Advanced Soil Mechanics
Fifth Edition

T0199558

Advanced Soil Mechanics
Fifth Edition

Braja M. Das

CRC Press
Taylor & Francis Group
Boca Raton London New York

CRC Press is an imprint of the
Taylor & Francis Group, an **informa** business

CRC Press
Taylor & Francis Group
6000 Broken Sound Parkway NW, Suite 300
Boca Raton, FL 33487-2742

First issued in paperback 2020

ISBN 13: 978-0-367-73010-9 (pbk)
ISBN 13: 978-0-8153-7913-3 (hbk)

Library of Congress Cataloging-in-Publication Data

Names: Das, Braja M., 1941- author.
Title: Advanced soil mechanics / Braja Das.
Description: 5th edition. | Boca Raton : Taylor & Francis, a CRC title, part
of the Taylor & Francis imprint, a member of the Taylor & Francis Group,
the academic division of T&F Informa, plc, [2019] | Includes
bibliographical references and index. |
Identifiers: LCCN 2018047522 (print) | LCCN 2018047957 (ebook) | ISBN
9781351215169 (ePub) | ISBN 9781351215152 (Mobipocket) | ISBN
9781351215176 (Adobe PDF) | ISBN 9780815379133 (hardback) | ISBN
9781351215183 (ebook)
Subjects: LCSH: Soil mechanics--Textbooks.
Classification: LCC TA710 (ebook) | LCC TA710 .D257 2019 (print) | DDC
624.1/5136--dc23
LC record available at https://lccn.loc.gov/2018047522

Visit the Taylor & Francis Web site at
http://www.taylorandfrancis.com

and the CRC Press Web site at
http://www.crcpress.com

To
Janice, Joe, Valerie, and Elizabeth

Contents

8   Consolidation

Preface

This textbook is intended for use in an introductory graduate level course that broadens (expands) the fundamental concepts acquired by students in their undergraduate work. The introductory graduate course can be followed by advanced courses dedicated to topics such as mechanical and chemical stabilization of soils, geoenvironmental engineering, finite element application to geotechnical engineering, critical state soil mechanics, geosynthetics, rock mechanics, and others.

The first edition of this book was published jointly by Hemisphere Publishing Corporation and McGraw-Hill Book Company of New York with a 1983 copyright. Taylor & Francis Group published the second, third, and fourth editions with 1997, 2008, and 2014 copyrights, respectively. The book has a total of 11 chapters and an appendix. SI units have been used throughout the text.

The following is a summary of additional materials given in this edition.

- Several new example problems have been added. The book now has more than 100 example problems which help the readers understand the theories presented. About 70 additional line drawings have been added to the text.
- In Chapter 1, "Soil aggregate, plasticity, and classification," relationships for determination of liquid limit by one-point method from test results of fall cone have been added to Section 1.8.1. Section 1.13.1 provides several correlations for estimation of the relative density of granular soil. This section also has correlations between uniformity coefficient, angularity, and maximum and minimum void ratios of granular soil. Effect of nonplastic fines on maximum and minimum void ratios of granular soils is given in Section 1.14.1.
- In Chapter 3, "Stresses and displacements in a soil mass—two-dimensional problems," stress determination for vertical line load located at the apex of an infinite wedge is presented in Section 3.3. Section 3.7 provides stress relationships for horizontal and inclined line loads

acting at the apex of an infinite wedge. Section 3.12 describes the stress distribution under a symmetrical vertical triangular strip load.

- In Chapter 4, "Stresses and displacements in a soil mass—three-dimensional problems," vertical stress calculation below flexible circular area with parabolic and conical loading are presented in Sections 4.10 and 4.11, respectively. Also a relationship for vertical stress under a uniformly loaded flexible elliptical area is given in Section 4.12.

- In Chapter 8, "Consolidation," Section 8.1 explains the fundamentals of the time-dependent settlement of saturated cohesive soil using the behavior of Kelvin model under load. Section 8.3.1 provides a simplified procedure developed by Hanna et al. (2013) to estimate the average degree of consolidation due to ramp loading. The log-log method proposed by Jose et al. (1989) and the Oikawa method (1987) to determine the preconsolidation pressure have been discussed in Section 8.5.1.1. A compilation of several correlations presently available in the literature for the recompression index of clay is given in Table 8.5. Design curves for prefabricated vertical drains have been elaborated upon in Section 8.17.

- In Chapter 9, "Shear strength of soils," the relevance of various laboratory test methods to field conditions has been briefly discussed in Section 9.4. In Section 9.7 a discussion has been provided to quantify the difference between the secant friction angle (ϕ_{secant}) and the ultimate friction angle (ϕ_{cv}) of granular soils based on the analysis of Bolten (1968). This section also includes the correlations for ϕ_{cv} for single mineral soil as discussed by Koerner (1970). Recently developed correlations for drained friction angle of normally consolidated clay (Sorensen and Okkels, 2013) are summarized in Section 9.12. Section 9.15 provides several correlations for the undrained shear strength of remolded clay. Relationships for determination of undrained shear strength using tapered vanes have been added to Section 9.22.

- In Chapter 10, "Elastic settlement of shallow foundations," the strain influence factor method for settlement calculation as provided by Terzaghi et al. (1996) has been discussed in Section 10.5.2. Settlement calculation based on the theory of elasticity as given in Section 10.6.1 has been substantially expanded. Elastic settlement in granular soil considering the change in soil modulus with elastic strain, and the effect of ground water table rise on elastic settlement of shallow foundations on granular soil have been discussed in two new sections: Sections 10.7 and 10.8.

- In Chapter 11, "Consolidation settlement of shallow foundations," discussion of Griffith's (1984) influence factor for determination of the average vertical stress increase in a soil layer has been added to Section 11.2.

Acknowledgments

Thanks are due to my wife, Janice, for her help in preparing the revised manuscript. I would also like to thank Tony Moore, senior editor, and Gabriella Williams, editorial assistant, Taylor & Francis Group, for working with me during the entire publication process of the book.

Thanks are due to Professor Nagaratnam Sivakugan of the College of Science, Technology, and Engineering, James Cook University, Queensland, Australia, for providing the cover page image.

Braja M. Das

Acknowledgments

Author

Braja M. Das is the Dean Emeritus of the College of Engineering and Computer Science at California State University, USA. He is a Fellow and Life Member of the American Society of Civil Engineers, a Life Member of the American Society for Engineering Education, and an Emeritus Member of the Transportation Research Board (TRB) AFS-80 Committee on Stabilization of Geomaterials and Stabilized Materials.

Professor Das is the author or coauthor of several textbooks and reference books in the area of geotechnical engineering, and he is Founder and Editor-in-Chief of the *International Journal of Geotechnical Engineering*.

Chapter I

Soil aggregate, plasticity, and classification

1.1 INTRODUCTION

Soils are aggregates of mineral particles; and together with air and/or water in the void spaces, they form three-phase systems. A large portion of the earth's surface is covered by soils, and they are widely used as construction and foundation materials. Soil mechanics is the branch of engineering that deals with the engineering properties of soils and their behavior under stress.

This book is divided into 11 chapters: "Soil Aggregate, Plasticity, and Classification," "Stresses and Strains: Elastic Equilibrium," "Stresses and Displacement in a Soil Mass: Two-Dimensional Problems," "Stresses and Displacement in a Soil Mass: Three-Dimensional Problems," "Pore Water Pressure due to Undrained Loading," "Permeability," "Seepage," "Consolidation," "Shear Strength of Soil," "Elastic Settlement of Shallow Foundations," and "Consolidation Settlement of Shallow Foundations." This chapter is a brief overview of some soil properties and their classification. It is assumed that the reader has been previously exposed to a basic soil mechanics course.

1.2 SOIL: SEPARATE SIZE LIMITS

A naturally occurring soil sample may have particles of various sizes. Over the years, various agencies have tried to develop the size limits of gravel, sand, silt, and clay. Some of these size limits are shown in Table 1.1.

Referring to Table 1.1, it is important to note that some agencies classify clay as particles smaller than 0.005 mm in size, and others classify it as particles smaller than 0.002 mm in size. However, it needs to be realized that particles defined as clay on the basis of their size are not necessarily clay minerals. Clay particles possess the tendency to develop plasticity when mixed with water; these are clay minerals. Kaolinite, illite, montmorillonite, vermiculite, and chlorite are examples of some clay minerals.

Table 1.1 Soil: separate size limits

Agency	Classification	Size limits (mm)
U.S. Department of Agriculture (USDA)	Gravel	>2
	Very coarse sand	2–1
	Coarse sand	1–0.5
	Medium sand	0.5–0.25
	Fine sand	0.25–0.1
	Very fine sand	0.1–0.05
	Silt	0.05–0.002
	Clay	<0.002
International Society of Soil Mechanics and Geotechnical Engineering (ISSMGE)	Gravel	>2
	Coarse sand	2–0.2
	Fine sand	0.2–0.02
	Silt	0.02–0.002
	Clay	<0.002
Federal Aviation Administration (FAA)	Gravel	>2
	Sand	2–0.075
	Silt	0.075–0.005
	Clay	<0.005
Massachusetts Institute of Technology (MIT)	Gravel	>2
	Coarse sand	2–0.6
	Medium sand	0.6–0.2
	Fine sand	0.2–0.06
	Silt	0.06–0.002
	Clay	<0.002
American Association of State Highway and Transportation Officials (AASHTO)	Gravel	76.2–2
	Coarse sand	2–0.425
	Fine sand	0.425–0.075
	Silt	0.075–0.002
	Clay	<0.002
Unified (U.S. Army Corps of Engineers, U.S. Bureau of Reclamation, and American Society for Testing and Materials)	Gravel	76.2–4.75
	Coarse sand	4.75–2
	Medium sand	2–0.425
	Fine sand	0.425–0.075
	Silt and clay (fines)	<0.075

Fine particles of quartz, feldspar, or mica may be present in a soil in the size range defined for clay, but these will not develop plasticity when mixed with water. It appears that it is more appropriate for soil particles with sizes <2 or 5 µm as defined under various systems to be called *clay-size particles* rather than *clay*. True clay particles are mostly of colloidal size range (<1 µm), and 2 µm is probably the upper limit.

1.3 CLAY MINERALS

Clay minerals are complex silicates of aluminum, magnesium, and iron. Two basic crystalline units form the clay minerals: (1) a silicon–oxygen tetrahedron, and (2) an aluminum or magnesium octahedron. A silicon–oxygen tetrahedron unit, shown in Figure 1.1a, consists of four oxygen atoms surrounding a silicon atom. The tetrahedron units combine to form a *silica sheet* as shown in Figure 1.2a. Note that the three oxygen atoms located at the base of each tetrahedron are shared by neighboring tetrahedra. Each silicon atom with a positive valence of 4 is linked to four oxygen atoms with a total negative valence of 8. However, each oxygen atom at the base of the tetrahedron is linked to two silicon atoms. This leaves one negative valence charge of the top oxygen atom of each tetrahedron to be counterbalanced. Figure 1.1b shows an octahedral unit consisting of six hydroxyl units surrounding an aluminum (or a magnesium) atom. The combination of the aluminum octahedral units forms a *gibbsite sheet* (Figure 1.2b). If the main metallic atoms in the octahedral units are magnesium, these sheets are referred to as *brucite sheets*. When the silica sheets are stacked over the octahedral sheets, the oxygen atoms replace the hydroxyls to satisfy their valence bonds. This is shown in Figure 1.2c.

Some clay minerals consist of repeating layers of two-layer sheets. A two-layer sheet is a combination of a silica sheet with a gibbsite sheet, or a combination of a silica sheet with a brucite sheet. The sheets are about 7.2 Å thick. The repeating layers are held together by hydrogen bonding and secondary valence forces. *Kaolinite* is the most important clay mineral belonging to this type (Figure 1.3). Figure 1.4 shows a scanning electron micrograph of kaolinite. Other common clay minerals that fall into this category are *serpentine* and *halloysite*.

The most common clay minerals with three-layer sheets are *illite* and *montmorillonite* (Figure 1.5). A three-layer sheet consists of an octahedral sheet in the middle with one silica sheet at the top and one at the bottom. Repeated layers of these sheets form the clay minerals. *Illite* layers

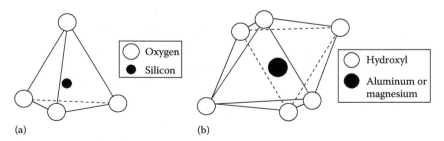

(a)   (b)

Figure 1.1 (a) Silicon–oxygen tetrahedron unit and (b) aluminum or magnesium octahedral unit.

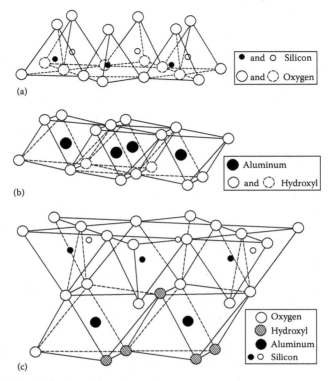

Figure 1.2 (a) Silica sheet, (b) gibbsite sheet, and (c) silica–gibbsite sheet. [After Grim, R. E., *J. Soil Mech. Found. Div.*, ASCE, 85(2), 1–17, 1959.]

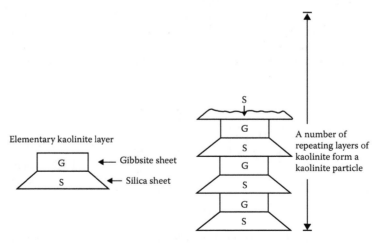

Figure 1.3 Symbolic structure for kaolinite.

Figure 1.4 Scanning electron micrograph of a kaolinite specimen. (Courtesy of David J. White, Ingios Geotechnics, Inc. Northfield, Minnesota.)

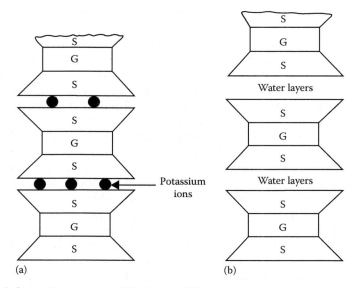

Figure 1.5 Symbolic structure of (a) illite and (b) montmorillonite.

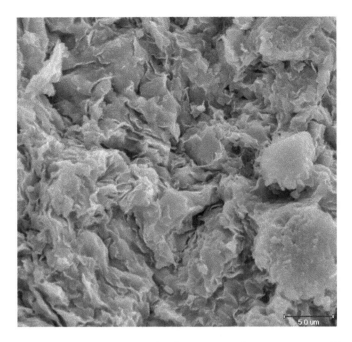

Figure 1.6 Scanning electron micrograph of a montmorillonite specimen. (Courtesy of David J, White, Ingios Geotechnics, Inc. Northfield, Minnesota.)

are bonded together by potassium ions. The negative charge to balance the potassium ions comes from the substitution of aluminum for some silicon in the tetrahedral sheets. Substitution of this type by one element for another without changing the crystalline form is known as *isomorphous substitution*. *Montmorillonite* has a similar structure to illite. However, unlike illite, there are no potassium ions present, and a large amount of water is attracted into the space between the three-sheet layers. Figure 1.6 shows a scanning electron micrograph of montmorillonite.

The surface area of clay particles per unit mass is generally referred to as *specific surface*. The lateral dimensions of kaolinite platelets are about 1,000–20,000 Å with thicknesses of 100–1,000 Å. Illite particles have lateral dimensions of 1000–5000 Å and thicknesses of 50–500 Å. Similarly, montmorillonite particles have lateral dimensions of 1000–5000 Å with thicknesses of 10–50 Å. If we consider several clay samples all having the same mass, the highest surface area will be in the sample in which the particle sizes are the smallest. So it is easy to realize that the specific surface of kaolinite will be small compared to that of montmorillonite. The specific surfaces of kaolinite, illite, and montmorillonite are about 15, 90, and 800 m²/g, respectively. Table 1.2 lists the specific surfaces of some clay minerals.

Table 1.2 Specific surface area and cation exchange capacity of some clay minerals

Clay mineral	Specific surface (m²/g)	Cation exchange capacity (me/100 g)
Kaolinite	10–20	3
Illite	80–100	25
Montmorillonite	800	100
Chlorite	5–50	20
Vermiculite	5–400	150
Halloysite (4H₂O)	40	12
Halloysite (2H₂O)	40	12

Clay particles carry a net negative charge. In an ideal crystal, the positive and negative charges would be balanced. However, isomorphous substitution and broken continuity of structures result in a net negative charge at the faces of the clay particles. (There are also some positive charges at the edges of these particles.) To balance the negative charge, the clay particles attract positively charged ions from salts in their pore water. These are referred to as exchangeable ions. Some are more strongly attracted than others, and the cations can be arranged in a series in terms of their affinity for attraction as follows:

$$Al^{3+} > Ca^{2+} > Mg^{2+} > NH_4^+ > K^+ > H^+ > Na^+ > Li^+$$

This series indicates that, for example, Al^{3+} ions can replace Ca^{2+} ions, and Ca^{2+} ions can replace Na^+ ions. The process is called *cation exchange*. For example,

$$Na_{clay} + CaCl_2 \rightarrow Ca_{clay} + NaCl$$

Cation exchange capacity (CEC) of a clay is defined as the amount of exchangeable ions, expressed in milliequivalents, per 100 g of dry clay. Table 1.2 gives the CEC of some clays.

1.4 NATURE OF WATER IN CLAY

The presence of exchangeable cations on the surface of clay particles was discussed in the preceding section. Some salt precipitates (cations in excess of the exchangeable ions and their associated anions) are also present on the surface of dry clay particles. When water is added to clay, these cations and anions float around the clay particles (Figure 1.7).

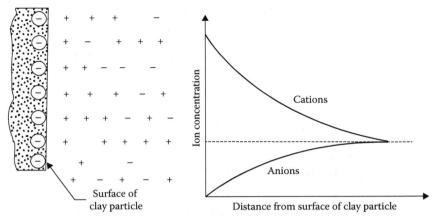

Figure 1.7 Diffuse double layer.

At this point, it must be pointed out that water molecules are dipolar, since the hydrogen atoms are not symmetrically arranged around the oxygen atoms (Figure 1.8a). This means that a molecule of water is like a rod with positive and negative charges at opposite ends (Figure 1.8b). There are three general mechanisms by which these dipolar water molecules, or *dipoles*, can be electrically attracted toward the surface of the clay particles (Figure 1.9):

a. Attraction between the negatively charged faces of clay particles and the positive ends of dipoles
b. Attraction between cations in the double layer and the negatively charged ends of dipoles. The cations are in turn attracted by the negatively charged faces of clay particles
c. Sharing of the hydrogen atoms in the water molecules by hydrogen bonding between the oxygen atoms in the clay particles and the oxygen atoms in the water molecules

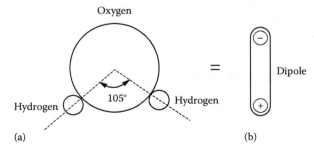

Figure 1.8 Dipolar nature of water: (a) unsymmetrical arrangement of hydrogen atoms; (b) dipole.

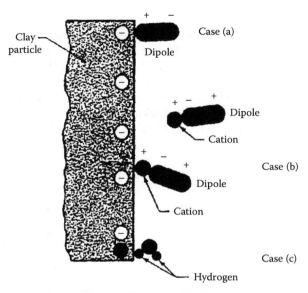

Figure 1.9 Dipolar water molecules in diffuse double layer.

The electrically attracted water that surrounds the clay particles is known as *double-layer water*. The plastic property of clayey soils is due to the existence of double-layer water. Thicknesses of double-layer water for typical kaolinite and montmorillonite crystals are shown in Figure 1.10. Since the innermost layer of double-layer water is very strongly held by a clay particle, it is referred to as *adsorbed water*.

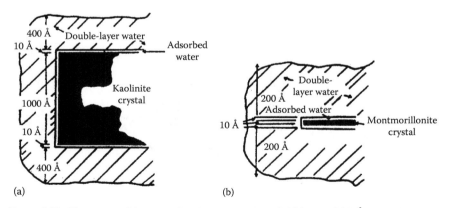

Figure 1.10 Clay water (a) typical kaolinite particle, 10,000 by 1,000 Å and (b) typical montmorillonite particle, 1,000 by 10 Å. [After Lambe, T. W., *Trans.* ASCE, 125, 682, 1960.]

1.5 REPULSIVE POTENTIAL

The nature of the distribution of ions in the diffuse double layer is shown in Figure 1.7. Several theories have been presented in the past to describe the ion distribution close to a charged surface. Of these, the Gouy–Chapman theory has received the most attention. Let us assume that the ions in the double layers can be treated as point charges, and that the surface of the clay particles is large compared to the thickness of the double layer. According to Boltzmann's theorem, we can write that (Figure 1.11)

$$n_+ = n_{+(0)} \exp \frac{-v_+ e\Phi}{KT} \tag{1.1}$$

$$n_- = n_{-(0)} \exp \frac{-v_- e\Phi}{KT} \tag{1.2}$$

where
n_+ is the local concentration of positive ions at a distance x
n_- is the local concentration of negative ions at a distance x
$n_{+(0)}$, $n_{-(0)}$ are the concentration of positive and negative ions away from the clay surface in the equilibrium liquid
Φ is the average electric potential at a distance x (Figure 1.12)
v_+, v_- are ionic valences
e is the unit electrostatic charge, 4.8×10^{-10} esu
K is the Boltzmann constant, 1.38×10^{-16} erg/K
T is the absolute temperature

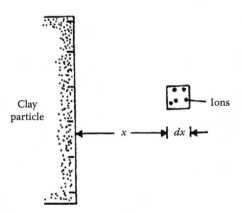

Figure 1.11 Derivation of repulsive potential equation.

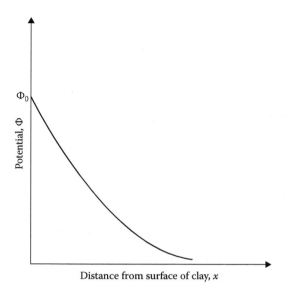

Figure 1.12 Nature of variation of potential Φ with distance from the clay surface.

The charge density ρ at a distance x is given by

$$\rho = v_+ \, en_+ - v_- en_- \tag{1.3}$$

According to Poisson's equation

$$\frac{d^2\Phi}{dx^2} = \frac{-4\pi\rho}{\lambda} \tag{1.4}$$

where λ is the dielectric constant of the medium.

Assuming $v_+ = v_-$ and $n_{+(0)} = n_{-(0)} = n_0$, and combining Equations 1.1 through 1.4, we obtain

$$\frac{d^2\Phi}{dx^2} = \frac{8\pi n_0 ve}{\lambda} \sinh \frac{ve\Phi}{KT} \tag{1.5}$$

It is convenient to rewrite Equation 1.5 in terms of the following nondimensional quantities

$$y = \frac{ve\Phi}{KT} \tag{1.6}$$

$$z = \frac{ve\Phi_0}{KT} \tag{1.7}$$

and

$$\xi = \kappa x \tag{1.8}$$

where Φ_0 is the potential at the surface of the clay particle and

$$\kappa^2 = \frac{8\pi n_0 e^2 v^2}{\lambda K T} \, (\text{cm}^{-2}) \tag{1.9}$$

Thus, from Equation 1.5

$$\frac{d^2 y}{d\xi^2} = \sinh y \tag{1.10}$$

The boundary conditions for solving Equation 1.10 are

1. At $\xi = \infty$, $y = 0$ and $dy/d\xi = 0$
2. At $\xi = 0$, $y = z$, that is, $\Phi = \Phi_0$

The solution yields the relation

$$e^{y/2} = \frac{(e^{z/2} + 1) + (e^{z/2} - 1)e^{-\xi}}{(e^{z/2} + 1) - (e^{z/2} - 1)e^{-\xi}} \tag{1.11}$$

Equation 1.11 gives an approximately exponential decay of potential. The nature of the variation of the nondimensional potential y with the nondimensional distance is given in Figure 1.13.

For a small surface potential (<25 mV), we can approximate Equation 1.5 as

$$\frac{d^2 \Phi}{dx^2} = \kappa^2 \Phi \tag{1.12}$$

$$\Phi = \Phi_0 e^{-\kappa x} \tag{1.13}$$

Equation 1.13 describes a purely exponential decay of potential. For this condition, the center of gravity of the diffuse charge is located at a distance of $x = 1/\kappa$. The term $1/\kappa$ is generally referred to as the double-layer *thickness*.

There are several factors that will affect the variation of the repulsive potential with distance from the surface of the clay layer. The effect of the cation concentration and ionic valence is shown in Figures 1.14 and 1.15, respectively. For a given value of Φ_0 and x, the repulsive potential Φ decreases with the increase of ion concentration n_0 and ionic valence v.

When clay particles are close and parallel to each other, the nature of variation of the potential will be as shown in Figure 1.16. Note for this case

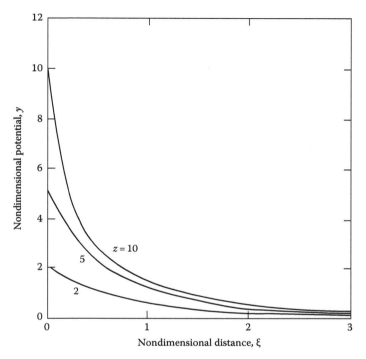

Figure 1.13 Variation of nondimensional potential with nondimensional distance.

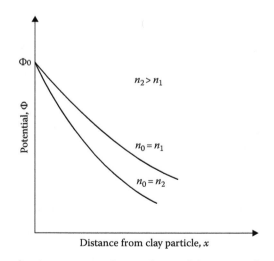

Figure 1.14 Effect of cation concentration on the repulsive potential.

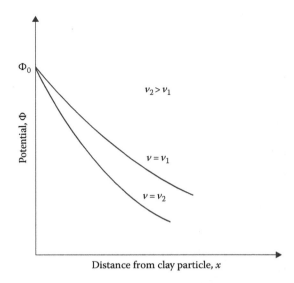

Figure 1.15 Effect of ionic valence on the repulsive potential.

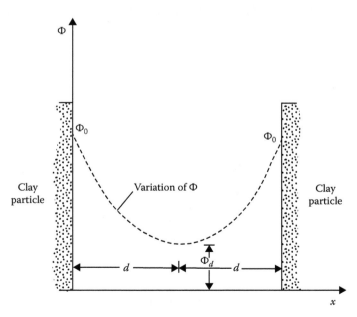

Figure 1.16 Variation of Φ between two parallel clay particles.

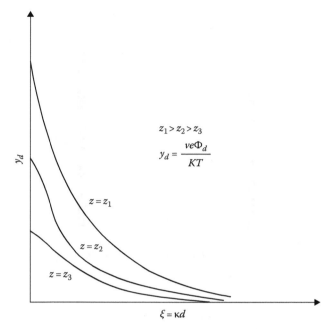

Figure 1.17 Nature of variation of the nondimensional midplane potential for two parallel plates.

that at $x = 0$, $\Phi = \Phi_0$, and at $x = d$ (midway between the plates), $\Phi = \Phi_d$ and $d\Phi/dx = 0$. Numerical solutions for the nondimensional potential $y = y_d$ (i.e., $\Phi = \Phi_d$) for various values of z and $\xi = \kappa d$ (i.e., $x = d$) are given by Verweg and Overbeek (1948) (see also Figure 1.17).

1.6 REPULSIVE PRESSURE

The repulsive pressure midway between two parallel clay plates (Figure 1.18) can be given by the Langmuir equation

$$p = 2n_0 KT \left(\cosh \frac{ve\Phi_d}{KT} - 1 \right) \tag{1.14}$$

where p is the repulsive pressure, that is, the difference between the osmotic pressure midway between the plates in relation to that in the equilibrium solution. Figure 1.19, which is based on the results of Bolt (1956), shows the theoretical and experimental variation of p between two clay particles.

Although the Guoy–Chapman theory has been widely used to explain the behavior of clay, there have been several important objections to this theory. A good review of these objections has been given by Bolt (1955).

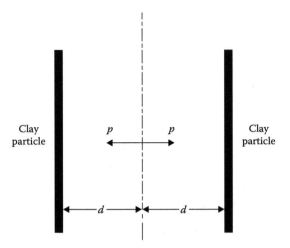

Figure 1.18 Repulsive pressure midway between two parallel clay plates.

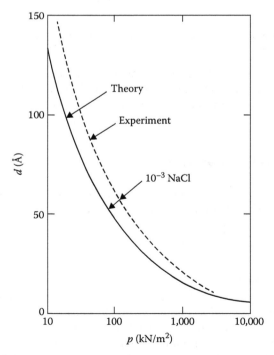

Figure 1.19 Repulsive pressure between sodium montmorillonite clay particles. [After Bolt, G. H., *Geotechnique*, 6(2), 86, 1956.]

1.7 FLOCCULATION AND DISPERSION OF CLAY PARTICLES

In addition to the repulsive force between the clay particles, there is an attractive force, which is largely attributed to the Van der Waal force. This is a secondary bonding force that acts between all adjacent pieces of matter. The force between two flat parallel surfaces varies inversely as $1/x^3$ to $1/x^4$, where x is the distance between the two surfaces. Van der Waal's force is also dependent on the dielectric constant of the medium separating the surfaces. However, if water is the separating medium, substantial changes in the magnitude of the force will not occur with minor changes in the constitution of water.

The behavior of clay particles in a suspension can be qualitatively visualized from our understanding of the attractive and repulsive forces between the particles and with the aid of Figure 1.20. Consider a dilute suspension of clay particles in water. These colloidal clay particles will undergo Brownian movement and, during this random movement, will come close to

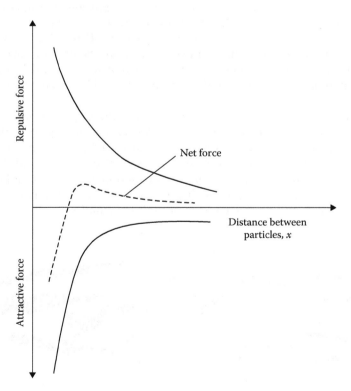

Figure 1.20 Dispersion and flocculation of clay in a suspension.

each other at distances within the range of interparticle forces. The forces of attraction and repulsion between the clay particles vary at different rates with respect to the distance of separation. The force of repulsion decreases exponentially with distance, whereas the force of attraction decreases as the inverse third or fourth power of distance, as shown in Figure 1.20. Depending on the distance of separation, if the magnitude of the repulsive force is greater than the magnitude of the attractive force, the net result will be repulsion. The clay particles will settle individually and form a dense layer at the bottom; however, they will remain separate from their neighbors (Figure 1.21a). This is referred to as the *dispersed state* of the soil. On the contrary, if the net force between the particles is attraction, flocs will be formed and these flocs will settle to the bottom. This is called *flocculated* clay (Figure 1.21b).

1.7.1 Salt flocculation and nonsalt flocculation

We saw in Figure 1.14 the effect of salt concentration, n_0, on the repulsive potential of clay particles. High salt concentration will depress the double layer of clay particles and hence the force of repulsion. We noted earlier in this section that the Van der Waal force largely contributes to the force of attraction between clay particles in suspension. If the clay particles are suspended in water with a high salt concentration, the flocs of the clay particles formed by dominant attractive forces will give them mostly an orientation approaching parallelism (face-to-face type). This is called a salt-type flocculation (Figure 1.22a).

Another type of force of attraction between the clay particles, which is not taken into account in colloidal theories, is that arising from the electrostatic attraction of the positive charges at the edge of the particles and the negative charges at the face. In a soil–water suspension with low salt concentration, this electrostatic force of attraction may produce a flocculation with an orientation approaching a perpendicular array. This is shown in Figure 1.22b and is referred to as nonsalt flocculation.

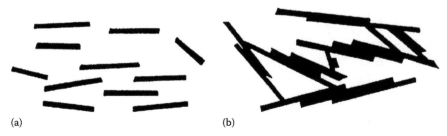

(a)   (b)

Figure 1.21 (a) Dispersion and (b) flocculation of clay.

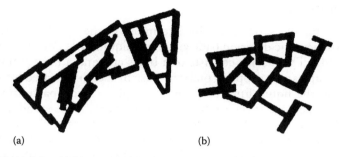

(a)   (b)

Figure 1.22 (a) Salt and (b) nonsalt flocculation of clay particles.

1.8 CONSISTENCY OF COHESIVE SOILS

The presence of clay minerals in a fine-grained soil will allow it to be remolded in the presence of some moisture without crumbling. If a clay slurry is dried, the moisture content will gradually decrease, and the slurry will pass from a liquid state to a plastic state. With further drying, it will change to a semisolid state and finally to a solid state, as shown in Figure 1.23. In 1911, A. Atterberg, a Swedish scientist, developed a method for describing the limit consistency of fine-grained soils on the basis of moisture content. These limits are the *liquid limit*, the *plastic limit*, and the *shrinkage limit*.

The liquid limit is defined as the moisture content, in percent, at which the soil changes from a liquid state to a plastic state. The moisture contents (in percent) at which the soil changes from a plastic to a semisolid state and from a semisolid to a solid state are defined as the plastic limit and the shrinkage limit, respectively. These limits are generally referred to as the *Atterberg limits*. The Atterberg limits of cohesive soil depend on several factors, such as the amount and type of clay minerals and the type of adsorbed cation.

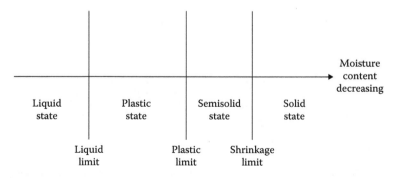

Figure 1.23 Consistency of cohesive soils.

1.8.1 Liquid limit

The liquid limit of a soil is generally determined by the Standard Casagrande device. A schematic diagram (side view) of a liquid limit device is shown in Figure 1.24a. This device consists of a brass cup and a hard rubber base. The brass cup can be dropped onto the base by a cam operated by a crank. To perform the liquid limit test, one must place a soil paste in the cup. A groove is then cut at the center of the soil pat with the standard grooving tool (Figure 1.24b). By using the crank-operated cam, the cup is lifted and dropped from a height of 10 mm. The moisture content, in percent, required to close a distance of 12.7 mm along the bottom of the groove (see Figure 1.24c and d) after 25 blows is defined as the *liquid limit*.

It is difficult to adjust the moisture content in the soil to meet the required 12.7 mm closure of the groove in the soil pat at 25 blows. Hence, at least

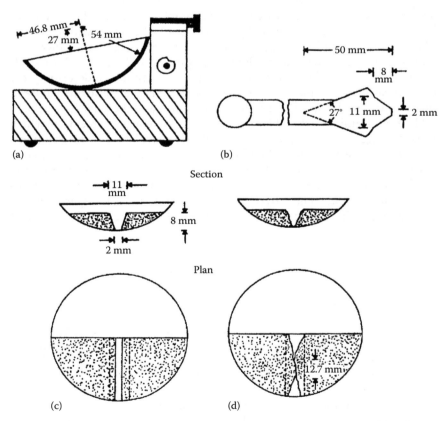

Figure 1.24 Schematic diagram of (a) liquid limit device, (b) grooving tool, (c) soil pat at the beginning of the test, and (d) soil pat at the end of the test.

three tests for the same soil are conducted at varying moisture contents, with the number of blows, N, required to achieve closure varying between 15 and 35. The moisture content of the soil, in percent, and the corresponding number of blows are plotted on semilogarithmic graph paper (Figure 1.25). The relationship between moisture content and log N is approximated as a straight line. This line is referred to as the *flow curve*. The moisture content corresponding to N = 25, determined from the flow curve, gives the liquid limit of the soil. The slope of the flow line is defined as the *flow index* and may be written as

$$I_F = \frac{w_1 - w_2}{\log(N_2/N_1)}$$

(1.15)

where
   I_F is the flow index
   w_1 is the moisture content of soil, in percent, corresponding to N_1 blows
   w_2 is the moisture content corresponding to N_2 blows

Note that w_2 and w_1 are exchanged to yield a positive value even though the slope of the flow line is negative. Thus, the equation of the flow line can be written in a general form as

$$w = -I_F \log N + C$$

(1.16)

where C is a constant.

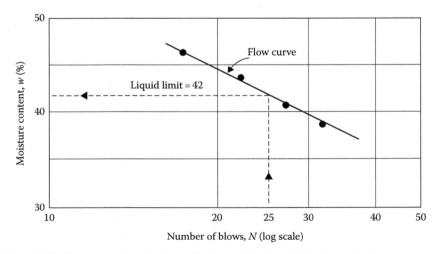

Figure 1.25 Flow curve for the determination of the liquid limit for a silty clay.

From the analysis of hundreds of liquid limit tests in 1949, the U.S. Army Corps of Engineers, at the Waterways Experiment Station in Vicksburg, Mississippi, proposed an empirical equation of the form

$$LL = w_N \left(\frac{N}{25} \right)^{\tan\beta} \tag{1.17}$$

where
> N is the number of blows in the liquid limit device for a 12.7 mm groove closure
> w_N is the corresponding moisture content
> $\tan\beta = 0.121$ (but note that $\tan\beta$ is not equal to 0.121 for all soils)

Equation 1.17 generally yields good results for the number of blows between 20 and 30. For routine laboratory tests, it may be used to determine the liquid limit when only one test is run for a soil. This procedure is generally referred to as the *one-point method* and was also adopted by ASTM under designation D-4318 (ASTM, 2014). The reason that the one-point method yields fairly good results is that a small range of moisture content is involved when $N = 20$–30.

Another method of determining the liquid limit, which is popular in Europe and Asia, is the *fall cone method* (British Standard—BS 1377). In this test, the liquid limit is defined as the moisture content at which a standard cone of apex angle 30° and weight of 0.78 N (80 gf) will penetrate a distance $d = 20$ mm in 5 s when allowed to drop from a position of point contact with the soil surface (Figure 1.26a). Due to the difficulty in achieving the liquid limit from a single test, four or more tests can be conducted at various moisture contents to determine the fall cone penetration, d, in 5 s. A semilogarithmic graph can then be plotted with moisture content w versus cone penetration d. The plot results in a straight line. The moisture content corresponding to $d = 20$ mm is the liquid limit (Figure 1.26b). From Figure 1.26b, the *flow index* can be defined as

$$I_{FC} = \frac{w_2(\%) - w_1(\%)}{\log d_2 - \log d_1} \tag{1.18}$$

where w_1, w_2 are the moisture contents at cone penetrations of d_1 and d_2, respectively.

As in the case of the percussion cup method (ASTM D4318), attempts have been made to develop the estimation of liquid limit by a one-point method. They are

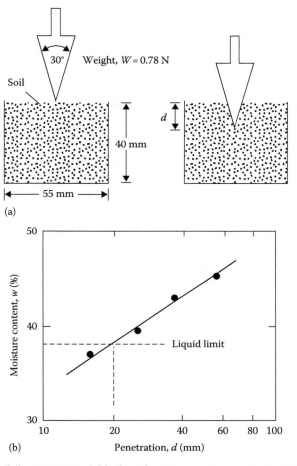

Figure 1.26 (a) Fall cone test and (b) plot of moisture content versus cone penetration for determination of liquid limit.

- Nagaraj and Jayadeva (1981)

$$LL = \frac{w}{0.77 \log d} \tag{1.19}$$

$$LL = \frac{w}{0.65 + 0.0175d} \tag{1.20}$$

- Feng (2001)

$$LL = w\left(\frac{20}{d}\right)^{0.33} \tag{1.21}$$

Table 1.3 Summary of main differences among fall cones (Summarized from Budhu, 1985)

Country	Cone details	Penetration for liquid limit (mm)
Russia	Cone angle = 30° Cone mass = 76 g	10
Britain, France	Cone angle = 30° Cone mass = 80 g	20
India	Cone angle = 31° Cone mass = 148 g	20.4
Sweden, Canada (Québec)	Cone angle = 60° Cone mass = 60 g	10

Note: Duration of penetration is 5s in all cases.

where w (%) is the moisture content for a cone penetration d (mm) falling between 15 mm to 25 mm.

The dimensions of the cone tip angle, cone weight, and the penetration (mm) at which the liquid limit is determined varies from country to country. Table 1.3 gives a summary of different fall cones used in various countries.

A number of major studies have shown that the undrained shear strength of the soil at liquid limit varies between 1.7 and 2.3 kN/m². Based on tests conducted on a large number of soil samples, Feng (2001) has given the following correlation between the liquid limits determined according to ASTM D4318 and British Standard BS1377.

$$LL_{(BS)} = 2.6 + 0.94[LL_{(ASTM)}] \tag{1.22}$$

Example 1.1

One liquid limit test was conducted on a soil using the fall cone. Following are the results: $w = 29.5\%$ at $d = 15$ mm. Estimate the liquid limit of the soil using Equations 1.19, 1.20, and 1.21.

Solution

From Equation 1.19,

$$LL = \frac{w}{0.77 \log d} = \frac{29.5}{(0.77)(\log 15)} = 32.58$$

From Equation 1.20,

$$LL = \frac{w}{0.65 + 0.0175d} = \frac{29.5}{0.65 + (0.0175)(15)} = 32.33$$

From Equation 1.21,

$$LL = w\left(\frac{20}{d}\right)^{0.33} = (29.5)\left(\frac{20}{15}\right)^{0.33} = 32.43$$

1.8.2 Plastic limit

The *plastic limit* is defined as the moist content, in percent, at which the soil crumbles when rolled into threads of 3.2 mm diameter. The plastic limit is the lower limit of the plastic stage of soil. The plastic limit test is simple and is performed by repeated rolling of an ellipsoidal size soil mass by hand on a ground glass plate. The procedure for the plastic limit test is given by ASTM Test Designation D-4318 (ASTM, 2014).

As in the case of liquid limit determination, the fall cone method can be used to obtain the plastic limit. This can be achieved by using a cone of similar geometry, but with a mass of 2.35 N (240 gf). Three to four tests at varying moist contents of soil are conducted, and the corresponding cone penetrations d are determined. The moisture content corresponding to a cone penetration of $d = 20$ mm is the plastic limit. Figure 1.27 shows the liquid and plastic limit determined by the fall cone test for Cambridge Gault clay reported by Wroth and Wood (1978).

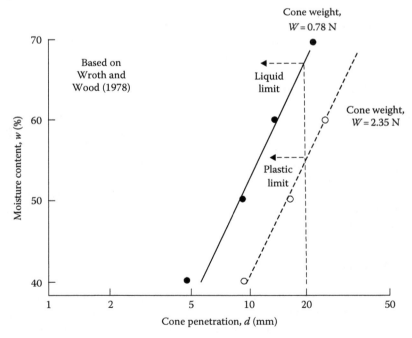

Figure 1.27 Liquid and plastic limits for Cambridge Gault clay determined by the fall cone test.

The difference between the liquid limit and the plastic limit of a soil is defined as the plasticity index, PI

$$PI = LL - PL \tag{1.23}$$

where
　　LL is the liquid limit
　　PL is the plastic limit

Sridharan et al. (1999) showed that the plasticity index can be correlated to the flow index as obtained from the liquid limit tests. According to their study

$$PI(\%) = 4.12 I_F(\%) \tag{1.24}$$

and

$$PI(\%) = 0.74 I_{FC}(\%) \tag{1.25}$$

1.9 LIQUIDITY INDEX

The relative consistency of a cohesive soil can be defined by a ratio called the *liquidity index* LI. It is defined as

$$LI = \frac{w_N - PL}{LL - PL} = \frac{w_N - PL}{PI} \tag{1.26}$$

where w_N is the natural moisture content. It can be seen from Equation 1.26 that, if $w_N = LL$, then the liquidity index is equal to 1. Again, if $w_N = PL$, the liquidity index is equal to 0. Thus, for a natural soil deposit which is in a plastic state (i.e., $LL \geq w_N \geq PL$), the value of the liquidity index varies between 1 and 0. A natural deposit with $w_N \geq LL$ will have a liquidity index greater than 1. In an undisturbed state, these soils may be stable; however, a sudden shock may transform them into a liquid state. Such soils are called *sensitive clays*.

1.10 ACTIVITY

Since the plastic property of soil is due to the adsorbed water that surrounds the clay particles, we can expect that the type of clay minerals and their proportional amounts in a soil will affect the liquid and plastic limits. Skempton (1953) observed that the plasticity index of a

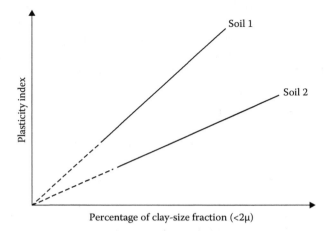

Figure 1.28 Relationship between plasticity index and percentage of clay-size fraction by weight.

soil linearly increases with the percent of clay-size fraction (percent finer than 2μ by weight) present in it. This relationship is shown in Figure 1.28. The average lines for all the soils pass through the origin. The correlations of PI with the clay-size fractions for different clays plot separate lines. This is due to the type of clay minerals in each soil. On the basis of these results, Skempton defined a quantity called activity, which is the slope of the line correlating PI and percent finer than 2μ. This activity *A* may be expressed as

$$A = \frac{PI}{(\text{percentage of clay-size fraction by weight})} \qquad (1.27)$$

Activity is used as an index for identifying the swelling potential of clay soils. Typical values of activities for various clay minerals are given in Table 1.4.

Table 1.4 Activities of clay minerals

Mineral	Activity (A)
Smectites	1–7
Illite	0.5–1
Kaolinite	0.5
Halloysite ($4H_2O$)	0.5
Halloysite ($2H_2O$)	0.1
Attapulgite	0.5–1.2
Allophane	0.5–1.2

Seed et al. (1964a) studied the plastic property of several artificially pre-pared mixtures of sand and clay. They concluded that, although the rela-tionship of the plasticity index to the percent of clay-size fraction is linear (as observed by Skempton), it may not always pass through the origin. This is shown in Figure 1.29. Thus, the activity can be redefined as

$$A = \frac{\text{PI}}{\text{percent of clay-size fraction} - C'} \tag{1.28}$$

where C' is a constant for a given soil. For the experimental results shown in Figure 1.29, $C' = 9$.

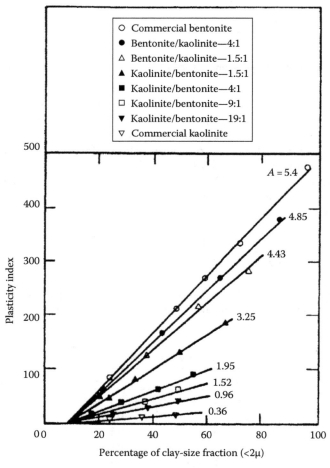

Figure 1.29 Relationship between plasticity index and clay-size fraction by weight for kaolinite/bentonite clay mixtures. [After Seed, H. B. et al., *J. Soil Mech. Found. Eng. Div.*, ASCE, 90(SM4), 107, 1964.]

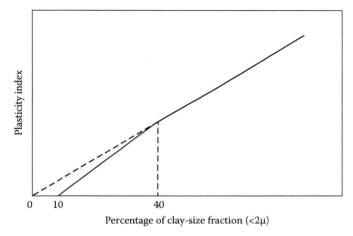

Figure 1.30 Simplified relationship between plasticity index and percentage of clay-size fraction by weight. [After Seed, H. B. et al., *J. Soil Mech. Found. Eng. Div.*, ASCE, 90(SM6), 75, 1964.]

Further works of Seed et al. (1964b) have shown that the relationship of the plasticity index to the percentage of clay-size fractions present in a soil can be represented by two straight lines. This is shown qualitatively in Figure 1.30. For clay-size fractions greater than 40%, the straight line passes through the origin when it is projected back.

1.11 GRAIN-SIZE DISTRIBUTION OF SOIL

For a basic understanding of the nature of soil, the distribution of the grain size present in a given soil mass must be known. The grain-size distribution of coarse-grained soils (gravelly and/or sandy) is determined by sieve analysis. Table 1.5 gives the opening size of some U.S. sieves.

The cumulative percent by weight of a soil passing a given sieve is referred to as the *percent finer*. Figure 1.31 shows the results of a sieve analysis for a sandy soil. The grain-size distribution can be used to determine some of the basic soil parameters, such as the effective size, the uniformity coefficient, and the coefficient of gradation.

The *effective size* of a soil is the diameter through which 10% of the total soil mass is passing and is referred to as D_{10}. The *uniformity coefficient* C_u is defined as

$$C_u = \frac{D_{60}}{D_{10}} \tag{1.29}$$

where D_{60} is the diameter through which 60% of the total soil mass is passing.

Table 1.5 U.S. standard sieves

Sieve no.	Opening size (mm)
3	6.35
4	4.75
6	3.36
8	2.38
10	2.00
16	1.19
20	0.84
30	0.59
40	0.425
50	0.297
60	0.25
70	0.21
100	0.149
140	0.105
200	0.075
270	0.053

The *coefficient of gradation* C_c is defined as

$$C_c = \frac{(D_{30})^2}{(D_{60})(D_{10})} \tag{1.30}$$

where D_{30} is the diameter through which 30% of the total soil mass is passing.

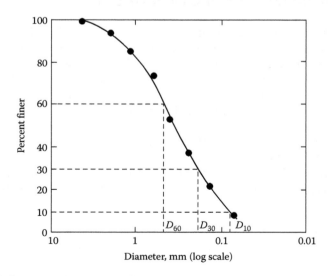

Figure 1.31 Grain-size distribution of a sandy soil.

A soil is called a *well-graded* soil if the distribution of the grain sizes extends over a rather large range. In that case, the value of the uniformity coefficient is large. Generally, a soil is referred to as well graded if C_u is larger than about 4–6 and C_c is between 1 and 3. When most of the grains in a soil mass are of approximately the same size—that is, C_u is close to 1—the soil is called *poorly graded*. A soil might have a combination of two or more well-graded soil fractions, and this type of soil is referred to as a *gap-graded* soil.

The sieve analysis technique described earlier is applicable for soil grains larger than No. 200 (0.075 mm) sieve size. For fine-grained soils, the procedure used for determination of the grain-size distribution is hydrometer analysis. This is based on the principle of sedimentation of soil grains.

1.12 WEIGHT–VOLUME RELATIONSHIPS

Figure 1.32a shows a soil mass that has a total volume V and a total weight W. To develop the weight–volume relationships, the three phases of the soil mass, that is, soil solids, air, and water, have been separated in Figure 1.32b. Note that

$$W = W_s + W_w \tag{1.31}$$

and, also

$$V = V_s + V_w + V_a \tag{1.32}$$

$$V_v = V_w + V_a \tag{1.33}$$

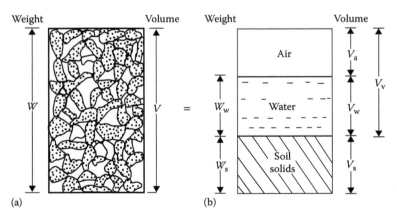

Figure 1.32 Weight–volume relationships for soil aggregate: (a) soil mass of volume V; (b) three phases of the soil mass.

where
  W_s is the weight of soil solids
  W_w is the weight of water
  V_s is the volume of the soil solids
  V_w is the volume of water
  V_a is the volume of air

The weight of air is assumed to be zero. The volume relations commonly used in soil mechanics are void ratio, porosity, and degree of saturation.

Void ratio e is defined as the ratio of the volume of voids to the volume of solids:

$$e = \frac{V_v}{V_s} \tag{1.34}$$

Porosity n is defined as the ratio of the volume of voids to the total volume:

$$n = \frac{V_v}{V} \tag{1.35}$$

  Also, $V = V_s + V_v$
and so

$$n = \frac{V_v}{V_s + V_v} = \frac{V_v/V_s}{(V_s/V_s)+(V_v/V_s)} = \frac{e}{1+e} \tag{1.36}$$

Degree of saturation S_r is the ratio of the volume of water to the volume of voids and is generally expressed as a percentage:

$$S_r(\%) = \frac{V_w}{V_v} \times 100 \tag{1.37}$$

The weight relations used are moisture content and unit weight. *Moisture content w* is defined as the ratio of the weight of water to the weight of soil solids, generally expressed as a percentage:

$$w(\%) = \frac{W_w}{W_s} \times 100 \tag{1.38}$$

Unit weight γ is the ratio of the total weight to the total volume of the soil aggregate:

$$\gamma = \frac{W}{V} \tag{1.39}$$

This is sometimes referred to as moist unit weight since it includes the weight of water and the soil solids. If the entire void space is filled with water (i.e., $V_a = 0$), it is a saturated soil; Equation 1.39 will then give us the saturated unit weight γ_{sat}.

The dry unit weight γ_d is defined as the ratio of the weight of soil solids to the total volume:

$$\gamma_d = \frac{W_s}{V} \tag{1.40}$$

Useful weight–volume relations can be developed by considering a soil mass in which the volume of soil solids is unity, as shown in Figure 1.33. Since $V_s = 1$, from the definition of void ratio given in Equation 1.34, the volume of voids is equal to the void ratio e. The weight of soil solids can be given by

$$W_s = G_s \gamma_w V_s = G_s \gamma_w \quad (\text{since} \, V_s = 1)$$

where
G_s is the specific gravity of soil solids
γ_w is the unit weight of water (9.81 kN/m³)

From Equation 1.38, the weight of water is $W_w = w W_s = w G_s \gamma_w$. So the moist unit weight is

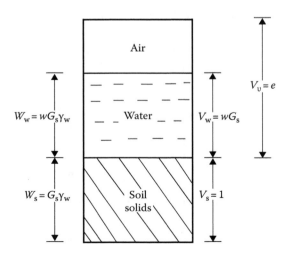

Figure 1.33 Weight–volume relationship for $V_s = 1$.

$$\gamma = \frac{W}{V} = \frac{W_s + W_w}{V_s + V_\upsilon} = \frac{G_s\gamma_w + wG_s\gamma_w}{1+e} = \frac{G_s\gamma_w(1+w)}{1+e} \qquad (1.41)$$

The dry unit weight can also be determined from Figure 1.33 as

$$\gamma_d = \frac{W_s}{V} = \frac{G_s\gamma_w}{1+e} \qquad (1.42)$$

The degree of saturation can be given by

$$S_r = \frac{V_w}{V_\upsilon} = \frac{W_w/\gamma_w}{V_\upsilon} = \frac{wG_s\gamma_w/\gamma_w}{e} = \frac{wG_s}{e} \qquad (1.43)$$

For saturated soils, $S_r = 1$. So, from Equation 1.43,

$$e = wG_s \qquad (1.44)$$

By referring to Figure 1.34, the relation for the unit weight of a saturated soil can be obtained as

$$\gamma_{sat} = \frac{W}{V} = \frac{W_s + W_w}{V} = \frac{G_s\gamma_w + e\gamma_w}{1+e} \qquad (1.45)$$

Basic relations for unit weight such as Equations 1.41, 1.42, and 1.45 in terms of porosity n can also be derived by considering a soil mass that has a total volume of unity as shown in Figure 1.35. In this case (for $V = 1$), from Equation 1.35, $V_\upsilon = n$. So, $V_s = V - V_v = 1 - n$.

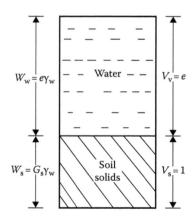

Figure 1.34 Weight–volume relation for saturated soil with $V_s = 1$.

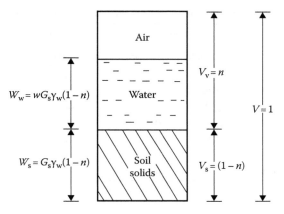

Figure 1.35 Weight–volume relationship for V = 1.

The weight of soil solids is equal to $(1 - n)G_s\gamma_w$, and the weight of water $W_w = wW_s = w(1 - n)G_s\gamma_w$. Thus, the moist unit weight is

$$\gamma = \frac{W}{V} = \frac{W_s + W_w}{V} = \frac{(1 - n)G_s\gamma_w + w(1 - n)G_s\gamma_w}{1}$$

$$= G_s\gamma_w(1 - n)(1 + w) \qquad (1.46)$$

The dry unit weight is

$$\gamma_d = \frac{W_s}{V} = (1 - n)G_s\gamma_w \qquad (1.47)$$

If the soil is saturated (Figure 1.36),

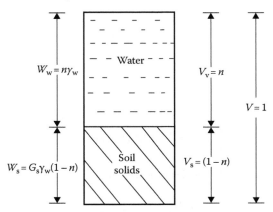

Figure 1.36 Weight–volume relationship for saturated soil with V = 1.

Table 1.6 Typical values of void ratios and dry unit weights for granular soils

Soil type	Void ratio, e		Dry unit weight, γ_d	
	Maximum	Minimum	Minimum (kN/m³)	Maximum (kN/m³)
Gravel	0.6	0.3	16	20
Coarse sand	0.75	0.35	15	19
Fine sand	0.85	0.4	14	19
Standard Ottawa sand	0.8	0.5	14	17
Gravelly sand	0.7	0.2	15	22
Silty sand	1	0.4	13	19
Silty sand and gravel	0.85	0.15	14	23

$$\gamma_{sat} = \frac{W_s + W_w}{V} = (1 - n)G_s\gamma_w + n\gamma_w = [G_s - n(G_s - 1)]\gamma_w \qquad (1.48)$$

Table 1.6 gives some typical values of void ratios and dry unit weights encountered in granular soils.

Example 1.2

For a soil in natural state, given $e = 0.8$, $w = 24\%$, and $G_s = 2.68$.

    a. Determine the moist unit weight, dry unit weight, and degree of saturation.
    b. If the soil is completely saturated by adding water, what would its moisture content be at that time? Also, find the saturated unit weight.

Solution

Part a:

From Equation 1.41, the moist unit weight is

$$\gamma = \frac{G_s\gamma_w(1 + w)}{1 + e}$$

Since $\gamma_w = 9.81$ kN/m³,

$$\gamma = \frac{(2.68)(9.81)(1 + 0.24)}{1 + 0.8} = 18.11\,\text{kN/m}^3$$

From Equation 1.42, the dry unit weight is

$$\gamma_d = \frac{G_s\gamma_w}{1 + e} = \frac{(2.68)(9.81)}{1 + 0.8} = 14.61\,\text{kN/m}^3$$

From Equation 1.43, the degree of saturation is

$$S_r(\%) = \frac{wG_s}{e} \times 100 = \frac{(0.24)(2.68)}{0.8} \times 100 = 80.4\%$$

Part b:

From Equation 1.44, for saturated soils, $e = wG_s$, or

$$w(\%) = \frac{e}{G_s} \times 100 = \frac{0.8}{2.68} \times 100 = 29.85\%$$

From Equation 1.45, the saturated unit weight is

$$\gamma_{sat} = \frac{G_s\gamma_w + e\gamma_w}{1+e} = \frac{9.81(2.68+0.8)}{1+0.8} = 18.97 \, \text{kN/m}^3$$

Example 1.3

In the natural state, a moist soil has a volume of 0.0093 m³ and weighs 177.6 N. The oven dry weight of the soil is 153.6 N. If $G_s = 2.71$, calculate the moisture content, moist unit weight, dry unit weight, void ratio, porosity, and degree of saturation.

Solution

Refer to Figure 1.37. The moisture content (Equation 1.38) is

$$w = \frac{W_w}{W_s} = \frac{W - W_s}{W_s} = \frac{177.6 - 153.6}{153.6} = \frac{24}{153.6} \times 100 = \mathbf{15.6\%}$$

The moist unit weight (Equation 1.39) is

$$\gamma = \frac{W}{V} = \frac{177.6}{0.0093} = 19{,}096 \, \text{N/m}^3 \approx \mathbf{19.1 \, kN/m^3}$$

For dry unit weight (Equation 1.40), we have

$$\gamma_d = \frac{W_s}{V} = \frac{153.6}{0.0093} = 16{,}516 \, \text{N/m}^3 \approx \mathbf{16.52 \, kN/m^3}$$

The void ratio (Equation 1.34) is found as follows:

$$e = \frac{V_v}{V_s}$$

$$V_s = \frac{W_s}{G_s\gamma_w} = \frac{0.1536}{2.71 \times 9.81} = 0.0058 \, \text{m}^3$$

$$V_v = V - V_s = 0.0093 - 0.0058 = 0.0035 \, \text{m}^3$$

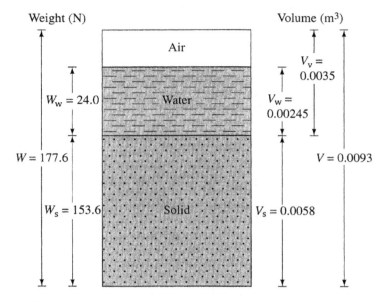

Figure 1.37 Three phases of a soil sample.

So

$$e = \frac{0.0035}{0.0058} \approx 0.60$$

For porosity (Equation 1.36), we have

$$n = \frac{e}{1+e} = \frac{0.60}{1+0.60} = 0.375$$

We find the degree of saturation (Equation 1.37) as follows:

$$S = \frac{V_w}{V_v}$$

$$V_w = \frac{W_w}{\gamma_w} = \frac{0.024}{9.81} = 0.00245 \text{ m}^3$$

So

$$S = \frac{0.00245}{0.0035} \times 100 = 70\%$$

Example 1.4

For a saturated soil, show that

$$\gamma_{sat} = \left(\frac{e}{w}\right)\left(\frac{1+w}{1+e}\right)\gamma_w.$$

Solution

From Equations 1.44 and 1.45,

$$\gamma_{sat} = \frac{(G_s + e)\gamma_w}{1+e} \tag{a}$$

and

$$e = wG_s$$

or

$$G_s = \frac{e}{w} \tag{b}$$

Combining Equations (a) and (b) gives

$$\gamma_{sat} = \frac{\left(\dfrac{e}{w} + e\right)\gamma_w}{1+e} = \left(\frac{e}{w}\right)\left(\frac{1+w}{1+e}\right)\gamma_w$$

1.13 RELATIVE DENSITY AND RELATIVE COMPACTION

Relative density is a term generally used to describe the degree of compaction of coarse-grained soils. Relative density D_r is defined as

$$D_r = \frac{e_{max} - e}{e_{max} - e_{min}} \tag{1.49}$$

where
    e_{max} is the maximum possible void ratio
    e_{min} is the minimum possible void ratio
    e is the void ratio in natural state of soil

Equation 1.49 can also be expressed in terms of dry unit weight of the soil:

$$\gamma_d(\text{max}) = \frac{G_s\gamma_w}{1+e_{min}} \quad \text{or} \quad e_{min} = \frac{G_s\gamma_w}{\gamma_d(\text{max})} - 1 \tag{1.50}$$

Similarly,

$$e_{max} = \frac{G_s \gamma_w}{\gamma_d(min)} - 1 \tag{1.51}$$

and

$$e = \frac{G_s \gamma_w}{\gamma_d} - 1 \tag{1.52}$$

where $\gamma_d(max)$, $\gamma_d(min)$, and γ_d are the maximum, minimum, and natural-state dry unit weights of the soil. Substitution of Equations 1.50 through 1.52 into Equation 1.49 yields

$$D_r = \left[\frac{\gamma_d(max)}{\gamma_d}\right]\left[\frac{\gamma_d - \gamma_d(min)}{\gamma_d(max) - \gamma_d(min)}\right] \tag{1.53}$$

Relative density is generally expressed as a percentage. It has been used by several investigators to correlate the angle of friction of soil, the soil liquefaction potential, etc.

Another term occasionally used in regard to the degree of compaction of coarse-grained soils is *relative compaction*, R_c, which is defined as

$$R_c = \frac{\gamma_d}{\gamma_d(max)} \tag{1.54a}$$

Comparing Equations 1.53 and 1.54a,

$$R_c = \frac{R_o}{1 - D_r(1 - R_o)} \tag{1.54b}$$

where $R_o = \gamma_d(min)/\gamma_d(max)$.

Lee and Singh (1971) reviewed 47 different soils and gave the approximate relation between relative compaction and relative density as

$$R_c = 80 + 0.2D_r \tag{1.54c}$$

where D_r is in percent.

1.13.1 Correlations for relative density of granular soil

Several correlations have been proposed for estimation of relative density from standard penetration test results obtained from field soil exploration programs. Some of those relationships are given below.

Kulhawy and Mayne (1990) modified an empirical relationship for relative density that was given by Marcuson and Bieganousky (1977), which can be expressed as

$$D_r(\%) = 12.2 + 0.75 \left[222N_{60} + 2311 - 711OCR - 779 \left(\frac{\sigma'_o}{p_a} \right) - 50C_u^2 \right]^{0.5}$$

$$(1.55)$$

where

D_r = relative density
N_{60} = standard penetration number for an energy ratio of 60%
σ'_o = effective overburden pressure
C_u = uniformity coefficient of sand

$$OCR = \frac{\text{preconsolidation pressure, } \sigma'_c}{\text{effective overburden pressure, } \sigma'_o}$$

p_a = atmospheric pressure ($\approx$100 kN/m²)

Meyerhof (1957) developed a correlation between D_r and N_{60} as

$$N_{60} = \left[17 + 24 \left(\frac{\sigma'_o}{p_a} \right) \right] D_r^2$$

or

$$D_r = \left\{ \frac{N_{60}}{\left[17 + 24 \left(\frac{\sigma'_o}{p_a} \right) \right]} \right\}^{0.5}$$

$$(1.56)$$

Equation 1.56 provides a reasonable estimate only for clean, medium fine sand.

Cubrinovski and Ishihara (1999) also proposed a correlation between N_{60} and the relative density of sand (D_r) that can be expressed as

$$D_r(\%) = \left[\frac{N_{60} \left(0.23 + \frac{0.06}{D_{50}} \right)^{1.7}}{9} \left(\frac{1}{\frac{\sigma'_o}{p_a}} \right) \right]^{0.5} (100)$$

$$(1.57)$$

where

p_a = atmospheric pressure ($\approx$100 kN/m²)
D_{50} = sieve size through which 50% of the soil will pass (mm)

Kulhawy and Mayne (1990) correlated the corrected standard penetration number and the relative density of sand in the form

$$D_r(\%) = \left[\frac{(N_1)_{60}}{C_p C_A C_{OCR}} \right]^{0.5} (100) \tag{1.58}$$

where

$$C_p = \text{grain-size correlations factor} = 60 + 25 \log D_{50} \tag{1.59}$$

$$C_A = \text{correlation factor for aging} = 1.2 + 0.05 \log \left(\frac{t}{100} \right) \tag{1.60}$$

$$C_{OCR} = \text{correlation factor for overconsolidation} = OCR^{0.18} \tag{1.61}$$

D_{50} = diameter through which 50% of the soil will pass (mm)
t = age of soil since deposition (years)
OCR = overconsolidation ratio

Skempton (1986) suggested that, for sand with a relative density greater than 35%,

$$\frac{(N_1)_{60}}{D_r^2} \approx 60 \tag{1.62}$$

where $(N_1)_{60}$ is N_{60} corrected to an effective overburden pressure of $p_a \approx 100$ kN/m². $(N_1)_{60}$ should be multiplied by 0.92 for coarse sands and 1.08 for fine sands.

More recently Mujtaba et al. (2017) have provided the following correlation for D_r:

$$D_r(\%) = 1.96 N_{60} - 19.2 \left(\frac{p_a}{\sigma_o'} \right)^{0.23} + 29.2 \tag{1.63}$$

For more details of N_{60} and $(N_1)_{60}$, the readers are referred to Das (2016).

1.14 RELATIONSHIP BETWEEN e_{max} AND e_{min}

The maximum and minimum void ratios for granular soils described in Section 1.13 depend on several factors such as

- Grain size
- Grain shape

- Nature of grain-size distribution
- Fine content F_c (i.e., fraction smaller than 0.075 mm)

Following are some of the correlations now available in the literature related to e_{max} and e_{min} of granular soils.

- Clean sand ($F_c = 0\%-5\%$)

Miura et al. (1997) conducted an extensive study of the physical characteristics of about 200 samples of granular material, which included mostly clean sand, some glass beads, and lightweight aggregates (LWA). Figure 1.38 shows a plot of e_{max} versus e_{min} obtained from that study, which shows that

$$e_{max} \approx 1.62 e_{min} \tag{1.64}$$

Cubrinovski and Ishihara (2002) analyzed a large number of clean sand samples based on which it was suggested that

$$e_{max} = 0.072 + 1.53 e_{min} \tag{1.65}$$

Based on best-fit linear regression lines, Cubrinovski and Ishihara (2002) also provided the following relationships for other soils:

- Sand with fines ($5\% < F_c \le 15\%$)

$$e_{max} = 0.25 + 1.37 e_{min} \tag{1.66}$$

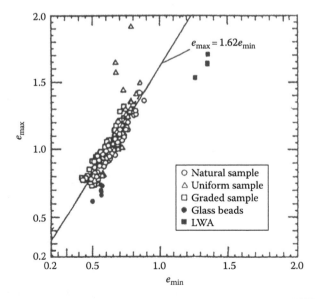

Figure 1.38 Plot of e_{max} versus e_{min} based on the results of Miura et al. (1997).

- Sand with fines and clay ($15\% < F_c \leq 30\%$; $P_c = 5\%\text{--}20\%$)

$$e_{max} = 0.44 + 1.21 e_{min} \tag{1.67}$$

- Silty soils ($30\% < F_c \leq 70\%$; $P_c = 5\%\text{--}20\%$)

$$e_{max} = 0.44 + 1.32 e_{min} \tag{1.68}$$

where
   F_c is the fine fraction for which grain size is smaller than 0.075 mm
   P_c is the clay-size fraction (<0.005 mm)

With a very large database, Cubrinovski and Ishihara (1999, 2002) developed a unique relationship between $e_{max} - e_{min}$ and median grain size D_{50}. The database included results from clean sand, sand with fines, and sand with clay, silty soil, gravelly sand, and gravel. This relationship is shown in Figure 1.39. In spite of some scatter, the average line can be given by the relation

$$e_{max} - e_{min} = 0.23 + \frac{0.06}{D_{50}(mm)} \tag{1.69}$$

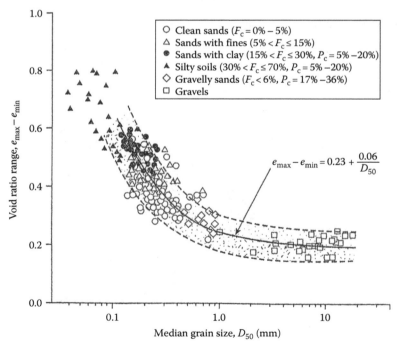

Figure 1.39 Plot of $e_{max} - e_{min}$ versus median grain size (D_{50}). [Redrawn after Cubrinovski and Ishihara, *Soils Found.*, 42(6), 65–78, 2002.]

It appears that the upper and lower limits of $e_{max} - e_{min}$ versus D_{50} as shown in Figure 1.39 can be approximated as

- Lower limit

$$e_{max} - e_{min} = 0.16 + \frac{0.045}{D_{50}(mm)} \qquad (1.70)$$

- Upper limit

$$e_{max} - e_{min} = 0.29 + \frac{0.079}{D_{50}(mm)} \qquad (1.71)$$

Youd (1973) analyzed the variation of e_{max} and e_{min} of several sand samples and provided relationships between angularity A of sand particles and the uniformity coefficient ($C_u = D_{60}/D_{10}$). The angularity, A, of a granular soil particle is defined as

$$A = \frac{\text{Average radius of corners and edges}}{\text{Radius of maximum inscribed sphere}} \qquad (1.72)$$

For further details on angularity the readers may refer to Das and Sobhan (2018).

The qualitative descriptions of sand particles with the range of angularity as provided by Youd (1973) are given below.

- **Very angular**—The particles that have unworn fractured surfaces and multiple sharp corners and edges. The value of A varies within a range of 0.12–0.17 with a mean value of 0.14.
- **Angular**—The particles with sharp corners having prismoidal or tetrahedral shapes with $A = 0.17$–0.25 with a mean value of 0.21.
- **Sub-angular**—The particles have blunted or slightly rounded corners and edges with $A = 0.25$–0.35 with a mean value of about 0.30.
- **Sub-rounded**—The particles have well-rounded edges and corners. The magnitude of A varies in the range of 0.35–0.49 with a mean value of 0.41.
- **Rounded**—The particles are irregularly shaped and rounded with no distinct corners or edges for which $A = 0.49$–0.79 with a mean value of 0.59.
- **Well-rounded**—The particles have a spherical or ellipsoidal shape with $A = 0.7$–1.0 with a mean value of about 0.48.

The variations of e_{max} and e_{min} with criteria described above are given in Figure 1.40. Note that, for a given value of C_u, the maximum and minimum void ratios increase with the decrease in angularity. Also, for a given value of A, the magnitudes of e_{max} and e_{min} decrease with an increase in C_u.

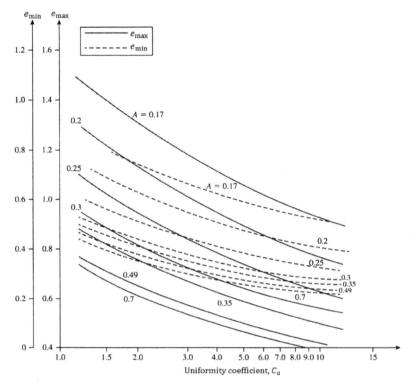

Figure 1.40 Variation of e_{max} and e_{min} with A and C_u. (Adapted from Youd, T. L., Factors controlling maximum and minimum densities of sand, *Evaluation of Relative Density and Its Role in Geotechnical Projects Involving Cohesionless Soils*, STP 523, ASTM, 98–122, 1973.)

1.14.1 Effect of nonplastic fines on e_{max} and e_{min}

The amount of *nonplastic fines* present in a given granular soil has a great influence on e_{max} and e_{min}. In order to visualize this, let us consider the study of McGeary (1961) related to the determination of the minimum void ratio (e_{min}) for idealized spheres (also see Lade et al., 1998). McGeary (1961) conducted tests with mixtures of two different sizes of steel spheres. The larger spheres had a diameter (D) of 3.15 mm. The diameter of the small spheres (d) varied from 0.91 mm to 0.15 mm. This provided a D/d ratio in the range of 3.45 to 19.69. Figure 1.41 shows the variation of e_{min} with the percent of small spheres in the mixture by volume for $D/d = 3.45$ and 4.77. For a given D/d value, the magnitude of e_{min} decreases with the increase in the volume of small spheres to an absolute minimum value $e_{min(min)}$. This occurs when the volume of small spheres in the mix is V_F. Beyond this point, the magnitude of e_{min} increases with the increase in the volume of smaller spheres.

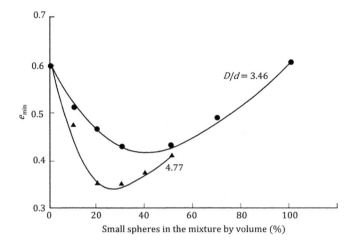

Figure 1.41 Test results of McGeary (1961)—Variation of minimum void ratio with percent of smaller steel spheres by volume.

Table 1.7 provides a summary of all of the test results of McGeary (1961). This is also shown in Figure 1.42, from which it can be concluded that (a) for $D/d \geq 7$, the magnitude of $e_{\text{min(min)}}$ remains approximately constant (≈ 0.2), and (b) at $e_{\text{min(min)}}$, the approximate magnitude of $V_F \approx 27\%$.

In order to compare the preceding experimental results with idealized spheres with the actual soil, we consider the study of Lade et al. (1998), which was conducted with two types: Nevada sand (retained on No. 200 U.S. sieve) and Nevada nonplastic fines (passing No. 200 U.S. sieve). Table 1.8 shows the D_{50} (size through 50% soil will pass) for the two sands and the nonplastic fines. Figure 1.43 shows the variation of e_{max} and e_{min} with percent of fines by volume. From this figure it can be seen that:

Table 1.7 Interpolated values of $e_{\text{min(min)}}$ from binary packing based on the tests of McGeary (1961)

D/d	$e_{\text{min(min)}}$	Approximate volume of small spheres at which $e_{\text{min(min)}}$ occurs, V_F (%)
3.46	0.426	41.3
4.77	0.344	26.2
6.56	0.256	25.0
11.25	0.216	27.5
16.58	0.213	26.3
19.69	0.192	27.5

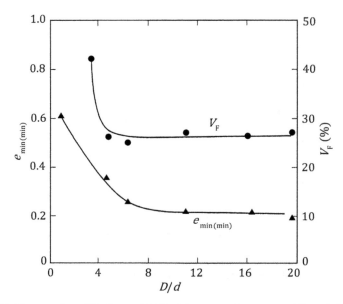

Figure 1.42 Test results of McGeary (1961)—Variation of $e_{min(min)}$ and V_F with D/d.

Table 1.8 $D_{50\text{-sand}}$ and $D_{50\text{-fines}}$ of the soils used by Lade et al. (1998)

Sand description	$D_{50\text{-sand}}$ (mm)	$D_{50\text{-fines}}$ (mm)	$\dfrac{D_{50\text{-sand}}}{D_{50\text{-fines}}}$
Nevada 50/80	0.211	0.050	4.22
Nevada 80/200	0.120	0.050	2.4

- For a given sand and fine mixture, the e_{max} and e_{min} decrease with the increase in the volume of fines from zero to about 30%. This is approximately similar to the behavior of ideal spheres shown in Figures 1.41 and 1.42. This is the *filling-of-the-void phase* where fines tend to fill the void spaces between the larger sand particles.
- There is a *transition zone*, where the percentage of fines is between 30 and 40%.
- For percentage of fines greater than about 40%, the magnitudes of e_{max} and e_{min} start increasing. This is the *replacement-of-solids phase*, where larger-sized solid particles are pushed out and gradually replaced by fines.

1.15 SOIL CLASSIFICATION SYSTEMS

Soil classification is the arrangement of soils into various groups or subgroups to provide a common language to express briefly the general usage characteristics

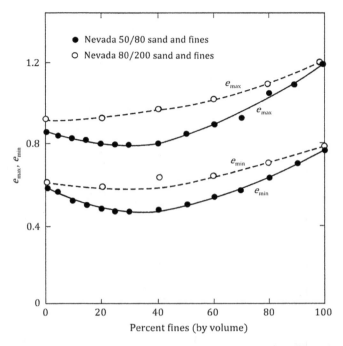

Figure 1.43 Variation of e_{max} and e_{min} with percent of nonplastic fines (Based on the test results of Lade et al., 1998). *Note*: For 50/80 sand and fines, $D_{50\text{-sand}}/D_{50\text{-fines}}$ = 4.22 and, for 80/200 sand and fines, $D_{50\text{-sand}}/D_{50\text{-fines}}$ = 2.4.

without detailed descriptions. At the present time, two major soil classification systems are available for general engineering use. They are the unified system and the American Association of State Highway and Transportation Officials (AASHTO) system. Both systems use simple index properties such as grain-size distribution, liquid limit, and plasticity index of soil.

1.15.1 Unified system

The unified system of soil classification was originally proposed by A. Casagrande in 1948 and was then revised in 1952 by the Corps of Engineers and the U.S. Bureau of Reclamation. In its present form [also see ASTM D-2487, ASTM (2014)], the system is widely used by various organizations, geotechnical engineers in private consulting business, and building codes.

Initially, there are two major divisions in this system. A soil is classified as a coarse-grained soil (gravelly and sandy) if more than 50% is retained on a No. 200 sieve and as a fine-grained soil (silty and clayey) if 50% or more is passing through a No. 200 sieve. The soil is then further classified by a number of subdivisions, as shown in Table 1.9.

Table 1.9 Unified soil classification system

Major divisions		Group symbols	Typical names	Criteria or classification[a]
Coarse-grained soils (<50% passing No. 200 sieve)[a]	Gravels (<50% of coarse fraction passing No. 4 sieve)			
	Gravels with few or no fines	GW	Well-graded gravels; gravel–sand mixtures (few or no fines)	$C_u = \dfrac{D_{60}}{D_{10}} > 4;\quad C_c = \dfrac{(D_{30})^2}{(D_{10})(D_{60})}$ Between 1 and 3
		GP	Poorly graded gravels; gravel–sand mixtures (few or no fines)	Not meeting the two criteria for GW
	Gravels with fines	GM	Silty gravels; gravel–sand–silt mixtures	Atterberg limits below A-line or plasticity index less than 4[b] (see Figure 1.44)
		GC	Clayey gravels; gravel–sand–clay mixtures	Atterberg limits above A-line with plasticity index greater than 7[b] (see Figure 1.44)
	Sands (≥50% of coarse fraction passing No. 4 sieve)			
	Clean sands (few or no fines)	SW	Well-graded sands; gravelly sands (few or no fines)	$C_u = \dfrac{D_{60}}{D_{10}} > 6;\quad C_c = \dfrac{(D_{30})^2}{(D_{10})(D_{60})}$ Between 1 and 3
		SP	Poorly graded sands; gravelly sands (few or no fines)	Not meeting the two criteria for SW

Sands with fines (appreciable amount of fines)	SM	Silty sands; sand–silt mixtures	Atterberg limits below A-line or plasticity index less than 4[b] (see Figure 1.44)
	SC	Clayey sands; sand–clay mixtures	Atterberg limits above A-line with plasticity index greater than 7[b] (see Figure 1.44)
Fine-grained soils (≥50% passing No. 200 sieve)			
Silts and clay (liquid limit less than 50)	ML	Inorganic silts; very fine sands; rock flour; silty or clayey fine sands	See Figure 1.44
	CL	Inorganic clays (low to medium plasticity); gravelly clays; sandy clays; silty clays; lean clays	See Figure 1.44
	OL	Organic silts; organic silty clays (low plasticity)	See Figure 1.44
Silts and clay (liquid limit greater than 50)	MH	Inorganic silts; micaceous or diatomaceous fine sandy or silty soils; elastic silt	See Figure 1.44
	CH	Inorganic clays (high plasticity); fat clays	See Figure 1.44
	OH	Organic clays (medium to high plasticity); organic silts	See Figure 1.44
Highly organic silts	Pt	Peat; mulch; and other highly organic soils	

Group symbols are G, gravel; W, well-graded; S, sand; P, poorly graded; C, clay; H, high plasticity; M, silt; L, low plasticity; O, organic silt or clay; Pt, peat and highly organic soil.

[a] Classification based on percentage of fines: <5% passing No. 200: GW. GP. SW. SP: >12% passing No. 200: GM. GC. SM. SC: 5%–12% passing No. 200: borderline—dual symbols required such as GW–GM. GW–GC. GP–GM. GP–SC. SW–SM. SW–SC. SP–SM. SP–SC.

[b] Atterberg limits above A-line and plasticity index between 4 and 7 are borderline cases. It needs dual symbols (see Figure 1.44).

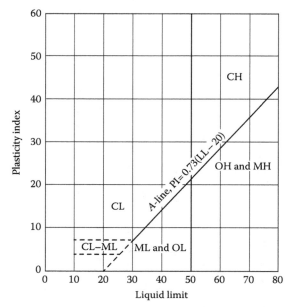

Figure 1.44 Plasticity chart.

Example 1.5

For a soil specimen, given the following,

Passing No. 4 sieve = 92%	Passing No. 40 sieve = 78%
Passing No. 10 sieve = 81%	Passing No. 200 sieve = 65%
Liquid limit = 48	Plasticity index = 32

classify the soil by the unified classification system.

Solution

Since more than 50% is passing through a No. 200 sieve, it is a fine-grained soil, that is, it could be ML, CL, OL, MH, CH, or OH. Now, if we plot LL = 48 and PI = 32 on the plasticity chart given in Figure 1.44, it falls in the zone CL. So the soil is classified as CL.

1.15.2 AASHTO classification system

This system of soil classification was developed in 1929 as the Public Road Administration Classification System. It has undergone several revisions, with the present version proposed by the Committee on Classification of Materials for Subgrades and Granular Type Roads of the Highway Research Board in 1945 [ASTM (2014) Test Designation D-3282].

The AASHTO classification system in present use is given in Table 1.10. According to this system, soil is classified into seven major

Table 1.10 Classification of highway subgrade materials

General classification	Granular materials (35% or less of total sample passing No. 200 sieve)						
	A-1		A-3	A-2			
Group classification	A-1-a	A-1-b	A-3	A-2-4	A-2-5	A-2-6	A-2-7
Sieve analysis (percent passing)							
No. 10	50 max.						
No. 40	30 max.	50 max.	50 min.				
No. 200	15 max.	25 max.	10 max.	35 max.	35 max.	35 max.	35 max.
Characteristics of fraction passing No. 40							
Liquid limit				40 max.	41 min.	40 max.	41 min.
Plasticity index	6 max.		NP	10 max.	10 max.	11 min.	11 min.
Usual types of significant constituent materials	Stone fragments, gravel, and sand		Fine sand	Silty or clayey gravel and sand			
General subgrade rating	Excellent to good						

General classification	Silt–clay materials (more than 35% or total sample passing No. 200 sieve)			
				A-7 A-7-5[a]
Group classification	A-4	A-5	A-6	A-7-6[b]
Sieve analysis (percent passing)				
No. 10				
No. 40				
No. 200	36 min.	36 min.	36 min.	36 min.
Characteristics of fraction passing No. 40				
Liquid limit	40 max.	41 min.	40 max.	41 min.
Plasticity index	10 max.	10 max.	11 min.	11 min.
Usual types of significant constituent materials	Silty soils		Clayey soils	
General subgrade rating	Fair to poor			

[a] For A-7-5, PI $\leq$ LL − 30.
[b] For A-7-6, PI > LL − 30.

groups: A-1 through A-7. Soils classified into Groups A-1, A-2, and A-3 are granular materials, where 35% or less of the particles pass through the No. 200 sieve. Soils where more than 35% pass through the No. 200 sieve are classified into groups A-4, A-5, A-6, and A-7. These are mostly silt and clay-type materials. The classification system is based on the following criteria:

1. Grain size
   Gravel: Fraction passing the 75 mm sieve and retained on No. 10 (2 mm) U.S. sieve
   Sand: Fraction passing the No. 10 (2 mm) U.S. sieve and retained on the No. 200 (0.075 mm) U.S. sieve
   Silt and clay: Fraction passing the No. 200 U.S. sieve
2. Plasticity: The term *silty* is applied when the fine fractions of the soil have a plasticity index of 10 or less. The term *clayey* is applied when the fine fractions have a plasticity index of 11 or more.
3. If cobbles and *boulders* (size larger than 75 mm) are encountered, they are excluded from the portion of the soil sample on which classification is made. However, the percentage of such material is recorded.

To classify a soil according to Table 1.10, the test data are applied from left to right. By the process of elimination, the first group from the left into which the test data will fit is the correct classification.

For the evaluation of the quality of a soil as a highway subgrade material, a number called the *group index* (GI) is also incorporated with the groups and subgroups of the soil. The number is written in parentheses after the group or subgroup designation. The group index is given by the equation

$$GI = (F - 35)[0.2 + 0.005(LL - 40)] + 0.01(F - 15)(PI - 10) \qquad (1.73)$$

where
   F is the percent passing the No. 200 sieve
   LL is the liquid limit
   PI is the plasticity index

The first term of Equation 1.73—that is, $(F - 35)[0.2 + 0.005(LL - 40)]$— is the partial group index determined from the liquid limit. The second term—that is, $0.01(F - 15) (PI - 10)$—is the partial group index determined from the plasticity index. Following are the rules for determining the group index:

1. If Equation 1.73 yields a negative value for GI, it is taken as 0.
2. The group index calculated from Equation 1.73 is rounded off to the nearest whole number (e.g., GI = 3.4 is rounded off to 3; GI = 3.5 is rounded off to 4).

3. There is no upper limit for the group index.
4. The group index of soils belonging to groups A-1-a, A-1-b, A-2-4, A-2-5, and A-3 is always 0.
5. When calculating the group index for soils that belong to groups A-2-6 and A-2-7, use the partial group index for PI, or

$$GI = 0.01(F-15)(PI-10) \tag{1.74}$$

In general, the quality of performance of a soil as a subgrade material is inversely proportional to the group index.

Example 1.6

Classify the following soil by the AASHTO classification system.

Passing No. 10 sieve: 100%
Passing No. 40 sieve: 92%
Passing No. 200 sieve: 86%
Liquid limit (LL): 70
Plasticity index (PI): 32

Solution

Percent passing the No. 200 sieve is 86%. So, it is a silty clay material (i.e., A-4, A-5, A-6, or A-7) as shown in Table 1.10. Proceeding from left to right, we see that it falls under A-7. For this case, PI = 32 < LL − 30. So, this is A-7-5. From Equation 1.73

$$GI = (F-35)[0.2 + 0.005(LL-40)] + 0.01(F-15)(PI-10)$$

Now, $F = 86$; LL = 70; PI = 32; so

$$GI = (86-35)[0.2 + 0.005(70-40)] + 0.01(86-15)(32-10)$$

$$= 33.47 \approx 33$$

Thus, the soil is A-7-5(33).

1.16 COMPACTION

Compaction of loose fills is a simple way of increasing the stability and load-bearing capacity of soils, and this is generally achieved by using smooth-wheel rollers, sheepsfoot rollers, rubber-tired rollers, and vibratory rollers. In order to write the specifications for field compaction, Proctor compaction tests are generally conducted in the laboratory. A brief description of the Proctor compaction test procedure is as follows:

1.16.1 Standard Proctor compaction test

A standard laboratory soil compaction test was first developed by Proctor (1933), and this is usually referred to as the *standard Proctor test* (ASTM designation D-698). The test is conducted by compaction of three layers of soil in a mold that is 944 cm³ in volume. Each layer of soil is subjected to 25 blows by a hammer weighing 24.6 N with a 304.8 mm drop. From the known volume of the mold, weight of moist compacted soil in the mold, and moisture content of the compacted soil, the dry unit weight of compaction can be determined as

$$\gamma_{moist} = \frac{\text{Weight of moist soil in the mold}}{\text{Volume of the mold}}$$

$$\gamma_d = \frac{\gamma_{moist}}{1 + w}$$

where
  γ_{moist} is the moist unit weight of compacted soil
  γ_d is the dry unit weight of compacted soil
  w is the moisture content of soil

The test can be repeated several times at various moist contents of soil. By plotting a graph of γ_d against the corresponding moisture content, the optimum moisture content w_{opt} and the maximum dry unit weight $\gamma_{d(max)}$ can be obtained (Figure 1.45). Also plotted in Figure 1.45 is the variation of

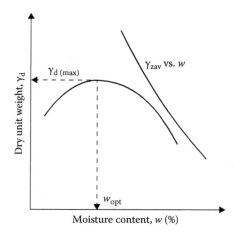

Figure 1.45 Nature of variation of γ_d versus w.

the dry unit weights, assuming the degree of saturation to be 100%. These are the *theoretical maximum* dry unit weights that can be attained for a given moisture content when there will be no air in the void spaces. With the degree of saturation as 100%

$$e = wG_s \tag{1.75}$$

The maximum dry unit weight at a given moisture content with zero air voids can be given by (Equation 1.42)

$$\gamma_{zav} = \frac{G_s\gamma_w}{1+e} = \frac{G_s\gamma_w}{1+wG_s} = \frac{\gamma_w}{(1/G_s)+w} \tag{1.76}$$

where γ_{zav} is the zero-air-void unit weight (dry).

For standard Proctor compaction test, the compaction energy E can be expressed as

$$E = \frac{(24.5\,\text{N/blow})(3\,\text{layers})(25\,\text{blows/layer})(0.3048\,\text{m})}{(944/10^6)\text{m}^3}$$

$$= 593,294\,\text{N-m/m}^3 \approx 593\,\text{kN-m/m}^3$$

1.16.2 Modified Proctor compaction test

With the development of heavier compaction equipment, the standard Proctor test has been modified for better representation of field conditions. In the *modified Proctor test* (ASTM designation D-1577), the same mold as in the standard Proctor test is used. However, the soil is compacted in 5 layers, with a 44.5 N hammer giving 25 blows to each layer. The height of drop of the hammer is 457.2 mm. Hence, the compactive effort in the modified Proctor test is equal to

$$E = \frac{(25\,\text{blows/layer})(5\,\text{layers})(44.5\,\text{N/blow})(0.4572\,\text{m})}{(944/10^6)\,\text{m}^3}$$

$$= 2,694,041\,\text{N-m/m}^3 \approx 2604\,\text{kN-m/m}^3$$

The maximum dry unit weight obtained from the modified Proctor test will be higher than that obtained from the standard Proctor test due to the application of higher compaction energy. It will also be accompanied by a lower optimum moisture content compared to that obtained from the standard Proctor compaction test.

1.17 EMPIRICAL RELATIONSHIPS FOR PROCTOR COMPACTION TESTS

Omar et al. (2003) presented the results of modified Proctor compaction tests on 311 soil samples. Of these samples, 45 were gravelly soil (GP, GP–GM, GW, GW–GM, and GM), 264 were sandy soil (SP, SP–SM, SW–SM, SW, SC–SM, SC, and SM), and two were clay with low plasticity (CL). Based on the tests, the following correlations were developed:

$$\rho_{d(max)} = [4,804,574G_s - 195.55(LL)^2 + 156,971(R\#4)^{0.5} - 9,527,830]^{0.5}$$

(1.77)

$$\ln(w_{opt}) = 1.195 \times 10^{-4}(LL)^2 - 1.964G_s - 6.617 \times 10^{-5}(R\#4) + 7.651$$

(1.78)

where
$\rho_{d(max)}$ is the maximum dry density
w_{opt} is the optimum moisture content (%)
G_s is the specific gravity of soil solids
LL is the liquid limit, in percent
R#4 is the percent retained on No. 4 sieve

For granular soils with less than 12% fines (i.e., finer than No. 200 sieve), relative density may be a better indicator for end product compaction specification in the field. Based on laboratory compaction tests on 55 clean sands (less than 5% finer than No. 200 sieve), Patra et al. (2010) provided the following relationships:

$$D_r = AD_{50}^{-B}$$

(1.79)

$$A = 0.216\ln E - 0.850$$

(1.80)

$$B = -0.03\ln E + 0.306$$

(1.81)

where
D_r is the maximum relative density of compaction achieved with compaction energy E, kN-m/m^3
D_{50} is the median grain size, mm

Gurtug and Sridharan (2004) proposed correlations for optimum moisture content and maximum dry unit weight with the plastic limit PL of cohesive soils. These correlations can be expressed as

$$w_{opt}(\%) = [1.95 - 0.38(\log E)](PL) \qquad (1.82)$$

$$\gamma_{d(max)}(kN/m^3) = 22.68e^{0.0183w_{opt}(\%)} \qquad (1.83)$$

where
   PL is the plastic limit, %
   E is the compaction energy, kN-m/m³

For modified Proctor test, $E \approx 2700$ kN/m³. Hence,

$$w_{opt}(\%) \approx 0.65(PL)$$

$$\gamma_{d(max)}(kN/m^3) \approx 22.68e^{-0.012(PL)}$$

Osman et al. (2008) analyzed a number of laboratory compaction test results on fine-grained (cohesive) soil, including those provided by Gurtug and Sridharan (2004). Based on this study, the following correlations were developed:

$$w_{opt}(\%) \approx (1.99 - 0.165\ln E)(PI) \qquad (1.84)$$

$$\gamma_{d(max)}(kN/m^3) \approx L - Mw_{opt}(\%) \qquad (1.85)$$

where

$$L = 14.34 + 1.195\ln E \qquad (1.86)$$

$$M = -0.19 + 0.073\ln E \qquad (1.87)$$

w_{opt} is the optimum moisture content, %
PI is the plasticity index, %
$\gamma_{d(max)}$ is the maximum dry unit weight, kN/m³
E is the compaction energy, kN-m/m³

DiMatteo et al. (2009) analyzed the results of 71 fine-grained soils and provided the following correlations for optimum moisture content w_{opt} and maximum dry unit weight $\gamma_{d(max)}$ for modified Proctor tests ($E = 2700$ kN-m/m³)

$$w_{opt}(\%) = -0.86(LL) + 3.04\left(\frac{LL}{G_s}\right) + 2.2 \qquad (1.88)$$

$$\gamma_{d(max)}(kN/m^3) = 40.316\left(w_{opt}^{-0.295}\right)(PI^{0.32}) - 2.4 \qquad (1.89)$$

where

    LL is the liquid limit, %
    PI is the plasticity index, %
    G_s is the specific gravity of soil solids

Mujtaba et al. (2013) conducted laboratory compaction tests on 110 sandy soil samples (SM, SP-SM, SP, SW-SM, and SW). Based on the test results the following correlations were provided for $\gamma_{d(max)}$ and w_{opt} (optimum moisture content):

$$\gamma_{d(max)}(kN/m^3) = 4.49 \log(C_u) + 151 \log(E) + 10.2 \tag{1.90}$$

$$\log w_{opt} = 1.67 - 0.193 \log(C_u) - 0.153 \log(E) \tag{1.91}$$

where

    C_u = uniformity coefficient
    E = compaction energy (kN-m/m³)

Example 1.7

For a sand with 4% finer than No. 200 sieve, estimate the maximum relative density of compaction that may be obtained from a modified Proctor test. Given D_{50} = 1.4 mm.

Solution

For the modified Proctor test, E = 2696 kN-m/m³.
From Equation 1.80

$$A = 0.216 \ln E - 0.850 = (0.216)(\ln 2696) - 0.850 = 0.856$$

From Equation 1.81

$$B = -0.03 \ln E + 0.306 = -(0.03)(\ln 2696) + 0.306 = 0.069$$

From Equation 1.79

$$D_r = A D_{50}^{-B} = (0.856)(1.4)^{-0.069} = 0.836 = 83.6\%$$

Example 1.8

For a silty clay soil given LL = 43 and PL = 18. Estimate the maximum dry unit weight of compaction that can be achieved by conducting modified Proctor test. Use Equation 1.85.

Solution

For the modified Proctor test, $E = 2696$ kN-m/m^3.
From Equations 1.86 and 1.87

$$L = 14.34 + 1.195 \ln E = 14.34 + 1.195 \ln(2696) = 23.78$$

$$M = -0.19 + 0.073 \ln E = -0.19 + 0.073 \ln(2696) = 0.387$$

From Equation 1.84

$$w_{opt}\,(\%) = (1.99 - 0.165 \ln E)(PI)$$

$$= [1.99 - 0.165 \ln(2696)](43 - 18)$$

$$= 17.16\%$$

From Equation 1.85

$$\gamma_{d(max)} = L - M w_{opt} = 23.78 - (0.387)(17.16) = 17.14 \text{ kN/m}^3$$

REFERENCES

American Society for Testing and Materials, *Annual Book of ASTM Standards*, sec. 4, vol. 04.08, ASTM, West Conshohocken, PA, 2014.

Atterberg, A., Uber die Physikalische Bodenuntersuschung und Uber die Plastizitat der Tone, *Int. Mitt. Bodenkunde*, 1, 5, 1911.

Bolt, G. H., Analysis of validity of Gouy-Chapman theory of the electric double layer, *J. Colloid Sci.*, 10, 206, 1955.

Bolt, G. H., Physical chemical analysis of compressibility of pure clay, *Geotechnique*, 6(2), 86, 1956.

Budhu, M., The effect of clay content on liquid limit from a fall cone and the British cup device, *Geotech. Testing J.*, ASTM, 8(2), 91–95, 1985.

Casagrande, A., Classification and identification of soils, *Trans.* ASCE, 113, 901–930, 1948.

Cubrinovski, M. and K. Ishihara, Empirical correlation between SPT N-value and relative density for sandy soils, *Soils Found.*, 39(5), 61–71, 1999.

Cubrinovski, M. and K. Ishihara, Maximum and minimum void ratio characteristics of sands, *Soils Found.*, 42(6), 65–78, 2002.

Das, B., *Principles of Foundation Engineering*, 8th edn., Cengage, Boston, MA, 2016.

Das, B. and K. Sobhan, *Principles of Geotechnical Engineering*, 9th edn., Cengage, Stamford, CT, 2018.

DiMatteo, L. D., F. Bigotti, and R. Rico, Best-fit model to estimate proctor properties of compacted soil, *J. Geotech. Geoenviron. Eng.*, ASCE, 135(7), 992–996, 2009.

Feng, T. W., A linear log d – log w model for the determination of consistency limits of soils, *Can. Geotech. J.*, 38(6), 1335–1342, 2001.

Grim, R. E., Physico-Chemical Properties of Soils, *J. Soil Mech. Found. Div.*, ASCE, 85(SM2), 1–17, 1959.

Gurtug, Y. and A. Sridharan, Compaction behaviour and prediction of its characteristics of fine grained soils with particular reference to compaction energy, *Soils Found.*, 44(5), 27–36, 2004.

Kulhawy, F. H. and P. W. Mayne, *Manual for Estimating Soil Properties for Foundation Design,* Final Report (EL-6800), Electric Power Research Institute (EPRI), Palo Alto, CA , 1990.

Lade, P. V., C. D. Liggio, and J. A. Yamamuro, Effects of non-plastic fines on minimum and maximum void ratios of sand, *Geotech. Testing J.*, ASTM, 21(4), 336–347, 1998.

Lambe, T. W., Compacted clay: Structure, *Trans.* ASCE, 125, 682–717, 1960.

Lee, K. L. and A. Singh, Relative density and relative compaction, *J. Soil Mech. Found. Div.*, ASCE, 97(SM7), 1049–1052, 1971.

McGeary, R. K., Mechanical packing of spherical particles, *J. Am. Ceramic Soc.*, 44(11), 513–522, 1961.

Meyerhof, G. G., Discussion on research on determining the density of sands by spoon penetration testing, *Proc.4th Int. Conf. Soil Mech. Found. Eng.*, 3(110), 1957.

Miura, K., K. Maeda, M. Furukama, and S. Toki, Physical characteristics of sands with different primary properties, *Soils Found.*, 37(3), 53–64, 1997.

Mujtaba, H., K. Farooq, N. Sivakugan, and B. M. Das, Correlation between gradation parameters and compaction characteristics of sandy soil, *Int. J. Geotech. Eng.*, Maney Publishing, U.K., 7(4), 395–401, 2013.

Mujtaba, H., K. Farooq, N. Sivakugan, and B. M. Das, Evaluation of relative density and friction angle based on SPT-N values, *J. Civ. Eng.*, KSCE, DOI10.1007/s 12205-017-1899-5, 2017.

Nagaraj, T. S. and M. S. Jayadeva, Re-examination of one-point methods of liquid limit determination, *Geotechnique*, 31(3), 413–425, 1981.

Omar, M., S. Abdallah, A. Basma, and S. Barakat, Compaction characteristics of granular soils in United Arab Emirates, *Geotech. Geol. Eng.*, 21(3), 283–295, 2003.

Osman, S., E. Togrol, and C. Kayadelen, Estimating compaction behavior of fine-grained soils based on compaction energy, *Can. Geotech. J.*, 45(6), 877–887, 2008.

Patra, C. R., N. Sivakugan, B. M. Das, and S. K. Rout, Correlation of relative density of clean sand with median grain size and compaction energy, *Int. J. Geotech. Eng.*, 4(2), 196–203, 2010.

Proctor, R. R., Design and construction of rolled earth dams, *Eng. News Record*, 3, 245–248, 286–289, 348–351, 372–376, 1933.

Seed, H. B., R. J. Woodward, and R. Lundgren, Clay mineralogical aspects of the Atterberg limits, *J. Soil Mech. Found. Eng. Div.*, ASCE, 90(SM4), 107–131, 1964a.

Seed, H. B., R. J. Woodward, and R. Lundgren, Fundamental aspects of Atterberg limits, *J. Soil Mech. Found. Eng. Div.*, ASCE, 90(SM6), 75–105, 1964b.

Skempton, A. W., The colloidal activity of clay, *Proc. 3rd Int. Conf. Soil Mech. Found. Eng.*, Zurich, Switzerland, Vol. 1, pp. 57–61, 1953.

Skempton, A. W., Standard penetration test procedures and the effects in sand of overburden pressure, relative density, particle size, aging, and overconsolidation, *Geotechnique*, 36(3), 425, 1986.

Sridharan, A., H. B. Nagaraj, and K. Prakash, Determination of the plasticity index from flow index, *Geotech. Testing J.*, ASTM, 22(2), 175–181, 1999.

Verweg, E. J. W. and J. Th. G. Overbeek, *Theory of Stability of Lyophobic Colloids*, Elsevier-North Holland, Amsterdam, the Netherlands, 1948.

Wroth, C. P. and D. M. Wood, The correlation of index properties with some basic engineering properties of soils, *Can. Geotech. J.*, 15(2), 137–145, 1978.

Youd, T. L., Factors controlling maximum and minimum densities of sand, *Evaluation of Relative Density and Its Role in Geotechnical Projects Involving Cohesionless Soils*, STP 523, ASTM, 98–122, 1973.

Chapter 2

Stresses and strains
Elastic equilibrium

2.1 INTRODUCTION

An important function in the study of soil mechanics is to predict the stresses and strains imposed at a given point in a soil mass due to certain loading conditions. This is necessary to estimate settlement and to conduct stability analysis of earth and earth-retaining structures, as well as to determine stress conditions on underground and earth-retaining structures.

An idealized stress–strain diagram for a material is shown in Figure 2.1. At low stress levels, the strain increases linearly with stress (branch *ab*), which is the elastic range of the material. Beyond a certain stress level, the material reaches a plastic state, and the strain increases with no further increase in stress (branch *bc*). The theories of stresses and strains presented in this chapter are for the elastic range only. In determining stress and strain in a soil medium, one generally resorts to the principles of the theory of elasticity, although soil in nature is not fully homogeneous, elastic, or isotropic. However, the results derived from the elastic theories can be judiciously applied to the problem of soil mechanics.

2.2 BASIC DEFINITION AND SIGN CONVENTIONS FOR STRESSES

An elemental soil mass with sides measuring dx, dy, and dz is shown in Figure 2.2. Parameters σ_x, σ_y, and σ_z are the normal stresses acting on the planes normal to the x, y, and z axes, respectively. The normal stresses are considered positive when they are directed onto the surface. Parameters τ_{xy}, τ_{yx}, τ_{yz}, τ_{zy}, τ_{zx}, and τ_{xz} are shear stresses. The notations for the shear stresses follow.

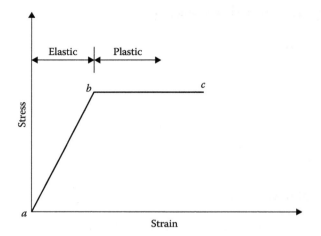

Figure 2.1 Idealized stress–strain diagram.

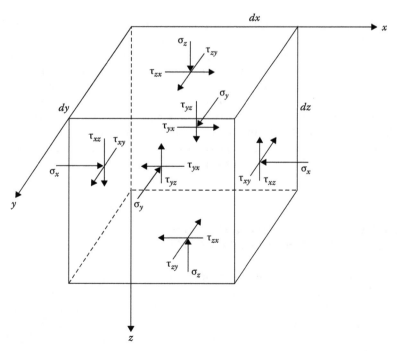

Figure 2.2 Notations for normal and shear stresses in a Cartesian coordinate system.

If τ_{ij} is a shear stress, it means the stress is acting on a plane normal to the i axis, and its direction is parallel to the j axis. A shear stress τ_{ij} is considered positive if it is directed in the negative j direction while acting on a plane whose outward normal is the positive i direction. For example, all shear stresses are positive in Figure 2.2. For equilibrium

$$\tau_{xy} = \tau_{yx} \tag{2.1}$$

$$\tau_{xz} = \tau_{zz} \tag{2.2}$$

$$\tau_{yz} = \tau_{zy} \tag{2.3}$$

Figure 2.3 shows the notations for the normal and shear stresses in a polar coordinate system (two-dimensional case). For this case, σ_r and σ_θ are the normal stresses, and $\tau_{r\theta}$ and $\tau_{\theta r}$ are the shear stresses. For equilibrium, $\tau_{r\theta} = \tau_{\theta r}$. Similarly, the notations for stresses in a cylindrical coordinate system are shown in Figure 2.4. Parameters σ_r, σ_θ, and σ_z are the normal stresses, and the shear stresses are $\tau_{r\theta} = \sigma_{\theta r}$, $\sigma_{\theta z} = \sigma_{z\theta}$, and $\tau_{rz} = \tau_{zr}$.

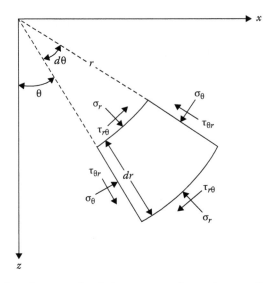

Figure 2.3 Notations for normal and shear stresses in a polar coordinate system.

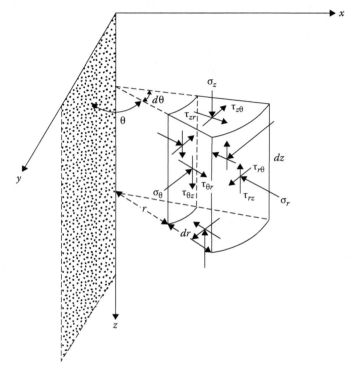

Figure 2.4 Notations for normal and shear stresses in cylindrical coordinates.

2.3 EQUATIONS OF STATIC EQUILIBRIUM

Figure 2.5 shows the stresses acting on an elemental soil mass with sides measuring dx, dy, and dz. Let γ be the unit weight of the soil. For equilibrium, summing the forces in the x direction

$$\Sigma F_x = \left[\sigma_x - \left(\sigma_x + \frac{\partial \sigma_x}{\partial x} dx \right) \right] dy\, dz + \left[\tau_{zx} - \left(\tau_{zx} + \frac{\partial \tau_{zx}}{\partial z} dz \right) \right] dx\, dy$$

$$+ \left[\tau_{yx} - \left(\tau_{yx} + \frac{\partial \tau_{yx}}{\partial y} dy \right) \right] dx\, dz = 0$$

or

$$\frac{\partial \sigma_x}{\partial x} + \frac{\partial \tau_{yx}}{\partial y} + \frac{\partial \tau_{zx}}{\partial z} = 0 \tag{2.4}$$

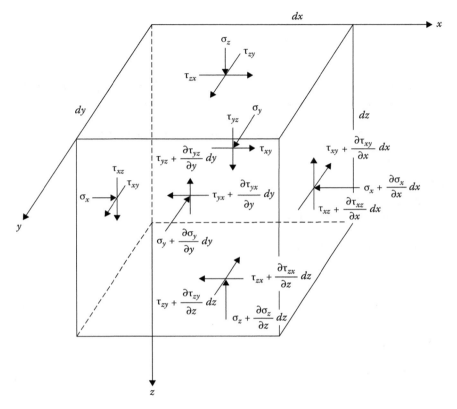

Figure 2.5 Derivation of equations of equilibrium.

Similarly, along the y direction, $\Sigma F_y = 0$, or

$$\frac{\partial \sigma_y}{\partial y} + \frac{\partial \tau_{xy}}{\partial x} + \frac{\partial \tau_{zy}}{\partial z} = 0 \tag{2.5}$$

Along the z direction

$$\Sigma F_z = \left[\sigma_z - \left(\sigma_z + \frac{\partial \sigma_z}{\partial z} dz\right)\right] dx\, dy + \left[\tau_{xz} - \left(\tau_{xz} + \frac{\partial \tau_{xz}}{\partial x} dx\right)\right] dy\, dz$$

$$+ \left[\tau_{yz} - \left(\tau_{yz} + \frac{\partial \tau_{yz}}{\partial y} dy\right)\right] dx\, dz + \gamma(dx\, dy\, dz) = 0$$

The last term of the preceding equation is the self-weight of the soil mass.

Thus

$$\frac{\partial \sigma_z}{\partial z} + \frac{\partial \tau_{xz}}{\partial x} + \frac{\partial \tau_{yz}}{\partial y} - \gamma = 0 \tag{2.6}$$

Equations 2.4 through 2.6 are the static equilibrium equations in the Cartesian coordinate system. These equations are written in terms of *total stresses*.

They may, however, be written in terms of *effective stresses* as

$$\sigma_x = \sigma'_x + u = \sigma'_x + \gamma_w h \tag{2.7}$$

where
  σ'_x is the effective stress
  u is the pore water pressure
  γ_w is the unit weight of water
  h is the pressure head

Thus

$$\frac{\partial \sigma_x}{\partial x} = \frac{\partial \sigma'_x}{\partial x} + \gamma_w \frac{\partial h}{\partial x} \tag{2.8}$$

Similarly

$$\frac{\partial \sigma_y}{\partial y} = \frac{\partial \sigma'_y}{\partial y} + \gamma_w \frac{\partial h}{\partial y} \tag{2.9}$$

and

$$\frac{\partial \sigma_z}{\partial z} = \frac{\partial \sigma'_z}{\partial z} + \gamma_w \frac{\partial h}{\partial z} \tag{2.10}$$

Substitution of the proper terms in Equations 2.4 through 2.6 results in

$$\frac{\partial \sigma'_x}{\partial x} + \frac{\partial \tau_{yx}}{\partial y} + \frac{\partial \tau_{zx}}{\partial z} + \gamma_w \frac{\partial h}{\partial x} = 0 \tag{2.11}$$

$$\frac{\partial \sigma'_y}{\partial y} + \frac{\partial \tau_{xy}}{\partial x} + \frac{\partial \tau_{zy}}{\partial z} + \gamma_w \frac{\partial h}{\partial y} = 0 \tag{2.12}$$

$$\frac{\partial \sigma'_z}{\partial z} + \frac{\partial \tau_{xz}}{\partial x} + \frac{\partial \tau_{yz}}{\partial y} + \gamma_w \frac{\partial h}{\partial z} - \gamma' = 0 \tag{2.13}$$

where γ' is the effective unit weight of soil. Note that the shear stresses will not be affected by the pore water pressure.

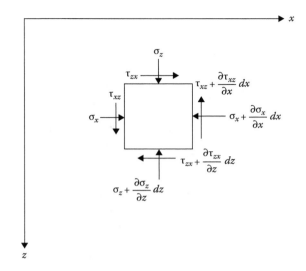

Figure 2.6 Derivation of static equilibrium equation for a two-dimensional problem in Cartesian coordinates.

In soil mechanics, a number of problems can be solved by two-dimensional stress analysis. Figure 2.6 shows the cross-section of an elemental soil prism of unit length with the stresses acting on its faces. The static equilibrium equations for this condition can be obtained from Equations 2.4 through 2.6 by substituting $\tau_{xy} = \tau_{yx} = 0$, $\tau_{yz} = \tau_{zy} = 0$, and $\partial\sigma_y/\partial y = 0$. Note that $\tau_{xz} = \tau_{zx}$. Thus

$$\frac{\partial\sigma_x}{\partial x} + \frac{\partial\tau_{xz}}{\partial z} = 0 \tag{2.14}$$

$$\frac{\partial\sigma_z}{\partial z} + \frac{\partial\tau_{xz}}{\partial x} - \gamma = 0 \tag{2.15}$$

Figure 2.7 shows an elemental soil mass in polar coordinates. Parameters σ_r and σ_θ are the normal components of stress in the radial and tangential directions, and $\tau_{\theta r}$ and $\tau_{r\theta}$ are the shear stresses. In order to obtain the static equations of equilibrium, the forces in the radial and tangential directions need to be considered. Thus

$$\Sigma F_r = \left[\sigma_r r\, d\theta - \left(\sigma_r + \frac{\partial\sigma_r}{\partial r}\, dr\right)(r + dr)\, d\theta\right]$$

$$+ \left[\sigma_\theta\, dr \sin d\theta/2 + \left(\sigma_\theta + \frac{\partial\sigma_\theta}{\partial\theta}\, d\theta\right) dr \sin d\theta/2\right]$$

$$+ \left[\tau_{\theta r}\, dr \cos d\theta/2 - \left(\tau_{\theta r} + \frac{\partial\tau_{\theta r}}{\partial\theta}\, d\theta\right) dr \cos d\theta/2\right] + \gamma(r\, d\theta\, dr)\cos\theta = 0$$

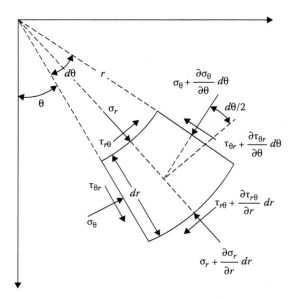

Figure 2.7 Derivation of static equilibrium equation for a two-dimensional problem in polar coordinates.

Taking sin $d\theta/2 \approx d\theta/2$ and cos $d\theta/2 \approx 1$, neglecting infinitesimally small quantities of higher order, and noting that $\partial(\sigma_r r)/\partial r = r(\partial\sigma_r/\partial r) + \sigma_r$ and $\tau_{\theta r} = \tau_{r\theta}$, the previous equation yields

$$\frac{\partial\sigma_r}{\partial r} + \frac{1}{r}\frac{\partial\tau_{r\theta}}{\partial\theta} + \frac{\sigma_r - \sigma_\theta}{r} - \gamma\cos\theta = 0 \tag{2.16}$$

Similarly, the static equation of equilibrium obtained by adding the components of forces in the tangential direction is

$$\frac{1}{r}\frac{\partial\sigma_\theta}{\partial\theta} + \frac{\partial\tau_{r\theta}}{\partial r} + \frac{2\tau_{r\theta}}{r} + \gamma\sin\theta = 0 \tag{2.17}$$

The stresses in the cylindrical coordinate system on a soil element are shown in Figure 2.8. Summing the forces in the radial, tangential, and vertical directions, the following relations are obtained:

$$\frac{\partial\sigma_r}{\partial r} + \frac{1}{r}\frac{\partial\tau_{r\theta}}{\partial\theta} + \frac{\partial\tau_{zr}}{\partial z} + \frac{\sigma_r - \sigma_\theta}{r} = 0 \tag{2.18}$$

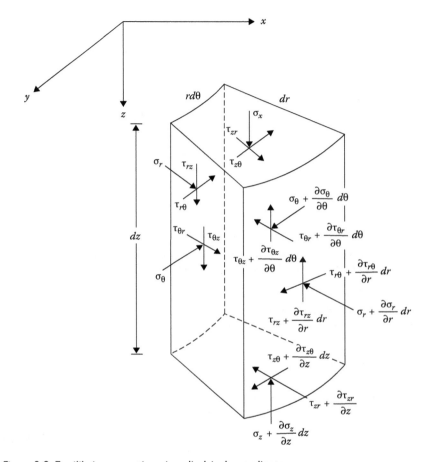

Figure 2.8 Equilibrium equations in cylindrical coordinates.

$$\frac{\partial \tau_{r\theta}}{\partial r} + \frac{1}{r}\frac{\partial \sigma_{\theta}}{\partial \theta} + \frac{\partial \tau_{\theta z}}{\partial z} + \frac{2\tau_{r\theta}}{r} = 0 \qquad (2.19)$$

$$\frac{\partial \tau_{zr}}{\partial r} + \frac{1}{r}\frac{\partial \tau_{\theta z}}{\partial \theta} + \frac{\partial \sigma_{z}}{\partial z} + \frac{\tau_{zr}}{r} - \gamma = 0 \qquad (2.20)$$

2.4 CONCEPT OF STRAIN

Consider an elemental volume of soil as shown in Figure 2.9a. Owing to the application of stresses, point A undergoes a displacement such that its components in the x, y, and z directions are u, v, and w, respectively. The adjacent point B undergoes displacements of $u + (\partial u/\partial x)dx$, $v + (\partial v/\partial x)dx$,

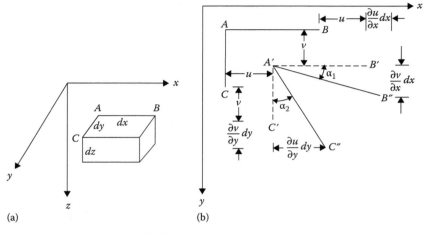

Figure 2.9 Concept of strain: (a) elemental volume of soil measuring dx dy dz; (b) rotation of sides AB and AC of the elemental volume.

and $w + (\partial w/\partial x)dx$ in the x, y, and z directions, respectively. So, the change in the length AB in the x direction is $u + (\partial u/\partial x)dx - u = (\partial u/\partial x)dx$. Hence, the strain in the x direction, ϵ_x, can be given as

$$\epsilon_x = \frac{1}{dx}\left(\frac{\partial u}{\partial x}dx\right) = \frac{\partial u}{\partial x} \qquad (2.21)$$

Similarly, the strains in the y and z directions can be written as

$$\epsilon_y = \frac{\partial v}{\partial y} \qquad (2.22)$$

$$\epsilon_z = \frac{\partial w}{\partial z} \qquad (2.23)$$

where ϵ_y and ϵ_z are the strains in the y and z directions, respectively.

Owing to stress application, sides AB and AC of the element shown in Figure 2.9a undergo a rotation as shown in Figure 2.9b (see $A'B''$ and $A'C''$). The small change in angle for side AB is α_1, the magnitude of which may be given as $[(\partial v/\partial x)dx](1/dx) = \partial v/\partial x$, and the magnitude of the change in angle α_2 for side AC is $[(\partial u/\partial y)dy](1/dy) = \partial u/\partial y$. The shear strain γ_{xy} is equal to the sum of the change in angles α_1 and α_2. Therefore

$$\gamma_{xy} = \frac{\partial u}{\partial y} + \frac{\partial v}{\partial x} \qquad (2.24)$$

Similarly, the shear strains γ_{xz} and γ_{yz} can be derived as

$$\gamma_{xz} = \frac{\partial u}{\partial z} + \frac{\partial w}{\partial x} \tag{2.25}$$

and

$$\gamma_{yz} = \frac{\partial v}{\partial z} + \frac{\partial w}{\partial y} \tag{2.26}$$

Generally, in soil mechanics, the compressive normal strains are considered positive. For shear strain, if there is an increase in the right angle BAC (Figure 2.9b), it is considered positive. As shown in Figure 2.9b, the shear strains are all negative.

2.5 HOOKE'S LAW

The axial strains for an ideal, elastic, isotropic material in terms of the stress components are given by Hooke's law as

$$\epsilon_x = \frac{\partial u}{\partial x} = \frac{1}{E}[\sigma_x - v(\sigma_y + \sigma_z)] \tag{2.27}$$

$$\epsilon_y = \frac{\partial v}{\partial y} = \frac{1}{E}[\sigma_y - v(\sigma_x + \sigma_z)] \tag{2.28}$$

and

$$\epsilon_z = \frac{\partial w}{\partial z} = \frac{1}{E}[\sigma_z - v(\sigma_x + \sigma_y)] \tag{2.29}$$

where
E is the Young's modulus
v is the Poisson's ratio

Form the relation given by Equations 2.27 through 2.29, the stress components can be expressed as

$$\sigma_x = \frac{vE}{(1+v)(1-2v)}(\epsilon_x + \epsilon_y + \epsilon_z) + \frac{E}{1+v}\epsilon_x \tag{2.30}$$

$$\sigma_y = \frac{vE}{(1+v)(1-2v)}(\epsilon_x + \epsilon_y + \epsilon_z) + \frac{E}{1+v}\epsilon_y \tag{2.31}$$

$$\sigma_z = \frac{vE}{(1+v)(1-2v)}(\epsilon_x + \epsilon_y + \epsilon_z) + \frac{E}{1+v}\epsilon_z \tag{2.32}$$

The shear strains in terms of the stress components are

$$\gamma_{xy} = \frac{\tau_{xy}}{G} \tag{2.33}$$

$$\gamma_{xz} = \frac{\tau_{xz}}{G} \tag{2.34}$$

and

$$\gamma_{yz} = \frac{\tau_{yz}}{G} \tag{2.35}$$

where shear modulus

$$G = \frac{E}{2(1+v)} \tag{2.36}$$

2.6 PLANE STRAIN PROBLEMS

A state of stress generally encountered in many problems in soil mechanics is the plane strain condition. Long retaining walls and strip foundations are examples where plane strain conditions are encountered. Referring to Figure 2.10, for the strip foundation, the strain in the y direction is zero (i.e., $\epsilon_y = 0$). The stresses at all sections in the xz plane are the same, and the shear stresses on these sections are zero (i.e., $\tau_{yx} = \tau_{xy} = 0$ and $\tau_{yz} = \tau_{zy} = 0$). Thus, from Equation 2.28

$$\epsilon_y = 0 = \frac{1}{E}[\sigma_y - v(\sigma_x + \sigma_z)]$$

$$\sigma_y = v(\sigma_x + \sigma_z) \tag{2.37}$$

Substituting Equation 2.37 into Equations 2.27 and 2.29

$$\epsilon_x = \frac{1-v^2}{E}\left[\sigma_x - \frac{v}{1-v}\sigma_z\right] \tag{2.38}$$

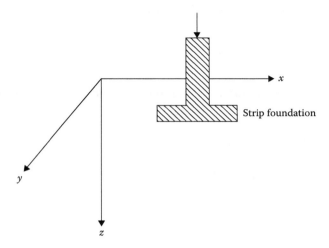

Figure 2.10 Strip foundation: plane strain problem.

and

$$\epsilon_z = \frac{1-v^2}{E}\left[\sigma_z - \frac{v}{1-v}\sigma_x\right] \tag{2.39}$$

Since $\tau_{xy} = 0$ and $\tau_{yz} = 0$

$$\gamma_{xy} = 0 \quad \gamma_{yz} = 0 \tag{2.40}$$

and

$$\gamma_{xz} = \frac{\tau_{xz}}{G} \tag{2.41}$$

2.6.1 Compatibility equation

The three strain components given by Equations 2.38, 2.39, and 2.41 are functions of the displacements u and w and are not independent of each other. Hence, a relation should exist such that the strain components give single-valued continuous solutions. It can be obtained as follows. From Equation 2.21, $\epsilon_x = \partial u/\partial x$. Differentiating twice with respect to z

$$\frac{\partial^2 \epsilon_x}{\partial z^2} = \frac{\partial^3 u}{\partial x\,\partial z^2} \tag{2.42}$$

From Equation 2.23, $\epsilon_z = \partial w/\partial z$. Differentiating twice with respect to x

$$\frac{\partial^2 \epsilon_z}{\partial x^2} = \frac{\partial^3 w}{\partial z\, \partial x^2} \tag{2.43}$$

Similarly, differentiating γ_{xz} (Equation 2.25) with respect to x and z

$$\frac{\partial^2 \gamma_{xz}}{\partial x\, \partial z} = \frac{\partial^3 u}{\partial x\, \partial z^2} + \frac{\partial^3 w}{\partial x^2\, \partial z} \tag{2.44}$$

Combining Equations 2.42 through 2.44, we obtain

$$\frac{\partial^2 \epsilon_x}{\partial z^2} + \frac{\partial^2 \epsilon_z}{\partial x^2} = \frac{\partial^2 \gamma_{xz}}{\partial x\, \partial z} \tag{2.45}$$

Equation 2.45 is the compatibility equation in terms of strain components. Compatibility equations in terms of the stress components can also be derived. Let $E' = E/(1 - v^2)$ and $v' = v/(1 - v)$. So, from Equation 2.38, $\epsilon_x = 1/E'(\sigma_x - v'\sigma_z)$. Hence

$$\frac{\partial^2 \epsilon_x}{\partial z^2} = \frac{1}{E'}\left(\frac{\partial^2 \sigma_x}{\partial z^2} - v'\frac{\partial^2 \sigma_z}{\partial z^2} \right) \tag{2.46}$$

Similarly, from Equation 2.39, $\epsilon_x = (1/E')(\sigma_z - v'\sigma_x)$. Thus

$$\frac{\partial^2 \epsilon_z}{\partial x^2} = \frac{1}{E'}\left(\frac{\partial^2 \sigma_z}{\partial x^2} - v'\frac{\partial^2 \sigma_x}{\partial x^2} \right) \tag{2.47}$$

Again, from Equation 2.41

$$\gamma_{xz} = \frac{\tau_{xz}}{G} = \frac{2(1+v)}{E}\tau_{xz} = \frac{2(1+v')}{E'}\tau_{xz} \tag{2.48}$$

$$\frac{\partial^2 \gamma_{xz}}{\partial x\, \partial z} = \frac{2(1+v')}{E'}\frac{\partial^2 \tau_{xz}}{\partial x\, \partial z}$$

Substitution of Equations 2.46 through 2.48 into Equation 2.45 yields

$$\frac{\partial^2 \sigma_x}{\partial z^2} + \frac{\partial^2 \sigma_z}{\partial x^2} - v'\left(\frac{\partial^2 \sigma_z}{\partial z^2} + \frac{\partial^2 \sigma_x}{\partial x^2} \right) = 2(1+v')\frac{\partial^2 \tau_{xz}}{\partial x\, \partial z} \tag{2.49}$$

From Equations 2.14 and 2.15

$$\frac{\partial}{\partial x}\left(\frac{\partial \sigma_x}{\partial x}+\frac{\partial \tau_{xz}}{\partial z}\right)+\frac{\partial}{\partial z}\left(\frac{\partial \sigma_z}{\partial z}+\frac{\partial \tau_{xz}}{\partial x}-\gamma\right)=0$$

or

$$2\frac{\partial^2 \tau_{xz}}{\partial x\,\partial z}=-\left(\frac{\partial^2 \sigma_x}{\partial x^2}+\frac{\partial^2 \sigma_z}{\partial z^2}\right)+\frac{\partial}{\partial z}(\gamma) \tag{2.50}$$

Combining Equations 2.49 and 2.50

$$\left(\frac{\partial^2}{\partial x^2}+\frac{\partial^2}{\partial z^2}\right)(\sigma_x+\sigma_z)=(1+v')\frac{\partial}{\partial z}(\gamma)$$

For weightless materials, or for a constant unit weight γ, the previous equation becomes

$$\left(\frac{\partial^2}{\partial x^2}+\frac{\partial^2}{\partial z^2}\right)(\sigma_x+\sigma_z)=0 \tag{2.51}$$

Equation 2.51 is the *compatibility equation* in terms of stress.

2.6.2 Stress function

For the plane strain condition, in order to determine the stress at a given point due to a given load, the problem reduces to solving the equations of equilibrium together with the compatibility equation (Equation 2.51) and the boundary conditions. For a weight-less medium (i.e., $\gamma = 0$), the equations of equilibrium are

$$\frac{\partial \sigma_x}{\partial x}+\frac{\partial \tau_{xz}}{\partial z}=0 \tag{2.14}$$

$$\frac{\partial \sigma_z}{\partial z}+\frac{\partial \tau_{xz}}{\partial x}=0 \tag{2.15}$$

The usual method of solving these problems is to introduce a stress function referred to as *Airy's stress function*. The stress function ϕ in terms of x and z should be such that

$$\sigma_x=\frac{\partial^2 \phi}{\partial z^2} \tag{2.52}$$

$$\sigma_z = \frac{\partial^2 \phi}{\partial x^2} \tag{2.53}$$

$$\tau_{xz} = -\frac{\partial^2 \phi}{\partial x\,\partial z} \tag{2.54}$$

The aforementioned equations will satisfy the equilibrium equations. When Equations 2.52 through 2.54 are substituted into Equation 2.51, we get

$$\frac{\partial^4 \phi}{\partial x^4} + 2\frac{\partial^4 \phi}{\partial x^2\,\partial z^2} + \frac{\partial^4 \phi}{\partial z^4} = 0 \tag{2.55}$$

So, the problem reduces to finding a function ϕ in terms of x and z such that it will satisfy Equation 2.55 and the boundary conditions.

2.6.3 Compatibility equation in polar coordinates

For solving plane strain problems in polar coordinates, assuming the soil to be weightless (i.e., $\gamma = 0$), the equations of equilibrium are (from Equations 2.16 and 2.17)

$$\frac{\partial \sigma_r}{\partial r} + \frac{1}{r}\frac{\partial \tau_{r\theta}}{\partial \theta} + \frac{\sigma_r - \sigma_\theta}{r} = 0$$

$$\frac{1}{r}\frac{\partial \sigma_\theta}{\partial \theta} + \frac{\partial \tau_{r\theta}}{\partial r} + \frac{2\tau_{r\theta}}{r} = 0$$

The compatibility equation in terms of stresses can be given by

$$\left(\frac{\partial^2}{\partial r^2} + \frac{1}{r}\frac{\partial}{\partial r} + \frac{1}{r^2}\frac{\partial^2}{\partial \theta^2} \right)(\sigma_r + \sigma_\theta) = 0 \tag{2.56}$$

The Airy stress function ϕ should be such that

$$\sigma_r = \frac{1}{r}\frac{\partial \phi}{\partial r} + \frac{1}{r^2}\frac{\partial^2 \phi}{\partial \theta^2} \tag{2.57}$$

$$\sigma_\theta = \frac{\partial^2 \phi}{\partial r^2} \tag{2.58}$$

$$\tau_{r\theta} = \frac{1}{r^2}\frac{\partial\phi}{\partial\theta} - \frac{1}{r}\frac{\partial^2\phi}{\partial r\,\partial\theta} = -\frac{\partial}{\partial r}\left(\frac{1}{r}\frac{\partial\phi}{\partial\theta}\right) \tag{2.59}$$

The previous equations satisfy the equilibrium equations. The compatibility equation in terms of stress function is

$$\left(\frac{\partial^2}{\partial r^2} + \frac{1}{r}\frac{\partial}{\partial r} + \frac{1}{r^2}\frac{\partial^2}{\partial\theta^2}\right)\left(\frac{\partial^2\phi}{\partial r^2} + \frac{1}{r}\frac{\partial\phi}{\partial r} + \frac{1}{r^2}\frac{\partial^2\phi}{\partial\theta^2}\right) = 0 \tag{2.60}$$

Similar to Equation 2.37, for the plane strain condition

$$\sigma_y = \upsilon(\sigma_r + \sigma_\theta)$$

Example 2.1

The stress at any point inside a semi-infinite medium due to a line load of intensity q per unit length (Figure 2.11) can be given by a stress function

$$\phi = Ax\,\tan^{-1}\left(\frac{z}{x}\right)$$

where A is a constant. This equation satisfies the compatibility equation (Equation 2.55). (a) Find σ_x, σ_z, σ_y, and τ_{xz}. (b) Applying proper boundary conditions, find A.

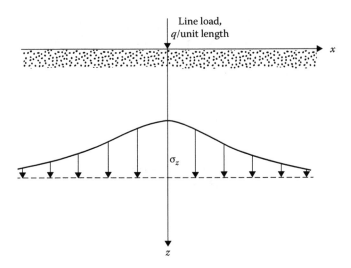

Figure 2.11 Stress at a point due to a line load.

Solution

Part a:

$$\phi = Ax\tan^{-1}\left(\frac{z}{x}\right)$$

The relations for σ_x, σ_z, σ_y, and τ_{xz} are given in Equations 2.52 through 2.54.

$$\sigma_x = \frac{\partial^2\phi}{\partial z^2}$$

$$\frac{\partial\phi}{\partial z} = Ax\frac{1}{1+(z/x)^2}\frac{1}{x} = \frac{A}{1+(z/x)^2}$$

$$\sigma_x = \frac{\partial^2\phi}{\partial z^2} = -\frac{2Azx^2}{(x^2+z^2)^2}$$

$$\sigma_x = \frac{\partial^2\phi}{\partial x^2}$$

$$\frac{\partial\phi}{\partial x} = A\tan^{-1}\frac{z}{x} - \frac{Az}{[1+(z/x)^2]x} = A\tan^{-1}\frac{z}{x} - \frac{Axz}{(x^2+z^2)}$$

$$\sigma_z = \frac{\partial^2\phi}{\partial x^2} = -\frac{A}{1+(z/x)^2}\frac{z}{x^2} - \frac{Az}{x^2+z^2} + \frac{2Ax^2z}{(x^2+z^2)^2}$$

$$= -\frac{Az}{x^2+z^2} - \frac{Az}{x^2+z^2} + \frac{2Ax^2z}{(x^2+z^2)^2} = -\frac{2Az^3}{(x^2+z^2)^2}$$

$$\tau_{zx} = -\frac{\partial^2\phi}{\partial x\,\partial z}$$

$$\frac{\partial\phi}{\partial x} = A\tan^{-1}\frac{z}{x} - \frac{Axz}{(x^2+z^2)}$$

$$\frac{\partial^2\phi}{\partial x\,\partial z} = \frac{A}{1+(z/x)^2}\frac{1}{x} - \frac{Ax}{x^2+z^2} + \frac{2Axz^2}{(x^2+z^2)^2}$$

or

$$\frac{\partial^2\phi}{\partial x\,\partial z} = \frac{2Axz^2}{(x^2+z^2)^2}$$

$$\tau_{xz} = -\frac{\partial^2 \phi}{\partial x \, \partial z} = -\frac{2Axz^2}{(x^2+z^2)^2}$$

$$\sigma_y = v(\sigma_x + \sigma_z) = v\left[-\frac{2Azx^2}{(x^2+z^2)^2} - \frac{2Az^3}{(x^2+z^2)^2} \right]$$

$$= -\frac{2Azv}{(x^2+z^2)^2}(x^2+z^2) = -\frac{2Azv}{(x^2+z^2)}$$

Part b: Consider a unit length along the *y* direction. We can write

$$q = \int_{-\infty}^{+\infty} (\sigma_z)(1)(dx) = \int_{-\infty}^{+\infty} -\frac{2Az^3}{(x^2+z^2)^2}\,dx$$

$$= -\frac{2Az^3}{2z^2}\left(\frac{x}{x^2+z^2} + \int \frac{dx}{x^2+z^2} \right)_{-\infty}^{+\infty}$$

$$= -Az\left(\frac{x}{x^2+z^2} + \frac{1}{z}\tan^{-1}\frac{x}{z} \right)_{-\infty}^{+\infty} = -A(\pi/2 + \pi/2) = -A\pi$$

$$A = -\frac{q}{\pi}$$

So

$$\sigma_x = \frac{2qx^2z}{\pi(x^2+z^2)^2} \quad \sigma_z = \frac{2qz^3}{\pi(x^2+z^2)^2} \quad \tau_{xz} = \frac{2qxz^2}{\pi(x^2+z^2)^2}$$

We can see that at $z = 0$ (i.e., at the surface) and for any value of $x \neq 0$, σ_x, σ_z, and τ_{xz} are equal to zero.

2.7 EQUATIONS OF COMPATIBILITY FOR THREE-DIMENSIONAL PROBLEMS

For three-dimensional problems in the Cartesian coordinate system as shown in Figure 2.2, the *compatibility equations in terms of stresses* are (assuming the body force to be zero or constant)

$$\nabla^2 \sigma_x + \frac{1}{1+v}\frac{\partial^2 \Theta}{\partial x^2} = 0 \tag{2.61}$$

$$\nabla^2 \sigma_y + \frac{1}{1+v} \frac{\partial^2 \Theta}{\partial y^2} = 0 \tag{2.62}$$

$$\nabla^2 \sigma_z + \frac{1}{1+v} \frac{\partial^2 \Theta}{\partial z^2} = 0 \tag{2.63}$$

$$\nabla^2 \tau_{xy} + \frac{1}{1+v} \frac{\partial^2 \Theta}{\partial x \partial y} = 0 \tag{2.64}$$

$$\nabla^2 \tau_{yz} + \frac{1}{1+v} \frac{\partial^2 \Theta}{\partial y \partial z} = 0 \tag{2.65}$$

$$\nabla^2 \tau_{xz} + \frac{1}{1+v} \frac{\partial^2 \Theta}{\partial x \partial z} = 0 \tag{2.66}$$

where

$$\nabla^2 = \frac{\partial^2}{\partial x^2} + \frac{\partial^2}{\partial y^2} + \frac{\partial^2}{\partial z^2}$$

and

$$\Theta = \sigma_x + \sigma_y + \sigma_z$$

The compatibility equations in terms of stresses for the cylindrical coordinate system (Figure 2.4) are as follows (for constant or zero body force):

$$\nabla^2 \sigma_z + \frac{1}{1+v} \frac{\partial^2 \Theta}{\partial z^2} = 0 \tag{2.67}$$

$$\nabla^2 \sigma_r + \frac{1}{1+v} \frac{\partial^2 \Theta}{\partial r^2} - \frac{4}{r^2} \frac{\partial \tau_{r\theta}}{\partial \theta} + \frac{2}{r^2}(\sigma_\theta + \sigma_r) = 0 \tag{2.68}$$

$$\nabla^2 \sigma_\theta + \frac{1}{1+v} \left(\frac{1}{r} \frac{\partial \Theta}{\partial r} + \frac{1}{r^2} \frac{\partial^2 \Theta}{\partial \theta^2} \right) + \frac{4}{r^2} \frac{\partial \tau_{r\theta}}{\partial \theta} - \frac{2}{r^2}(\sigma_\theta + \sigma_r) = 0 \tag{2.69}$$

$$\nabla^2 \tau_{rz} + \frac{1}{1+v} \frac{\partial^2 \Theta}{\partial r \partial z} - \frac{\tau_{rz}}{r^2} - \frac{2}{r^2} \frac{\partial \tau_{\theta z}}{\partial \theta} = 0 \tag{2.70}$$

$$\nabla^2 \tau_{r\theta} + \frac{1}{1+v} \frac{\partial}{\partial r} \left(\frac{1}{r} \frac{\partial \Theta}{\partial \theta} \right) - \frac{4}{r^2} \tau_{r\theta} - \frac{2}{r^2} \frac{\partial}{\partial \theta} (\sigma_\theta - \sigma_r) = 0 \tag{2.71}$$

$$\nabla^2 \tau_{z\theta} + \frac{1}{1+v} \frac{1}{r} \frac{\partial^2 \Theta}{\partial \theta \partial z} + \frac{2}{r} \frac{\partial \tau_{rz}}{\partial \theta} - \frac{\tau_{z\theta}}{r^2} = 0 \tag{2.72}$$

2.8 STRESSES ON AN INCLINED PLANE AND PRINCIPAL STRESSES FOR PLANE STRAIN PROBLEMS

The fundamentals of plane strain problems are explained in Section 2.5. For these problems, the strain in the y direction is zero (i.e., $\tau_{yx} = \tau_{xy} = 0$; $\tau_{yz} = \tau_{zy} = 0$) and σ_y is constant for all sections in the plane.

If the stresses at a point in a soil mass (i.e., σ_x, σ_y, σ_z, $\tau_{xz}(= \tau_{zx})$) are known (as shown in Figure 2.12a), the normal stress σ and the shear stress τ on an inclined plane BC can be determined by considering a soil prism of unit length in the direction of the y axis. Summing the components of all forces in the n direction (Figure 2.12b) gives

$\Sigma F_n = 0$

$\sigma \, dA = (\sigma_x \cos \theta)(dA \cos \theta) + (\sigma_z \sin \theta)(dA \sin \theta)$
$\qquad + (\tau_{xz} \sin \theta)(dA \cos \theta) + (\tau_{xz} \cos \theta)(dA \sin \theta)$

where dA is the area of the inclined face of the prism. Thus

$$\sigma = \sigma_x \cos^2 \theta + \sigma_z \sin^2 \theta + 2\tau_{xz} \sin \theta \cos \theta$$

$$= \left(\frac{\sigma_x + \sigma_z}{2} \right) + \left(\frac{\sigma_x - \sigma_z}{2} \right) \cos 2\theta + \tau_{xz} \sin 2\theta \tag{2.73}$$

Similarly, summing the forces in the s direction gives

$\Sigma F_s = 0$

$\tau \, dA = -(\sigma_x \sin \theta)(dA \cos \theta) + (\sigma_z \cos \theta)(dA \sin \theta)$
$\qquad + (\tau_{xz} \cos \theta)(dA \cos \theta) - (\tau_{xz} \sin \theta)(dA \sin \theta)$

$\tau = -\sigma_x \sin \theta \cos \theta + \sigma_z \sin \theta \cos \theta + \tau_{xz}(\cos^2 \theta - \sin^2 \theta)$

$$= \tau_{xz} \cos 2\theta - \left(\frac{\sigma_x - \sigma_z}{2} \right) \sin 2\theta \tag{2.74}$$

Note that σ_y has no contribution to σ or τ.

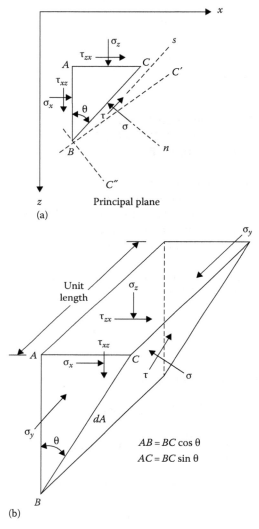

Figure 2.12 (a) Stresses on an inclined plane for the plane strain case; (b) soil prism of unit length in the direction of y-axis.

2.8.1 Transformation of stress components from polar to Cartesian coordinate system

In some instances, it is helpful to know the relations for transformation of stress components in a polar coordinate system to a Cartesian coordinate system. This can be done by a principle similar to that demonstrated earlier for finding the stresses on an inclined plane. Comparing Figures 2.12 and 2.13,

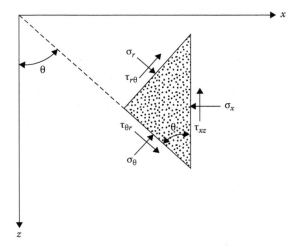

Figure 2.13 Transformation of stress components from polar to Cartesian coordinate system.

it is obvious that we can substitute σ_r for σ_z, σ_θ for σ_x, and $\tau_{r\theta}$ for τ_{xz} in Equations 2.73 and 2.74 to obtain σ_x and τ_{xz} as shown in Figure 2.13. So

$$\sigma_x = \sigma_r \sin^2\theta + \sigma_\theta \cos^2\theta + 2\tau_{r\theta} \sin\theta\cos\theta \tag{2.75}$$

$$\tau_{xz} = -\sigma_\theta \sin\theta\cos\theta + \sigma_r \sin\theta\cos\theta + \tau_{r\theta}(\cos^2\theta - \sin^2\theta) \tag{2.76}$$

Similarly, it can be shown that

$$\sigma_z = \sigma_r \cos^2\theta + \sigma_\theta \sin^2\theta - 2\tau_{r\theta} \sin\theta\cos\theta \tag{2.77}$$

2.8.2 Principal stress

A plane is defined as a *principal plane* if the shear stress acting on it is zero. This means that the only stress acting on it is a normal stress. The normal stress on a principal plane is referred to as the *principal stress*. In a plane strain case, σ_y is a *principal stress*, and the xz plane is a principal plane. The orientation of the other two principal planes can be determined by considering Equation 2.74. On an inclined plane, if the shear stress is zero, it follows that

$$\tau_{xz} \cos 2\theta = \left(\frac{\sigma_x - \sigma_z}{2}\right)\sin 2\theta$$

$$\tan 2\theta = \frac{2\tau_{xz}}{\sigma_x - \sigma_z} \tag{2.78}$$

From Equation 2.78, it can be seen that there are two values of θ at right angles to each other that will satisfy the relation. These are the directions of the two principal planes BC' and BC'' as shown in Figure 2.12. Note that there are now three principal planes that are at right angles to each other. Besides σ_y, the expressions for the two other principal stresses can be obtained by substituting Equation 2.78 into Equation 2.73, which gives

$$\sigma_{p(1)} = \frac{\sigma_x + \sigma_z}{2} + \sqrt{\left(\frac{\sigma_x - \sigma_z}{2}\right)^2 + \tau_{xz}^2} \tag{2.79}$$

$$\sigma_{p(3)} = \frac{\sigma_x + \sigma_z}{2} - \sqrt{\left(\frac{\sigma_x - \sigma_z}{2}\right)^2 + \tau_{xz}^2} \tag{2.80}$$

where $\sigma_{p(1)}$ and $\sigma_{p(3)}$ are the principal stresses. Also

$$\sigma_{p(1)} + \sigma_{p(3)} = \sigma_x + \sigma_z \tag{2.81}$$

Comparing the magnitude of the principal stresses, $\sigma_{p(1)} > \sigma_y = \sigma_{p(2)} > \sigma_{p(3)}$. Thus $\sigma_{p(1)}$, $\sigma_{p(2)}$, and $\sigma_{p(3)}$ are referred to as the major, intermediate, and minor principal stresses. From Equations 2.37 and 2.81, it follows that

$$\sigma_y = v[\sigma_{p(1)} + \sigma_{p(3)}] \tag{2.82}$$

2.8.3 Mohr's circle for stresses

The shear and normal stresses on an inclined plane (Figure 2.12) can also be determined graphically by using Mohr's circle. The procedure to construct Mohr's circle is explained later.

The sign convention for normal stress is positive for compression and negative for tension. The shear stress on a given plane is positive if it tends to produce a clockwise rotation about a point outside the soil element, and it is negative if it tends to produce a counterclockwise rotation about a point outside the element (Figure 2.14). Referring to plane AB in Figure 2.12a, the normal stress is $+\sigma_x$ and the shear stress is $+\tau_{xz}$. Similarly, on plane AC, the stresses are $+\sigma_z$ and $-\tau_{xz}$. The stresses on planes AB and AC can be plotted on a graph with normal stresses along the abscissa and shear stresses along the ordinate. Points B and C in Figure 2.15 refer to the stress conditions on planes AB and AC, respectively. Now, if points B and C are

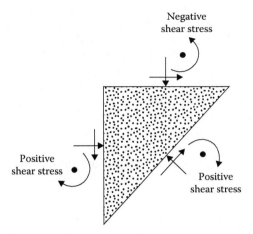

Figure 2.14 Sign convention for shear stress used for the construction of Mohr's circle.

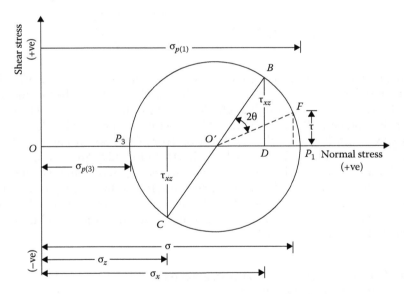

Figure 2.15 Mohr's circle.

joined by a straight line, it will intersect the normal stress axis at O'. With O' as the center and $O'B$ as the radius, if a circle $BP_1 CP_3$ is drawn, it will be Mohr's circle. The radius of Mohr's circle is

$$O'B = \sqrt{O'D^2 + BD^2} = \sqrt{\left(\frac{\sigma_x - \sigma_z}{2}\right)^2 + \tau_{xz}^2} \qquad (2.83)$$

Any radial line in Mohr's circle represents a given plane, and the coordinates of the points of intersection of the radial line and the circumference of Mohr's circle give the stress condition on that plane. For example, let us find the stresses on plane BC. If we start from plane AB and move an angle θ in the clockwise direction in Figure 2.12, we reach plane BC. In Mohr's circle in Figure 2.15, the radial line $O'B$ represents the plane AB. We move an angle 2θ in the clockwise direction to reach point F. Now the radial line $O'F$ in Figure 2.15 represents plane BC in Figure 2.12. The coordinates of point F will give us the stresses on the plane BC.

Note that the ordinates of points P_1 and P_3 are zero, which means that $O'P_1$ and $O'P_3$ represent the major and minor principal planes, and $OP_1 = \sigma_{p(1)}$ and $OP_3 = \sigma_{p(3)}$:

$$\sigma_{p(1)} = OP_1 = OO' + O'P_1 = \frac{\sigma_x + \sigma_z}{2} + \sqrt{\left(\frac{\sigma_x - \sigma_z}{2}\right)^2 + \tau_{xz}^2}$$

$$\sigma_{p(3)} = OP_3 = OO' - O'P_3 = \frac{\sigma_x + \sigma_z}{2} - \sqrt{\left(\frac{\sigma_x - \sigma_z}{2}\right)^2 + \tau_{xz}^2}$$

The previous two relations are the same as Equations 2.79 and 2.80. Also note that the principal plane $O'P_1$ in Mohr's circle can be reached by moving clockwise from $O'B$ through angle $BO'P_1 = \tan^{-1}[2\tau_{xz}/(\sigma_x - \sigma_z)]$. The other principal plane $O'P_3$ can be reached by moving through angle $180° + \tan^{-1}[2\tau_{xz}/(\sigma_x - \sigma_z)]$ in the clockwise direction from $O'B$. So, in Figure 2.12, if we move from plane AB through angle $(1/2)\tan^{-1}[2\tau_{xz}/(\sigma_x - \sigma_z)]$, we will reach plane BC', on which the principal stress $\sigma_{p(1)}$ acts. Similarly, moving clockwise from plane AB through angle $1/2\{180° + \tan^{-1}[2\tau_{xz}/(\sigma_x - \sigma_z)]\} = 90° + (1/2)\tan^{-1}[2\tau_{xz}/(\sigma_x - \sigma_z)]$ in Figure 2.12, we reach plane BC'', on which the principal stress $\sigma_{p(3)}$ acts. These are the same conclusions as derived from Equation 2.78.

2.8.4 Pole method for finding stresses on an inclined plane

A pole is a unique point located on the circumference of Mohr's circle. If a line is drawn through the pole parallel to a given plane, the point of intersection of this line and Mohr's circle will give the stresses on the plane. The procedure for finding the pole is shown in Figure 2.16.

Figure 2.16a shows the same stress element as Figure 2.12. The corresponding Mohr's circle is given in Figure 2.16b. Point B on Mohr's circle represents the stress conditions on plane AB (Figure 2.16a). If a line is drawn through B parallel to AB, it will intersect Mohr's circle at P. Point P is the

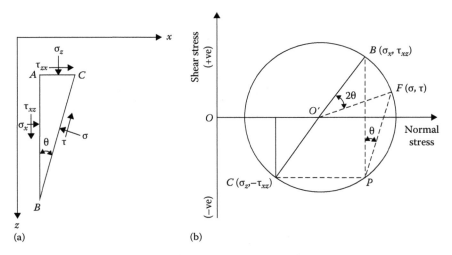

Figure 2.16 Pole method of finding stresses on an inclined plane: (a) stress element; (b) corresponding Mohr's circle.

pole for Mohr's circle. We could also have found pole P by drawing a line through C parallel to plane AC. To find the stresses on plane BC, we draw a line through P parallel to BC. It will intersect Mohr's circle at F, and the coordinates of point F will give the normal and shear stresses on plane AB.

Example 2.2

The stresses at a point in a soil mass are shown in Figure 2.17 (plane strain case). Determine the principal stresses and show their directions. Use $v = 0.35$.

Solution

Based on the sign conventions explained in Section 2.2,

$$\sigma_z = +100 \text{ kN/m}^2, \quad \sigma_x = +50 \text{ kN/m}^2, \quad \text{and} \quad \tau_{xz} = -25 \text{ kN/m}^2$$

$$\sigma_p = \frac{\sigma_x + \sigma_z}{2} \pm \sqrt{\left(\frac{\sigma_x - \sigma_z}{2}\right)^2 + \tau_{xz}^2}$$

$$= \frac{50 + 100}{2} \pm \sqrt{\left(\frac{50 - 100}{2}\right)^2 + (-25)^2} = (75 \pm 35.36) \text{ kN/m}^2$$

$$\sigma_{p(1)} = 110.36 \text{ kN/m}^2 \quad \sigma_{p(3)} = 39.64 \text{ kN/m}^2$$

$$\sigma_{p(2)} = v[\sigma_{p(1)} + \sigma_{p(3)}] = (0.35)(110.36 + 39.34) = 52.5 \text{ kN/m}^2$$

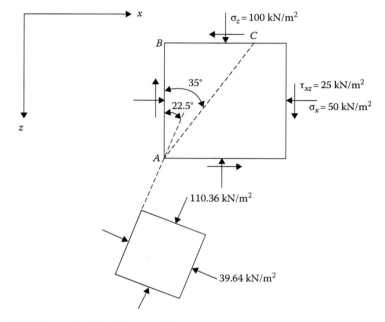

Figure 2.17 Determination of principal stresses at a point.

From Equation 2.78

$$\tan 2\theta = \frac{2\tau_{xz}}{\sigma_x - \sigma_z} = \frac{(2)(-25)}{(50-100)} = 1$$

$$2\theta = \tan^{-1}(1) = 45° \text{ and } 225° \text{ so } \theta = 22.5° \text{ and } 112.5°$$

Parameter $\sigma_{p(2)}$ is acting on the xz plane. The directions of $\sigma_{p(1)}$ and $\sigma_{p(3)}$ are shown in Figure 2.17.

Example 2.3

Refer to Example 2.2.

a. Determine the magnitudes of $\sigma_{p(1)}$ and $\sigma_{p(3)}$ by using Mohr's circle.
b. Determine the magnitudes of the normal and shear stresses on plane AC shown in Figure 2.17.

Solution

Part a: For Mohr's circle, on plane AB, $\sigma_x = 50\,\text{kN/m}^2$ and $\tau_{xz} = -25\,\text{kN/m}^2$. On plane BC, $\sigma_z = +100$ and $\tau_{xz} +25\,\text{kN/m}^2$. For the stresses, Mohr's circle is plotted in Figure 2.18. The radius of the circle is

$$O'H = \sqrt{(O'I)^2 + (HI)^2} = \sqrt{25^2 + 25^2} = 35.36\,\text{kN/m}^2$$

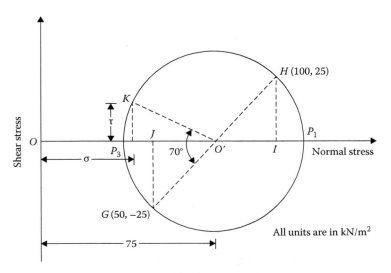

Figure 2.18 Mohr's circle for stress determination.

$$\sigma_{p(1)} = OO' + O'P_1 = 75 + 35.36 = 110.36 \text{ kN/m}^2$$

$$\sigma_{p(3)} = OO' + O'P_1 = 75 - 35.36 = 39.64 \text{ kN/m}^2$$

The angle $GO'P_3 = 2\theta = \tan^{-1}(JG/O'J) = \tan^{-1}(25/25) = 45°$. So, we move an angle $\theta = 22.5°$ clockwise from plane AB to reach the minor principal plane, and an angle $\theta = 22.5 + 90 = 112.5°$ clockwise from plane AB to reach the major principal plane. The orientation of the major and minor principal stresses is shown in Figure 2.17.

Part b: Plane AC makes an angle $35°$, measured clockwise, with plane AB. If we move through an angle of $(2)(35°) = 70°$ from the radial line $O'G$ (Figure 2.18), we reach the radial line $O'K$. The coordinates of K will give the normal and shear stresses on plane AC. So

$$\tau = O'K \sin 25° = 35.36 \sin 25° = 14.94 \text{ kN/m}^2$$

$$\sigma = OO' - O'K \cos 25° = 75 - 35.36 \cos 25° = 42.95 \text{ kN/m}^2$$

Note: This could also be solved using Equations 2.73 and 2.74:

$$\tau = \tau_{xz} \cos 2\theta - \left(\frac{\sigma_x - \sigma_z}{2}\right) \sin 2\theta$$

where
$\tau_{xz} = -25 \text{ kN/m}^2$
$\theta = 35°$
$\sigma_x = +50 \text{ kN/m}^2$
$\sigma_z = +100 \text{ kN/m}^2$ (watch the sign conventions)

So

$$\tau = -25\cos 70 - \left(\frac{50-100}{2}\right)\sin 70 = -8.55 - (-23.49)$$

$$= 14.94 \text{ kN/m}^2$$

$$\sigma = \left(\frac{\sigma_x + \sigma_z}{2}\right) + \left(\frac{\sigma_x - \sigma_z}{2}\right)\cos 2\theta + \tau_{xz}\sin 2\theta$$

$$= \left(\frac{50+100}{2}\right) + \left(\frac{50-100}{2}\right)\cos 70 + (-25)\sin 70$$

$$= 75 - 8.55 - 23.49 = 42.96 \text{ kN/m}^2$$

2.9 STRAINS ON AN INCLINED PLANE AND PRINCIPAL STRAIN FOR PLANE STRAIN PROBLEMS

Consider an elemental soil prism $ABDC$ of unit length along the y direction (Figure 2.19). The lengths of the prism along the x and z directions are $AB = dx$ and $AC = dz$, respectively. When subjected to stresses, the soil prism is deformed and displaced. The length in the y direction still remains unity. $A'B''D''C''$ is the deformed shape of the prism in the displaced position.

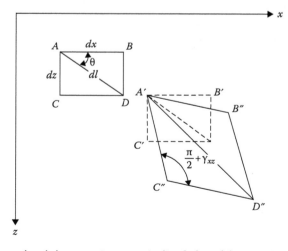

Figure 2.19 Normal and shear strains on an inclined plane (plane strain case).

If the normal strain on an inclined plane AD making an angle θ with the x axis is equal to ϵ,

$$A'D'' = AD(1+\epsilon) = dl(1 +\epsilon) \tag{2.84}$$

where $AD = dl$.

Note that the angle $B''A'C''$ is equal to $(\pi/2 - \gamma_{xz})$. So the angle $A'C''D''$ is equal to $+(\pi/2 + \gamma_{xz})$. Now

$$(A'D'')^2 = (A'C'')^2 + (C'D'')^2 - 2(A'C'')(C'D'')\cos(\pi/2 + \gamma_{xz}) \tag{2.85}$$

$$A'C'' = AC(1+\epsilon_z) = dz(1+\epsilon_z) = dl(\sin\theta)(1+\epsilon_z) \tag{2.86}$$

$$C''D'' = A'B'' = dx(1 +\epsilon_x) = dl(\cos\theta)(1+\epsilon_x) \tag{2.87}$$

Substitution of Equations 2.84, 2.86, and 2.87 into Equation 2.85 gives

$$(1+\epsilon)^2(dl)^2 = [dl(\sin\theta)(1+\epsilon_z)]^2 + [dl(\cos\theta)(1+\epsilon_x)]^2$$
$$+ 2(dl)^2(\sin\theta)(\cos\theta)(1+\epsilon_x)(1+\epsilon_z)\sin\gamma_{xz} \tag{2.88}$$

Taking $\sin\gamma_{xz} \approx \gamma_{xz}$ and neglecting the higher order terms of strain such as $\epsilon^2, \epsilon_x^2, \epsilon_z^2, \epsilon_x\gamma_{xz}, \epsilon_z\gamma_{xz}, \epsilon_x\epsilon_z\gamma_{xz}$, Equation 2.88 can be simplified to

$$1+2\epsilon = (1+2\epsilon_z)\sin^2\theta + (1+2\epsilon_x)\cos^2\theta + 2\gamma_{xz}\sin\theta\cos\theta$$

$$\epsilon = \epsilon_x \cos^2\theta + \epsilon_z \sin^2\theta + \frac{\gamma_{xz}}{2}\sin 2\theta \tag{2.89}$$

or

$$\epsilon = \frac{\epsilon_x + \epsilon_z}{2} + \frac{\epsilon_x - \epsilon_z}{2}\cos 2\theta + \frac{\gamma_{xz}}{2}\sin 2\theta \tag{2.90}$$

Similarly, the shear strain on plane AD can be derived as

$$\gamma = \gamma_{xz}\cos 2\theta - (\epsilon_x - \epsilon_z)\sin 2\theta \tag{2.91}$$

Comparing Equations 2.90 and 2.91 with Equations 2.73 and 2.74, it appears that they are similar except for a factor of 1/2 in the last terms of the equations.

The principal strains can be derived by substituting zero for shear strain in Equation 2.91. Thus

$$\tan 2\theta = \frac{\gamma_{xz}}{\epsilon_x - \epsilon_y} \tag{2.92}$$

There are two values of θ that will satisfy the aforementioned relation. Thus, from Equations 2.90 and 2.92, we obtain

$$\epsilon_p = \frac{\epsilon_x + \epsilon_z}{2} \pm \sqrt{\left(\frac{\epsilon_x - \epsilon_z}{2}\right)^2 + \left(\frac{\gamma_{xz}}{2}\right)^2} \tag{2.93}$$

where ϵ_p = principal strain. Also note that Equation 2.93 is similar to Equations 2.79 and 2.80.

2.10 STRESS COMPONENTS ON AN INCLINED PLANE, PRINCIPAL STRESS, AND OCTAHEDRAL STRESSES: THREE-DIMENSIONAL CASE

2.10.1 Stress on an inclined plane

Figure 2.20 shows a tetrahedron $AOBC$. The face AOB is on the xy plane with stresses σ_z, τ_{zy}, and τ_{zx} acting on it. The face AOC is on the yz plane subjected to stresses σ_x, τ_{xy}, and τ_{xz}. Similarly, the face BOC is on the xz

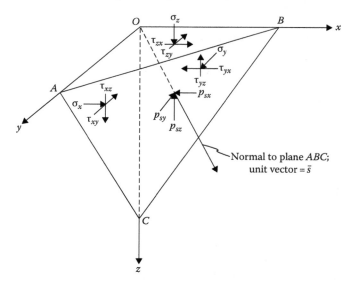

Figure 2.20 Stresses on an inclined plane—three-dimensional case.

plane with stresses σ_y, τ_{yx}, and τ_{yz}. Let it be required to find the x, y, and z components of the stresses acting on the inclined plane ABC.

Let $\mathbf{i}$, $\mathbf{j}$, and $\mathbf{k}$ be the unit vectors in the x, y, and z directions, and let s be the unit vector in the direction perpendicular to the inclined plane ABC:

$$s = \cos(s,x)\mathbf{i} + \cos(s,y)\mathbf{j} + \cos(s,z)\mathbf{k} \tag{2.94}$$

If the area of ABC is dA, then the area of AOC can be given as $dA(s \cdot i) = dA\cos(s, x)$. Similarly, the area of $BOC = dA(s \cdot j) = dA\cos(s, y)$, and the area of $AOB = dA(s \cdot k) = dA\cos(s, z)$.

For equilibrium, summing the forces in the x direction, $\Sigma F_x = 0$:

$$p_{sx}\,dA = [\sigma_x \cos(s,x) + \tau_{yx}\cos(s,y) + \tau_{zx}\cos(s,z)]dA$$

or

$$p_{sx} = \sigma_x \cos(s,x) + \tau_{yx}\cos(s,y) + \tau_{zx}\cos(s,z) \tag{2.95}$$

where p_{sx} is the stress component on plane ABC in the x direction.

Similarly, summing the forces in the y and z directions

$$p_{sy} = \tau_{xy}\cos(s,x) + \sigma_y \cos(s,y) + \tau_{zy}\cos(s,z) \tag{2.96}$$

$$p_{sz} = \tau_{xz}\cos(s,x) + \tau_{yz}\cos(s,y) + \sigma_z \cos(s,z) \tag{2.97}$$

where p_{sy} and p_{sz} are the stress components on plane ABC in the y and z directions, respectively. Equations 2.95 through 2.97 can be expressed in matrix form as

$$\begin{vmatrix} p_{sx} \\ p_{sy} \\ p_{sz} \end{vmatrix} = \begin{vmatrix} \sigma_x & \tau_{yx} & \tau_{xz} \\ \tau_{xy} & \sigma_y & \tau_{zy} \\ \tau_{xz} & \tau_{yz} & \sigma_z \end{vmatrix} \begin{vmatrix} \cos(s,x) \\ \cos(s,y) \\ \cos(s,z) \end{vmatrix} \tag{2.98}$$

The normal stress on plane ABC can now be determined as

$$\sigma = p_{sx}\cos(s,x) + p_{sy}\cos(s,y) + p_{sz}\cos(s,z)$$

$$= \sigma_x \cos^2(s,x) + \sigma_y \cos^2(s,y) + \sigma_z \cos^2(s,z) + 2\tau_{xy}\cos(s,x)\cos(s,y)$$

$$+ 2\tau_{yz}\cos(s,y)\cos(s,z) + 2\tau_{zx}\cos(s,x)\cos(s,z) \tag{2.99}$$

The shear stress τ on the plane can be given as

$$\tau = \sqrt{\left(p_{sx}^2 + p_{sy}^2 + p_{sz}^2\right) - \sigma^2} \tag{2.100}$$

2.10.2 Transformation of axes

Let the stresses in a soil mass in the Cartesian coordinate system be given. If the stress components in a new set of orthogonal axes (x_1, y_1, z_1) as shown in Figure 2.21 are required, they can be determined in the following manner. The direction cosines of the x_1, y_1, and z_1 axes with respect to the x, y, and z axes are shown:

	x	y	z
x_1	l_1	m_1	n_1
y_1	l_2	m_2	n_2
z_1	l_3	m_3	n_3

Following the procedure adopted to obtain Equation 2.98, we can write

$$\begin{vmatrix} p_{x_1x} \\ p_{x_1y} \\ p_{x_1z} \end{vmatrix} = \begin{vmatrix} \sigma_x & \tau_{yx} & \tau_{zx} \\ \tau_{xy} & \sigma_y & \tau_{zy} \\ \tau_{xz} & \tau_{yz} & \sigma_z \end{vmatrix} \begin{vmatrix} l_1 \\ m_1 \\ n_1 \end{vmatrix} \tag{2.101}$$

where p_{x_1x}, p_{x_1y}, and p_{x_1z} are stresses parallel to the x, y, and z axes and are acting on the plane perpendicular to the x_1 axis (i.e., y_1z_1 plane).

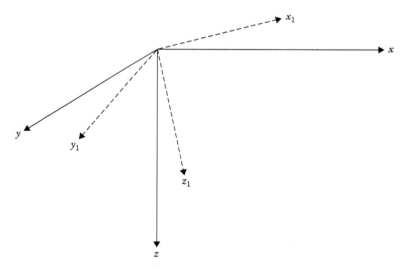

Figure 2.21 Transformation of stresses to a new set of orthogonal axes.

We can now take the components of p_{x_1x}, p_{x_1y}, and p_{x_1z} to determine the normal and shear stresses on the y_1z_1 plane, or

$$\sigma_{x_1} = l_1 p_{x_1x} + m_1 p_{x_1y} + n_1 p_{x_1z}$$

$$\tau_{x_1y_1} = l_2 p_{x_1x} + m_2 p_{x_1y} + n_2 p_{x_1z}$$

$$\tau_{x_1z_1} = l_3 p_{x_1x} + m_3 p_{x_1y} + n_3 p_{x_1z}$$

In a matrix form, the previous three equations may be expressed as

$$\begin{vmatrix} \sigma_{x_1} \\ \tau_{x_1y_1} \\ \tau_{x_1z_1} \end{vmatrix} = \begin{vmatrix} l_1 & m_1 & n_1 \\ l_2 & m_2 & n_2 \\ l_3 & m_3 & n_3 \end{vmatrix} \begin{vmatrix} p_{x_1x} \\ p_{x_1y} \\ p_{x_1z} \end{vmatrix} \tag{2.102}$$

In a similar manner, the normal and shear stresses on the x_1z_1 plane (i.e., $\sigma_{y_1}, \tau_{y_1x_1}$, and $\tau_{y_1z_1}$) and on the x_1y_1 plane (i.e., $\sigma_{z_1}, \tau_{z_1x_1}$, and $\tau_{z_1y_1}$) can be determined. Combining these terms, we can express the stresses in the new set of orthogonal axes in a matrix form. Thus

$$\begin{vmatrix} \sigma_{x_1} & \tau_{y_1x_1} & \tau_{z_1x_1} \\ \tau_{x_1y_1} & \sigma_{y_1} & \tau_{z_1y_1} \\ \tau_{x_1z_1} & \tau_{y_1z_1} & \sigma_{z_1} \end{vmatrix} = \begin{vmatrix} l_1 & m_1 & n_1 \\ l_2 & m_2 & n_2 \\ l_3 & m_2 & n_2 \end{vmatrix} \begin{vmatrix} \sigma_x & \tau_{yx} & \tau_{zx} \\ \tau_{xy} & \sigma_y & \tau_{zy} \\ \tau_{xz} & \tau_{yz} & \sigma_z \end{vmatrix} \begin{vmatrix} l_1 & l_2 & l_3 \\ m_1 & m_2 & m_3 \\ n_1 & n_2 & n_3 \end{vmatrix}$$

$$\tag{2.103}$$

Note: $\tau_{xy} = \tau_{yx}, \tau_{zy} = \tau_{yz}$, and $\tau_{zx} = \tau_{xz}$.

Solution of Equation 2.103 gives the following relations:

$$\sigma_{x_1} = l_1^2 \sigma_x + m_1^2 \sigma_y + n_1^2 \sigma_z + 2m_1n_1\tau_{yz} + 2n_1l_1\tau_{zx} + 2l_1m_1\tau_{xy} \tag{2.104}$$

$$\sigma_{y_1} = l_2^2 \sigma_x + m_2^2 \sigma_y + n_2^2 \sigma_z + 2m_2n_2\tau_{yz} + 2n_2l_2\tau_{zx} + 2l_2m_2\tau_{xy} \tag{2.105}$$

$$\sigma_{z_1} = l_3^2 \sigma_x + m_3^2 \sigma_y + n_3^2 \sigma_z + 2m_3n_3\tau_{yz} + 2n_3l_3\tau_{zx} + 2l_3m_3\tau_{xy} \tag{2.106}$$

$$\tau_{x_1y_1} = \tau_{y_1x_1} = l_1l_2\sigma_x + m_1m_2\sigma_y + n_1n_2\sigma_z + (m_1n_2 + m_2n_1)\tau_{yz}$$

$$+ (n_1l_2 + n_2l_1)\tau_{zx} + (l_1m_2 + l_2m_1)\tau_{xy} \tag{2.107}$$

$$\tau_{x_1 z_1} = \tau_{z_1 x_1} = l_1 l_3 \sigma_x + m_1 m_3 \sigma_y + n_1 n_3 \sigma_z + (m_1 n_3 + m_3 n_1)\tau_{yz}$$
$$+ (n_1 l_3 + n_3 l_1)\tau_{zx} + (l_1 m_3 + l_3 m_1)\tau_{xy} \tag{2.108}$$

$$\tau_{y_1 z_1} = \tau_{z_1 y_1} = l_2 l_3 \sigma_x + m_2 m_3 \sigma_y + n_2 n_3 \sigma_z + (m_2 n_3 + m_3 n_2)\tau_{yz}$$
$$+ (n_2 l_3 + n_3 l_2)\tau_{zx} + (l_2 m_3 + l_3 m_2)\tau_{xy} \tag{2.109}$$

2.10.3 Principal stresses

The preceding procedure allows the determination of the stresses on any plane from the known stresses based on a set of orthogonal axes. As discussed earlier, a plane is defined as a principal plane if the shear stresses acting on it are zero, which means that the only stress acting on it is a normal stress. This normal stress on a principal plane is referred to as a *principal stress*. In order to determine the principal stresses, refer to Figure 2.20, in which x, y, and z are a set of orthogonal axes. Let the stresses on planes OAC, BOC, and AOB be known, and let ABC be a principal plane. The direction cosines of the normal drawn to this plane are l, m, and n with respect to the x, y, and z axes, respectively. Note that

$$l^2 + m^2 + n^2 = 1 \tag{2.110}$$

If ABC is a principal plane, then the only stress acting on it will be a normal stress σ_p. The x, y, and z components of σ_p are $\sigma_p l$, $\sigma_p m$, and $\sigma_p n$. Referring to Equations 2.95 through 2.97, we can write

$$\sigma_p l = \sigma_x l + \tau_{yx} m + \tau_{zx} n$$

or

$$(\sigma_x - \sigma_p)l + \tau_{yx} m + \tau_{zx} n = 0 \tag{2.111}$$

Similarly

$$\tau_{xy} l + (\sigma_y - \sigma_p)m + \tau_{zy} n = 0 \tag{2.112}$$

$$\tau_{xz} l + \tau_{yz} m + (\sigma_z - \sigma_p)n = 0 \tag{2.113}$$

From Equations 2.110 through 2.113, we note that l, m, and n cannot all be equal to zero at the same time. So

$$\begin{vmatrix} (\sigma_x - \sigma_p) & \tau_{yx} & \tau_{zx} \\ \tau_{xy} & (\sigma_y - \sigma_p) & \tau_{zy} \\ \tau_{xz} & \tau_{yz} & (\sigma_z - \sigma_p) \end{vmatrix} = 0 \qquad (2.114)$$

or

$$\sigma_p^3 - I_1\sigma_p^2 + I_2\sigma_p - I_3 = 0 \qquad (2.115)$$

where

$$I_1 = \sigma_x + \sigma_y + \sigma_z \qquad (2.116)$$

$$I_2 = \sigma_x\sigma_y + \sigma_y\sigma_z + \sigma_x\sigma_z - \tau_{xy}^2 - \tau_{yz}^2 - \tau_{xz}^2 \qquad (2.117)$$

$$I_3 = \sigma_x\sigma_y\sigma_z + 2\tau_{xy}\tau_{yz}\tau_{xz} - \sigma_x\tau_{yz}^2 - \sigma_y\tau_{xz}^2 - \sigma_z\tau_{xy}^2 \qquad (2.118)$$

I_1, I_2, and I_3 defined in Equations 2.116 through 2.118 are independent of direction cosines and hence independent of the choice of axes. So, they are referred to as stress invariants.

Solution of Equation 2.115 gives three real values of σ_p. So there are three principal planes and they are mutually perpendicular to each other. The directions of these planes can be determined by substituting each σ_p in Equations 2.111 through 2.113 and solving for l, m, and n, and observing the direction cosine condition for $l^2 + m^2 + n^2 = 1$. Note that these values for l, m, and n are the direction cosines for the normal drawn to the plane on which σ_p is acting. The maximum, intermediate, and minimum values of $\sigma_{p(i)}$ are referred to as the major principal stress, intermediate principal stress, and minor principal stress, respectively.

2.10.4 Octahedral stresses

The octahedral stresses at a point are the normal and shear stresses acting on the planes of an imaginary octahedron surrounding that point. The normals to these planes have direction cosines of $\pm 1\sqrt{3}$ with respect to the direction of the principal stresses (Figure 2.22). The axes marked 1, 2, and 3 are the directions of the principal stresses $\sigma_{p(1)}$, $\sigma_{p(2)}$, and $\sigma_{p(3)}$. The expressions for the octahedral normal stress σ_{oct} can be obtained using

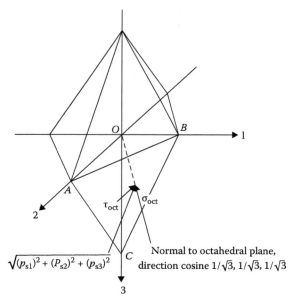

Figure 2.22 Octahedral stress.

Equations 2.95 through 2.97 and 2.99. Now, compare planes ABC in Figures 2.20 and 2.22. For the octahedral plane ABC in Figure 2.22

$$p_{s1} = \sigma_{p(1)}l \tag{2.119}$$

$$p_{s2} = \sigma_{p(2)}m \tag{2.120}$$

$$p_{s3} = \sigma_{p(3)}n \tag{2.121}$$

where p_{s1}, p_{s2}, and p_{s3} are stresses acting on plane ABC parallel to the principal stress axes 1, 2, and 3, respectively. Parameters l, m, and n are the direction cosines of the normal drawn to the octahedral plane and are all equal to $1/\sqrt{3}$. Thus, from Equation 2.99

$$\sigma_{\text{oct}} = l_1^2 \sigma_{p(1)} + m_1^2 \sigma_{p(2)} + n_1^2 \sigma_{p(3)}$$

$$= \frac{1}{3}[\sigma_{p(1)} + \sigma_{p(2)} + \sigma_{p(3)}] \tag{2.122}$$

The shear stress on the octahedral plane is

$$\tau_{oct} = \sqrt{[(p_{s1})^2 + (p_{s2})^2 + (p_{s3})^2] - \sigma_{oct}^2} \qquad (2.123)$$

where τ_{oct} is the octahedral shear stress, or

$$\tau_{oct} = \frac{1}{3}\sqrt{[\sigma_{p(1)} - \sigma_{p(2)}]^2 + [\sigma_{p(2)} - \sigma_{p(3)}]^2 + [\sigma_{p(3)} - \sigma_{p(1)}]^2} \qquad (2.124)$$

The octahedral normal and shear stress expressions can also be derived as a function of the stress components for any set of orthogonal axes x, y, z. From Equation 2.116

$$I_1 = \text{const} = \sigma_x + \sigma_y + \sigma_z = \sigma_{p(1)} + \sigma_{p(2)} + \sigma_{p(3)} \qquad (2.125)$$

So

$$\sigma_{oct} = \frac{1}{3}[\sigma_{p(1)} + \sigma_{p(2)} + \sigma_{p(3)}] = \frac{1}{3}(\sigma_x + \sigma_y + \sigma_z) \qquad (2.126)$$

Similarly, from Equation 2.117

$$I_2 = \text{const} = (\sigma_x\sigma_y + \sigma_y\sigma_z + \sigma_z\sigma_x) - \tau_{xy}^2 - \tau_{yz}^2 - \tau_{xz}^2$$

$$= \sigma_{p(1)}\sigma_{p(2)} + \sigma_{p(2)}\sigma_{p(3)} + \sigma_{p(3)}\sigma_{p(1)} \qquad (2.127)$$

Combining Equations 2.124, 2.125, and 2.127 gives

$$\tau_{oct} = \frac{1}{3}\sqrt{(\sigma_x - \sigma_y)^2 + (\sigma_y - \sigma_z)^2 + (\sigma_z - \sigma_x)^2 + 6\tau_{xy}^2 + 6\tau_{yz}^2 + 6\tau_{xz}^2} \qquad (2.128)$$

Example 2.4

The stresses at a point in a soil mass are as follows:

$\sigma_x = 50 \text{ kN/m}^2$ $\tau_{xy} = 30 \text{ kN/m}^2$
$\sigma_y = 40 \text{ kN/m}^2$ $\tau_{yz} = 25 \text{ kN/m}^2$
$\sigma_z = 80 \text{ kN/m}^2$ $\tau_{xz} = 25 \text{ kN/m}^2$

Determine the normal and shear stresses on a plane with direction cosines $l = 2/3$, $m = 2/3$, and $n = 1/3$.

Solution

From Equation 2.98

$$\begin{vmatrix} p_{sx} \\ p_{sy} \\ p_{sz} \end{vmatrix} = \begin{vmatrix} \sigma_x & \tau_{xy} & \tau_{xz} \\ \tau_{xy} & \sigma_y & \tau_{yz} \\ \tau_{xz} & \tau_{yz} & \sigma_z \end{vmatrix} \begin{vmatrix} l \\ m \\ n \end{vmatrix}$$

The normal stress on the inclined plane (Equation 2.99) is

$$\begin{aligned}
\sigma &= p_{sx}l + p_{sy}m + p_{sz}n \\
&= \sigma_x l^2 + \sigma_y m^2 + \sigma_z n^2 + 2\tau_{xy}lm + 2\tau_{yz}mn + 2\tau_{xz}ln \\
&= 50(2/3)^2 + 40(2/3)^2 + 80(1/3)^2 + 2(30)(2/3)(2/3) \\
&\quad + 2(25)(2/3)(1/3) + 2(25)(2/3)(1/3) = 97.78 \text{ kN/m}^2
\end{aligned}$$

$$\begin{aligned}
p_{sx} &= \sigma_x l + \tau_{xy}m + \tau_{xz}n = 50(2/3) + 30(2/3) + 25(1/3) \\
&= 33.33 + 20 + 8.33 = 61.66 \text{ kN/m}^2
\end{aligned}$$

$$\begin{aligned}
p_{sy} &= \tau_{xy}l + \sigma_y m + \tau_{yz}n = 30(2/3) + 40(2/3) + 25(1/3) \\
&= 20 + 26.67 + 8.33 = 55 \text{ kN/m}^2
\end{aligned}$$

$$\begin{aligned}
p_{sz} &= \tau_{xz}l + \tau_{yz}m + \sigma_z n = 25(2/3) + 25(2/3) + 80(1/3) \\
&= 16.67 + 16.67 + 26.67 = 60.01 \text{ kN/m}^2
\end{aligned}$$

The resultant stress is

$$p = \sqrt{p_{sx}^2 + p_{sy}^2 + p_{sz}^2} = \sqrt{61.66^2 + 55^2 + 60.01^2} = 102.2 \text{ kN/m}^2$$

The shear stress on the plane is

$$\tau = \sqrt{p^2 - \sigma^2} = \sqrt{102.2^2 - 97.78^2} = 29.73 \text{ kN/m}^2$$

Example 2.5

At a point in a soil mass, the stresses are as follows:

$$\sigma_x = 25 \text{ kN/m}^2 \qquad \tau_{xy} = 30 \text{ kN/m}^2$$
$$\sigma_y = 40 \text{ kN/m}^2 \qquad \tau_{yz} = -6 \text{ kN/m}^2$$
$$\sigma_z = 17 \text{ kN/m}^2 \qquad \tau_{xz} = -10 \text{ kN/m}^2$$

Determine the principal stresses and also the octahedral normal and shear stresses.

Solution

From Equation 2.114

$$
\begin{vmatrix}
(\sigma_x - \sigma_p) & \tau_{yx} & \tau_{zx} \\
\tau_{xy} & (\sigma_y - \sigma_p) & \tau_{zy} \\
\tau_{xz} & \tau_{yz} & (\sigma_z - \sigma_p)
\end{vmatrix} = 0
$$

$$
\begin{vmatrix}
(25 - \sigma_p) & 30 & -10 \\
30 & (40 - \sigma_p) & -6 \\
-10 & -6 & (17 - \sigma_p)
\end{vmatrix} = \sigma_p^3 - 82\sigma_p^2 + 1069\sigma_p - 800 = 0
$$

The three roots of the equation are

$$\sigma_{p(1)} = 65.9 \text{ kN/m}^2$$

$$\sigma_{p(2)} = 15.7 \text{ kN/m}^2$$

$$\sigma_{p(3)} = 0.4 \text{ kN/m}^2$$

$$\sigma_{oct} = \frac{1}{3}[\sigma_{p(1)} + \sigma_{p(2)} + \sigma_{p(3)}]$$

$$= \frac{1}{3}(65.9 + 15.7 + 0.4) = 27.33 \text{ kN/m}^2$$

$$\tau_{oct} = \frac{1}{3}\sqrt{[\sigma_{p(1)} - \sigma_{p(2)}]^2 + [\sigma_{p(2)} - \sigma_{p(3)}]^2 + [\sigma_{p(3)} - \sigma_{p(1)}]^2}$$

$$= \frac{1}{3}\sqrt{(65.9 - 15.7)^2 + (15.7 - 0.4)^2 + (0.4 - 65.9)^2} = 27.97 \text{ kN/m}^2$$

2.11 STRAIN COMPONENTS ON AN INCLINED PLANE, PRINCIPAL STRAIN, AND OCTAHEDRAL STRAIN: THREE-DIMENSIONAL CASE

We have seen the analogy between the stress and strain equations derived in Sections 2.7 and 2.8 for the plane strain case. Referring to Figure 2.20, let the strain components at a point in a soil mass be represented by ϵ_x, ϵ_y, ϵ_z, γ_{xy}, γ_{yz}, and γ_{zx}. The normal strain on plane ABC (the normal to plane ABC has direction cosines of l, m, and n) can be given by

$$\epsilon = l^2\epsilon_x + m^2\epsilon_y + n^2\epsilon_z + lm\gamma_{xy} + mn\gamma_{yz} + ln\gamma_{zx} \qquad (2.129)$$

This equation is similar in form to Equation 2.99 derived for normal stress. When we replace ϵ_x, ϵ_y, ϵ_z, $\gamma_{xy}/2$, $\gamma_{yz}/2$, and $\gamma_{zx}/2$, respectively, for σ_x, σ_y, σ_z, τ_{xy}, τ_{yz}, and τ_{zx} in Equation 2.99, Equation 2.129 is obtained.

If the strain components at a point in the Cartesian coordinate system (Figure 2.21) are known, the components in a new set of orthogonal axes can be given by (similar to Equation 2.103)

$$
\begin{vmatrix}
\epsilon_{x1} & \dfrac{1}{2}\gamma_{x1y1} & \dfrac{1}{2}\gamma_{x1z1} \\[2mm]
\dfrac{1}{2}\gamma_{x1y1} & \epsilon_{y1} & \dfrac{1}{2}\gamma_{y1z1} \\[2mm]
\dfrac{1}{2}\gamma_{x1z1} & \dfrac{1}{2}\gamma_{y1z1} & \epsilon_{z1}
\end{vmatrix}
$$

$$
=
\begin{vmatrix}
l_1 & m_1 & n_1 \\
l_2 & m_2 & n_2 \\
l_3 & m_3 & n_3
\end{vmatrix}
\begin{vmatrix}
\epsilon_x & \dfrac{1}{2}\gamma_{xy} & \dfrac{1}{2}\gamma_{xz} \\[2mm]
\dfrac{1}{2}\gamma_{xy} & \epsilon_y & \dfrac{1}{2}\gamma_{yz} \\[2mm]
\dfrac{1}{2}\gamma_{xz} & \dfrac{1}{2}\gamma_{yz} & \epsilon_z
\end{vmatrix}
\begin{vmatrix}
l_1 & l_2 & l_3 \\
m_1 & m_2 & m_3 \\
n_1 & m_2 & n_3
\end{vmatrix}
\tag{2.130}
$$

The equations for principal strains at a point can also be written in a form similar to that given for stress (Equation 2.115) as

$$
\epsilon_p^3 - J_1\,\epsilon_p^2 + J_2\,\epsilon_p - J_3 = 0 \tag{2.131}
$$

where ϵ_p is the principal strain

$$
J_1 = \epsilon_x + \epsilon_y + \epsilon_z \tag{2.132}
$$

$$
J_2 = \epsilon_x\epsilon_y + \epsilon_y\epsilon_z + \epsilon_z\epsilon_x - \left(\frac{\gamma_{xy}}{2}\right)^2 - \left(\frac{\gamma_{yz}}{2}\right)^2 - \left(\frac{\gamma_{xz}}{2}\right)^2 \tag{2.133}
$$

$$
J_3 = \epsilon_x\epsilon_y\epsilon_z + \frac{\gamma_{xy}\gamma_{yz}\gamma_{zx}}{4} - \epsilon_x\left(\frac{\gamma_{yz}}{2}\right)^2 - \epsilon_y\left(\frac{\gamma_{xz}}{2}\right)^2 - \epsilon_z\left(\frac{\gamma_{xy}}{2}\right)^2 \tag{2.134}
$$

J_1, J_2, and J_3 are the strain invariants and are not functions of the direction cosines.

The normal and shear strain relations for the octahedral planes are

$$\epsilon_{\text{oct}} = \frac{1}{3}[\epsilon_{p(1)} + \epsilon_{p(2)} + \epsilon_{p(3)}] \tag{2.135}$$

$$\gamma_{\text{oct}} = \frac{2}{3}\sqrt{[\epsilon_{p(1)} - \epsilon_{p(2)}]^2 + [\epsilon_{p(2)} - \epsilon_{p(3)}]^2 + [\epsilon_{p(3)} - \epsilon_{p(1)}]^2} \tag{2.136}$$

where
ϵ_{oct} is the octahedral normal strain
γ_{oct} is the octahedral shear strain
$\epsilon_{p(1)}$, $\epsilon_{p(2)}$, $\epsilon_{p(3)}$ are the major, intermediate, and minor principal strains, respectively

Equations 2.135 and 2.136 are similar to the octahedral normal and shear stress relations given by Equations 2.126 and 2.128.

Chapter 3

Stresses and displacements in a soil mass

Two-dimensional problems

3.1 INTRODUCTION

Estimating the increase in stress at various points and the associated displacement caused in a soil mass due to external loading using the theory of elasticity is an important component in the safe design of the foundations of structures. The ideal assumption of the theory of elasticity, namely that the medium is homogeneous, elastic, and isotropic, is not quite true for most natural soil profiles. It does, however, provide a close estimation of geotechnical engineers and, using proper safety factors, safe designs can be developed.

This chapter deals with two-dimensional problems (plane strain cases) involving stresses and displacements induced by various types of loading. The expressions for stresses and displacements are obtained on the assumption that soil is a perfectly elastic material. Problems relating to plastic equilibrium are not treated in this chapter.

Stresses and displacements related to three-dimensional problems are treated in Chapter 4.

3.2 VERTICAL LINE LOAD ON THE SURFACE

Figure 3.1 shows the case where a line load of q per unit length is applied at the surface of a homogeneous, elastic, and isotropic soil mass. The stresses at a point P defined by r and θ can be determined by using the stress function

$$\phi = \frac{q}{\pi} r\theta \sin\theta \qquad (3.1)$$

In the polar coordinate system, the expressions for the stresses are as follows:

$$\sigma_r = \frac{1}{r}\frac{\partial\phi}{\partial r} + \frac{1}{r^2}\frac{\partial^2\phi}{\partial\theta^2} \qquad (2.57)$$

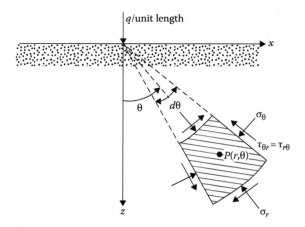

Figure 3.1 Vertical line load on the surface of a semi-infinite mass.

$$\sigma_\theta = \frac{\partial^2 \phi}{\partial r^2} \tag{2.58}$$

and

$$\tau_{r\theta} = -\frac{\partial}{\partial r}\left(\frac{1}{r}\frac{\partial \phi}{\partial \theta}\right) \tag{2.59}$$

Substituting the values of ϕ in the previous equations, we get

$$\sigma_r = \frac{1}{r}\left(\frac{q}{\pi}\theta\sin\theta\right) + \frac{1}{r^2}\left(\frac{q}{\pi}r\cos\theta + \frac{q}{\pi}r\cos\theta - \frac{q}{\pi}r\theta\sin\theta\right)$$

$$= \frac{2q}{\pi r}\cos\theta \tag{3.2}$$

Similarly

$$\sigma_\theta = 0 \tag{3.3}$$

and

$$\tau_{r\theta} = 0 \tag{3.4}$$

The stress function assumed in Equation 3.1 will satisfy the *compatibility equation*:

$$\left(\frac{\partial^2}{\partial r^2}+\frac{1}{r}\frac{\partial}{\partial r}+\frac{1}{r^2}\frac{\partial^2}{\partial \theta^2}\right)\left(\frac{\partial^2 \phi}{\partial r^2}+\frac{1}{r}\frac{\partial \phi}{\partial r}+\frac{1}{r^2}\frac{\partial^2 \phi}{\partial \theta^2}\right)=0 \tag{2.60}$$

Also, it can be seen that the stresses obtained in Equations 3.2 through 3.4 satisfy the boundary conditions. For $\theta = 90°$, $r > 0$, $\sigma_r = 0$, and at $r = 0$, σ_r is theoretically equal to infinity, which signifies that plastic flow will occur locally. Note that σ_r and σ_θ are the major and minor principal stresses at point P.

Using the earlier expressions for σ_r, σ_θ, and $\tau_{r\theta}$, we can derive the stresses in the rectangular coordinate system (Figure 3.2):

$$\sigma_z = \sigma_r \cos^2 \theta + \sigma_\theta \sin^2 \theta - 2\tau_{r\theta}\sin\theta\cos\theta \tag{2.77}$$

or,

$$\sigma_z = \frac{2q}{\pi r}\cos^3 \theta = \frac{2q}{\pi\sqrt{x^2+z^2}}\left(\frac{z}{\sqrt{x^2+z^2}}\right)^3 = \frac{2qz^3}{\pi(x^2+z^2)^2} \tag{3.5}$$

Similarly

$$\sigma_x = \sigma_r \sin^2 \theta + \sigma_\theta \cos^2 \theta + 2\tau_{r\theta}\sin\theta\cos\theta \tag{2.75}$$

or,

$$\sigma_x = \frac{2qx^2z}{\pi(x^2+z^2)^2} \tag{3.6}$$

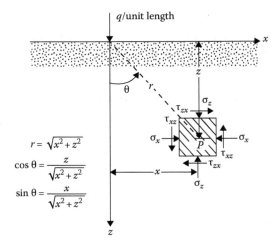

Figure 3.2 Stresses due to a vertical line load in rectangular coordinates.

Table 3.1 Values of $\sigma_z/(q/z)$, $\sigma_x/(q/z)$, and $\tau_{xz}/(q/z)$
(Equations 3.5 through 3.7)

x/z	$\sigma_z/(q/z)$	$\sigma_x/(q/z)$	$\tau_{xz}/(q/z)$
0	0.637	0	0
0.1	0.624	0.006	0.062
0.2	0.589	0.024	0.118
0.3	0.536	0.048	0.161
0.4	0.473	0.076	0.189
0.5	0.407	0.102	0.204
0.6	0.344	0.124	0.207
0.7	0.287	0.141	0.201
0.8	0.237	0.151	0.189
0.9	0.194	0.157	0.175
1.0	0.159	0.159	0.159
1.5	0.060	0.136	0.090
2.0	0.025	0.102	0.051
3.0	0.006	0.057	0.019

and

$$\tau_{xz} = -\sigma_\theta \sin\theta \cos\theta + \sigma_r \sin\theta \cos\theta + \tau_{r\theta}(\cos^2\theta - \sin^2\theta) \tag{2.76}$$

or,

$$\tau_{xz} = \frac{2qxz^2}{\pi(x^2 + z^2)^2} \tag{3.7}$$

For the plane strain case

$$\sigma_y = v(\sigma_x + \sigma_z) \tag{3.8}$$

The variation of the values for σ_x, σ_z, and τ_{xz} with x/z in a nondimensional form are given in Table 3.1. Figure 3.3 shows plots of σ_z, σ_x, τ_{xz} with x/z in a nondimensional form.

3.2.1 Displacement on the surface (z = 0)

By relating displacements to stresses via strain, the vertical displacement w at the *surface* (i.e., $z = 0$) can be obtained as

$$w = \frac{2}{\pi} \frac{1-v^2}{E} q \ln|x| + C \tag{3.9}$$

where
   E is the modulus of elasticity
   v is Poisson's ratio
   C is a constant

$$\sigma_z/(q/z),\ \sigma_x/(q/z),\ \tau_{xz}/(q/z)$$

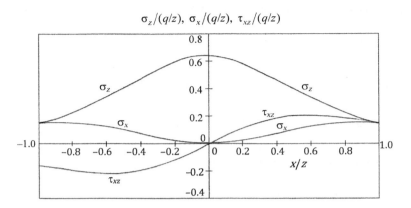

Figure 3.3 Plot of $\sigma_z/(q/z)$, $\sigma_x/(q/z)$, and $\tau_{xz}/(q/z)$ vs. x/z (Equations 3.5 through 3.7).

The magnitude of the constant can be determined if the vertical displacement at a point is specified.

Example 3.1

For the point A in Figure 3.4, calculate the increase of vertical stress σ_z due to the two line loads.

Solution

The increase of vertical stress at A due to the line load $q_1 = 20$ kN/m:

$$\text{Given,}\ \frac{x}{z} = \frac{2\,m}{2\,m} = 1$$

From Table 3.1, for $x/z = 1$, $\sigma_z/(q/z) = 0.159$. So

$$\sigma_{z(1)} = 0.159\left(\frac{q_1}{z}\right) = 0.159\left(\frac{20}{2}\right) = 1.59 \text{ kN/m}^2$$

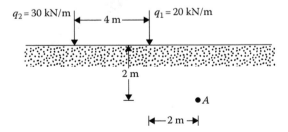

Figure 3.4 Two line loads acting on the surface.

The increase of vertical stress at A due to the line load $q_2 = 30$ kN/m:

Given, $\dfrac{x}{z} = \dfrac{6\,\text{m}}{2\,\text{m}} = 3$

From Table 3.1, for $x/z = 3$, $\sigma_z/(q/z) = 0.006$. Thus

$$\sigma_{z(2)} = 0.006\left(\frac{q_2}{z}\right) = 0.006\left(\frac{30}{2}\right) = 0.09 \text{ kN/m}^2$$

So, the total increase of vertical stress is

$$\sigma_z = \sigma_{z(1)} + \sigma_{z(2)} = 1.59 + 0.09 = 1.68 \text{ kN/m}^2$$

3.3 VERTICAL LINE LOAD AT THE APEX OF AN INFINITE WEDGE

Figure 3.5 shows an infinite wedge with its apex at O. A vertical line load of q/unit length is applied at O. The stress at a point P located at a radial distance r can be given as (Michell, 1900):

$$\sigma_r = \left(\frac{2q}{r}\right)\frac{\cos\theta}{2\alpha + \sin 2\alpha} \tag{3.10}$$

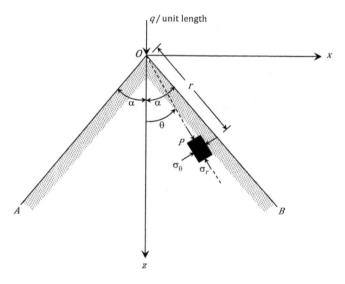

Figure 3.5 Infinite wedge with vertical line load.

$$\sigma_\theta = 0 \tag{3.11}$$

$$\tau_{r\theta} = 0 \tag{3.12}$$

Note that, if $\alpha = \pi/2$, which is the case in Figure 3.1, then Equation 3.10 becomes $\sigma_r = (2q/\pi r)\cos\theta$, which is the same as Equation 3.2.

Example 3.2

Refer to Figure 3.5. Given: $\alpha = 40°$; $q = 30$ kN/m. Determine σ_r at P (2.5 m, 4 m).

Solution

$$\theta = \tan^{-1}\left(\frac{x}{z}\right) = \tan^{-1}\left(\frac{2.5}{4}\right) = 32^c$$

$$r = \left[(2.5)^2 + (4)^2\right]^{0.5} = 4.717 \text{ m}$$

$$\alpha = \left(\frac{\pi}{180}\right)(40) = 0.698 \text{ radians}$$

From Equation 3.10,

$$\sigma_r = \left(\frac{2q}{r}\right)\frac{\cos\theta}{2\alpha + \sin 2\alpha} = \left[\frac{(2)(30)}{4.717}\right]\left[\frac{\cos 32}{(2)(0.698) + \sin 80}\right] = 4.53 \text{ kN/m}^2$$

3.4 VERTICAL LINE LOAD ON THE SURFACE OF A FINITE LAYER

Equations 3.5 through 3.7 were derived with the assumption that the homogeneous soil mass extends to a great depth. However, in many practical cases, a stiff layer such as rock or highly incompressible material may be encountered at a shallow depth (Figure 3.6). At the interface of the top soil layer and the lower incompressible layer, the shear stresses will modify the pattern of stress distribution. Poulos (1966) and Poulos and Davis (1974) expressed the vertical stress σ_z and vertical displacement at the surface (w at $z = 0$) in the forms:

$$\sigma_z = \frac{q}{\pi h}I_1 \tag{3.13}$$

$$w_{z=0} = \frac{q}{\pi E}I_2 \tag{3.14}$$

where I_1 and I_2 are influence values.

I_1 is a function of z/h, x/h, and v. Similarly, I_2 is a function of x/h and v. The variations of I_1 and I_2 are given in Tables 3.2 and 3.3, respectively, for $v = 0$.

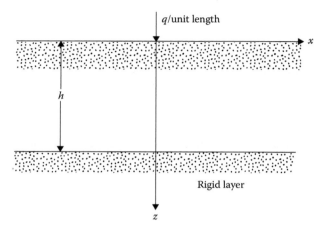

Figure 3.6 Vertical line load on a finite elastic layer.

Table 3.2 Variation of I_1 ($v = 0$)

x/h	z/h				
	0.2	0.4	0.6	0.8	1.0
0	9.891	5.157	3.641	2.980	2.634
0.1	5.946	4.516	3.443	2.885	2.573
0.2	2.341	3.251	2.948	2.627	2.400
0.3	0.918	2.099	2.335	2.261	2.144
0.4	0.407	1.301	1.751	1.857	1.840
0.5	0.205	0.803	1.265	1.465	1.525
0.6	0.110	0.497	0.889	1.117	1.223
0.8	0.032	0.185	0.408	0.592	0.721
1.0	0.000	0.045	0.144	0.254	0.357
1.5	−0.019	−0.035	−0.033	−0.018	0.010
2.0	−0.013	−0.025	−0.035	−0.041	−0.042
4.0	0.009	0.009	0.008	0.007	0.006
8.0	0.002	0.002	0.002	0.002	0.002

3.5 VERTICAL LINE LOAD INSIDE A SEMI-INFINITE MASS

Equations 3.5 through 3.7 were also developed on the basis of the assumption that the line load is applied on the surface of a semi-infinite mass. However, in some cases, the line load may be embedded. Melan (1932) gave the solution of stresses at a point P due to a vertical line load of

Table 3.3 Variation of I_2 ($\nu = 0$)

x/h	I_2
0.1	3.756
0.2	2.461
0.3	1.730
0.4	1.244
0.5	0.896
0.6	0.643
0.7	0.453
0.8	0.313
1.0	0.126
1.5	−0.012
2.0	−0.017
4.0	−0.002
8.0	0

q per unit length applied inside a semi-infinite mass (at point A, Figure 3.7). The final equations are given as follows:

$$\sigma_z = \frac{q}{\pi}\left(\frac{1}{2(1-\nu)}\left\{ \frac{(z-d)^3}{r_1^4} + \frac{(z+d)[(z+d)^2 + 2dz]}{r_2^4} - \frac{8dz(d+z)x^2}{r_2^6} \right\} \right.$$

$$\left. + \frac{1-2\nu}{4(1-\nu)}\left(\frac{z-d}{r_1^2} + \frac{3z+d}{r_2^4} - \frac{4zx^2}{r_2^4} \right) \right)$$

(3.15)

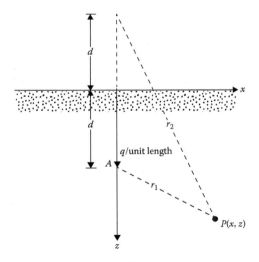

Figure 3.7 Vertical line load inside a semi-infinite mass.

$$\sigma_x = \frac{q}{\pi}\left\{\frac{1}{2(1-v)}\left[\frac{(z-d)x^2}{r_1^4}+\frac{(z+d)(x^2+2d^2)-2dx^2}{r_2^4}+\frac{8dz(d+z)x^2}{r_2^6}\right]\right.$$

$$\left.+\frac{1-2v}{4(1-v)}\left(\frac{d-z}{r_1^2}+\frac{z+3d}{r_2^2}+\frac{4zx^2}{r_2^4}\right)\right\} \tag{3.16}$$

$$\tau_{xz} = \frac{qx}{\pi}\left\{\frac{1}{2(1-v)}+\left[\frac{(z-d)^2}{r_1^4}+\frac{z^2-2dz-d^2}{r_2^4}+\frac{8dz(d+z)^2}{r_2^6}\right]\right.$$

$$\left.+\frac{1-2v}{4(1-v)}\left[\frac{1}{r_1^2}-\frac{1}{r_2^2}+\frac{4z(d+z)}{r_2^4}\right]\right\} \tag{3.17}$$

Figure 3.8 shows a plot of $\sigma_z/(q/d)$ based on Equation 3.15.

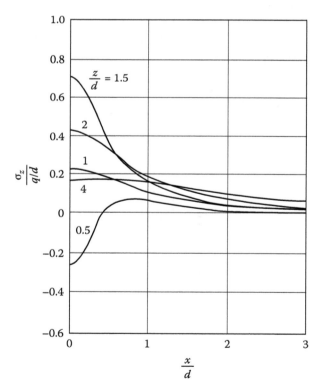

Figure 3.8 Plot of $\sigma_z/(q/d)$ versus x/d for various values of z/d (Equation 3.15).

3.6 HORIZONTAL LINE LOAD ON THE SURFACE

The stresses due to a horizontal line load of q per unit length (Figure 3.9) can be evaluated by a stress function of the form

$$\phi = \frac{q}{\pi} r\theta \cos\theta \tag{3.18}$$

Proceeding in a similar manner to that shown in Section 3.2 for the case of vertical line load, we obtain the stresses at a point P defined by r and θ as

$$\sigma_r = \frac{2q}{\pi r}\sin\theta \tag{3.19}$$

$$\sigma_\theta = 0 \tag{3.20}$$

$$\tau_{r\theta} = 0 \tag{3.21}$$

In the rectangular coordinate system,

$$\sigma_z = \frac{2q}{\pi}\frac{xz^2}{(x^2 + z^2)^2} \tag{3.22}$$

$$\sigma_x = \frac{2q}{\pi}\frac{x^3}{(x^2 + z^2)^2} \tag{3.23}$$

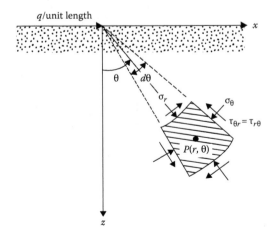

Figure 3.9 Horizontal line load on the surface of a semi-infinite mass.

$$\tau_{xz} = \frac{2q}{\pi} \frac{x^2 z}{(x^2 + z^2)^2} \tag{3.24}$$

For the plane strain case, $\sigma_y = v(\sigma_x + \sigma_z)$.

Some values of σ_x, σ_z, and τ_{xz} in a nondimensional form are given in Table 3.4. Figure 3.10 shows plots of σ_z, σ_x, and τ_{xz} with x/z in a nondimensional form.

Table 3.4 Values of $\sigma_z/(q/z)$, $\sigma_x/(q/z)$, and $\tau_{xz}/(q/z)$ (Equations 3.22 through 3.24)

x/z	$\sigma_z/(q/z)$	$\sigma_x/(q/z)$	$\tau_{xz}/(q/z)$
0	0	0	0
0.1	0.062	0.0006	0.006
0.2	0.118	0.0049	0.024
0.3	0.161	0.0145	0.048
0.4	0.189	0.0303	0.076
0.5	0.204	0.0509	0.102
0.6	0.207	0.0743	0.124
0.7	0.201	0.0984	0.141
0.8	0.189	0.1212	0.151
0.9	0.175	0.1417	0.157
1.0	0.159	0.1591	0.159
1.5	0.090	0.2034	0.136
2.0	0.051	0.2037	0.102
3.0	0.019	0.1719	0.057

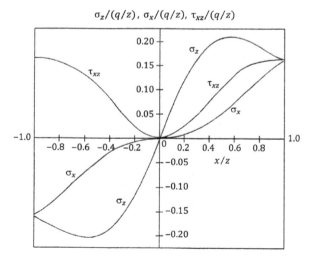

Figure 3.10 Plot of $\sigma_z/(q/z)$, $\sigma_x/(q/z)$, and $\tau_{xz}/(q/z)$ vs. x/z (Equations 3.22 through 3.24).

Example 3.3

Refer to Figure 3.11. Given: $\alpha_1 = 30°$; $\alpha_2 = 45°$; $a = 2$ m; $a_1 = 3$ m; $a_2 = 5$ m; $b = 2$ m; $q_1 = 40$ kN/m; $q_2 = 30$ kN/m. Determine σ_z at M and N.

Solution

Components of $q_1 \rightarrow$ Vertical: $q_1 \sin 30$
                                   Horizontal: $q_1 \cos 30$

Components of $q_2 \rightarrow$ Vertical: $q_2 \sin 45$
                                   Horizontal: $q_2 \cos 45$

From Equations 3.5 and 3.22,

At point N:

$$\sigma_z = \frac{2(q_1 \sin 30)(2)^3}{\pi[(5)^2 + (2)^2]^2} + \frac{2(-q_1 \cos 30)(5)(2)^2}{\pi[(5)^2 + (2)^2]^2} + \frac{2(q_2 \sin 45)(2)^3}{\pi[(5+2)^2 + (2)^2]^2}$$

$$+ \frac{2(q_2 \cos 45)(5+2)(2)^2}{\pi[(5+2)^2 + (2)^2]^2}$$

$$= 0.121 - 0.5245 + 0.0385 + 0.1346$$

$$\approx -0.23 \text{ kN/m}^2$$

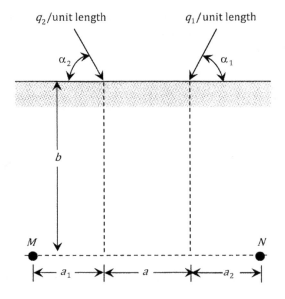

Figure 3.11 Two inclined line loads.

At point M:

$$\sigma_z = \frac{2(q_1 \sin 30)(2)^3}{\pi[(-5)^2 + (2)^2]^2} + \frac{2(-q_1 \cos 30)(-5)(2)^2}{\pi[(-5)^2 + (2)^2]^2} + \frac{2(q_2 \sin 45)(2)^3}{\pi[(-3)^2 + (2)^2]^2}$$

$$+ \frac{2(q_2 \cos 45)(-3)(2)^2}{\pi[(-3)^2 + (2)^2]^2}$$

$$= 0.121 + 0.5245 + 0.639 - 0.959$$

$$\approx 0.327 \text{ kN/m}^2$$

3.7 HORIZONTAL AND INCLINED LINE LOAD AT THE APEX OF AN INFINITE WEDGE

Figure 3.12(a) shows a horizontal line load of q/unit length at the apex of an infinite wedge. Similar to Equations 3.10 to 3.12, the stress at P can be given as:

$$\sigma_r = \frac{2q}{r} \frac{\sin\theta}{2\alpha - \sin2\alpha} \tag{3.25}$$

$$\sigma_\theta = 0 \tag{3.26}$$

$$\tau_{r\theta} = 0 \tag{3.27}$$

Figure 3.12(b) shows an *inclined* line load at the apex of an infinite wedge, which is a combination of the loadings shown in Figures 3.5 and 3.12(a). For this condition

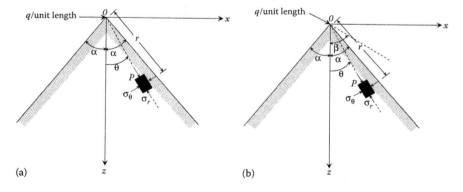

(a)   (b)

Figure 3.12 (a) Horizontal line load; (b) inclined line load at the apex of an infinite load.

$$\sigma_r = \frac{2q}{r}\left[\frac{\cos\beta\cos\theta}{2\alpha+\sin2\alpha}+\frac{\sin\beta\sin\theta}{2\alpha-\sin2\alpha}\right]$$

(3.28)

$$\sigma_\theta = 0$$

(3.29)

$$\tau_{r\theta} = 0$$

(3.30)

3.8 HORIZONTAL LINE LOAD INSIDE A SEMI-INFINITE MASS

If the horizontal line load acts inside a semi-infinite mass as shown in Figure 3.13, Melan's solutions for stresses at a point $P(x, z)$ may be given as follows:

$$\sigma_z = \frac{qx}{\pi}\left\{\frac{1}{2(1-v)}\left[\frac{(z-d)^2}{r_1^4}-\frac{d^2-z^2+6dz}{r_2^4}+\frac{8dz\,x^2}{r_2^6}\right]\right.$$

$$\left.-\frac{1-2v}{4(1-v)}\left[\frac{1}{r_1^2}-\frac{1}{r_2^2}-\frac{4z(d+z)}{r_2^4}\right]\right\}$$

(3.31)

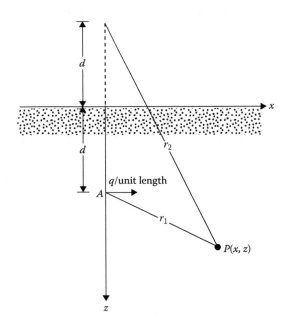

Figure 3.13 Horizontal line load inside a semi-infinite mass.

$$\sigma_x = \frac{qx}{\pi}\left\{\frac{1}{2(1-v)}+\left[\frac{x^2}{r_1^4}-\frac{x^2+8dz+6d^2}{r_2^4}+\frac{8dz(d+z)^2}{r_2^6}\right]\right.$$

$$\left.+\frac{1-2v}{4(1-v)}\left[\frac{1}{r_1^2}+\frac{3}{r_2^2}-\frac{4z(d+z)}{r_2^4}\right]\right\} \tag{3.32}$$

$$\tau_{xz} = \frac{q}{\pi}\left\{\frac{1}{2(1-v)}\left[\frac{(z-d)x^2}{r_1^4}+\frac{(2dz+x^2)(d+z)}{r_2^4}+\frac{8dz(d+z)x^2}{r_2^6}\right]\right.$$

$$\left.+\frac{1-2v}{4(1-v)}\left(\frac{z-d}{r_1^2}+\frac{3z+d}{r_2^2}+\frac{4z(d+z)^2}{r_2^4}\right)\right\} \tag{3.33}$$

3.9 UNIFORM VERTICAL LOADING ON AN INFINITE STRIP ON THE SURFACE

Figure 3.14 shows the case where a uniform vertical load of q per unit area is acting on a flexible infinite strip on the surface of a semi-infinite elastic mass. To obtain the stresses at a point $P(x, z)$, we can consider an elementary strip of width ds located at a distance s from the centerline of the load. The load per unit length of this elementary strip is $q \cdot ds$, and it can be approximated as a line load.

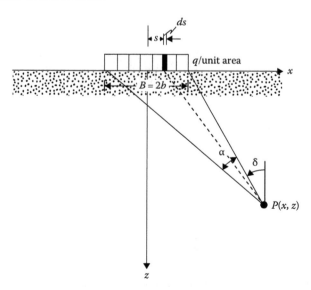

Figure 3.14 Uniform vertical loading on an infinite strip.

The increase of vertical stress, σ_z, at P due to the elementary strip loading can be obtained by substituting $x - s$ for x and $q \cdot ds$ for q in Equation 3.5, or

$$d\sigma_z = \frac{2q\,ds}{\pi} \frac{z^3}{[(x-s)^2 + z^2]^2} \tag{3.34}$$

The total increase of vertical stress, σ_z, at P due to the loaded strip can be determined by integrating Equation 3.34 with limits of $s = b$ to $s = -b$; so

$$\sigma_z = \int d\sigma_z = \frac{2q}{\pi} \int_{-b}^{+b} \frac{z^3}{[(x-s)^2 + z^2]^2}\,ds$$

$$= \frac{q}{\pi}\left[\tan^{-1}\frac{z}{x-b} - \tan^{-1}\frac{z}{x+b} - \frac{2bz(x^2 - z^2 - b^2)}{(x^2 + z^2 - b^2)^2 + 4b^2z^2} \right] \tag{3.35}$$

In a similar manner, referring to Equations 3.6 and 3.7,

$$\sigma_x = \int d\sigma_x = \frac{2q}{\pi} \int_{-b}^{+b} \frac{(x-s)^2 z}{[(x-s)^2 + z^2]^2}\,ds$$

$$= \frac{q}{\pi}\left[\tan^{-1}\frac{z}{x-b} - \tan^{-1}\frac{z}{x+b} + \frac{2bz(x^2 - z^2 - b^2)}{(x^2 + z^2 - b^2)^2 + 4b^2z^2} \right] \tag{3.36}$$

$$\tau_{xz} = \frac{2q}{\pi} \int_{-b}^{+b} \frac{(x-s)z^2}{[(x-s)^2 + z^2]^2}\,ds = \frac{4bqxz^2}{\pi[(x^2 + z^2 - b^2)^2 + 4b^2z^2]} \tag{3.37}$$

Equations 3.35 and 3.36 are for $x \geq b$. However, for $x = 0$ to $x < b$, the term $\tan^{-1}[z/(x - b)]$ becomes negative. For such cases, replace $\tan^{-1}[z/(x - b)]$ with $\pi + \tan^{-1}[z/(x - b)]$. Also note that, due to symmetry, the *magnitudes* of σ_z, σ_x, and τ_{xz} are the same at $\pm x$ for a given value of z.

The expressions for σ_z, σ_x, and τ_{xz} given in Equations 3.35 through 3.37 can be presented in a simplified form:

$$\sigma_z = \frac{q}{\pi}[\alpha + \sin\alpha\cos(\alpha + 2\delta)] \tag{3.38}$$

$$\sigma_x = \frac{q}{\pi}[\alpha - \sin\alpha\cos(\alpha + 2\delta)] \tag{3.39}$$

$$\tau_{xz} = \frac{q}{\pi}[\sin\alpha\sin(\alpha + 2\delta)] \tag{3.40}$$

where α and δ are the angles shown in Figure 3.14.

(**Note:** The angle δ is positive measured counterclockwise from the vertical drawn at *P*.)

Tables 3.5 through 3.7 give the values of σ_z/q, σ_x/q, τ_{xz}/q for various values of x/b and z/b.

The variation of σ_z/q with $\pm x/B$ for z/B = 0.25, 0.5, 1.0 and 2.0 is shown in Figure 3.15.

Using the relationships given in Equation 3.35, isobars for σ_z/q can be drawn. This is shown in Figure 3.16.

3.9.1 Vertical displacement at the surface (z = 0)

The vertical surface displacement *relative* to the center of the strip load can be expressed as

$$w_{z=0}(x) - w_{z=0}(x=0) = \frac{2q(1-v^2)}{\pi E} \left\{ \begin{array}{l} (x-b)\ln|x-b| - \\ (x+b)\ln|x-b| + 2b\ln b \end{array} \right\} \tag{3.41}$$

Example 3.4

Refer to Figure 3.14. Given *B* = 4 m. For point *P*, *z* = 1 m and *x* = 1 m. Determine σ_z/q, σ_x/q, and τ_{xz}/q at *P*. Use Equations 3.35 through 3.37.

Solution

σ_z/q *Calculation*

Given *b* = *B*/2 = 4/2 = 2

$$z = 1 \text{ m}$$

From Equation 3.35, since $x < b$

$$\frac{\sigma_z}{q} = \frac{1}{\pi}\left[\pi + \tan^{-1}\frac{z}{x-b} - \tan^{-1}\frac{z}{x+b} - \frac{2bz(x^2 - z^2 - b^2)}{(x^2 + z^2 - b^2)^2 + 4b^2z^2} \right]$$

$$\tan^{-1}\frac{z}{x-b} = \tan^{-1}\frac{1}{1-2} = -45° = -0.785 \text{ rad}$$

$$\tan^{-1}\frac{z}{x+b} = \tan^{-1}\frac{1}{1+2} = 18.43° = 0.322 \text{ rad}$$

$$\frac{2bz(x^2 - z^2 - b^2)}{(x^2 + z^2 - b^2)^2 + 4b^2z^2} = \frac{(2)(2)(1)(1^2 - 1^2 - 2^2)}{(1^2 + 1^2 - 2^2)^2 + (4)(2^2)(1^2)} = -0.08$$

Table 3.5 Values of σ_z/q (Equation 3.35)

z/b	x/b										
	0.0	0.1	0.2	0.3	0.4	0.5	0.6	0.7	0.8	0.9	1.0
0.00	1.000	1.000	1.000	1.000	1.000	1.000	1.000	1.000	1.000	1.000	0.000
0.10	1.000	1.000	0.999	0.999	0.999	0.998	0.997	0.993	0.980	0.909	0.500
0.20	0.997	0.997	0.996	0.995	0.992	0.988	0.979	0.959	0.909	0.775	0.500
0.30	0.990	0.989	0.987	0.984	0.978	0.967	0.947	0.908	0.833	0.697	0.499
0.40	0.977	0.976	0.973	0.966	0.955	0.937	0.906	0.855	0.773	0.651	0.498
0.50	0.959	0.958	0.953	0.943	0.927	0.902	0.864	0.808	0.727	0.620	0.497
0.60	0.937	0.935	0.928	0.915	0.896	0.866	0.825	0.767	0.691	0.598	0.495
0.70	0.910	0.908	0.899	0.885	0.863	0.831	0.788	0.732	0.662	0.581	0.492
0.80	0.881	0.878	0.869	0.853	0.829	0.797	0.755	0.701	0.638	0.566	0.489
0.90	0.850	0.847	0.837	0.821	0.797	0.765	0.724	0.675	0.617	0.552	0.485
1.00	0.818	0.815	0.805	0.789	0.766	0.735	0.696	0.650	0.598	0.540	0.480
1.10	0.787	0.783	0.774	0.758	0.735	0.706	0.670	0.628	0.580	0.529	0.474
1.20	0.755	0.752	0.743	0.728	0.707	0.679	0.646	0.607	0.564	0.517	0.468
1.30	0.725	0.722	0.714	0.699	0.679	0.654	0.623	0.588	0.548	0.506	0.462
1.40	0.696	0.693	0.685	0.672	0.653	0.630	0.602	0.569	0.534	0.495	0.455
1.50	0.668	0.666	0.658	0.646	0.629	0.607	0.581	0.552	0.519	0.484	0.448
1.60	0.642	0.639	0.633	0.621	0.605	0.586	0.562	0.535	0.506	0.474	0.440
1.70	0.617	0.615	0.608	0.598	0.583	0.565	0.544	0.519	0.492	0.463	0.433
1.80	0.593	0.591	0.585	0.576	0.563	0.546	0.526	0.504	0.479	0.453	0.425
1.90	0.571	0.569	0.564	0.555	0.543	0.528	0.510	0.489	0.467	0.443	0.417
2.00	0.550	0.548	0.543	0.535	0.524	0.510	0.494	0.475	0.455	0.433	0.409
2.10	0.530	0.529	0.524	0.517	0.507	0.494	0.479	0.462	0.443	0.423	0.401
2.20	0.511	0.510	0.506	0.499	0.490	0.479	0.465	0.449	0.432	0.413	0.393
2.30	0.494	0.493	0.489	0.483	0.474	0.464	0.451	0.437	0.421	0.404	0.385
2.40	0.477	0.476	0.473	0.467	0.460	0.450	0.438	0.425	0.410	0.395	0.378
2.50	0.462	0.461	0.458	0.452	0.445	0.436	0.426	0.414	0.400	0.386	0.370
2.60	0.447	0.446	0.443	0.439	0.432	0.424	0.414	0.403	0.390	0.377	0.363
2.70	0.433	0.432	0.430	0.425	0.419	0.412	0.403	0.393	0.381	0.369	0.355
2.80	0.420	0.419	0.417	0.413	0.407	0.400	0.392	0.383	0.372	0.360	0.348
2.90	0.408	0.407	0.405	0.401	0.396	0.389	0.382	0.373	0.363	0.352	0.341
3.00	0.396	0.395	0.393	0.390	0.385	0.379	0.372	0.364	0.355	0.345	0.334
3.10	0.385	0.384	0.382	0.379	0.375	0.369	0.363	0.355	0.347	0.337	0.327
3.20	0.374	0.373	0.372	0.369	0.365	0.360	0.354	0.347	0.339	0.330	0.321
3.30	0.364	0.363	0.362	0.359	0.355	0.351	0.345	0.339	0.331	0.323	0.315
3.40	0.354	0.354	0.352	0.350	0.346	0.342	0.337	0.331	0.324	0.316	0.308
3.50	0.345	0.345	0.343	0.341	0.338	0.334	0.329	0.323	0.317	0.310	0.302
3.60	0.337	0.336	0.335	0.333	0.330	0.326	0.321	0.316	0.310	0.304	0.297
3.70	0.328	0.328	0.327	0.325	0.322	0.318	0.314	0.309	0.304	0.298	0.291

(Continued)

Table 3.5 (Continued) Values of σ_z/q (Equation 3.35)

					x/b					
z/b	0.0	0.1	0.2	0.3	0.4	0.5	0.6	0.7	0.8	0.9	1.0
3.80	0.320	0.320	0.319	0.317	0.315	0.311	0.307	0.303	0.297	0.292	0.285
3.90	0.313	0.313	0.312	0.310	0.307	0.304	0.301	0.296	0.291	0.286	0.280
4.00	0.306	0.305	0.304	0.303	0.301	0.298	0.294	0.290	0.285	0.280	0.275
4.10	0.299	0.299	0.298	0.296	0.294	0.291	0.288	0.284	0.280	0.275	0.270
4.20	0.292	0.292	0.291	0.290	0.288	0.285	0.282	0.278	0.274	0.270	0.265
4.30	0.286	0.286	0.285	0.283	0.282	0.279	0.276	0.273	0.269	0.265	0.260
4.40	0.280	0.280	0.279	0.278	0.276	0.274	0.271	0.268	0.264	0.260	0.256
4.50	0.274	0.274	0.273	0.272	0.270	0.268	0.266	0.263	0.259	0.255	0.251
4.60	0.268	0.268	0.268	0.266	0.265	0.263	0.260	0.258	0.254	0.251	0.247
4.70	0.263	0.263	0.262	0.261	0.260	0.258	0.255	0.253	0.250	0.246	0.243
4.80	0.258	0.258	0.257	0.256	0.255	0.253	0.251	0.248	0.245	0.242	0.239
4.90	0.253	0.253	0.252	0.251	0.250	0.248	0.246	0.244	0.241	0.238	0.235
5.00	0.248	0.248	0.247	0.246	0.245	0.244	0.242	0.239	0.237	0.234	0.231
	1.1	1.2	1.3	1.4	1.5	1.6	1.7	1.8	1.9	2.0	
0.00	0.000	0.000	0.000	0.000	0.000	0.000	0.000	0.000	0.000	0.000	
0.10	0.091	0.020	0.007	0.003	0.002	0.001	0.001	0.000	0.000	0.000	
0.20	0.225	0.091	0.040	0.020	0.011	0.007	0.004	0.003	0.002	0.002	
0.30	0.301	0.165	0.090	0.052	0.031	0.020	0.013	0.009	0.007	0.005	
0.40	0.346	0.224	0.141	0.090	0.059	0.040	0.027	0.020	0.014	0.011	
0.50	0.373	0.267	0.185	0.128	0.089	0.063	0.046	0.034	0.025	0.019	
0.60	0.391	0.298	0.222	0.163	0.120	0.088	0.066	0.050	0.038	0.030	
0.70	0.403	0.321	0.250	0.193	0.148	0.113	0.087	0.068	0.053	0.042	
0.80	0.411	0.338	0.273	0.218	0.173	0.137	0.108	0.086	0.069	0.056	
0.90	0.416	0.351	0.291	0.239	0.195	0.158	0.128	0.104	0.085	0.070	
1.00	0.419	0.360	0.305	0.256	0.214	0.177	0.147	0.122	0.101	0.084	
1.10	0.420	0.366	0.316	0.271	0.230	0.194	0.164	0.138	0.116	0.098	
1.20	0.419	0.371	0.325	0.282	0.243	0.209	0.178	0.152	0.130	0.111	
1.30	0.417	0.373	0.331	0.291	0.254	0.221	0.191	0.166	0.143	0.123	
1.40	0.414	0.374	0.335	0.298	0.263	0.232	0.203	0.177	0.155	0.135	
1.50	0.411	0.374	0.338	0.303	0.271	0.240	0.213	0.188	0.165	0.146	
1.60	0.407	0.373	0.339	0.307	0.276	0.248	0.221	0.197	0.175	0.155	
1.70	0.402	0.370	0.339	0.309	0.281	0.254	0.228	0.205	0.183	0.164	
1.80	0.396	0.368	0.339	0.311	0.284	0.258	0.234	0.212	0.191	0.172	
1.90	0.391	0.364	0.338	0.312	0.286	0.262	0.239	0.217	0.197	0.179	
2.00	0.385	0.360	0.336	0.311	0.288	0.265	0.243	0.222	0.203	0.185	
2.10	0.379	0.356	0.333	0.311	0.288	0.267	0.246	0.226	0.208	0.190	
2.20	0.373	0.352	0.330	0.309	0.288	0.268	0.248	0.229	0.212	0.195	
2.30	0.366	0.347	0.327	0.307	0.288	0.268	0.250	0.232	0.215	0.199	

(Continued)

Table 3.5 (Continued) Values of σ_z/q (Equation 3.35)

	x/b									
z/b	1.1	1.2	1.3	1.4	1.5	1.6	1.7	1.8	1.9	2.0
2.40	0.360	0.342	0.323	0.305	0.287	0.268	0.251	0.234	0.217	0.202
2.50	0.354	0.337	0.320	0.302	0.285	0.268	0.251	0.235	0.220	0.205
2.60	0.347	0.332	0.316	0.299	0.283	0.267	0.251	0.236	0.221	0.207
2.70	0.341	0.327	0.312	0.296	0.281	0.266	0.251	0.236	0.222	0.208
2.80	0.335	0.321	0.307	0.293	0.279	0.265	0.250	0.236	0.223	0.210
2.90	0.329	0.316	0.303	0.290	0.276	0.263	0.249	0.236	0.223	0.211
3.00	0.323	0.311	0.299	0.286	0.274	0.261	0.248	0.236	0.223	0.211
3.10	0.317	0.306	0.294	0.283	0.271	0.259	0.247	0.235	0.223	0.212
3.20	0.311	0.301	0.290	0.279	0.268	0.256	0.245	0.234	0.223	0.212
3.30	0.305	0.296	0.286	0.275	0.265	0.254	0.243	0.232	0.222	0.211
3.40	0.300	0.291	0.281	0.271	0.261	0.251	0.241	0.231	0.221	0.211
3.50	0.294	0.286	0.277	0.268	0.258	0.249	0.239	0.229	0.220	0.210
3.60	0.289	0.281	0.273	0.264	0.255	0.246	0.237	0.228	0.218	0.209
3.70	0.284	0.276	0.268	0.260	0.252	0.243	0.235	0.226	0.217	0.208
3.80	0.279	0.272	0.264	0.256	0.249	0.240	0.232	0.224	0.216	0.207
3.90	0.274	0.267	0.260	0.253	0.245	0.238	0.230	0.222	0.214	0.206
4.00	0.269	0.263	0.256	0.249	0.242	0.235	0.227	0.220	0.212	0.205
4.10	0.264	0.258	0.252	0.246	0.239	0.232	0.225	0.218	0.211	0.203
4.20	0.260	0.254	0.248	0.242	0.236	0.229	0.222	0.216	0.209	0.202
4.30	0.255	0.250	0.244	0.239	0.233	0.226	0.220	0.213	0.207	0.200
4.40	0.251	0.246	0.241	0.235	0.229	0.224	0.217	0.211	0.205	0.199
4.50	0.247	0.242	0.237	0.232	0.226	0.221	0.215	0.209	0.203	0.197
4.60	0.243	0.238	0.234	0.229	0.223	0.218	0.212	0.207	0.201	0.195
4.70	0.239	0.235	0.230	0.225	0.220	0.215	0.210	0.205	0.199	0.194
4.80	0.235	0.231	0.227	0.222	0.217	0.213	0.208	0.202	0.197	0.192
4.90	0.231	0.227	0.223	0.219	0.215	0.210	0.205	0.200	0.195	0.190
5.00	0.227	0.224	0.220	0.216	0.212	0.207	0.203	0.198	0.193	0.188

Table 3.6 Values of σ_x/q (Equation 3.36)

	x/b					
z/b	0	0.5	1.0	1.5	2.0	2.5
0	1.000	1.000	0	0	0	0
0.5	0.450	0.392	0.347	0.285	0.171	0.110
1.0	0.182	0.186	0.225	0.214	0.202	0.162
1.5	0.080	0.099	0.142	0.181	0.185	0.165
2.0	0.041	0.054	0.091	0.127	0.146	0.145
2.5	0.230	0.033	0.060	0.089	0.126	0.121

Table 3.7 Values of τ_{xz}/q (Equation 3.37)

z/b	x/b					
	0	0.5	1.0	1.5	2.0	2.5
0	—	—	—	—	—	—
0.5	—	0.127	0.300	0.147	0.055	0.025
1.0	—	0.159	0.255	0.210	0.131	0.074
1.5	—	0.128	0.204	0.202	0.157	0.110
2.0	—	0.096	0.159	0.175	0.157	0.126
2.5	—	0.072	0.124	0.147	0.144	0.127

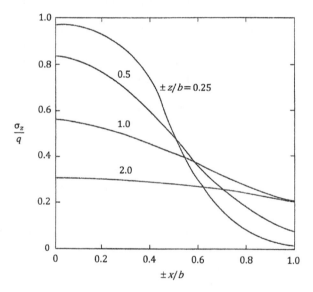

Figure 3.15 Variation of σ_z/q with x/B and z/B (Equation 3.35).

Hence

$$\frac{\sigma_z}{q} = \frac{1}{\pi}[\pi - 0.785 - 0.322 - (-0.8)] = 0.902$$

σ_x/q *Calculation*
From Equation 3.36

$$\frac{\sigma_x}{q} = \frac{1}{\pi}\left[\pi + \tan^{-1}\frac{z}{x-b} - \tan^{-1}\frac{z}{x+b} + \frac{2bz(x^2 - z^2 - b^2)}{(x^2 + z^2 - b^2)^2 + 4b^2 z^2}\right]$$

$$= \frac{1}{\pi}[\pi - 0.785 - 0.322 + (-0.8)] = 0.392$$

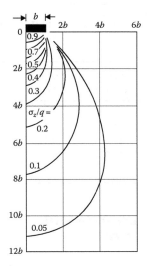

Figure 3.16 Isobars for σ_z/q below a strip load.

τ_{xz}/q *Calculation*

From Equation 3.37

$$\frac{\tau_{xz}}{q} = \frac{4bqxz^2}{\pi[(x^2 + z^2 - b^2)^2 + 4b^2z^2]} = \frac{(4)(2)(1)(1^2)}{\pi[(1^2 + 1^2 - 2^2)^2 + (4)(2^2)(1^2)]} = 0.127$$

Example 3.5

For the infinite strip load shown in Figure 3.14, given: $B = 4$ m; $q = 105$ kN/m²; $\nu = 0.3$. Determine the variation of σ_z, σ_x, τ_{xz}, $\sigma_{p(1)}$ (major principal stress), $\sigma_{p(2)}$ (intermediate principal stress), and $\sigma_{p(3)}$ (minor principal stress) at $x = 0$, 2 m, 4 m, 6 m, and 8 m at $z = 3$.

Solution

Given: $B = 4$ m; $b = 4/2 = 2$ m; $q = 105$ kN/m²; $\nu = 0.3$; $z = 3$ m

x (m)	x/b	z/b	σ_z/q^a	σ_x/q^b	τ_{xz}/q^c	σ_z (kN/m²)	σ_x (kN/m²)	τ_{xz} (kN/m²)
0	0	1.5	0.668	0.08	0	70.14	8.4	0
2	1	1.5	0.448	0.142	0.204	47.07	14.91	21.42
4	2	1.5	0.146	0.185	0.157	15.33	19.43	16.49
6	3	1.5	0.042	0.139	0.080	4.41	14.6	8.4
8	4	1.5	0.015	0.095	0.038	1.58	9.98	4.0

[a] Table 3.5
[b] Table 3.6
[c] Table 3.7

$$\sigma_{p(1)}, \ \sigma_{p(3)} = \frac{\sigma_z + \sigma_x}{2} \pm \sqrt{\left(\frac{\sigma_z - \sigma_x}{2}\right)^2 + \tau_{xz}^2}; \quad \sigma_{p(2)} = \nu(\sigma_x + \sigma_z); \quad \nu = 0.3$$

x (m)	$\sigma_{p(1)}$ (kN/m²)	$\sigma_{p(2)}$ (kN/m²)	$\sigma_{p(3)}$ (kN/m²)
0	70.1	23.6	8.4
2	57.7	18.6	4.3
4	33.9	10.4	0.76
6	19.3	5.7	−0.25
8	11.5	3.4	−0.03

3.10 UNIFORM STRIP LOAD INSIDE A SEMI-INFINITE MASS

Strip loads can be located inside a semi-infinite mass as shown in Figure 3.17. The distribution of vertical stress σ_z due to this type of loading can be determined by integration of Melan's solution (Equation 3.15). This has been given by Kezdi and Rethati (1988). The magnitude of σ_z at a point P *along the centerline of the load* (i.e., $x = 0$) can be given as

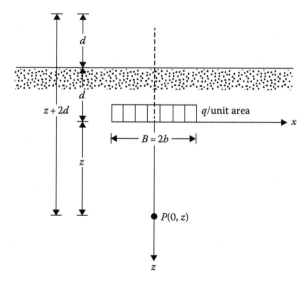

Figure 3.17 Strip load inside a semi-infinite mass.

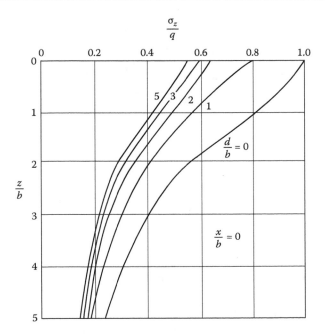

$$\frac{\sigma_z}{q}$$

Figure 3.18 Plot of σ_z/q versus z/b (Equation 3.42).

$$\sigma_z = \frac{q}{\pi}\left\{\frac{b(z+2d)}{(z+2d)^2+b^2} + \tan^{-1}\frac{b}{z+2d} + \frac{bz}{z^2+b^2}\right.$$

$$+ \tan^{-1}\frac{b}{z} - \frac{v-1}{2v}(z+2d)\left[\frac{b}{(z+2d)^2+b^2} - \frac{b}{z^2+b^2}\right]$$

$$\left. + \frac{v+1}{2v}\frac{2(z+2d)db(z+d)}{(z^2+b^2)^2}\right\} \quad \text{(for } x=0\text{)} \tag{3.42}$$

Figure 3.18 shows the influence of d/b on the variation of σ_z/q.

3.11 UNIFORM HORIZONTAL LOADING ON AN INFINITE STRIP ON THE SURFACE

If a uniform horizontal load is applied on an infinite strip of width $2b$ as shown in Figure 3.19, the stresses at a point inside the semi-infinite mass can be determined by using a similar procedure of superposition as outlined in Section 3.9 for vertical loading. For an elementary strip of width ds, the load

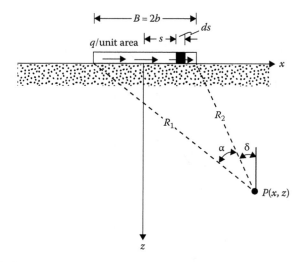

Figure 3.19 Uniform horizontal loading on an infinite strip.

per unit length is $q \cdot ds$. Approximating this as a line load, we can substitute $q \cdot ds$ for q and $x - s$ for x in Equations 3.22 through 3.24. Thus, at a point $P(x, z)$

$$\sigma_z = \int d\sigma_z = \frac{2q}{\pi} \int_{s=-b}^{s=+b} \frac{(x-s)z^2}{[(x-s)^2 + z^2]^2}\, ds = \frac{4bqxz^2}{\pi\,[(x^2 + z^2 - b^2)^2 + 4b^2z^2]} \quad (3.43)$$

$$\sigma_x = \int d\sigma_x = \frac{2q}{\pi} \int_{s=-b}^{s=+b} \frac{(x-s)^3}{[(x-s)^2 + z^2]^2}\, ds$$

$$= \frac{q}{\pi}\left[2.303\log\frac{(x+b)^2 + z^2}{(x+b)^2 + z^2} - \frac{4bxz^2}{(x^2 + z^2 - b^2)^2 + 4b^2z^2} \right] \quad (3.44)$$

$$\tau_{xz} = \int d\tau_{xz} = \frac{2q}{\pi} \int_{s=-b}^{s=+b} \frac{(x-s)^2 z}{[(x-s)^2 + z^2]^2}\, ds$$

$$= \frac{q}{\pi}\left[\tan^{-1}\frac{z}{x-b} - \tan^{-1}\frac{z}{x+b} - \frac{2bz(x^2 - z^2 - b^2)}{(x^2 + z^2 - b^2)^2 + 4b^2z^2} \right] \quad (3.45)$$

For $x = 0$ to $x < b$, the term $\tan^{-1}(z/(x - b))$ in Equation 3.45 will be negative. So, it should be replaced by $\pi + \tan^{-1}(z/(x - b))$. For a given value of z, the magnitude of τ_{xz} is the same at $\pm x$.

The expressions for stresses given by Equations 3.43 through 3.45 may also be simplified as follows:

$$\sigma_z = \frac{q}{\pi}[\sin\alpha\sin(\alpha + 2\delta)] \tag{3.46}$$

$$\sigma_x = \frac{q}{\pi}\left[2.303\log\frac{R_1^2}{R_2^2} - \sin\alpha\sin(\alpha + 2\delta)\right] \tag{3.47}$$

$$\tau_{xz} = \frac{q}{\pi}[\alpha - \sin\alpha\cos(\alpha + 2\delta)] \tag{3.48}$$

where R_1, R_2, α, and δ are as defined in Figure 3.19 The angle δ is positive measured counterclockwise from the vertical drawn at P.

The variations of σ_z, σ_x, and τ_{xz} in a nondimensional form are given in Tables 3.8 through 3.10.

3.11.1 Horizontal displacement at the surface (z = 0)

The horizontal displacement u at a point on the surface ($z = 0$) relative to the center of the strip loading is of the form

$$u_{z=0}(x) - u_{z=0}(x = 0) = \frac{2q(1 - v^2)}{\pi E}\left\{\begin{array}{l}(x - b)\ln|x - b| - \\ (x + b)\ln|x - b| + 2b\ln b\end{array}\right\} \tag{3.49}$$

Table 3.8 Values of σ_z/q (Equation 3.43)

z/b	0	0.5	1.0	1.5	2.0	2.5
				x/b		
0	—	—	—	—	—	—
0.5	—	0.127	0.300	0.147	0.055	0.025
1.0	—	0.159	0.255	0.210	0.131	0.074
1.5	—	0.128	0.204	0.202	0.157	0.110
2.0	—	0.096	0.159	0.175	0.157	0.126
2.5	—	0.072	0.124	0.147	0.144	0.127

Table 3.9 Values of σ_x/q (Equation 3.44)

x/b	z/b																		
	0	0.1	0.2	0.3	0.4	0.5	0.6	0.7	1.0	1.5	2.0	2.5	3.0	3.5	4.0	4.5	5.0	5.5	6.0
0	0	0	0	0	0	0	0	0	0	0	0	0	0	0	0	0	0	0	0
0.1	0.1287	0.1252	0.1180	0.1073	0.0946	0.0814	0.0687	0.0572	0.0317	0.0121	0.0051	0.0024	0.0013	0.0007	0.0004	0.0003	0.0002	0.00013	0.0001
0.25	0.3253	0.3181	0.2982	0.2693	0.2357	0.2014	0.1692	0.2404	0.0780	0.0301	0.0129	0.0062	0.0033	0.00019	0.0012	0.0007	0.0005	0.00034	0.00025
0.5	0.6995	0.6776	0.6195	0.5421	0.4608	0.3851	0.3188	0.2629	0.1475	0.0598	0.0269	0.0134	0.0073	0.0042	0.0026	0.0017	0.00114	0.00079	0.00057
0.75	1.2390	1.1496	0.9655	0.7855	0.6379	0.5210	0.4283	0.3541	0.2058	0.0899	0.0429	0.0223	0.0124	0.0074	0.0046	0.0030	0.00205	0.00144	0.00104
1.0	—	1.5908	1.1541	0.9037	0.7312	0.6024	0.5020	0.4217	0.2577	0.1215	0.0615	0.0333	0.0191	0.0116	0.0074	0.0049	0.00335	0.00236	0.00171
1.25	1.3990	1.3091	1.1223	0.9384	0.7856	0.6623	0.5624	0.4804	0.3074	0.1548	0.0825	0.0464	0.0275	0.0170	0.0110	0.0074	0.00510	0.00363	0.00265
1.5	1.0248	1.0011	0.9377	0.8517	0.7591	0.6697	0.5881	0.5157	0.3489	0.1874	0.1049	0.0613	0.0373	0.0236	0.0155	0.0105	0.00736	0.00528	0.00387
1.75	0.8273	0.8170	0.7876	0.7437	0.6904	0.6328	0.5749	0.5190	0.3750	0.2162	0.1271	0.0770	0.0483	0.0313	0.0209	0.0144	0.01013	0.00732	0.00541
2.0	0.6995	0.6939	0.6776	0.6521	0.6195	0.5821	0.5421	0.5012	0.3851	0.2386	0.1475	0.0928	0.0598	0.0396	0.0269	0.0188	0.01339	0.00976	0.00727
2.5	0.5395	0.5372	0.5304	0.5194	0.5047	0.4869	0.4667	0.4446	0.3735	0.2627	0.1788	0.1211	0.0826	0.0572	0.0403	0.0289	0.02112	0.01569	0.01185
3.0	0.4414	0.4402	0.4366	0.4303	0.4229	0.4132	0.4017	0.3889	0.3447	0.2658	0.1962	0.1421	0.1024	0.0741	0.0541	0.0400	0.02993	0.02269	0.01742
4.0	0.3253	0.3248	0.3235	0.3212	0.3181	0.3143	0.3096	0.3042	0.2846	0.2443	0.2014	0.1616	0.1276	0.0999	0.0780	0.0601	0.04789	0.03781	0.03006
5.0	0.2582	0.2580	0.2573	0.2562	0.2547	0.2527	0.2504	0.2477	0.2375	0.2151	0.1888	0.1618	0.1362	0.1132	0.0934	0.0767	0.06285	0.05156	0.04239
6.0	0.2142	0.2141	0.2137	0.2131	0.2123	0.2112	0.2098	0.2083	0.2023	0.1888	0.1712	0.1538	0.1352	0.1173	0.1008	0.0861	0.07320	0.06207	0.05259

Table 3.10 Values of τ_{xz}/q (Equation 3.45)

| z/b | \multicolumn{5}{c}{x/b} |
	0	0.5	1.0	1.5	2.0
0	1.000	1.000	0	0	0
0.5	0.959	0.902	0.497	0.089	0.019
1.0	0.818	0.735	0.480	0.214	0.084
1.5	0.688	0.607	0.448	0.271	0.146
2.0	0.550	0.510	0.409	0.288	0.185
2.5	0.462	0.436	0.370	0.285	0.205

Example 3.6

Refer to Figure 3.19. Given $B = 4$ m. For point P, $z = 1$ m. Determine σ_z/q, σ_x/q, and τ_{xz}/q at $x = \pm 1$ m.

Solution

Calculation for σ_z/q

Given $b = B/2 = 2$ m

$z = 1$ m

$x = \pm 1$ m

From Equation 3.43

$$\frac{\sigma_z}{q} = \frac{4bxz^2}{\pi[(x^2 + z^2 - b^2)^2 + 4b^2z^2]} = \frac{(4)(2)(\pm 1)(1^2)}{\pi\{(\pm 1)^2 + [1^2 - 2^2] + (4)(2^2)(1^2)\}}$$

$$= \begin{cases} 0.127 \text{ at } x = 1\,\text{m} \\ -0.127 \text{ at } x = -1\,\text{m} \end{cases}$$

Calculation for σ_x/q

From Equation 3.44

$$\frac{\sigma_x}{q} = \frac{1}{\pi}\left[2.303\log\frac{(x+b)^2 + z^2}{(x-b)^2 + z^2} - \frac{4bxz^2}{(x^2 + z^2 - b^2)^2 + 4b^2z^2}\right]$$

At $x = +1$ m

$$\frac{\sigma_x}{q} = \frac{1}{\pi}\left[2.303\log\frac{(1+2)^2 + 1^2}{(1-2)^2 + 1^2} - \frac{(4)(2)(1)(1^2)}{[(1^2) + 1^2 - 2^2]^2 + (4)(2^2)(1^2)}\right] = 0.385$$

At $x = -1$ m

$$\frac{\sigma_x}{q} = \frac{1}{\pi}\left[2.303\log\frac{(-1+2)^2 + 1^2}{(-1-2)^2 + 1^2} - \frac{(4)(2)(-1)(1^2)}{[(-1)^2 + 1^2 - 2^2]^2 + (4)(2^2)(1^2)}\right] = -0.385$$

Calculation of τ_{xz} *at* $x = \pm 1$ m

Note: $x < b$. From Equation 3.45

$$
\frac{\tau_{xz}}{q} = \frac{1}{\pi}\left[\pi + \tan^{-1}\frac{z}{x-b} - \tan^{-1}\frac{z}{x+b} - \frac{2bz(x^2 - z^2 - b^2)}{(x^2 + z^2 - b^2)^2 + 4b^2 z^2}\right]
$$

$$
= \frac{1}{\pi}\left[\pi + \tan^{-1}\frac{1}{1-2} - \tan^{-1}\frac{1}{1+2} - \frac{(2)(2)(1)(1^2 - 1^2 - 2^2)}{(1^2 + 1^2 - 2^2)^2 + (4)(2^2)(1^2)}\right]
$$

$$
= \frac{1}{\pi}[\pi - 0.785 - 0.322 + (-0.8)] = 0.902
$$

3.12 SYMMETRICAL VERTICAL TRIANGULAR STRIP LOAD ON THE SURFACE

Figure 3.20 shows a symmetrical vertical triangular strip load on the surface of a semi-infinite mass. The stresses at a point $P(x, z)$ due to the loading can be expressed by the following relationships (Gray, 1936):

$$
\sigma_z = \frac{q}{\pi}\left[(\alpha_1 + \alpha_2) - \frac{x}{b}(\alpha_1 - \alpha_2)\right] \tag{3.50}
$$

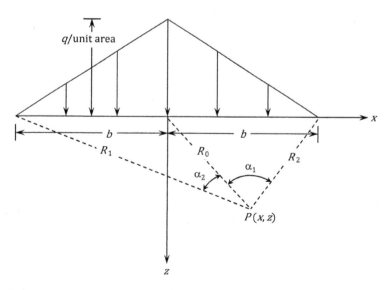

Figure 3.20 Symmetrical vertical triangular strip loading.

Table 3.11 Variation of σ_z/q, σ_x/q, and τ_{xz}/q with x/b and z/b (Equations 3.50 through 3.52)

x/b	z/b	σ_z/q	σ_x/q	τ_{xz}/q
0	0	1.0	1.0	0
	0.5	0.705	0.193	0
	1.0	0.50	0.059	0
	1.5	0.374	0.023	0
	2.0	0.195	0.011	0
0.25	0	0.750	0.75	0
	0.5	0.634	0.203	0.115
	1.0	0.471	0.068	0.076
0.5	0	0.5	0.5	0
	0.5	0.471	0.215	0.176
	1.0	0.396	0.091	0.130
0.75	0	0.250	0.250	0
	0.5	0.288	0.216	0.181
	1.0	0.298	0.117	0.153
1.0	0	0	0	0
	0.5	0.139	0.201	0.141
	1.0	0.205	0.134	0.148

$$\sigma_x = \frac{q}{\pi}\left[(\alpha_1 + \alpha_2) - \frac{x}{b}(\alpha_1 - \alpha_2) - \frac{2z}{b}\log_e\frac{R_1 R_2}{R_0^2}\right] \qquad (3.51)$$

$$\tau_{xz} = \frac{q}{\pi}\left(\frac{z}{b}\right)(\alpha_1 - \alpha_2) \qquad (3.52)$$

Table 3.11 gives some values of σ_z/q, σ_x/q, and τ_{xz}/q for various combinations of x/b and z/b.

Example 3.7

Refer to Figure 3.21. Determine σ_z, σ_x, and τ_{xz} at P (1.5 m, 1.5 m) using Equations 3.50 through 3.52.

Solution

Given: $q = 100$ kN/m²; $b = 3$ m; $x/b = 1.5/3 = 0.5$; $z/b = 1.5/3 = 0.5$

Angle $CPD = \tan^{-1}\left(\frac{1.5}{1.5}\right) = 45° = 0.785$ radians

Angle $BPD = \tan^{-1}\left(\frac{1.5}{1.5}\right) = 45° = 0.785$ radians

Angle $APD = \tan^{-1}\left(\frac{4.5}{1.5}\right) = 71.57° = 1.25$ radians

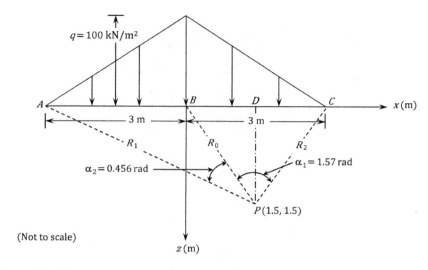

Figure 3.21 Calculation of stress due to vertical triangular strip load.

So,

$$\alpha_1 = 0.785 + 0.785 = 1.57 \text{ radians}$$

$$\alpha_2 = 1.25 - 0.785 = 0.465 \text{ radians}$$

$$R_0 = \frac{PD}{\cos 45} = \frac{1.5}{\cos 45} = 2.12 \text{ m}$$

$$R_2 = \frac{PD}{\cos 45} = 2.12 \text{ m}$$

$$R_1 = \sqrt{\left(\overline{PD}\right)^2 + \left(\overline{AD}\right)^2} = \sqrt{(1.5)^2 + (4.5)^2} = 4.74 \text{ m}$$

Equation 3.50:

$$\sigma_z = \frac{q}{\pi} + \left[(\alpha_1 + \alpha_2) - \frac{x}{b}(\alpha_1 - \alpha_2) \right]$$

$$= \frac{100}{\pi} + \left[(1.57 + 0.465) - \frac{1.5}{3}(1.57 - 0.465) \right] = 47.19 \text{ kN/m}^2$$

From Equation 3.51:

$$\sigma_x = \frac{q}{\pi}\left[(\alpha_1 + \alpha_2) - \frac{x}{b}(\alpha_1 - \alpha_2) - \frac{2z}{b}\log_e\frac{R_1 R_2}{R_0^2}\right]$$

$$= \frac{100}{\pi}\left[(1.57 + 0.465) - \left(\frac{1.5}{3}\right)(1.57 - 0.465) - (2)\left(\frac{1.5}{3}\right)\right.$$

$$\left. \times \log_e\left(\frac{4.74 \times 2.12}{2.12^2}\right)\right]$$

$$= \frac{100}{\pi}(2.035 - 0.5525 - 0.805) = 21.57 \text{ kN/m}^2$$

From Equation 3.52,

$$\tau_{xz} = \frac{q}{\pi}\left(\frac{z}{b}\right)(\alpha_1 - \alpha_2) = \frac{100}{\pi}\left(\frac{1.5}{3}\right)(1.57 - 0.465) = 17.59 \text{ kN/m}^2$$

Note: From Table 3.11,

$$\sigma_z = (0.471)(100) = 47.1 \text{ kN/m}^2$$

$$\sigma_x = (0.215)(100) = 21.5 \text{ kN/m}^2$$

$$\tau_{xz} = (0.176)(100) = 17.6 \text{ kN/m}^2$$

Example 3.8

For the triangular loading shown in Figure 3.22, determine σ_z, σ_x and τ_{xz} at point P (3 m, 1.5 m).

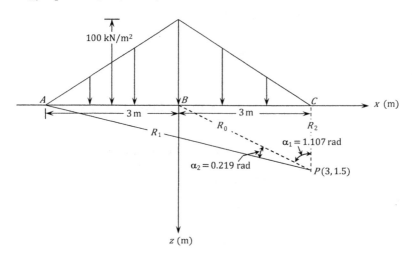

Figure 3.22 Symmetrical vertical triangular strip load—stress at P.

Solution

Refer to Figure 3.22.

$$\text{Angle } BPC = \alpha_1 = \tan^{-1}\left(\frac{3}{1.5}\right) = 63.43^\circ = 1.107 \text{ radians}$$

$$\text{Angle } APC = \tan^{-1}\left(\frac{6}{1.5}\right) = 75.96^\circ = 1.326 \text{ radians}$$

$$\alpha_2 = 1.326 - 1.107 = 0.219 \text{ radians}$$

$$R_2 = 1.5 \text{ m}$$
$$R_0 = \left[(3)^2 + (1.5)^2\right]^{0.5} = 3.39 \text{ m}$$
$$R_1 = \left[(6)^2 + (1.5)^2\right]^{0.5} = 6.16 \text{ m}$$
$$\sigma_z = \frac{q}{\pi}\left[(\alpha_1 + \alpha_2) - \left(\frac{x}{b}\right)(\alpha_1 - \alpha_2)\right]$$
$$= \frac{100}{\pi}\left[(1.107 + 0.219) - \left(\frac{3}{3}\right)(1.107 - 0.219)\right] = 13.94 \text{ kN/m}^2$$

$$\sigma_x = \frac{q}{\pi}\left[(\alpha_1 + \alpha_2) - \frac{x}{b}(\alpha_1 - \alpha_2) - \frac{2z}{b}\log_e\left(\frac{R_1 R_2}{R_0^2}\right)\right]$$
$$= \frac{100}{\pi}\left[(1.107 + 0.219) - \left(\frac{3}{3}\right)(1.107 - 0.219)\right.$$
$$\left. - \left(\frac{2 \times 1.5}{3}\right)\log_e\left(\frac{6.16 \times 1.5}{3.39^2}\right)\right]$$
$$= \frac{100}{\pi}[1.326 - 0.888 - (-0.218)] = 20.88 \text{ kN/m}^2$$

$$\tau_{xz} = \frac{q}{\pi}\left(\frac{z}{b}\right)(\alpha_1 - \alpha_2) = \frac{100}{\pi}\left(\frac{1.5}{3}\right)(1.107 - 0.219) = 14.14 \text{ kN/m}^2$$

Note: From Table 3.11: For $x/b = 3/3 = 1$ and $z/b = 1.5/3 = 0.5$,

$$\sigma_z = (0.139)(100) = 13.9 \text{ kN/m}^2$$

$$\sigma_x = (0.201)(100) = 20.1 \text{ kN/m}^2$$

$$\tau_{xz} = (0.141)(100) = 14.1 \text{ kN/m}^2$$

3.13 TRIANGULAR NORMAL LOADING ON AN INFINITE STRIP ON THE SURFACE

Figure 3.23 shows a vertical loading on an infinite strip on width $2b$. The load increases from zero to q across the width. For an elementary strip of width ds, the load per unit length can be given as $(q/2b)s \cdot ds$. Approximating this as a line load, we can substitute $(q/2b)s \cdot ds$ for q and $x - s$ for x in Equations 3.5 through 3.7 to determine the stresses at a point (x, z) inside the semi-infinite mass. Thus

$$\sigma_z = \int d\sigma_z = \left(\frac{1}{2b}\right)\left(\frac{2q}{\pi}\right)\int_{s=0}^{s=2b} \frac{z^3 s\, ds}{[(x-s)^2 + z^2]^2}$$

$$= \frac{q}{2\pi}\left(\frac{x}{b}\alpha - \sin 2\delta\right) \tag{3.53}$$

$$\sigma_x = \int d\sigma_x = \left(\frac{1}{2b}\right)\left(\frac{2q}{\pi}\right)\int_0^{2b} \frac{(x-s)^2 z s\, ds}{[(x-s)^2 + z^2]^2}$$

$$= \frac{q}{2\pi}\left(\frac{x}{b}\alpha - 2.303\frac{z}{b}\log\frac{R_1^2}{R_2^2} + \sin 2\delta\right) \tag{3.54}$$

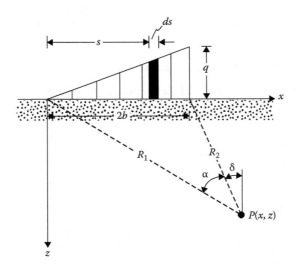

Figure 3.23 Linearly increasing vertical loading on an infinite strip.

Table 3.12 Values of σ_z/q (Equation 3.53)

x/b	0	0.5	1.0	1.5	2.0	2.5	3.0	4.0	5.0
					z/b				
−3	0	0.0003	0.0018	0.00054	0.0107	0.0170	0.0235	0.0347	0.0422
−2	0	0.0008	0.0053	0.0140	0.0249	0.0356	0.0448	0.0567	0.0616
−1	0	0.0041	0.0217	0.0447	0.0643	0.0777	0.0854	0.0894	0.0858
0	0	0.0748	0.1273	0.1528	0.1592	0.1553	0.1469	0.1273	0.1098
1	0.5	0.4797	0.4092	0.3341	0.2749	0.2309	0.1979	0.1735	0.1241
2	0.5	0.4220	0.3524	0.2952	0.2500	0.2148	0.1872	0.1476	0.1211
3	0	0.0152	0.0622	0.1010	0.1206	0.1268	0.1258	0.1154	0.1026
4	0	0.0019	0.0119	0.0285	0.0457	0.0596	0.0691	0.0775	0.0776
5	0	0.0005	0.0035	0.0097	0.0182	0.0274	0.0358	0.0482	0.0546

$$\tau_{xz} = \int d\tau_{xz} = \left(\frac{1}{2b}\right)\left(\frac{2q}{\pi}\right)\int_0^{2b} \frac{(x-s)z^2 ds}{[(x-s)^2 + z^2]^2}$$

$$= \frac{q}{2\pi} = \left(1 + \cos 2\delta - \frac{z}{b}\alpha\right) \qquad (3.55)$$

For Equations 3.53 through 3.55, the angle δ is positive in the counter-clockwise direction measured from the vertical drawn at P.

Nondimensional values of σ_z (Equation 3.53) are given in Table 3.12.

3.13.1 Vertical deflection at the surface

For this condition, the vertical deflection at the surface $(z = 0)$ can be expressed as

$$w_{z=0} = \left(\frac{q}{b\pi}\right)\left(\frac{1-\nu^2}{E}\right)\left[2b^2 \ln|2b - x| - \frac{x^2}{2}\ln\left|\frac{2b - x}{x}\right| - b(b + x)\right] \qquad (3.56)$$

Example 3.9

Refer to Figure 3.24. For a linearly increasing vertical loading on an infinite strip, given $b = 1$ m; $q = 100$ kN/m². Determine the vertical stress $\Delta\sigma_z$ at P (−1 m, 1.5 m).

Solution

Refer to Figure 3.24. Also note that $2b = 2$ m.

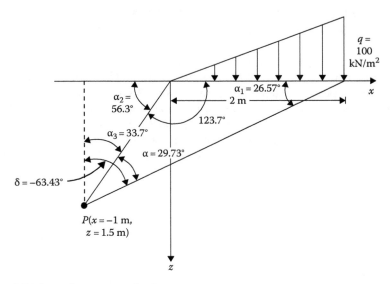

Figure 3.24 Linearly increasing load.

$$\alpha_1 = \tan^{-1}\left(\frac{1.5}{3}\right) = 26.57°$$

$$\alpha_2 = \tan^{-1}\left(\frac{1.5}{1}\right) = 56.3°$$

$$\alpha = \alpha_2 - \alpha_1 = 56.3 - 26.57 = 29.73°$$

$$\alpha_3 = 90 - \alpha_2 = 90 - 56.3 = 33.7°$$

$$\delta = -(\alpha_3 + \alpha) = -(33.7 + 29.73) = -63.43°$$

$$2\delta = -126.86°$$

From Equation 3.53

$$\frac{\sigma_z}{q} = \frac{1}{2\pi}\left(\frac{x}{b}\alpha - \sin 2\delta\right) = \frac{1}{2\pi}\left[\frac{(-1)}{1}\left(\frac{\pi}{180} \times 29.73\right) - \sin(-126.86)\right]$$

$$= \frac{1}{2\pi}[-0.519 - (-0.8)] = 0.0447$$

$$\sigma_z = (0.0447)(q) = (0.0447)(100) = 4.47 \text{ kN/m}^2$$

3.14 VERTICAL STRESS IN A SEMI-INFINITE MASS DUE TO EMBANKMENT LOADING

In several practical cases, it is necessary to determine the increase of vertical stress in a soil mass due to embankment loading. This can be done by the method of superposition as shown in Figure 3.25 and described later.

The stress at A due to the embankment loading as shown in Figure 3.25a is equal to the stress at A due to the triangular loading shown in Figure 3.25b minus the stress at A due to the triangular loading shown in Figure 3.25c.

Referring to Equation 3.53, the vertical stress at A due to the loading shown in Figure 3.25b is

$$\frac{q + (b/a)q}{\pi}(\alpha_1 + \alpha_2)$$

Similarly, the stress at A due to the loading shown in Figure 3.25c is

$$\left(\frac{b}{a}q\right)\frac{1}{\pi}\alpha_2$$

Thus the stress at A due to embankment loading (Figure 3.25a) is

$$\sigma_z = \frac{q}{\pi}\left[\left(\frac{a+b}{a}\right)(\alpha_1 + \alpha_2) - \frac{b}{a}\alpha_2\right]$$

or

$$\sigma_z = I_3 q \qquad\qquad (3.57)$$

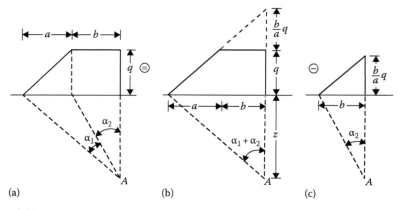

Figure 3.25 Vertical stress due to embankment loading: (a) embankment loading with an angle of $\alpha_1 + \alpha_2$ at A; (b) triangular loading with an angle $\alpha_1 + \alpha_2$ at A; (c) triangular loading with angle α_2 at A.

where I_3 is the influence factor,

$$I_3 = \frac{1}{\pi}\left[\left(\frac{a+b}{a}\right)(\alpha_1 + \alpha_2) - \frac{b}{a}\alpha_2\right] = \frac{1}{\pi}f\left(\frac{a}{z}, \frac{b}{z}\right)$$

The values of the influence factor for various a/z and b/z are given in Figure 3.26.

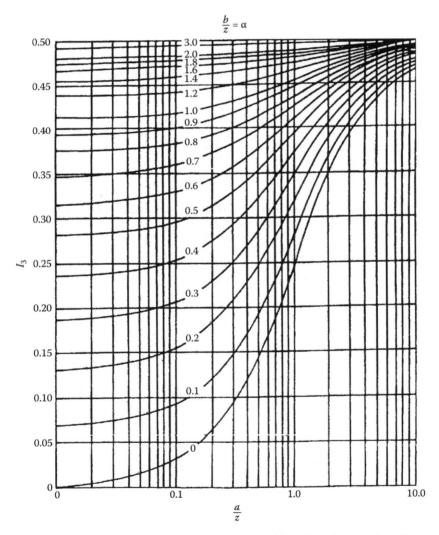

Figure 3.26 Influence factors for embankment load. (After Osterberg, J. O., Influence values for vertical stresses in semi-infinite mass due to embankment loading, Proc. 4th Int. Conf. Soil Mech. Found. Eng., vol. I, p. 393, 1957.)

Example 3.10

A 5 m-high embankment is to be constructed as shown in Figure 3.27. If the unit weight of compacted soil is 18.5 kN/m³, calculate the vertical stress due solely to the embankment at A, B, and C.

Solution

Vertical Stress at A: $q = \gamma H = 18.5 \times 5 = 92.5$ kN/m³ using the method of superposition and referring to Figure 3.28a.

$$\sigma_{zA} = \sigma_{z(1)} + \sigma_{z(2)}$$

For the left-hand section, $b/z = 2.5/5 = 0.5$ and $a/z = 5/5 = 1$. From Figure 3.26, $I_3 = 0.396$. For the right-hand section, $b/z = 7.5/5 = 1.5$ and $a/z = 5/5 = 1$. From Figure 3.26, $I_3 = 0.477$. So

$$\sigma_{zA} = (0.396 + 0.477)(92.5) = 80.75 \text{ kN/m}^2$$

Vertical stress at B: Using Figure 3.28b

$$\sigma_{zB} = \sigma_{z(1)} + \sigma_{z(2)} - \sigma_{z(3)}$$

For the left-hand section, $b/z = 0/10 = 0$, $a/z = 2.5/5 = 0.5$. So, from Figure 3.26, $I_3 = 0.14$. For the middle section, $b/z = 12.5/5 = 2.5$, $a/z = 5/5 = 1$. Hence, $I_3 = 0.493$. For the right-hand section, $I_3 = 0.14$ (same as the left-hand section). So

$$\sigma_{zB} = (0.14)(18.5 \times 2.5) + (0.493)(18.5 \times 5) - (0.14)(18.5 \times 2.5)$$

$$= (0.493)(92.5) = 45.5 \text{ kN/m}^2$$

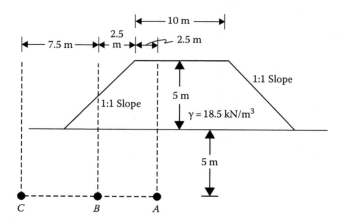

Figure 3.27 Stress increase due to embankment loading (not to scale).

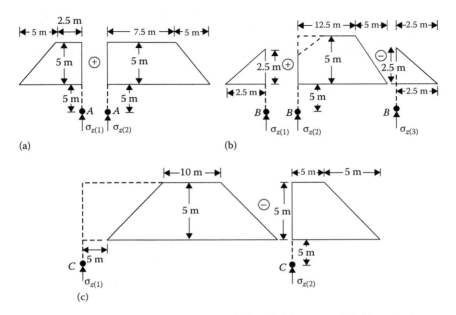

Figure 3.28 Calculation of stress increase at *A*, *B*, and *C* (not to scale): (a) vertical stress at *A*, (b) vertical stress at *B*, (c) vertical stress at *C*.

Vertical stress at C: Referring to Figure 3.28c

$$\sigma_{zC} = \sigma_{z(1)} - \sigma_{z(2)}$$

For the left-hand section, $b/z = 20/5 = 4$, $a/z = 5/5 = 1$. So, $I_3 = 0.498$.
For the right-hand section, $b/z = 5/5 = 1$, $a/z = 5/5 = 1$. So, $I_3 = 0.456$.
Hence

$$\sigma_{zC} = (0.498 - 0.456)(92.5) = 3.89 \text{ kN/m}^2$$

REFERENCES

Gray, H., Stress distribution in elastic solids, *Proc. 1st Int. Conf. Soil Mech. Found. Eng.*, Boston, 2, 157–168, 1936.

Kezdi, A. and L. Rethati, *Handbook of Soil Mechanics*, vol. 3, Elsevier, Amsterdam, the Netherlands, 1988.

Melan, E., Der Spannungzustand der durch eine Einzelkraft im Innern beanspruchten Halbschiebe, *Z. Angew. Math. Mech.*, 12, 343, 1932.

Michell, J. H., The stress in an aeolotropic elastic solid with an infinite plane boundary, *Proc. London Math. Soc.*, 32, 247–258, 1900.

Osterberg, J. O., Influence values for vertical stresses in semi-infinite mass due to embankment loading, *Proc. 4th Int. Conf. Soil Mech. Found. Eng.*, London, vol. 1, 393, 1957.

Poulos, H. G., Stresses and displacements in an elastic layer underlain by a rough rigid base, *Civ. Eng., Res. Rep. No. R63*, University of Sydney, Sydney, New South Wales, Australia, 1966.

Poulos, H. G. and E. H. Davis, *Elastic Solutions for Soil and Rock Mechanics*, Wiley, New York, 1974.

Chapter 4

Stresses and displacements in a soil mass

Three-dimensional problems

4.1 INTRODUCTION

In Chapter 3, the procedure for estimating stress and displacement for plane strain cases was discussed. This chapter relates to the calculation of stress and displacement for three-dimensional problems based on the theory of elasticity.

4.2 STRESSES DUE TO A VERTICAL POINT LOAD ON THE SURFACE

Boussinesq (1883) solved the problem for stresses inside a semi-infinite mass due to a point load acting on the surface. In rectangular coordinates, the stresses at a point $P(x,y,z)$ may be expressed as follows (Figure 4.1):

$$\sigma_z = \frac{3Qz^3}{2\pi R^5} \tag{4.1}$$

$$\sigma_x = \frac{3Q}{2\pi}\left\{\frac{x^2 z}{R^5} + \frac{1-2v}{3}\left[\frac{1}{R(R+z)} - \frac{(2R+z)x^2}{R^3(R+z)^2} - \frac{z}{R^3}\right]\right\} \tag{4.2}$$

$$\sigma_y = \frac{3Q}{2\pi}\left\{\frac{y^2 z}{R^5} + \frac{1-2v}{3}\left[\frac{1}{R(R+z)} - \frac{(2R+z)y^2}{R^3(R+z)^2} - \frac{z}{R^3}\right]\right\} \tag{4.3}$$

$$\tau_{xy} = \frac{3Q}{2\pi}\left[\frac{xyz}{R^5} + \frac{1-2v}{3}\frac{(2R+z)xy}{R^3(R+z)^2}\right] \tag{4.4}$$

$$\tau_{xz} = \frac{3Q}{2\pi}\frac{yz^2}{R^5} \tag{4.5}$$

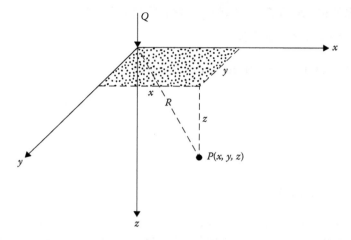

Figure 4.1 Concentrated point load on the surface (rectangular coordinates).

$$\tau_{yz} = \frac{3Q}{2\pi} \frac{yz^2}{R^5}$$ (4.6)

where

Q is the point load

$R = \sqrt{z^2 + r^2}$

$r = \sqrt{x^2 + y^2}$

ν is Poisson's ratio

In cylindrical coordinates, the stresses may be expressed as follows (Figure 4.2):

$$\sigma_z = \frac{3Qz^3}{2\pi R^5}$$ (4.7)

$$\sigma_r = \frac{Q}{2\pi}\left[\frac{3zr^2}{R^5} - \frac{1-2\nu}{R(R+z)}\right]$$ (4.8)

$$\sigma_\theta = \frac{Q}{2\pi}(1-2\nu)\left[\frac{1}{R(R+z)} - \frac{z}{R^3}\right]$$ (4.9)

$$\tau_{rz} = \frac{3Qrz^2}{2\pi R^5}$$ (4.10)

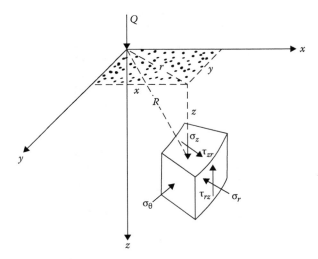

Figure 4.2 Concentrated point load (vertical) on the surface (cylindrical coordinates).

Equation 4.1 (or [4.7]) can be expressed as

$$\sigma_z = I_4 \frac{Q}{z^2} \tag{4.11}$$

where I_4 is the nondimensional influence factor:

$$I_4 = \frac{3}{2\pi}\left[1+\left(\frac{r}{z}\right)^2\right]^{-5/2} \tag{4.12}$$

Table 4.1 gives the values of I_4 for various values of r/z.
    Equation 4.8 for the radial stress can be expressed as

$$\sigma_r = \frac{Q}{2\pi z^2}\left\{\frac{3\left(\frac{r}{z}\right)^2}{\left[\left(\frac{r}{z}\right)^2+1\right]^{5/2}} - \frac{1-2v}{\left[\left(\frac{r}{z}\right)^2+1\right]+\left[\left(\frac{r}{z}\right)^2+1\right]^{1/2}}\right\} \tag{4.13}$$

or

$$\sigma_r = \frac{Q}{z^2}I_5 \tag{4.14}$$

Table 4.1 Value of I_4 (Equation 4.12)

r/z	I_4
0	0.4775
0.2	0.4329
0.4	0.3294
0.6	0.2214
0.8	0.1386
1.0	0.0844
1.2	0.0513
1.4	0.0317
1.6	0.0200
1.8	0.0129
2.0	0.0085
2.5	0.0034

where

$$I_5 = \frac{1}{2\pi}\left\{ \frac{3\left(\dfrac{r}{z}\right)^2}{\left[\left(\dfrac{r}{z}\right)^2+1\right]^{5/2}} - \frac{1-2v}{\left[\left(\dfrac{r}{z}\right)^2+1\right]+\left[\left(\dfrac{r}{z}\right)^2+1\right]^{1/2}} \right\} \qquad (4.15)$$

The variation of I_5 with r/z and v is given in Table 4.2.

Table 4.2 Variation of I_5 with r/z and v (Equation 4.15)

r/z	v			
	0	0.1	0.3	0.5
0	0	0	0	0
0.25	0	0	0	0.026
0.50	0.001	0.015	0.041	0.068
1.00	0.031	0.043	0.065	0.088
1.25	0.033	0.040	0.056	0.071
1.50	0.025	0.031	0.044	0.056
1.75	0.018	0.023	0.034	0.044
2.00	0.012	0.017	0.025	0.034
2.25	0.008	0.019	0.019	0.027
2.50	0.005	0.008	0.015	0.021
2.75	0.003	0.006	0.011	0.017
3.00	0.002	0.004	0.009	0.014

4.3 DEFLECTION DUE TO A CONCENTRATED POINT LOAD AT THE SURFACE

The deflections at a point due to a concentrated point load located at the surface are as follows (Figure 4.1):

$$u = \int \varepsilon_x dx = \frac{Q(1+v)}{2\pi E}\left[\frac{xz}{R^3} - \frac{(1-2v)x}{R(R+z)}\right] \tag{4.16}$$

$$v = \int \varepsilon_y dy = \frac{Q(1+v)}{2\pi E}\left[\frac{yz}{R^3} - \frac{(1-2v)y}{R(R+z)}\right] \tag{4.17}$$

$$w = \int \varepsilon_z dz = \frac{1}{E}\left[\sigma_z - v(\sigma_\gamma + \sigma_\theta)\right] = \frac{Q(1+v)}{2\pi E}\left[\frac{z^2}{R^3} - \frac{2(1-v)}{R}\right] \tag{4.18}$$

4.4 HORIZONTAL POINT LOAD ON THE SURFACE

Figure 4.3 shows a horizontal point load Q acting on the surface of a semi-infinite mass. This is generally referred to as Cerutti's problem. The stresses at a point $P(x, y, z)$ are as follows:

$$\sigma_z = \frac{3Qxz^2}{2\pi R^5} \tag{4.19}$$

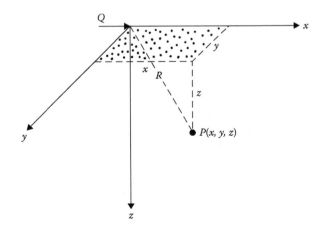

Figure 4.3 Horizontal point load on the surface.

$$\sigma_x = \frac{Q}{2\pi}\frac{x}{R^3}\left\{\frac{3x^2}{R^2}-(1-2v)+\frac{(1-2v)R^2}{(R+z)^2}\left[3-\frac{x^2(3R+z)}{R^2(R+z)}\right]\right\} \tag{4.20}$$

$$\sigma_y = \frac{Q}{2\pi}\frac{x}{R^3}\left\{\frac{3y^2}{R^2}-(1-2v)+\frac{(1-2v)R^2}{(R+z)^2}\left[3-\frac{y^2(3R+z)}{R^2(R+z)}\right]\right\} \tag{4.21}$$

$$\tau_{xy} = \frac{Q}{2\pi}\frac{y}{R^3}\left\{\frac{3x^2}{R^2}+\frac{(1-2v)R^2}{(R+z)^2}\left[1-\frac{x^2(3R+z)}{R^2(R+z)}\right]\right\} \tag{4.22}$$

$$\tau_{xz} = \frac{3Q}{2\pi}\frac{x^2z}{R^5} \tag{4.23}$$

$$\tau_{yz} = \frac{3Q}{2\pi}\frac{xyz}{R^5} \tag{4.24}$$

Also, the displacements at point P can be given as

$$u = \frac{Q}{2\pi}\frac{(1+v)}{E}\frac{1}{R}\left[\frac{x^2}{R^2}+1+\frac{(1-2v)R}{(R+z)}\left(1-\frac{x^2}{R(R+z)}\right)\right] \tag{4.25}$$

$$v = \frac{Q}{2\pi}\frac{(1+v)}{E}\frac{xy}{R^3}\left[1-\frac{(1-2v)R^2}{(R+z)^2}\right] \tag{4.26}$$

$$w = \frac{Q}{2\pi}\frac{(1+v)}{E}\frac{x}{R^2}\left[\frac{z}{R}+\frac{(1-2v)R}{(R+z)}\right] \tag{4.27}$$

Example 4.1

Refer to Figure 4.4. An inclined point load $Q = 400$ kN is applied at O. The load is in the xz plane. Determine the vertical stress σ_z at points A (3 m, 0, 2 m) and B (-3 m, 0, 2 m).

Solution

For the point load Q the horizontal component,

$$Q_h = Q\cos 30 = 400\cos 30 = 346.41 \text{ kN}$$

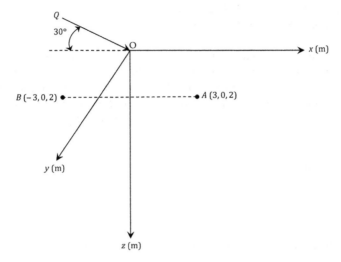

Figure 4.4 An inclined load applied at O.

Also the vertical component,

$$Q_v = 400 \sin 30 = 200 \text{ kN}$$

The components Q_h and Q_v are in the xz plane. Hence, combining Equations 4.1 and 4.19,

$$\sigma_z = \frac{3Q_v z^3}{2\pi R^5} + \frac{3Q_h x z^2}{2\pi R^5}$$

For point A:

$$R = (x^2 + y^2 + z^2)^{0.5} = [(3)^2 + (0)^2 + (2)^2]^{0.5} = 3.6 \text{ m}$$

$$\sigma_z = \frac{(3)(200)(2)^3}{2\pi(3.6)^5} + \frac{(3)(346.41)(3)(2)^2}{2\pi(3.6)^5} = 1.26 + 3.28 = 4.54 \text{ kN/m}^2$$

For point B:

$$R = 3.6 \text{ m}$$

$$\sigma_z = \frac{(3)(200)(2)^3}{2\pi(3.6)^5} + \frac{(3)(346.41)(-3)(2)^2}{2\pi(3.6)^5} = 1.26 - 3.28 = -2.02 \text{ kN/m}^2$$

4.5 VERTICAL STRESS DUE TO A LINE LOAD OF FINITE LENGTH

Figure 4.5 shows a flexible line load of length L, and the load per unit length is equal to q. In order to determine the vertical stress at P due to the line load, we consider an elementary length dy of the line load. The load on the elementary length is then equal to $q \cdot dy$. The vertical stress increase $d\sigma_z$ due to the elemental load at P can be obtained using Equation 4.1, or

$$d\sigma_z = \frac{3(q \cdot dy)z^3}{2\pi R^5} \tag{4.28}$$

where

$$R^5 = (a^2 + y^2 + z^2)^{5/2} \tag{4.29}$$

Thus, the total stress increase σ_z at P due to the entire line load of length L can be given as

$$\sigma_z = \int d\sigma_z = \int_0^L \frac{3(q \cdot dy)z^3}{2\pi(a^2 + y^2 + z^2)^{5/2}} = \frac{q}{z} I_6 \tag{4.30}$$

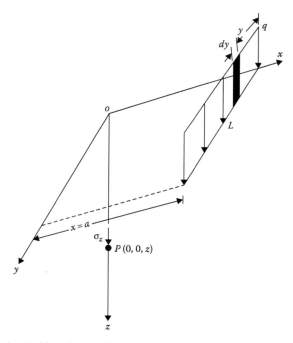

Figure 4.5 Line load of length L on the surface of a semi-infinite soil mass.

where

$$I_6 = \frac{1}{2\pi\left(m_1^2 + 1\right)^2}\left[\frac{3n_1}{\sqrt{m_1^2 + n_1^2 + 1}} - \left(\frac{n_1}{\sqrt{m_1^2 + n_1^2 + 1}}\right)^3\right] \tag{4.31}$$

$$m_1 = \frac{a}{z} \tag{4.32}$$

$$n_1 = \frac{L}{z} \tag{4.33}$$

Figure 4.6 shows a plot of the variation of I_6 with m_1 and n_1.

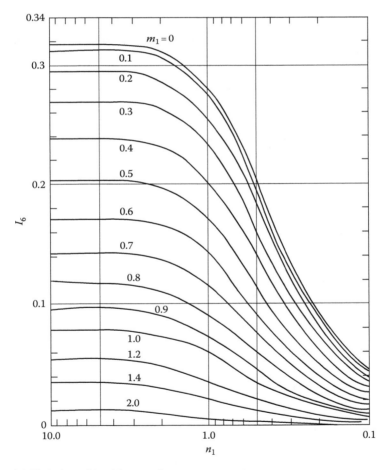

Figure 4.6 Variation of I_6 with m_1 and n_1.

Example 4.2

Refer to Figure 4.5. Given $a = 3$ m, $L = 4.8$ m, $q = 50$ kN/m. Determine the increase in stress, σ_z, due to the line load at

    a. Point with coordinates (0, 0, 6 m)
    b. Point with coordinates (0, 2.4 m, 6 m)

Solution

Part a:

$$m_1 = \frac{a}{z} = \frac{3}{6} = 0.5$$

$$n_1 = \frac{L}{z} = \frac{4.8}{6} = 0.8$$

From Figure 4.6, for $m_1 = 0.5$ and $n_1 = 0.8$, the value of I_6 is about 0.158. So

$$\sigma_z = \frac{q}{z}(I_6) = \frac{50}{6}(0.158) = 1.32\,\text{kN/m}^2$$

Part b:

As shown in Figure 4.7, the method of superposition can be used. Referring to Figure 4.7,

$$\sigma_z = \sigma_{z(1)} + \sigma_{z(2)}$$

For obtaining $\sigma_{z(1)}$ (Figure 4.7a),

$$m_1 = \frac{3}{6} = 0.5$$

$$n_1 = \frac{L_1}{z} = \frac{2.4}{6} = 0.4$$

From Figure 4.5, $I_{6(1)} \approx 0.1$. Similarly, for $\sigma_{z(2)}$ (Figure 4.7b)

$$m_1 = 0.5$$

$$n_1 = \frac{L_2}{z} = \frac{2.4}{6} = 0.4$$

So, $I_{6(2)} \approx 0.1$. Hence

$$\sigma_z = \frac{q}{z}[I_{6(1)} + I_{6(2)}] = \frac{50}{6}(0.1 + 0.1) = 1.67\,\text{kN/m}^2$$

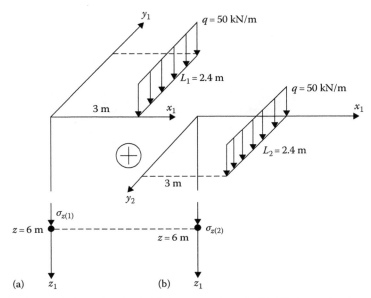

Figure 4.7 Stress due to a line load: (a) determination of $z(1)$; (b) determination of $z(2)$.

4.6 STRESSES BELOW A CIRCULARLY LOADED FLEXIBLE AREA (UNIFORM VERTICAL LOAD)

Integration of the Boussinesq's equation given in Section 4.2 can be adopted to obtain the stresses below the center of a circularly loaded flexible area. Figure 4.8 shows a circular area of radius b being subjected to a uniform load of q per unit area. Consider an elementary area dA. The load over the area is equal to $q \cdot dA$, and this can be treated as a point load. To determine the vertical stress due to the elementary load at a point P, we can substitute $q \cdot dA$ for Q and $\sqrt{r^2 + z^2}$ for R in Equation 4.1. Thus

$$d\sigma_z = \frac{(3q \cdot dA)z^3}{2\pi(r^2 + z^2)^{5/2}} \tag{4.34}$$

Since $dA = r d\theta \, dr$, the vertical stress at P due to the entire loaded area may now be obtained by substituting for dA in Equation 4.34 and then integrating

$$\sigma_z = \int_{\theta=0}^{\theta=2\pi} \int_{r=0}^{r=b} \frac{3q}{2\pi} \frac{z^3 r \, d\theta \, dr}{(r^2 + z^2)^{5/2}} = q\left[1 - \frac{z^3}{(b^2 + z^2)^{3/2}}\right] \tag{4.35}$$

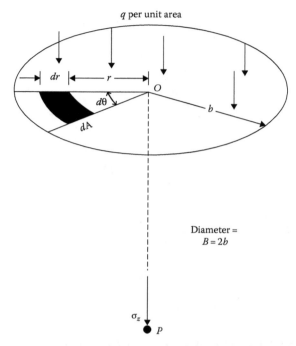

Figure 4.8 Stresses below the center of a circularly loaded area due to uniform vertical load.

Proceeding in a similar manner, we can also determine σ_r and σ_θ at point P as

$$\sigma_r = \sigma_\theta = \frac{q}{2}\left[1 + 2v - \frac{2(1+v)z}{(b^2 + z^2)^{1/2}} + \frac{z^3}{(b^2 + z^2)^{3/2}}\right] \qquad (4.36)$$

Equation (4.35) can be rewritten as

$$\frac{\sigma_z}{q} = 1 - \frac{1}{\left[\left(\frac{b}{z}\right)^2 + 1\right]^{3/2}} \qquad (4.37)$$

Table 4.3 gives the variation of σ_z/q with z/b.

Figure 4.9 shows a plot of σ_z/q vs. z/b.

A detailed tabulation of stresses below a uniformly loaded flexible circular area was given by Ahlvin and Ulery (1962). Referring to Figure 4.10, the stresses at point P may be given by

$$\sigma_z = q(A' + B') \qquad (4.38)$$

Table 4.3 Variation of σ_z/q with z/b
Equation 4.37)

z/b	σ_z/q
0.00	1.0000
0.02	0.9999
0.05	0.9998
0.10	0.9990
0.20	0.9925
0.40	0.9488
0.50	0.9106
0.80	0.7562
1.00	0.6465
1.50	0.4240
2.00	0.2845
2.50	0.1996
3.00	0.1436
4.00	0.0869
5.00	0.0571

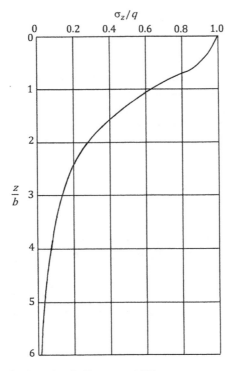

Figure 4.9 Variation of σ_z/q with z/b (Equation 4.37).

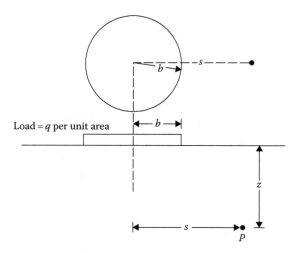

Figure 4.10 Stresses at any point below a circularly loaded area.

$$\sigma_r = q[2\nu A' + C + (1 - 2\nu)F] \qquad\qquad (4.39)$$

$$\sigma_\theta = q[2\nu A' - D + (1 - 2\nu)E] \qquad\qquad (4.40)$$

$$\tau_{rz} = \tau_{zr} = qG \qquad\qquad (4.41)$$

where A', B', C, D, E, F, and G are functions of s/b and z/b; the values of these are given in Tables 4.4 through 4.10.

Note that σ_θ is a principal stress, due to symmetry. The remaining two principal stresses can be determined as

$$\sigma_P = \frac{(\sigma_z + \sigma_r) \pm \sqrt{(\sigma_z - \sigma_r)^2 + (2\tau_{rz})^2}}{2} \qquad\qquad (4.42)$$

Example 4.3

Refer to Figure 4.10. Given that $q = 100$ kN/m^2, $B = 2b = 5$ m, and $\nu = 0.45$, determine the principal stresses at a point defined by $s = 3.75$ m and $z = 5$ m.

Solution

$s/b = 3.75/2.5 = 1.5$; $z/b = 5/2.5 = 2$. From Tables 4.4 through 4.10

$\quad A' = 0.06275$

$\quad B' = 0.06371$

$\quad C = -0.00782$

Table 4.4 Function A'

z/b	s/b																	
	0	0.2	0.4	0.6	0.8	1	1.2	1.5	2	3	4	5	6	7	8	10	12	14
0	1.0	1.0	1.0	1.0	1.0	0.5	0	0	0	0	0	0	0	0	0	0	0	0
0.1	0.90050	0.89748	0.88679	0.86126	0.78797	0.43015	0.09645	0.02787	0.00856	0.00211	0.00084	0.00042	0	0	0	0	0	0
0.2	0.80388	0.79824	0.77884	0.73483	0.63014	0.38269	0.15433	0.05251	0.01680	0.00419	0.00167	0.00083	0.00048	0.00030	0.00020	0	0	0
0.3	0.71265	0.70518	0.68316	0.62690	0.52081	0.34375	0.17964	0.07199	0.02440	0.00622	0.00250							
0.4	0.62861	0.62015	0.59241	0.53767	0.44329	0.31048	0.18709	0.08593	0.03118									
0.5	0.55279	0.54403	0.51622	0.46448	0.38390	0.28156	0.18556	0.09499	0.03701	0.01013	0.00407	0.00209	0.00118	0.00071	0.00053	0.00025	0.00014	0.00009
0.6	0.48550	0.47691	0.45078	0.40427	0.33676	0.25588	0.17952	0.10010										
0.7	0.42654	0.41874	0.39491	0.35428	0.29833	0.21727	0.17124	0.10228	0.04558									
0.8	0.37531	0.36832	0.34729	0.31243	0.26581	0.21297	0.16206	0.10236										
0.9	0.33104	0.32492	0.30669	0.27707	0.23832	0.19488	0.15253	0.10094										
1	0.29289	0.28763	0.27005	0.24697	0.21468	0.17868	0.14329	0.09849	0.05185	0.01742	0.00761	0.00393	0.00226	0.00143	0.00097	0.00050	0.00029	0.00018
1.2	0.23178	0.22795	0.21662	0.19890	0.17626	0.15101	0.12570	0.09192	0.05260	0.01935	0.00871	0.00459	0.00269	0.00171	0.00115			
1.5	0.16795	0.16552	0.15877	0.14804	0.13436	0.11892	0.10296	0.08048	0.05116	0.02142	0.01013	0.00548	0.00325	0.00210	0.00141	0.00073	0.00043	0.00027
2	0.10557	0.10453	0.10140	0.09647	0.09011	0.08269	0.07471	0.06275	0.04496	0.02221	0.01160	0.00659	0.00399	0.00264	0.00180	0.00094	0.00056	0.00036
2.5	0.07152	0.07098	0.06947	0.06698	0.06373	0.05974	0.05555	0.04880	0.03787	0.02143	0.01221	0.00732	0.00463	0.00308	0.00214	0.00115	0.00068	0.00043
3	0.05132	0.05101	0.05022	0.04886	0.04707	0.04487	0.04241	0.03839	0.03150	0.01980	0.01220	0.00770	0.00505	0.00346	0.00242	0.00132	0.00079	0.00051
4	0.02986	0.02976	0.02907	0.02802	0.02832	0.02749	0.02651	0.02490	0.02193	0.01592	0.01109	0.00768	0.00536	0.00384	0.00282	0.00160	0.00099	0.00065
5	0.01942	0.01938				0.01835			0.01573	0.01249	0.00949	0.00708	0.00527	0.00394	0.00298	0.00179	0.00113	0.00075
6	0.01361					0.01307			0.01168	0.00983	0.00795	0.00628	0.00492	0.00384	0.00299	0.00188	0.00124	0.00084
7	0.01005					0.00976			0.00894	0.00784	0.00661	0.00548	0.00445	0.00360	0.00291	0.00193	0.00130	0.00091
8	0.00772					0.00755			0.00703	0.00635	0.00554	0.00472	0.00398	0.00332	0.00276	0.00189	0.00134	0.00094
9	0.00612					0.00600			0.00566	0.00520	0.00466	0.00409	0.00353	0.00301	0.00256	0.00184	0.00133	0.00096
10								0.00477	0.00465	0.00438	0.00397	0.00352	0.00326	0.00273	0.00241			

Source: After Ahlvin, R.G. and Ulery, H.H., Tabulated values for determining the complete pattern of stresses, strains and deflections beneath a uniform load on a homogeneous half space, *Bull. 342*, Highway Research Record, pp. 1–13, 1962.

Table 4.5 Function B'

z/b	s/b																
	0	0.2	0.4	0.6	0.8	1	1.2	1.5	2	3	4	5	6	7	8	10	12	14
0	0	0	0	0	0	0	0	0	0	0	0	0	0	0	0	0	0	0
0.1	0.09852	0.10140	0.11138	0.13424	0.18796	0.05388	−0.07899	−0.02672	−0.00845	−0.00210	−0.00084	−0.00042	0	0	0	0	0	0
0.2	0.18857	0.19306	0.20772	0.23524	0.25983	0.08513	−0.07759	−0.04448	−0.01593	−0.00412	−0.00166	−0.00083	0	0	0	0	0	0
0.3	0.26362	0.26787	0.28018	0.29483	0.27257	0.10757	−0.04316	−0.04999	−0.02166	−0.00599	−0.00245		−0.00024	−0.00015	−0.00010			
0.4	0.32016	0.32259	0.32748	0.32273	0.26925	0.12404	−0.00766	−0.04535	−0.02522									
0.5	0.35777	0.35752	0.35323	0.33106	0.26236	0.13591	0.02165	−0.03455	−0.02651	−0.00991	−0.00388	−0.00199	−0.00116	−0.00073	−0.00049	−0.00025	−0.00014	−0.00009
0.6	0.37831	0.37531	0.36308	0.32822	0.25411	0.14440	0.04457	−0.02101	−0.02329									
0.7	0.38487	0.37962	0.36072	0.31929	0.24638	0.14986	0.06209	−0.00702										
0.8	0.38091	0.37408	0.35133	0.30699	0.23779	0.15292	0.07530	0.00614										
0.9	0.36962	0.36275	0.33734	0.29299	0.22891	0.15404	0.08507	0.01795										
1	0.35355	0.34553	0.32075	0.27819	0.21978	0.15355	0.09210	0.02814	−0.01005	−0.01115	−0.00608	−0.00344	−0.00210	−0.00135	−0.00092	−0.00048	−0.00028	−0.00018
1.2	0.31485	0.30730	0.28481	0.24836	0.20113	0.14915	0.10002	0.04378	0.00023	−0.00995	−0.00632	−0.00378	−0.00236	−0.00156	−0.00107	−0.00068	−0.00040	−0.00026
1.5	0.25602	0.25025	0.23338	0.20694	0.17368	0.13732	0.10193	0.05745	0.01385	−0.00669	−0.00600	−0.00401	−0.00265	−0.00181	−0.00126	−0.00084	−0.00050	−0.00033
2	0.17889	0.18144	0.16644	0.15198	0.13375	0.11331	0.09254	0.06371	0.02836	0.00028	−0.00410	−0.00371	−0.00278	−0.00202	−0.00148	−0.00094	−0.00059	−0.00039
2.5	0.12807	0.12633	0.12126	0.11327	0.10298	0.09130	0.07869	0.06022	0.03429	0.00661	−0.00130	−0.00271	−0.00250	−0.00201	−0.00156	−0.00099	−0.00065	−0.00046
3	0.09487	0.09394	0.09099	0.08635	0.08033	0.07325	0.06551	0.05354	0.03511	0.01112	0.00157	−0.00134	−0.00192	−0.00179	−0.00151	−0.00094	−0.00068	−0.00050
4	0.05707	0.05666	0.05562	0.05383	0.05145	0.04773	0.04532	0.03995	0.03066	0.01515	0.00595	0.00155	−0.00029	−0.00094	−0.00109	−0.00094	−0.00061	−0.00049
5	0.03772	0.03760				0.03384			0.02474	0.01522	0.00810	0.00371	0.00132	0.00013	−0.00043	−0.00037	−0.00047	−0.00045
6	0.02666					0.02468			0.01968	0.01380	0.00867	0.00496	0.00254	0.00110	0.00028	−0.00002	−0.00029	−0.00037
7	0.01980					0.01868			0.01577	0.01204	0.00842	0.00547	0.00332	0.00185	0.00093	0.00035	−0.00002	−0.00029
8	0.01526					0.01459			0.01279	0.01034	0.00779	0.00554	0.00372	0.00236	0.00141	0.00066	0.00035	−0.00025
9	0.01212					0.01170			0.01054	0.00888	0.00705	0.00533	0.00386	0.00265	0.00178		0.00066	−0.00012
10						0.00924			0.00879	0.00764	0.00631	0.00501	0.00382	0.00281	0.00199		0.00012	

Source: After Ahlvin, R.G. and Ulery, H.H., Tabulated values for determining the complete pattern of stresses, strains and deflections beneath a uniform load on a homogeneous half space, Bull. 342, Highway Research Record, pp. 1–13, 1962.

Table 4.6 Function C

z/b	s/b																	
	0	0.2	0.4	0.6	0.8	1	1.2	1.5	2	3	4	5	6	7	8	10	12	14
0	0	0	0	0	0	0	0	0	0	0	0	0	0	0	0	0	0	0
0.1	-0.04926	-0.05142	-0.05903	-0.07708	-0.12108	0.02247	0.12007	0.04475	0.01536	0.00403	0.00164	0.00082	0	0	0	0	0	0
0.2	-0.09429	-0.09775	-0.10872	-0.12977	-0.14552	0.02419	0.14896	0.07892	0.02951	0.00796	0.00325	0.00164	0.00094	0.00059	0.00039	0	0	0
0.3	-0.13181	-0.13484	-0.14415	-0.15023	-0.12990	0.01988	0.13394	0.09816	0.04148	0.01169	0.00483							
0.4	-0.16008	-0.16188	-0.16519	-0.15985	-0.11168	0.01292	0.11014	0.10422	0.05067									
0.5	-0.17889	-0.17835	-0.17497	-0.15625	-0.09833	0.00483	0.08730	0.10125	0.05690	0.01824	0.00778	0.00399	0.00231	0.00146	0.00098	0.00050	0.00029	0.00018
0.6	-0.18915	-0.18664	-0.17336	-0.14934	-0.08967	-0.00304	0.06731	0.09313	0.06129									
0.7	-0.19244	-0.18831	-0.17393	-0.14147	-0.08409	-0.01061	0.05028	0.08253										
0.8	-0.19046	-0.18481	-0.16784	-0.13393	-0.08066	-0.01744	0.03582	0.07114										
0.9	-0.18481	-0.17841	-0.16024	-0.12664	-0.07828	-0.02337	0.02359	0.05993										
1	-0.17678	-0.17050	-0.15188	-0.11995	-0.07634	-0.02843	0.01331	0.04939	0.05429	0.02726	0.01333	0.00726	0.00433	0.00278	0.00188	0.00098	0.00057	0.00036
1.2	-0.15742	-0.15117	-0.13467	-0.10763	-0.07289	-0.03575	-0.00245	0.03107	0.04552	0.02791	0.01467	0.00824	0.00501	0.00324	0.00221	0.00141	0.00083	0.00039
1.5	-0.12801	-0.12277	-0.11101	-0.09145	-0.06711	-0.04124	-0.01702	0.01088	0.03154	0.02652	0.01570	0.00933	0.00585	0.00386	0.00266	0.00179	0.00107	0.00069
2	-0.08944	-0.08491	-0.07976	-0.06925	-0.05560	-0.04144	-0.02687	-0.00782	0.01267	0.02070	0.01527	0.01013	0.00689	0.00462	0.00327	0.00209	0.00128	0.00083
2.5	-0.06403	-0.06068	-0.05839	-0.05259	-0.04522	-0.03605	-0.02800	-0.01536	0.00103	0.01384	0.01314	0.00987	0.00707	0.00506	0.00369	0.00232	0.00145	0.00096
3	-0.04744	-0.04560	-0.04339	-0.04089	-0.03642	-0.03130	-0.02587	-0.01748	-0.00528	0.00792	0.01030	0.00888	0.00689	0.00520	0.00392	0.00254	0.00168	0.00115
4	-0.02854	-0.02737	-0.02562	-0.02585	-0.02421	-0.02112	-0.01964	-0.01586	-0.00956	0.00038	0.00492	0.00602	0.00561	0.00476	0.00389	0.00250	0.00177	0.00127
5	-0.01886	-0.01810				-0.01568			-0.00939	-0.00293	-0.00128	0.00329	0.00391	0.00380	0.00341	0.00227	0.00173	0.00130
6	-0.01333					-0.01118			-0.00819	-0.00405	-0.00079	0.00129	0.00234	0.00272	0.00272	0.00193	0.00161	0.00128
7	-0.00990					-0.00902			-0.00678	-0.00417	-0.00180	-0.00004	0.00113	0.00174	0.00200	0.00157	0.00143	0.00128
8	-0.00763					-0.00699			-0.00552	-0.00393	-0.00225	-0.00077	0.00029	0.00096	0.00134	0.00134	0.00122	0.00120
9	-0.00607					-0.00423			-0.00452	-0.00353	-0.00235	-0.00118	-0.00027	0.00037	0.00082	0.00124	0.00110	0.00110
10								-0.00381	-0.00373	-0.00314	-0.00233	-0.00137	-0.00063	-0.00030	0.00040			

Source: After Ahlvin, R.G. and Ulery, H.H., Tabulated values for determining the complete pattern of stresses, strains and deflections beneath a uniform load on a homogeneous half space, Bull. 342, Highway Research Record, pp. 1–13, 1962.

Table 4.7 Function D

z/b	s/b																	
	0	0.2	0.4	0.6	0.8	1	1.2	1.5	2	3	4	5	6	7	8	10	12	14
0	0	0	0	0	0	0	0	0	0	0	0	0	0	0	0	0	0	0
0.1	0.04926	0.04998	0.05235	0.05716	0.06687	0.07635	0.04108	0.01803	0.00691	0.00193	0.00080	0.00041	0	0	0	0	0	0
0.2	0.09429	0.09552	0.09900	0.10546	0.11431	0.10932	0.07139	0.03444	0.01359	0.00384	0.00159	0.00081	0.00047	0.00029	0.00020	0	0	0
0.3	0.13181	0.13305	0.14051	0.14062	0.14267	0.12745	0.09078	0.04817	0.01982	0.00927	0.00238							
0.4	0.16008	0.16070	0.16229	0.16288	0.15756	0.13696	0.10248	0.05887	0.02545									
0.5	0.17889	0.17917	0.17826	0.17481	0.16403	0.14074	0.10894	0.06670	0.03039	0.00921	0.00390	0.00200	0.00116	0.00073	0.00049	0.00025	0.00015	0.00009
0.6	0.18915	0.18867	0.18573	0.17887	0.16489	0.14137	0.11186	0.07212										
0.7	0.19244	0.19132	0.18679	0.17782	0.16229	0.13926	0.11237	0.07551	0.03801									
0.8	0.19046	0.18927	0.18348	0.17306	0.15714	0.13548	0.11115	0.07728										
0.9	0.18481	0.18349	0.17709	0.16635	0.15063	0.13067	0.10866	0.07788										
1	0.17678	0.17503	0.16886	0.15824	0.14344	0.12513	0.10540	0.07753	0.04456	0.01611	0.00725	0.00382	0.00224	0.00142	0.00096	0.00050	0.00029	0.00018
1.2	0.15742	0.15618	0.15014	0.14073	0.12823	0.11340	0.09757	0.07484	0.04575	0.01796	0.00835	0.00446	0.00264	0.00169	0.00114	0.00073	0.00043	0.00027
1.5	0.12801	0.12754	0.12237	0.11549	0.10657	0.09608	0.08491	0.06833	0.04539	0.01983	0.00970	0.00532	0.00320	0.00205	0.00140	0.00095	0.00056	0.00036
2	0.08944	0.09080	0.08668	0.08273	0.07814	0.07187	0.06566	0.05589	0.04103	0.02098	0.01117	0.00643	0.00398	0.00260	0.00179	0.00115	0.00068	0.00044
2.5	0.06403	0.06565	0.06284	0.06068	0.05777	0.05525	0.05069	0.04486	0.03532	0.02045	0.01183	0.00717	0.00457	0.00306	0.00213	0.00133	0.00080	0.00052
3	0.04744	0.04834	0.04760	0.04548	0.04391	0.04195	0.03963	0.03606	0.02983	0.01904	0.01187	0.00755	0.00497	0.00341	0.00242	0.00160	0.00100	0.00065
4	0.02854	0.02928	0.02996	0.02798	0.02724	0.02661	0.02568	0.02408	0.02110	0.01552	0.01087	0.00757	0.00533	0.00382	0.00280	0.00180	0.00114	0.00077
5	0.01886	0.01950				0.01816			0.01535	0.01230	0.00939	0.00700	0.00523	0.00392	0.00299	0.00190	0.00124	0.00086
6	0.01333					0.01351			0.01149	0.00976	0.00788	0.00625	0.00488	0.00381	0.00301	0.00192	0.00130	0.00092
7	0.00990					0.00966			0.00899	0.00787	0.00662	0.00542	0.00445	0.00360	0.00292	0.00192	0.00131	0.00096
8	0.00763					0.00759			0.00727	0.00641	0.00554	0.00477	0.00402	0.00332	0.00275	0.00187	0.00133	0.00099
9	0.00607					0.00746			0.00601	0.00533	0.00470	0.00415	0.00358	0.00303	0.00260			
10								0.00542	0.00506	0.00450	0.00398	0.00364	0.00319	0.00278	0.00239			

Source: After Ahlvin, R. G. and Ulery, H.H., Tabulated values for determining the complete pattern of stresses, strains and deflections beneath a uniform load on a homogeneous half space, Bull. 342, Highway Research Record, pp. 1–13, 1962.

Table 4.8 Function E

z/b	s/b																	
	0	0.2	0.4	0.6	0.8	1	1.2	1.5	2	3	4	5	6	7	8	10	12	14
0	0.5	0.5	0.5	0.5	0.5	0.5	0.34722	0.22222	0.12500	0.05556	0.03125	0.02000	0.01389	0.01020	0.00781	0.00500	0.00347	0.00255
0.1	0.45025	0.449494	0.44698	0.44173	0.43008	0.39198	0.30445	0.20399	0.11806	0.05362	0.03045	0.01959	0.01342	0.00991	0.00762			
0.2	0.40194	0.400434	0.39591	0.38660	0.36798	0.32802	0.26598	0.18633	0.11121	0.05170	0.02965	0.01919	0.01272	0.00946	0.00734			
0.3	0.35633	0.35428	0.33809	0.33674	0.31578	0.28003	0.23311	0.16967	0.10450	0.04979	0.02886	0.01800						
0.4	0.31431	0.31214	0.30541	0.29298	0.27243	0.24200	0.20526	0.15428	0.09801	0.04608	0.02727							
0.5	0.27639	0.27407	0.26732	0.25511	0.23639	0.21119	0.18168	0.14028	0.09180									
0.6	0.24275	0.24247	0.23411	0.22289	0.20634	0.18520	0.16155	0.12759	0.08027									
0.7	0.21327	0.21112	0.20535	0.19525	0.18093	0.16356	0.14421	0.11620										
0.8	0.18765	0.18550	0.18049	0.17190	0.15977	0.14523	0.12928	0.10602										
0.9	0.16552	0.16337	0.15921	0.15179	0.14168	0.12954	0.11634	0.09686										
1	0.14645	0.14483	0.14610	0.13472	0.12618	0.11611	0.10510	0.08865	0.06552	0.03736	0.02352	0.01602	0.01157	0.00874	0.00683	0.00475	0.00332	0.00246
1.2	0.11589	0.11435	0.11201	0.10741	0.10140	0.09431	0.08657	0.07476	0.05728	0.03425	0.02208	0.01527	0.01113	0.00847	0.00664	0.00450	0.00318	0.00237
1.5	0.08398	0.08356	0.08159	0.07885	0.07517	0.07088	0.06611	0.05871	0.04703	0.03003	0.02008	0.01419	0.01049	0.00806	0.00636	0.00425	0.00304	0.00228
2	0.05279	0.05105	0.05146	0.05034	0.04850	0.04675	0.04442	0.04078	0.03454	0.02410	0.01706	0.01248	0.00943	0.00738	0.00590	0.00401	0.00290	0.00219
2.5	0.03576	0.03426	0.03489	0.03435	0.03360	0.03211	0.03150	0.02953	0.02599	0.01945	0.01447	0.01096	0.00850	0.00674	0.00546	0.00378	0.00276	0.00210
3	0.02566	0.02519	0.02470	0.02491	0.02444	0.02389	0.02330	0.02216	0.02007	0.01585	0.01230	0.00962	0.00763	0.00617	0.00505	0.00355	0.00263	0.00201
4	0.01493	0.01452	0.01495	0.01526	0.01446	0.01418	0.01395	0.01356	0.01281	0.01084	0.00900	0.00742	0.00612	0.00511	0.00431	0.00313	0.00237	0.00185
5	0.00971	0.00927				0.00929			0.00873	0.00774	0.00673	0.00579	0.00495	0.00425	0.00364	0.00275	0.00213	0.00168
6	0.00680					0.00632			0.00629	0.00574	0.00517	0.00457	0.00404	0.00354	0.00309	0.00241	0.00192	0.00154
7	0.00503					0.00493			0.00466	0.00438	0.00404	0.00370	0.00330	0.00296	0.00264	0.00213	0.00172	0.00140
8	0.00386					0.00377			0.00354	0.00344	0.00325	0.00297	0.00273	0.00250	0.00228	0.00185	0.00155	0.00127
9	0.00306					0.00227			0.00275	0.00273	0.00264	0.00246	0.00229	0.00212	0.00194	0.00163	0.00139	0.00116
10									0.00220	0.00225	0.00221	0.00203	0.00200	0.00181	0.00171			

Source: After Ahlvin, R.G. and Ulery, H.H., Tabulated values for determining the complete pattern of stresses, strains and deflections beneath a uniform load on a homogeneous half space. Bull. 342, Highway Research Record, pp. 1–13, 1962.

Table 4.9 Function F

z/b	s/b																	
	0	0.2	0.4	0.6	0.8	1	1.2	1.5	2	3	4	5	6	7	8	10	12	14
0	0.5	0.5	0.5	0.5	0.5	0	-0.34722	-0.22222	-0.12500	-0.05556	-0.03125	-0.02000	-0.01389	-0.01020	-0.00781	-0.00500	-0.00347	-0.00255
0.1	0.45025	0.44794	0.43981	0.41954	0.35789	0.03817	-0.20800	-0.17612	-0.10950	-0.05151	-0.02961	-0.01917						
0.2	0.40194	0.39781	0.38294	0.34823	0.26215	0.05466	-0.11165	-0.13381	-0.09441	-0.04750	-0.02798	-0.01835	-0.01295	-0.00961	-0.00742			
0.3	0.35633	0.35094	0.34508	0.29016	0.20503	0.06372	-0.05346	-0.09768	-0.08010	-0.04356	-0.02636							
0.4	0.31431	0.30801	0.28681	0.24469	0.17086	0.06848	-0.01818	-0.06835	-0.06684									
0.5	0.27639	0.26997	0.24890	0.20937	0.14752	0.07037	0.00388	-0.04529	-0.05479	-0.03595	-0.02320	-0.01590	-0.01154	-0.00875	-0.00681	-0.00450	-0.00318	-0.00237
0.6	0.24275	0.23444	0.21667	0.18138	0.13042	0.07068	0.01797	-0.02749										
0.7	0.21327	0.20762	0.18956	0.15903	0.11740	0.06963	0.02704	-0.01392	-0.03469									
0.8	0.18765	0.18287	0.16679	0.14053	0.10604	0.06774	0.03277	-0.00365										
0.9	0.16552	0.16158	0.14747	0.12528	0.09664	0.06533	0.03619	0.00408										
1	0.14645	0.14280	0.12395	0.11225	0.08850	0.06256	0.03819	0.00984	-0.01367	-0.01994	-0.01591	-0.01209	-0.00931	-0.00731	-0.00587	-0.00400	-0.00289	-0.00219
1.2	0.11589	0.11360	0.10460	0.09449	0.07486	0.05670	0.03913	0.01716	-0.00452	-0.01491	-0.01337	-0.01068	-0.00844	-0.00676	-0.00550	-0.00353	-0.00261	-0.00201
1.5	0.08398	0.08196	0.07719	0.06918	0.05919	0.04804	0.03686	0.02177	0.00413	-0.00879	-0.00995	-0.00870	-0.00723	-0.00596	-0.00495	-0.00307	-0.00233	-0.00183
2	0.05279	0.05348	0.04994	0.04614	0.04162	0.03593	0.03029	0.02197	0.01043	-0.00189	-0.00546	-0.00589	-0.00544	-0.00474	-0.00410	-0.00263	-0.00208	-0.00166
2.5	0.03576	0.03673	0.03459	0.03263	0.03014	0.02762	0.02406	0.01927	0.01188	0.00198	-0.00226	-0.00364	-0.00386	-0.00366	-0.00332	-0.00223	-0.00183	-0.00150
3	0.02566	0.02586	0.02255	0.02395	0.02263	0.02097	0.01911	0.01623	0.01144	0.00396	-0.00010	-0.00192	-0.00258	-0.00271	-0.00263	-0.00153	-0.00137	-0.00120
4	0.01493	0.01536	0.01412	0.01259	0.01386	0.01331	0.01256	0.01134	0.00912	0.00508	-0.00209	0.00026	-0.00076	-0.00127	-0.00148	-0.00096	-0.00099	-0.00093
5	0.00971	0.01011				0.00905			0.00700	0.00475	0.00277	0.00129	0.00031	-0.00030	-0.00066	-0.00053	-0.00066	-0.00070
6	0.00680					0.00675			0.00538	0.00409	0.00278	0.00170	0.00088	0.00030	-0.00010	-0.00020	-0.00041	-0.00049
7	0.00503					0.00483			0.00428	0.00346	0.00258	0.00178	0.00114	0.00064	0.00027	0.00003	-0.00020	-0.00033
8	0.00386					0.00380			0.00350	0.00291	0.00229	0.00174	0.00125	0.00082	0.00048	0.00020	0.00003	-0.00019
9	0.00306					0.00374			0.00291	0.00247	0.00203	0.00163	0.00124	0.00089	0.00062	0.00020	-0.00005	
10						0.00267			0.00246	0.00213	0.00176	0.00149	0.00126	0.00092	0.00070			

Source: After Ahlvin, R.G. and Ulery, H.H., Tabulated values for determining the complete pattern of stresses, strains and deflections beneath a uniform load on a homogeneous half space, *Bull. 342*, Highway Research Record, pp. 1–13, 1962.

Table 4.10 Function G

s/b

z/b	0	0.2	0.4	0.6	0.8	1	1.2	1.5	2	3	4	5	6	7	8	10	12	14
0	0	0	0	0	0	0.31831	0	0	0	0	0	0	0	0	0	0	0	0
0.1	0	0.00315	0.00802	0.01951	0.06682	0.31405	0.05555	0.00865	0.00159	0.00023	0.00007	0.00003	0	0	0	0	0	0
0.2	0	0.01163	0.02877	0.06441	0.16214	0.30474	0.13592	0.03060	0.00614	0.00091	0.00026	0.00010	0.00005	0.00003	0.00002	0	0	0
0.3	0	0.02301	0.05475	0.11072	0.21465	0.29228	0.18216	0.05747	0.01302	0.00201	0.00059							
0.4	0	0.03460	0.07883	0.14477	0.23442	0.27779	0.20195	0.08233	0.02138									
0.5	0	0.04429	0.09618	0.16426	0.23652	0.26216	0.20731	0.10185	0.03033	0.00528	0.00158	0.00063	0.00030	0.00016	0.00009	0.00004	0.00002	0.00001
0.6	0	0.04966	0.10729	0.17192	0.22949	0.24574	0.20496	0.11541										
0.7	0	0.05484	0.11256	0.17126	0.21772	0.22924	0.19840	0.12373	0.04718									
0.8	0	0.05590	0.11225	0.16534	0.20381	0.21295	0.18953	0.12855										
0.9	0	0.05496	0.10856	0.15628	0.18904	0.19712	0.17945	0.12881										
1	0	0.05266	0.10274	0.14566	0.17419	0.18198	0.16884	0.12745	0.06434	0.01646	0.00555	0.00233	0.00113	0.00062	0.00036	0.00015	0.00007	0.00004
1.2	0	0.04585	0.08831	0.12323	0.14615	0.15408	0.14755	0.12038	0.06967	0.02077	0.00743	0.00320	0.00159	0.00087	0.00051			
1.5	0	0.03483	0.06688	0.09293	0.11071	0.11904	0.11830	0.10477	0.07075	0.02599	0.01021	0.00460	0.00233	0.00130	0.00078	0.00033	0.00016	0.00009
2	0	0.02102	0.04069	0.05721	0.06948	0.07738	0.08067	0.07804	0.06275	0.03062	0.01409	0.00692	0.00369	0.00212	0.00129	0.00055	0.00027	0.00015
2.5	0	0.01293	0.02534	0.03611	0.04484	0.05119	0.05509	0.05668	0.05117	0.03099	0.01650	0.00886	0.00499	0.00296	0.00185	0.00082	0.00041	0.00023
3	0	0.00840	0.01638	0.02376	0.02994	0.03485	0.03843	0.04124	0.04039	0.02886	0.01745	0.01022	0.00610	0.00376	0.00241	0.00110	0.00057	0.00032
4	0	0.00382	0.00772	0.01149	0.01480	0.01764	0.02004	0.02271	0.02475	0.02215	0.01639	0.01118	0.00745	0.00499	0.00340	0.00167	0.00090	0.00052
5	0	0.00214				0.00992		0.01343	0.01551	0.01601	0.01364	0.01105	0.00782	0.00560	0.00404	0.00216	0.00122	0.00073
6	0					0.00602		0.00845	0.01014	0.01148	0.01082	0.00917	0.00733	0.00567	0.00432	0.00243	0.00150	0.00092
7	0					0.00396			0.00687	0.00830	0.00842	0.00770	0.00656	0.00539	0.00432	0.00272	0.00171	0.00110
8	0					0.00270			0.00481	0.00612	0.00656	0.00631	0.00568	0.00492	0.00413	0.00278	0.00185	0.00124
9	0					0.00177			0.00347	0.00459	0.00513	0.00515	0.00485	0.00438	0.00381	0.00274	0.00192	0.00133
10	0							0.00199	0.00258	0.00351	0.00407	0.00420	0.00411	0.00382	0.00346			

Source: After Ahlvin, R.G. and Ulery, H.H., Tabulated values for determining the complete pattern of stresses, strains and deflections beneath a uniform load on a homogeneous half space, Bull. 342, Highway Research Record, pp. 1–13, 1962.

$D = 0.05589$

$E = 0.04078$

$F = 0.02197$

$G = 0.07804$

So

$\sigma_z = q(A' + B') = 100(0.06275 + 0.06371) = 12.65\,\text{kN/m}^2$

$\sigma_\theta = q[2vA' - D + (1 - 2v)E]$

$\quad = 100\{2(0.45)(0.06275) - 0.05589 + [1 - (2)(0.45)]0.04078\}$

$\quad = 0.466\,\text{kN/m}^2$

$\sigma_r = q[2vA' + C + (1 - 2v)F]$

$\quad = 100[0.9(0.06275) - 0.00782 + 0.1(0.02197)] = 5.09\,\text{kN/m}^2$

$\tau_{rz} = qG = (100)(0.07804) = 7.8\,\text{kN/m}^2$

$\sigma_\theta = 0.466\,\text{kN/m}^2 = \sigma_2 \text{ (intermediate principal stress)}$

$\sigma_P = \dfrac{(12.65 + 5.09) \pm \sqrt{(12.65 - 5.09)^2 + (2 \times 7.8)^2}}{2}$

$\quad = \dfrac{17.74 \pm 17.34}{2}$

$\sigma_{P(1)} = 17.54\ \text{kN/m}^2 \text{ (major principal stress)}$

$\sigma_{P(3)} = 0.2\ \text{kN/m}^2 \text{(minor principal stress)}$

4.7 VERTICAL DISPLACEMENT DUE TO UNIFORMLY LOADED CIRCULAR AREA AT THE SURFACE

The vertical displacement due to a uniformly loaded circular area at a point (Figure 4.11) can be determined by using the same procedure we used previously for a point load, which involves determination of the strain ϵ_z from the equation

$$\epsilon_z = \frac{1}{E}[\sigma_z - v(\sigma_r + \sigma_\theta)] \tag{4.43}$$

and determination of the settlement by integration with respect to z.

The relations for σ_z, σ_r, and σ_θ are given in Equations 4.38 through 4.40. Substitution of the relations for σ_z, σ_r, and σ_θ in the preceding equation for strain and simplification gives (Ahlvin and Ulery, 1962)

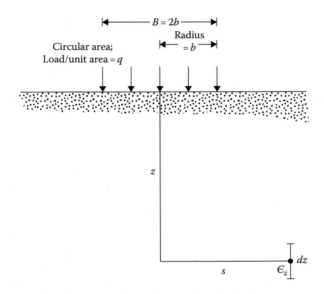

Figure 4.11 Elastic settlement due to a uniformly loaded circular area.

$$\epsilon_z = q\frac{1+\nu}{E}[(1-2\nu)A' + B']$$ (4.44)

where q is the load per unit area. A' and B' are nondimensional and are functions of z/b and s/b; their values are given in Tables 4.4 and 4.5.

The vertical deflection at a depth z can be obtained by integration of Equation 4.44 as

$$w = q\frac{1+\nu}{E}b\left[\frac{z}{b}I_7 + (1-\nu)I_8\right]$$ (4.45)

where
$I_7 = A'$ (Table 4.4)
b is the radius of the circular loaded area

The numerical values and I_8 (which is a function of z/b and s/b) are given in Table 4.11.

From Equation 4.45, it follows that the settlement at the surface (i.e., at $z = 0$) is

$$w_{(z=0)} = qb\frac{1-\nu^2}{E}I_8$$ (4.46)

Table 4.11 Values of I_8

z/b	s/b																	
	0	0.2	0.4	0.6	0.8	1	1.2	1.5	2	3	4	5	6	7	8	10	12	14
0		1.97987	1.91751	1.80575	1.62553	1.27319	0.93676	0.71185	0.51671	0.33815	0.25200	0.20045	0.16626	0.14315	0.12576	0.09918	0.08346	0.07023
0.1	1.80998	1.79018	1.72886	1.61961	1.44711	1.18107	0.92670	0.70888	0.51627	0.33794	0.25184	0.20081						
0.2	1.63961	1.62068	1.56242	1.46001	1.30614	1.09996	0.90098	0.70074	0.51382	0.33726	0.25162	0.20072	0.16688	0.14288	0.12512			
0.3	1.48806	1.47044	1.40979	1.32442	1.19210	1.02740	0.86726	0.68823	0.50966	0.33638	0.25124							
0.4	1.35407	1.33802	1.28963	1.20822	1.09555	0.96202	0.83042	0.67238	0.50412									
0.5	1.23607	1.22176	1.17894	1.10830	1.01312	0.90298	0.79308	0.65429	0.49728	0.33293	0.24996	0.19982	0.16668	0.14273	0.12493	0.09996	0.08295	0.07123
0.6	1.13238	1.11998	1.08350	1.02154	0.94120	0.84917	0.75653	0.63469										
0.7	1.04131	1.03037	0.99794	0.91049	0.87742	0.80030	0.72143	0.61442	0.48061									
0.8	0.96125	0.95175	0.92386	0.87928	0.82136	0.75571	0.68809	0.59398										
0.9	0.89072	0.88251	0.85856	0.82616	0.77950	0.71495	0.65677	0.57361										
1	0.82843	0.85005	0.80465	0.76809	0.72587	0.67769	0.62701	0.55364	0.45122	0.31877	0.24386	0.19673	0.16516	0.14182	0.12394	0.09952	0.08292	0.07104
1.2	0.72410	0.71882	0.70370	0.67937	0.64814	0.61187	0.57329	0.51552	0.43013	0.31162	0.24070	0.19520	0.16369	0.14099	0.12350			
1.5	0.60555	0.60233	0.57246	0.57633	0.55559	0.53138	0.50496	0.46379	0.39872	0.29945	0.23495	0.19053	0.16199	0.14058	0.12281	0.09876	0.8270	0.07064
2	0.47214	0.47022	0.44512	0.45656	0.44502	0.43202	0.41702	0.39242	0.35054	0.27740	0.22418	0.18618	0.15846	0.13762	0.12124	0.09792	0.08196	0.07026
2.5	0.38518	0.38403	0.38098	0.37608	0.36940	0.36155	0.35243	0.33698	0.30913	0.25550	0.21208	0.17898	0.15395	0.13463	0.11928	0.09700	0.08115	0.06980
3	0.32457	0.32403	0.32184	0.31887	0.31464	0.30969	0.30381	0.29364	0.27453	0.23487	0.19977	0.17154	0.14919	0.13119	0.11694	0.09558	0.08061	0.06897
4	0.24620	0.24588	0.24820	0.25128	0.24168	0.23932	0.23668	0.23164	0.22188	0.19908	0.17640	0.15596	0.13864	0.12396	0.11172	0.09300	0.07864	0.06848
5	0.19805	0.19785				0.19455			0.18450	0.17080	0.15575	0.14130	0.12785	0.11615	0.10585	0.08915	0.07675	0.06695
6	0.16554					0.16326			0.15750	0.14868	0.13842	0.12792	0.11778	0.10836	0.09990	0.08562	0.07452	0.06522
7	0.14217					0.14077			0.13699	0.13097	0.12404	0.11620	0.10843	0.10101	0.09387	0.08197	0.07210	0.06377
8	0.12448					0.12352			0.12112	0.11680	0.11176	0.10600	0.09976	0.09400	0.08848	0.07800	0.06928	0.06200
9	0.11079					0.10989			0.10854	0.10548	0.10161	0.09702	0.09234	0.08784	0.08298	0.07407	0.06678	0.05976
10								0.09900	0.09820	0.09510	0.09290	0.08980	0.08300	0.08180	0.07710			

Source: After Ahlvin, R.G. and Ulery, H.H., Tabulated values for determining the complete pattern of stresses, strains and deflections beneath a uniform load on a homogeneous half space, Bull. 342, Highway Research Record, pp. 1–13, 1962.

Example 4.4

Consider a uniformly loaded flexible circular area on the surface of a sand layer 9 m thick as shown in Figure 4.12. The circular area has a diameter of 3 m. Also given q = 100 kN/m²; for sand, E = 21,000 kN/m² and ν = 0.3.

  a. Use Equation 4.45 and determine the deflection of the center of the circular area (z = 0).
  b. Divide the sand layer into these layers of equal thickness of 3 m each. Use Equation 4.44 to determine the deflection at the center of the circular area.

Solution

Part a:

From Equation 4.45

$$w = \frac{q(1+\nu)}{E} b \left[\frac{z}{b} I_7 + (1-\nu)I_8 \right]$$

$$w_{\text{net}} = w_{(z=0,s=0)} - w_{(z=9m,s=0)}$$

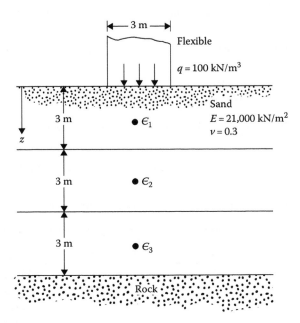

Figure 4.12 Elastic settlement calculation for a layer of finite thickness.

For $z/b = 0$ and $s/b = 0$, $I_7 = 1$ and $I_8 = 2$; so

$$w_{(z=0,\, s=0)} = \frac{100(1+0.3)}{21,000}(1.5)[(1-0.3)2] = 0.013\,\text{m} = 13\,\text{mm}$$

For $z/b = 9/1.5 = 6$ and $s/b = 0$, $I_7 = 0.01361$ and $I_8 = 0.16554$; so

$$w_{(z=9m,\, s=0)} = \frac{100(1+0.3)(1.5)}{21,000}[6(0.01361)+(1-0.3)0.16554]$$

$$= 0.00183\,\text{m} = 1.83\,\text{mm}$$

Hence, $w_{\text{net}} = 13 - 1.83 = 11.17$ mm.

Part b:

From Equation 4.44

$$\epsilon_z = \frac{q(1+\nu)}{E}[(1-2\nu)A' + B']$$

Layer 1: From Tables 4.4 and 4.5, for $z/b = 1.5/1.5 = 1$ and $s/b = 0$, $A' = 0.29289$ and $B' = 0.35355$

$$\epsilon_{z(1)} = \frac{100(1+0.3)}{21,000}[(1-0.6)(0.29289)+0.35355] = 0.00291$$

Layer 2: For $z/b = 4.5/1.5 = 3$ and $s/b = 0$, $A' = 0.05132$ and $B' = 0.09487$

$$\epsilon_{z(2)} = \frac{100(1+0.3)}{21,000}[(1-0.6)(0.05132)+0.09487] = 0.00071$$

Layer 3: For $z/b = 7.5/1.5 = 5$ and $s/b = 0$, $A' = 0.01942$ and $B' = 0.03772$

$$\epsilon_{z(3)} = \frac{100(1+0.3)}{21,000}[(1-0.6)(0.01942)+0.0.3772] = 0.00028$$

The final stages in the calculation are tabulated as follows:

Layer i	Layer thickness Δz_i (m)	Strain at the center of the layer $\epsilon_{z(i)}$	$\epsilon_{z(i)}\,\Delta z_i$ (m)
1	3	0.00291	0.00873
2	3	0.00071	0.00213
3	3	0.00028	0.00084
			$\Sigma 0.0117$ m
			= 11.7 mm

Example 4.5

A circular raft that is 5 m in diameter rests on a 10-m-thick saturated clay layer as shown in Figure 4.13a. An approximation of the variation of the unconfined compression strength of clay (q_u) with depth is also shown in Figure 4.13b. Calculate the elastic settlement (assuming a flexible area) at the center of the raft by dividing the 10-m-thick clay layer below the raft into ten 1-m-thick layers. Assume $E = 200q_u$ and $v = 0.5$.

Solution

Equation 4.44:

$$\epsilon_z = q\frac{1+v}{E}[(1-2v)A' + B']$$

Table 4.4 gives A' and Table 4.5 gives B'.

$q = 200 \text{ kN/m}^2$; $v = 0.5$; $B = 5$ m; $b = 2.5$ m; $E = 200q_u$.

$$\epsilon_z = \frac{(200)(1.5)}{E}\left\{[1-(2\times0.5)]A' + B'\right\} = \frac{300}{E}B'$$

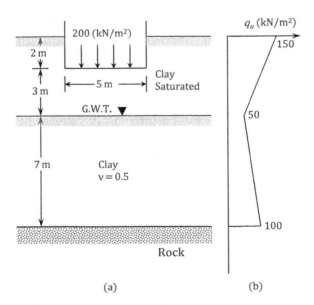

Figure 4.13 A circular raft on a saturated clay layer.

Layer No.	Depth to the middle of layer from bottom of foundation, z (m)	z/b	B'	E at the middle of the layer (kN/m²)	ϵ_z
1	0.5	0.2	0.18857	20,000	0.00283
2	1.5	0.6	0.37831	16,000	0.00709
3	2.5	1.0	0.35355	12,000	0.00884
4	3.5	1.4	0.27563	10,714	0.00772
5	4.5	1.8	0.20974	12,142	0.00523
6	5.5	2.2	0.15856	13,570	0.00351
7	6.5	2.6	0.12143	14,998	0.00243
8	7.5	3.0	0.09487	16,426	0.00173
9	8.5	3.4	0.07985	17,854	0.00134
10	9.5	3.8	0.06466	19,282	0.00101
					Σ0.04173

$$S_e = (0.04173)(1000 \text{ mm}) = 41.73 \text{ mm}$$

4.8 VERTICAL STRESS BELOW A RECTANGULAR LOADED AREA ON THE SURFACE

The stress at a point P at a depth z below the corner of a uniformly loaded (vertical) flexible rectangular area (Figure 4.14) can be determined by integration of Boussinesq's equations given in Section 4.2. The vertical load over the elementary area $dx\ dy$ may be treated as a point load of magnitude

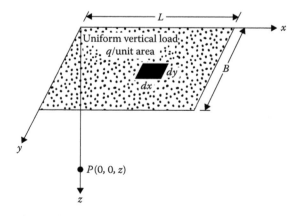

Figure 4.14 Vertical stress below the corner of a uniformly loaded (normal) rectangular area.

$q \cdot dx \cdot dy$. The vertical stress at P due to this elementary load can be evaluated with the aid of Equation 4.1:

$$d\sigma_z = \frac{3q \, dx \, dy \, z^3}{2\pi(x^2 + y^2 + z^2)^{5/2}}$$

The total increase of vertical stress at P due to the entire loaded area may be determined by integration of the previous equation with horizontal limits of $x = 0$ to $x = L$ and $y = 0$ to $y = B$. Newmark (1935) gave the results of the integration in the following form:

$$\sigma_z = qI_9 \tag{4.47}$$

$$I_9 = \frac{1}{4\pi} \left[\frac{2mn(m^2 + n^2 + 1)^{1/2}}{m^2 + n^2 + m^2n^2 + 1} \frac{m^2 + n^2 + 2}{m^2 + n^2 + 1} + \tan^{-1} \frac{2mn(m^2 + n^2 + 1)^{1/2}}{m^2 + n^2 - m^2n^2 + 1} \right] \tag{4.48}$$

where

$m = B/z$
$n = L/z$

The values of I_9 for various values of m and n are given in Table 4.12. Figure 4.15 shows a plot of I_9 with m and n.

The arctangent term in Equation 4.48 must be a positive angle in radians. When $m^2 + n^2 + 1 < m^2n^2$, it becomes a negative angle. So, a term π should be added to that angle.

For equations concerning the determination of σ_x, σ_y, τ_{xz}, τ_{yz}, and τ_{xy}, the reader is referred to the works of Holl (1940) and Giroud (1970).

The use of Table 4.12 for determination of the vertical stress at any point below a rectangular loaded area is shown in Example 4.6.

In most cases, the vertical stress below the center of a rectangular area is of importance. This can be given by the relationship

$$\Delta\sigma = qI_{10}$$

where

$$I_{10} = \frac{2}{\pi} \left[\frac{m_1'n_1'}{\sqrt{1 + m_1'^2 + n_1'^2}} \frac{1 + m_1'^2 + 2n_1'^2}{(1 + n_1'^2)(m_1'^2 + n_1'^2)} + \sin^{-1} \frac{m_1'}{\sqrt{m_1'^2 + n_1'^2} \sqrt{1 + n_1'^2}} \right] \tag{4.49}$$

$$m_1' = \frac{L}{B} \tag{4.50}$$

Table 4.12 Variation of I_9 with m and n

n \ m	0.1	0.2	0.3	0.4	0.5	0.6	0.7	0.8	0.9	1.0	1.2	1.4	1.6	1.8	2.0	2.5	3.0	4.0	5.0	6.0
0.1	0.0047	0.0092	0.0132	0.0168	0.0198	0.0222	0.0242	0.0258	0.0270	0.0279	0.0293	0.0301	0.0306	0.0309	0.0311	0.0314	0.0315	0.0316	0.0316	0.0316
0.2	0.0092	0.0179	0.0259	0.0328	0.0387	0.0435	0.0474	0.0504	0.0528	0.0547	0.0573	0.0589	0.0599	0.0606	0.0610	0.0616	0.0618	0.0619	0.0620	0.0620
0.3	0.0132	0.0259	0.0374	0.0474	0.0559	0.0629	0.0686	0.0731	0.0766	0.0794	0.0832	0.0856	0.0871	0.0880	0.0887	0.0895	0.0898	0.0901	0.0901	0.0902
0.4	0.0168	0.0328	0.0474	0.0602	0.0711	0.0801	0.0873	0.0931	0.0977	0.1013	0.1063	0.1094	0.1114	0.1126	0.1134	0.1145	0.1150	0.1153	0.1154	0.1154
0.5	0.0198	0.0387	0.0559	0.0711	0.0840	0.0947	0.1034	0.1104	0.1158	0.1202	0.1263	0.1300	0.1324	0.1340	0.1350	0.1363	0.1368	0.1372	0.1374	0.1374
0.6	0.0222	0.0435	0.0629	0.0801	0.0947	0.1069	0.1168	0.1247	0.1311	0.1361	0.1431	0.1475	0.1503	0.1521	0.1533	0.1548	0.1555	0.1560	0.1561	0.1562
0.7	0.0242	0.0474	0.0686	0.0873	0.1034	0.1169	0.1277	0.1365	0.1436	0.1491	0.1570	0.1620	0.1652	0.1672	0.1686	0.1704	0.1711	0.1717	0.1719	0.1719
0.8	0.0258	0.0504	0.0731	0.0931	0.1104	0.1247	0.1365	0.1461	0.1537	0.1598	0.1684	0.1739	0.1774	0.1797	0.1812	0.1832	0.1841	0.1847	0.1849	0.1850
0.9	0.0270	0.0528	0.0766	0.0977	0.1158	0.1311	0.1436	0.1537	0.1619	0.1684	0.1777	0.1836	0.1874	0.1899	0.1915	0.1938	0.1947	0.1954	0.1956	0.1957
1.0	0.0279	0.0547	0.0794	0.1013	0.1202	0.1361	0.1491	0.1598	0.1684	0.1752	0.1851	0.1914	0.1955	0.1981	0.1999	0.2024	0.2034	0.2042	0.2044	0.2045
1.2	0.0293	0.0573	0.0832	0.1063	0.1263	0.1431	0.1570	0.1684	0.1777	0.1851	0.1958	0.2028	0.2073	0.2103	0.2124	0.2151	0.2163	0.2172	0.2175	0.2176
1.4	0.0301	0.0589	0.0856	0.1094	0.1300	0.1475	0.1620	0.1739	0.1836	0.1914	0.2028	0.2102	0.2151	0.2184	0.2206	0.2236	0.2250	0.2260	0.2263	0.2264
1.6	0.0306	0.0599	0.0871	0.1114	0.1324	0.1503	0.1652	0.1774	0.1874	0.1955	0.2073	0.2151	0.2203	0.2237	0.2261	0.2294	0.2309	0.2320	0.2323	0.2325
1.8	0.0309	0.0606	0.0880	0.1126	0.1340	0.1521	0.1672	0.1797	0.1899	0.1981	0.2103	0.2183	0.2237	0.2274	0.2299	0.2333	0.2350	0.2362	0.2366	0.2367
2.0	0.0311	0.0610	0.0887	0.1134	0.1350	0.1533	0.1686	0.1812	0.1915	0.1999	0.2124	0.2206	0.2261	0.2299	0.2325	0.2361	0.2378	0.2391	0.2395	0.2397
2.5	0.0314	0.0616	0.0895	0.1145	0.1363	0.1548	0.1704	0.1832	0.1938	0.2024	0.2151	0.2236	0.2294	0.2333	0.2361	0.2401	0.2420	0.2434	0.2439	0.2441
3.0	0.0315	0.0618	0.0898	0.1150	0.1368	0.1555	0.1711	0.1841	0.1947	0.2034	0.2163	0.2250	0.2309	0.2350	0.2378	0.2420	0.2439	0.2455	0.2461	0.2463
4.0	0.0316	0.0619	0.0901	0.1153	0.1372	0.1560	0.1717	0.1847	0.1954	0.2042	0.2172	0.2260	0.2320	0.2362	0.2391	0.2434	0.2455	0.2472	0.2479	0.2481
5.0	0.0316	0.0620	0.0901	0.1154	0.1374	0.1561	0.1719	0.1849	0.1956	0.2044	0.2175	0.2263	0.2324	0.2366	0.2395	0.2439	0.2460	0.2479	0.2486	0.2489
6.0	0.0316	0.0620	0.0902	0.1154	0.1374	0.1562	0.1719	0.1850	0.1957	0.2045	0.2176	0.2264	0.2325	0.2367	0.2397	0.2441	0.2463	0.2482	0.2489	0.2492

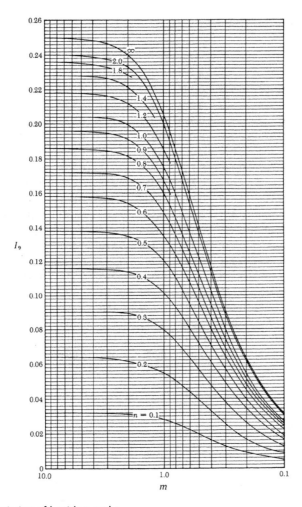

Figure 4.15 Variation of I_9 with m and n.

$$n_1' = \frac{z}{(B/2)} \tag{4.51}$$

The variation of I_{10} with m_1 and n_1 is given in Table 4.13.

Example 4.6

A distributed load of 50 kN/m² is acting on the flexible rectangular area 6 × 3 m as shown in Figure 4.16. Determine the vertical stress at point A, which is located at a depth of 3 m below the ground surface.

Table 4.13 Variation of I_{10} with m_1' and n_1'

n_1'	m_1'									
	1	2	3	4	5	6	7	8	9	10
0.20	0.994	0.997	0.997	0.997	0.997	0.997	0.997	0.997	0.997	0.997
0.40	0.960	0.976	0.977	0.977	0.977	0.977	0.977	0.977	0.977	0.977
0.60	0.892	0.932	0.936	0.936	0.937	0.937	0.937	0.937	0.937	0.937
0.80	0.800	0.870	0.878	0.880	0.881	0.881	0.881	0.881	0.881	0.881
1.00	0.701	0.800	0.814	0.817	0.818	0.818	0.818	0.818	0.818	0.818
1.20	0.606	0.727	0.748	0.753	0.754	0.755	0.755	0.755	0.755	0.755
1.40	0.522	0.658	0.685	0.692	0.694	0.695	0.695	0.696	0.696	0.696
1.60	0.449	0.593	0.627	0.636	0.639	0.640	0.641	0.641	0.641	0.642
1.80	0.388	0.534	0.573	0.585	0.590	0.591	0.592	0.592	0.593	0.593
2.00	0.336	0.481	0.525	0.540	0.545	0.547	0.548	0.549	0.549	0.549
3.00	0.179	0.293	0.348	0.373	0.384	0.389	0.392	0.393	0.394	0.395
4.00	0.108	0.190	0.241	0.269	0.285	0.293	0.298	0.301	0.302	0.303
5.00	0.072	0.131	0.174	0.202	0.219	0.229	0.236	0.240	0.242	0.244
6.00	0.051	0.095	0.130	0.155	0.172	0.184	0.192	0.197	0.200	0.202
7.00	0.038	0.072	0.100	0.122	0.139	0.150	0.158	0.164	0.168	0.171
8.00	0.029	0.056	0.079	0.098	0.113	0.125	0.133	0.139	0.144	0.147
9.00	0.023	0.045	0.064	0.081	0.094	0.105	0.113	0.119	0.124	0.128
10.00	0.019	0.037	0.053	0.067	0.079	0.089	0.097	0.103	0.108	0.112

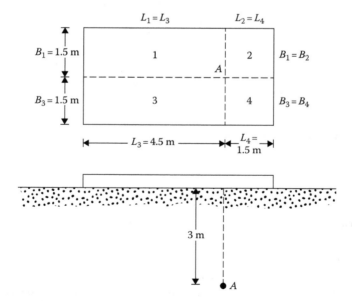

Figure 4.16 Distributed load on a flexible rectangular area.

Solution

The total increase of stress at A may be evaluated by summing the stresses contributed by the four rectangular loaded areas shown in Figure 4.16. Thus

$$\sigma_z = q[I_{9(1)} + I_{9(2)} + I_{9(3)} + I_{9(4)}]$$

$$n_{(1)} = \frac{L_1}{z} = \frac{4.5}{3} = 1.5 \quad m_{(1)} = \frac{B_1}{z} = \frac{1.5}{3} = 0.5$$

From Table 4.12, $I_{9(1)} = 0.131$. Similarly

$$n_{(2)} = \frac{L_2}{z} = \frac{1.5}{3} = 0.5 \quad m_{(2)} = \frac{B_2}{z} = 0.5 \quad I_{9(2)} = 0.084$$

$$n_{(3)} = 1.5 \quad m_{(3)} = 0.5 \quad I_{9(3)} = 0.131$$

$$n_{(4)} = 0.5 \quad m_{(4)} = 0.5 \quad I_{9(4)} = 0.085$$

So

$$\sigma_z = 50(0.131 + 0.084 + 0.131 + 0.084) = 21.5 \text{ kN/m}^2$$

Example 4.7

Refer to the flexible area on the surface of a clay layer shown in Figure 4.17. The uniformly distributed vertical load on the rectangular area is $q_1 = 100$ kN/m^2, and the uniformly distributed vertical load on the semicircular area is $q_2 = 200$ kN/m^2. Determine the vertical stress increase due to the loaded area located at a depth of 3 m below points A and B.

Solution

Stress below point A—The entire area can be divided into three parts (Figure 4.18):

- For rectangular area 1:

$$m = \frac{B}{z} = \frac{3}{3} = 1; \quad n = \frac{L}{z} = \frac{4}{3} = 1.33$$

  From Figure 4.15: $I_9 = 0.196$
- For rectangular area 2:

$$m = \frac{3}{3} = 1; \quad n = \frac{3}{3} = 1; \quad I_9 = 0.178$$

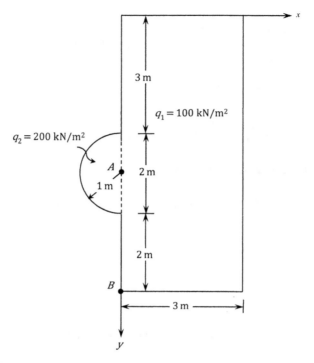

Figure 4.17 Flexible area on the surface of a clay layer.

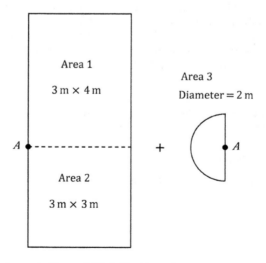

Figure 4.18 Area shown in Figure 4.17 divided into three parts.

- For half circle area 3 (see Equation 4.35):

$$\sigma_z = \frac{q_2}{2}\left[1 - \frac{z^3}{(b^2 + z^2)^{1.5}}\right]$$

Hence,

$$\sigma_z = q_1(0.196 + 0.178) + \frac{q_2}{2}\left[1 - \frac{3^3}{(1^2 + 3^2)^{1.5}}\right]$$

With $q_1 = 100$ kN/m² and $q_2 = 200$ kN/m²,

$$\sigma_z = 52.02 \text{ kN/m}^2$$

Stress below point B—The entire area can be divided into two parts (Figure 4.19).

- For rectangular area 1:

$$m = \frac{B}{z} = \frac{3}{3} = 1; \quad n = \frac{L}{z} = \frac{7}{3} = 2.33$$

From Figure 4.15: $I_9 = 0.202$

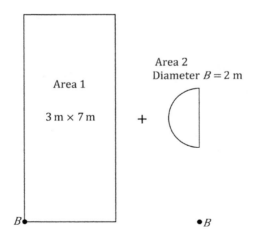

Figure 4.19 Area shown in Figure 4.17 divided into two parts.

- For semicircular area 2:

$$\frac{s}{b} = \frac{(2+1)}{1} = 3; \quad \frac{z}{b} = \frac{3}{1} = 3$$

From Equation 4.38,

$$\sigma_z = \frac{q_2}{2}(A' + B') = \frac{q_2}{2}(0.0198 + 0.01112)$$

Thus, total

$$\sigma_z = 100(0.202) + \frac{200}{2}(0.0198 + 0.01112) = \mathbf{23.29\ kN/m^2}$$

4.9 DEFLECTION DUE TO A UNIFORMLY LOADED FLEXIBLE RECTANGULAR AREA

The elastic deformation in the vertical direction at the corner of a uniformly loaded rectangular area of size $L \times B$ (Figure 4.14) can be obtained by proper integration of the expression for strain. The deflection at a depth z below the corner of the rectangular area can be expressed in the form (Harr, 1966)

$$w(\text{corner}) = \frac{qB}{2E}(1 - v^2)\left[I_{11} - \left(\frac{1-2v}{1-v}\right)I_{12}\right] \tag{4.52}$$

where

$$I_{11} = \frac{1}{\pi}\left[\ln\left(\frac{\sqrt{1 + m_1''^2 + n_1''^2} + m_1''}{\sqrt{1 + m_1''^2 + n_1''^2} - m_1''}\right) + m_1''\ln\left(\frac{\sqrt{1 + m_1''^2 + n_1''^2} + 1}{\sqrt{1 + m_1''^2 + n_1''^2} - 1}\right)\right] \tag{4.53}$$

$$I_{12} = \frac{n_1''}{\pi}\tan^{-1}\left(\frac{m_1''}{n_1''\sqrt{1 + m_1''^2 + n_1''^2}}\right) \tag{4.54}$$

$$m_1'' = \frac{L}{B} \tag{4.55}$$

$$n_1'' = \frac{z}{B} \tag{4.56}$$

Values of I_{11} and I_{12} are given in Tables 4.14 and 4.15.

Table 4.14 Variation of I_{11}

n″	Value of m_i''									
	1	2	3	4	5	6	7	8	9	10
0.00	1.122	1.532	1.783	1.964	2.105	2.220	2.318	2.403	2.477	2.544
0.25	1.095	1.510	1.763	1.944	2.085	2.200	2.298	2.383	2.458	2.525
0.50	1.025	1.452	1.708	1.890	2.032	2.148	2.246	2.331	2.406	2.473
0.75	0.933	1.371	1.632	1.816	1.959	2.076	2.174	2.259	2.334	2.401
1.00	0.838	1.282	1.547	1.734	1.878	1.995	2.094	2.179	2.255	2.322
1.25	0.751	1.192	1.461	1.650	1.796	1.914	2.013	2.099	2.175	2.242
1.50	0.674	1.106	1.378	1.570	1.717	1.836	1.936	2.022	2.098	2.166
1.75	0.608	1.026	1.299	1.493	1.641	1.762	1.862	1.949	2.025	2.093
2.00	0.552	0.954	1.226	1.421	1.571	1.692	1.794	1.881	1.958	2.026
2.25	0.504	0.888	1.158	1.354	1.505	1.627	1.730	1.817	1.894	1.963
2.50	0.463	0.829	1.095	1.291	1.444	1.567	1.670	1.758	1.836	1.904
2.75	0.427	0.776	1.037	1.233	1.386	1.510	1.613	1.702	1.780	1.850
3.00	0.396	0.728	0.984	1.179	1.332	1.457	1.561	1.650	1.729	1.798
3.25	0.369	0.686	0.935	1.128	1.281	1.406	1.511	1.601	1.680	1.750
3.50	0.346	0.647	0.889	1.081	1.234	1.359	1.465	1.555	1.634	1.705
3.75	0.325	0.612	0.848	1.037	1.189	1.315	1.421	1.511	1.591	1.662
4.00	0.306	0.580	0.809	0.995	1.147	1.273	1.379	1.470	1.550	1.621
4.25	0.289	0.551	0.774	0.957	1.107	1.233	1.339	1.431	1.511	1.582
4.50	0.274	0.525	0.741	0.921	1.070	1.195	1.301	1.393	1.474	1.545
4.75	0.260	0.501	0.710	0.887	1.034	1.159	1.265	1.358	1.438	1.510
5.00	0.248	0.479	0.682	0.855	1.001	1.125	1.231	1.323	1.404	1.477
5.25	0.237	0.458	0.655	0.825	0.969	1.093	1.199	1.291	1.372	1.444
5.50	0.227	0.440	0.631	0.797	0.939	1.062	1.167	1.260	1.341	1.413
5.75	0.217	0.422	0.608	0.770	0.911	1.032	1.137	1.230	1.311	1.384
6.00	0.208	0.406	0.586	0.745	0.884	1.004	1.109	1.201	1.282	1.355
6.25	0.200	0.391	0.566	0.722	0.858	0.977	1.082	1.173	1.255	1.328
6.50	0.193	0.377	0.547	0.699	0.834	0.952	1.055	1.147	1.228	1.301
6.75	0.186	0.364	0.529	0.678	0.810	0.927	1.030	1.121	1.203	1.275
7.00	0.179	0.352	0.513	0.658	0.788	0.904	1.006	1.097	1.178	1.251
7.25	0.173	0.341	0.497	0.639	0.767	0.881	0.983	1.073	1.154	1.227
7.50	0.168	0.330	0.482	0.621	0.747	0.860	0.960	1.050	1.131	1.204
7.75	0.162	0.320	0.468	0.604	0.728	0.839	0.939	1.028	1.109	1.181
8.00	0.158	0.310	0.455	0.588	0.710	0.820	0.918	1.007	1.087	1.160
8.25	0.153	0.301	0.442	0.573	0.692	0.801	0.899	0.987	1.066	1.139
8.50	0.148	0.293	0.430	0.558	0.676	0.783	0.879	0.967	1.046	1.118
8.75	0.144	0.285	0.419	0.544	0.660	0.765	0.861	0.948	1.027	1.099
9.00	0.140	0.277	0.408	0.531	0.644	0.748	0.843	0.930	1.008	1.080
9.25	0.137	0.270	0.398	0.518	0.630	0.732	0.826	0.912	0.990	1.061
9.50	0.133	0.263	0.388	0.506	0.616	0.717	0.810	0.895	0.972	1.043
9.75	0.130	0.257	0.379	0.494	0.602	0.702	0.794	0.878	0.955	1.026
10.00	0.126	0.251	0.370	0.483	0.589	0.688	0.778	0.862	0.938	1.009

Table 4.15 Variation of I_{12}

n_1''	Value of m_1''									
	1	2	3	4	5	6	7	8	9	10
0.25	0.098	0.103	0.104	0.105	0.105	0.105	0.105	0.105	0.105	0.105
0.50	0.148	0.167	0.172	0.174	0.175	0.175	0.175	0.176	0.176	0.176
0.75	0.166	0.202	0.212	0.216	0.218	0.219	0.220	0.220	0.220	0.220
1.00	0.167	0.218	0.234	0.241	0.244	0.246	0.247	0.248	0.248	0.248
1.25	0.160	0.222	0.245	0.254	0.259	0.262	0.264	0.265	0.265	0.266
1.50	0.149	0.220	0.248	0.261	0.267	0.271	0.274	0.275	0.276	0.277
1.75	0.139	0.213	0.247	0.263	0.271	0.277	0.280	0.282	0.283	0.284
2.00	0.128	0.205	0.243	0.262	0.273	0.279	0.283	0.286	0.288	0.289
2.25	0.119	0.196	0.237	0.259	0.272	0.279	0.284	0.288	0.290	0.292
2.50	0.110	0.186	0.230	0.255	0.269	0.278	0.284	0.288	0.291	0.293
2.75	0.102	0.177	0.223	0.250	0.266	0.277	0.283	0.288	0.291	0.294
3.00	0.096	0.168	0.215	0.244	0.262	0.274	0.282	0.287	0.291	0.294
3.25	0.090	0.160	0.208	0.238	0.258	0.271	0.279	0.285	0.290	0.293
3.50	0.084	0.152	0.200	0.232	0.253	0.267	0.277	0.283	0.288	0.292
3.75	0.079	0.145	0.193	0.226	0.248	0.263	0.273	0.281	0.287	0.291
4.00	0.075	0.138	0.186	0.219	0.243	0.259	0.270	0.278	0.285	0.289
4.25	0.071	0.132	0.179	0.213	0.237	0.254	0.267	0.276	0.282	0.287
4.50	0.067	0.126	0.173	0.207	0.232	0.250	0.263	0.272	0.280	0.285
4.75	0.064	0.121	0.167	0.201	0.227	0.245	0.259	0.269	0.277	0.283
5.00	0.061	0.116	0.161	0.195	0.221	0.241	0.255	0.266	0.274	0.281
5.25	0.059	0.111	0.155	0.190	0.216	0.236	0.251	0.263	0.271	0.278
5.50	0.056	0.107	0.150	0.185	0.211	0.232	0.247	0.259	0.268	0.276
5.75	0.054	0.103	0.145	0.179	0.206	0.227	0.243	0.255	0.265	0.273
6.00	0.052	0.099	0.141	0.174	0.201	0.223	0.239	0.252	0.262	0.270
6.25	0.050	0.096	0.136	0.170	0.197	0.218	0.235	0.248	0.259	0.267
6.50	0.048	0.093	0.132	0.165	0.192	0.214	0.231	0.245	0.256	0.265
6.75	0.046	0.089	0.128	0.161	0.188	0.210	0.227	0.241	0.252	0.262
7.00	0.045	0.087	0.124	0.156	0.183	0.205	0.223	0.238	0.249	0.259
7.25	0.043	0.084	0.121	0.152	0.179	0.201	0.219	0.234	0.246	0.256
7.50	0.042	0.081	0.117	0.149	0.175	0.197	0.216	0.231	0.243	0.253
7.75	0.040	0.079	0.114	0.145	0.171	0.193	0.212	0.227	0.240	0.250
8.00	0.039	0.077	0.111	0.141	0.168	0.190	0.208	0.224	0.236	0.247
8.25	0.038	0.074	0.108	0.138	0.164	0.186	0.205	0.220	0.233	0.244
8.50	0.037	0.072	0.105	0.135	0.160	0.182	0.201	0.217	0.230	0.241
8.75	0.036	0.070	0.103	0.132	0.157	0.179	0.198	0.214	0.227	0.238
9.00	0.035	0.069	0.100	0.129	0.154	0.176	0.194	0.210	0.224	0.235
9.25	0.034	0.067	0.098	0.126	0.151	0.172	0.191	0.207	0.221	0.233
9.50	0.033	0.065	0.095	0.123	0.147	0.169	0.188	0.204	0.218	0.230
9.75	0.032	0.064	0.093	0.120	0.145	0.166	0.185	0.201	0.215	0.227
10.00	0.032	0.062	0.091	0.118	0.142	0.163	0.182	0.198	0.212	0.224

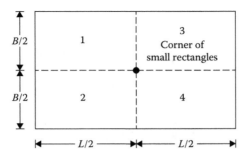

Figure 4.20 Determination of settlement at the center of a rectangular area of dimensions $L \times B$.

For surface deflection at the *corner of a rectangular area*, we can substitute $z/B = n_1'' = 0$ in Equation 4.52 and make the necessary calculations; thus

$$w(\text{corner}) = \frac{qB}{2E}(1 - v^2)I_{11} \qquad (4.57)$$

The deflection at the surface for the *center of a rectangular area* (Figure 4.20) can be found by adding the deflection for the corner of four rectangular areas of dimension $L/2 \times B/2$. Thus, from Equation 4.52

$$w(\text{center}) = 4\left[\frac{q(B/2)}{2E}\right](1 - v^2)I_{11} = \frac{qB}{E}(1 - v^2)I_{11} \qquad (4.58)$$

Example 4.8

Consider a flexible rectangular area measuring 3 m × 6 m ($B \times L$) on the ground surface. The flexible area is subjected to a loading $q = 100$ kN/m². A rock layer is located 6 m below the ground surface. Determine the deflection at the surface below the center of the loaded area. Use $E = 18,000$ kN/m² and $v = 0.3$.

Solution

$$w(\text{center}) = w(\text{center at } z = 0) - w(\text{center at } z = 6 \text{ m})$$

From Equation 4.58

$$w(\text{center at } z = 0) = \frac{qB}{E}(1 - v^2)I_{11}$$

$$m_1'' = \frac{L}{B} = \frac{6}{3} = 2$$

$$n_1'' = \frac{z}{B} = \frac{0}{3} = 0$$

From Table 4.14, for $m_1'' = 2$ and $n_1'' = 0$, the value of I_{11} is 1.532. Hence

$$w\,(\text{center at } z = 0) = \frac{(100)(3)}{18,000}(1-0.3^2)(1.532) = 0.0232\,\text{m} = 23.2\,\text{mm}$$

w(center at $z = 6$ m) = (4)[w(corner) at $z = 6$ m of a rectangular area measuring $B' \times L' = B/2 \times L/2 = 1.5$ m × 3 m

For this case

$$m_1'' = \frac{L'}{B'} = \frac{3\,\text{m}}{1.5\,\text{m}} = 2$$

$$n_1'' = \frac{z}{B'} = \frac{6}{1.5} = 4$$

From Table 4.14, $I_{11} = 0.580$; and from Table 4.15, $I_{12} = 0.138$. For one of the rectangular areas measuring $B' \times L'$, from Equation 4.52

$$w\,(\text{corner}) = \frac{qB'}{2E}(1-v^2)\left[I_{11} - \left(\frac{1-2v}{1-v}\right)I_{12}\right]$$

$$= \frac{(100(1.5)}{(2)(18,000)} \times (1-0.3^2)\left[0.58 - \left(\frac{1-2\times0.3}{1-0.3}\right)0.138\right]$$

$$= 0.0019\,\text{m} = 1.9\,\text{mm}$$

So, for the center of the rectangular area measuring $B \times L$
   w(center at $z = 6$ m) = (4)(1.9) = 7.6 mm
Hence
   w(center) = 23.2 − 7.6 = 15.6 mm

Example 4.9

The plan of a loaded flexible area is shown in Figure 4.21. If load is applied on the ground surface of a thick deposit of sand ($v = 0.25$), calculate the surface elastic settlement at A and B in terms of the modulus of elasticity of soil, E.

Solution

Settlement at A:
   For rectangular area (surface settlement) (from Equation 4.57):

$$S_{e(\text{corner})} = 2\left[\frac{q_1B'}{2E}[(1-v^2)I_{11}]\right]$$

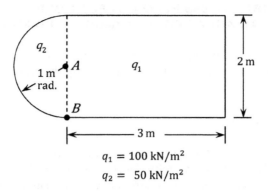

$$q_1 = 100 \text{ kN/m}^2$$
$$q_2 = 50 \text{ kN/m}^2$$

Figure 4.21 Plan of a loaded flexible area.

For half a circle (from Equation 4.46):

$$S_e = q_2 b \left(\frac{1 - v^2}{E} \right) \frac{I_8}{2}$$

Rectangle: $L = 3$ m; $B' = 1$ m; $m_1'' = 3/1 = 3$; $n_1'' = 0$. So $I_{11} = 1.783$.
Half circle: $z/b = 0$; $s/b = 0$; $I_8 = 2$.
So

$$
\begin{aligned}
\text{Total } S_e &= 2\left[\frac{q_1 B}{2E}(1 - v^2)I_{11} \right] + q_2 b \left(\frac{1 - v^2}{E} \right) \frac{I_8}{2} \\
&= 2\left[\frac{(100)(1)}{2E}(1 - 0.25^2)(1.783) \right] + (50)(1)\left(\frac{1 - 0.25^2}{E} \right) \frac{2}{2} \\
&= \frac{167.16}{E} + \frac{46.88}{E} = \frac{214.04}{E}
\end{aligned}
$$

Settlement at B:
 Rectangle: $L = 3$ m; $B = 1$ m; $m_1'' = 3/2 = 1.5$; $n_1'' = 0$. So $I_{11} = 1.358$.
 Half circle: $z/b = 0$; $s/b = 1$; $I_8 = 1.27$.

$$
\begin{aligned}
S_e &= \frac{q_1 B}{2E}(1 - v^2)I_{11} + q_2 b \left(\frac{1 - v^2}{E} \right) \frac{I_8}{2} \\
&= \frac{(100)(2)}{2E}(1 - 0.25^2)(1.358) + (50)(1)\left(\frac{1 - 0.25^2}{E} \right) \frac{1.27}{2} \\
&= \frac{127.3}{E} + \frac{29.77}{E} = \frac{157.07}{E}
\end{aligned}
$$

4.10 VERTICAL STRESS BELOW A FLEXIBLE CIRCULAR AREA WITH PARABOLIC LOADING

Figure 4.22 shows a flexible circular area subjected to parabolic type of loading. The loading pattern can be described by the relation

$$q' = q\left[1 - \left(\frac{r}{b}\right)^2\right] \tag{4.59}$$

Harr and Lovell (1963) studied this case and provided a solution for the variation of vertical stress (σ_z) with depth (z) in a nondimensional form, as shown in Figure 4.23. The vertical stress below the center line of the loaded area ($r/b = 0$) can be expressed as

$$\frac{\sigma_z}{q} = \left[\frac{1}{(z/b) + \sqrt{1 + (z/b)^2}}\right]^2 \left[1 + \frac{2(z/b)}{\sqrt{1 + (z/b)^2}}\right] \tag{4.60}$$

Schiffman (1963) evaluated the variation of radial stress (σ_r) with depth below the center line of the loaded area, which is a function of Poisson's ratio ν. Based on this analysis, the plot of σ_r/q vs. z/b for $\nu = 0$ is shown in Figure 4.24.

4.11 VERTICAL STRESS BELOW A FLEXIBLE CIRCULAR AREA WITH CONICAL LOADING

A flexible circular area with conical loading is shown in Figure 4.25. The variation of vertical stress, σ_z, below the center line ($r/b = 0$) of the area was obtained by Harr and Lovell (1963) and can be expressed as

$$\frac{\sigma_z}{q} = \left\{1 - \frac{1}{[(b/z)^2 + 1]^{0.5}}\right\} \tag{4.61}$$

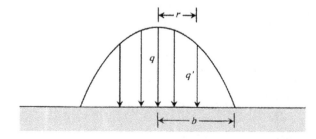

Figure 4.22 Vertical parabolic loading on a flexible circular area.

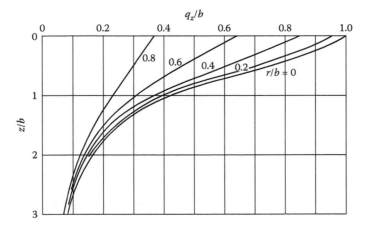

Figure 4.23 Variation of q_z/b with z/b and r/b. (Based on Harr, M. D. and C. W. Lovell, Jr., *Highway Res. Rec.*, 39, 68–77, 1963.)

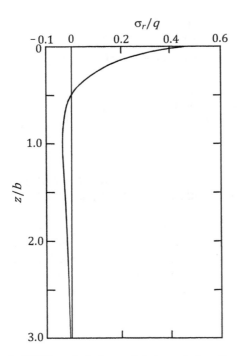

Figure 4.24 Schiffman's (1963) analysis for radial stress below the center of a flexible circular area with parabolic loading ($v = 0$).

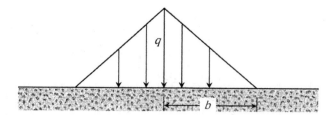

Figure 4.25 Vertical conical loading on a flexible circular area.

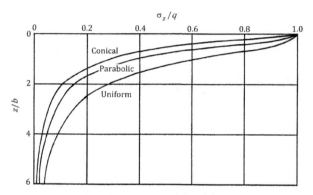

Figure 4.26 Variation of σ_z/q with z/b at $r/b = 0$ for circular, parabolic, and conical loading.

Figure 4.26 shows a comparison of the variation of σ_z/q with z/b at $r/b = 0$ for circular, parabolic, and conical loading (Equations 4.37, 4.60, and 4.61). Similar to Figure 4.24, Figure 4.27 shows the variation of σ_r with depth below the center line of the loaded area (Schiffman, 1963).

4.12 VERTICAL STRESS UNDER A UNIFORMLY LOADED FLEXIBLE ELLIPTICAL AREA

Deresiewicz (1960) developed the relationship for the variation of vertical stress along the center line of a uniformly loaded elliptical area. The plan of an elliptical area is shown in Figure 4.28, over which the magnitude of the uniformly distributed load per unit area is q. Figure 4.29 shows the variation of σ_z/q vs. z/a for various values of e. The term e is defined as

$$e = \left[1 - \left(\frac{a}{b}\right)^2\right] \tag{4.62}$$

Note that, for a circular area, $e = 0$.

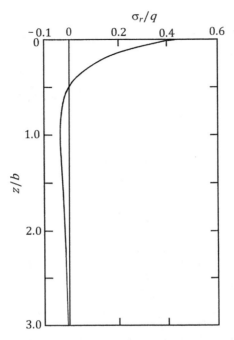

Figure 4.27 Schiffman's (1963) analysis for radial stress below the center of a flexible circular area with conical loading ($v = 0$).

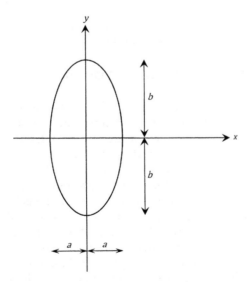

Figure 4.28 Uniformly loaded elliptical area.

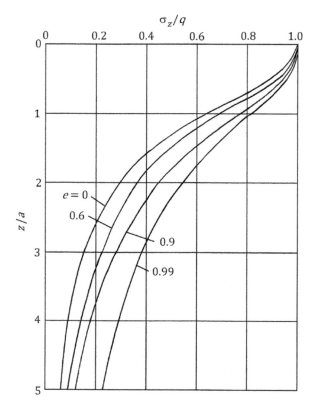

Figure 4.29 Variation of σ_z/q with z/b (Deresiewicz, 1966 analysis) along the center line of a uniformly loaded flexible elliptical area.

4.13 STRESSES IN A LAYERED MEDIUM

In the preceding sections, we discussed the stresses inside a homogeneous elastic medium due to various loading conditions. In actual cases of soil deposits, it is possible to encounter layered soils, each with a different modulus of elasticity. A case of practical importance is that of a stiff soil layer on top of a softer layer, as shown in Figure 4.30a. For a given loading condition, the effect of the stiff layer will be to reduce the stress concentration in the lower layer. Burmister (1943) worked on such problems involving two- and three-layer flexible systems. This was later developed by Fox (1948), Burmister (1958), Jones (1962), and Peattie (1962).

The effect of the reduction of stress concentration due to the presence of a stiff top layer is demonstrated in Figure 4.30b. Consider a flexible circular area of radius b subjected to a loading of q per unit area at the surface

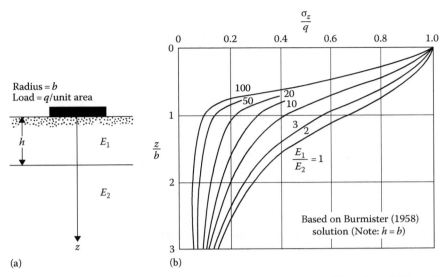

Figure 4.30 (a) Uniformly loaded circular area in a two-layered soil $E_1 > E_2$ and (b) vertical stress below the centerline of a uniformly loaded circular area.

of a two-layered system. E_1 and E_2 are the moduli of elasticity of the top and the bottom layer, respectively, with $E_1 > E_2$; and h is the thickness of the top layer. For $h = b$, the elasticity solution for the vertical stress σ_z at various depths below the center of the loaded area can be obtained from Figure 4.30b. The curves of σ_z/q against z/b for $E_1/E_2 = 1$ give the simple Boussinesq case, which is obtained by solving Equation 4.35. However, for $E_1/E_2 > 1$, the value of σ_z/q for a given z/b decreases with the increase of E_1/E_2. It must be pointed out that in obtaining these results it is assumed that there is *no slippage at the interface*.

The study of the stresses in a flexible layered system is of importance in highway pavement design.

4.14 VERTICAL STRESS AT THE INTERFACE OF A THREE-LAYER FLEXIBLE SYSTEM

Peattie (1962) prepared a number of graphs for determination of the vertical stress σ_z at the interfaces of three-layer systems (Figure 4.31) below the center of a uniformly loaded flexible circular area. These graphs are presented in Figures A.1 through A.32 (see the Appendix). In the determination of these stresses, it is assumed that Poisson's ratio for all layers is 0.5. The following parameters have been used in the graphs:

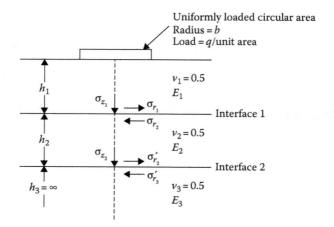

Figure 4.31 Uniformly loaded circular area on a three-layered medium.

$$K_1 = \frac{E_1}{E_2} \tag{4.63}$$

$$K_2 = \frac{E_2}{E_3} \tag{4.64}$$

$$A = \frac{b}{h_2} \tag{4.65}$$

$$H = \frac{h_1}{h_2} \tag{4.66}$$

For determination of the stresses σ_{z_1} and σ_{z_2} (vertical stresses at interfaces 1 and 2, respectively), we first obtain ZZ_1 and ZZ_2 from the graphs. The stresses can then be calculated from

$$\sigma_{z_1} = q(ZZ_1) \tag{4.67a}$$

and

$$\sigma_{z_2} = q(ZZ_2) \tag{4.67b}$$

Typical use of these graphs is shown in Example 4.6.

Example 4.10

A flexible circular area is subjected to a uniformly distributed load of 100 kN/m² as shown in Figure 4.32. Determine the vertical stress σ_{z_1} at the interface of the stiff and medium-stiff clay.

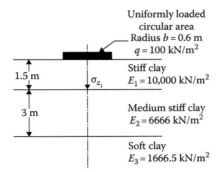

Figure 4.32 Flexible circular load on layered soil.

Solution

$$K_1 = \frac{E_1}{E_2} = \frac{10,000}{6,666} = 1.5$$

$$K_2 = \frac{E_2}{E_3} = \frac{6,666}{1666.5} = 4$$

$$A = \frac{b}{h_2} = \frac{0.6}{3} = 0.2$$

$$H = \frac{h_1}{h_2} = \frac{1.5}{3} = 0.5$$

From the figures given in the Appendix, we can prepare the following table:

	ZZ_1		
K_1	$K_2 = 0.2$	$K_2 = 2.0$	$K_2 = 20$
0.2	0.29	0.27	0.25
2.0	0.16	0.15	0.15
20.0	0.054	0.042	0.037

Based on the results of this table, a graph of ZZ_1 against K_2 for various values of K_1 is plotted (Figure 4.33). For this problem, $K_2 = 4$. So, the values of ZZ_1 for $K_2 = 4$ and $K_1 = 0.2$, 2.0, and 20.0 are obtained from Figure 4.17 and then plotted as in Figure 4.34. From this graph, $ZZ_1 = 0.16$ for $K_1 = 1.5$. Thus

$$\sigma_{z1} = 100(0.16) = 16\,\text{kN/m}^2$$

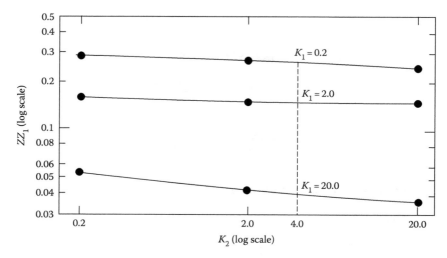

Figure 4.33 Plot of ZZ_1 vs. K_2.

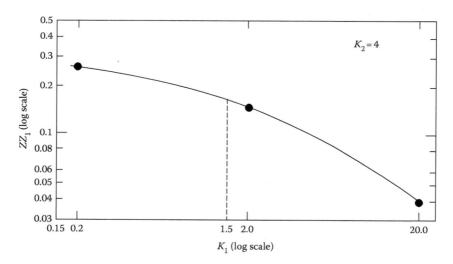

Figure 4.34 Plot of ZZ_1 vs. K_1.

4.15 VERTICAL STRESS IN WESTERGAARD MATERIAL DUE TO A VERTICAL POINT LOAD

Westergaard (1938) proposed a solution for the determination of the vertical stress due to a point load Q in an elastic solid medium in which there exist alternating layers with thin rigid reinforcements (Figure 4.35a). This type of assumption may be an idealization of a clay layer with thin

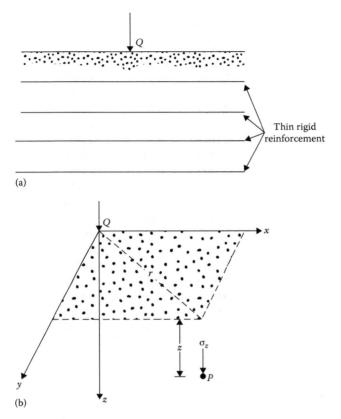

(a)

(b)

Figure 4.35 Westergaard's solution for vertical stress due to a point load. (Note: ν = Poisson's ratio of soil between the rigid layers.) (a) Westergaard type material; (b) Vertical stress at P due to a point load Q.

seams of sand. For such an assumption, the vertical stress increase at a point P (Figure 4.35b) can be given as

$$\sigma_z = \frac{Q\eta}{2\pi z^2}\left[\frac{1}{\eta^2 + (r/z)^2}\right]^{3/2} \tag{4.68}$$

where

$$\eta = \sqrt{\frac{1-2\nu}{2-2\nu}} \tag{4.69}$$

ν = Poisson's ratio of the solid between the rigid reinforcements

$$r = \sqrt{x^2 + y^2}$$

Table 4.16 Variation of I_{13} (Equation 4.71)

r/z	I_{13}		
	$v = 0$	$v = 0.2$	$v = 0.4$
0	0.3183	0.4244	0.9550
0.1	0.3090	0.4080	0.8750
0.2	0.2836	0.3646	0.6916
0.3	0.2483	0.3074	0.4997
0.4	0.2099	0.2491	0.3480
0.5	0.1733	0.1973	0.2416
0.6	0.1411	0.1547	0.1700
0.7	0.1143	0.1212	0.1221
0.8	0.0925	0.0953	0.0897
0.9	0.0751	0.0756	0.0673
1.0	0.0613	0.0605	0.0516
1.5	0.0247	0.0229	0.0173
2.0	0.0118	0.0107	0.0076
2.5	0.0064	0.0057	0.0040
3.0	0.0038	0.0034	0.0023
4.0	0.0017	0.0015	0.0010
5.0	0.0009	0.0008	0.0005

Equation 4.68 can be rewritten as

$$\sigma_z = \left(\frac{Q}{z^2} \right) I_{13} \tag{4.70}$$

where

$$I_{13} = \frac{1}{2\pi\eta^2} \left[\left(\frac{r}{\eta z} \right)^2 + 1 \right]^{-3/2} \tag{4.71}$$

Table 4.16 gives the variation of I_{13} with v.

In most practical problems of geotechnical engineering, Boussinesq's solution (Section 4.2) is preferred over Westergaard's solution.

4.16 SOLUTIONS FOR VERTICAL STRESS IN WESTERGAARD MATERIAL

The Westergaard material was explained in Section 4.15, in which the semi-infinite mass is assumed to be homogeneous, but reinforced internally so that no horizontal displacement can occur. Following are some solutions to obtain stress at a point due to surface loading on Westergaard material.

a. Vertical Stress (σ_z) due to a Line Load of Finite Length
Referring to Figure 4.5, the stress at P

$$\sigma_z = \frac{q}{z}\frac{\eta}{2\pi}\left[\frac{n_1}{m_1^2+\eta^2}\cdot\frac{1}{\left(m_1^2+n_1^2+\eta^2\right)^{0.5}}\right] \tag{4.72}$$

where

$$\eta = \sqrt{\frac{1-2v}{2-2v}}$$

$$m_1 = \frac{a}{z}$$

$$n_1 = \frac{L}{z}$$

b. Vertical Stress (σ_z) due to a Circularly Loaded Area
Referring to Figure 4.8, the vertical stress at P

$$\sigma_z = q\left\{1-\frac{\eta}{[\eta^2+(b/z)^2]^{0.5}}\right\} \tag{4.73}$$

Table 4.17 gives the variation of σ_z/q for $v = 0$.

Table 4.17 Variation of σ_z/q for $v = 0$
(Equation 4.73)

b/z	σ_z/q
0	0
0.1	0.0099
0.2	0.0378
0.3	0.0794
0.4	0.1296
0.5	0.1835
0.6	0.2375
0.7	0.2893
0.8	0.3377
0.9	0.3822
1.0	0.4227
2.0	0.6667
3.0	0.7706
4.0	0.8259
5.0	0.8599

c. Vertical Stress (σ_z) due to a Rectangularly Loaded Area
Referring to Figure 4.14, the vertical stress at P

$$\sigma_z = \frac{q}{2\pi}\left\{\cot^{-1}\left[\eta^2\left(\frac{1}{m^2}+\frac{1}{n^2}\right)+\eta^4\left(\frac{1}{m^2 n^2}\right)\right]^{0.5}\right\} \tag{4.74}$$

where

$$m = \frac{B}{z}$$

$$n = \frac{L}{z}$$

Figure 4.36 shows the variation of σ_z/q with m and n.

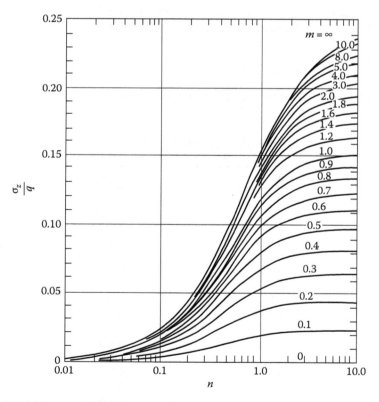

Figure 4.36 Variation of σ_z/q (Equation 4.74) with m and n. (*Note:* $\nu = 0$.)

4.17 DISTRIBUTION OF CONTACT STRESS OVER FOOTINGS

In calculation vertical stress, we generally assume that the foundation of a structure is flexible. In practice, this is not the case; no foundation is perfectly flexible, nor is it infinitely rigid. The actual nature of the distribution of contact stress will depend on the elastic properties of the foundation and the soil on which the foundation is resting.

Borowicka (1936, 1938) analyzed the problem of distribution of contact stress over uniformly loaded strip and circular rigid foundations resting on a semi-infinite elastic mass. The shearing stress at the base of the foundation was assumed to be zero. The analysis shows that the distribution of contact stress is dependent on a nondimensional factor K_r of the form

$$K_r = \frac{1}{6}\left(\frac{1-\nu_s^2}{1-\nu_f^2}\right)\left(\frac{E_f}{E_s}\right)\left(\frac{T}{b}\right)^3 \tag{4.75}$$

where
ν_s is the Poisson's ratio for soil
ν_f is the Poisson's ratio for foundation material
E_f, E_s are the Young's modulus of foundation material and soil, respectively

$$b = \begin{cases} \text{Half-width for strip foundation} \\ \text{Radius for circular foundation} \end{cases}$$

T is the thickness of foundation

Figure 4.37 shows the distribution of contact stress for a circular foundation. Note that $K_r = 0$ indicates a perfectly flexible foundation, and $K_r = \infty$ means a perfectly rigid foundation.

4.17.1 Foundations of clay

When a flexible foundation resting on a saturated clay ($\phi = 0$) is loaded with a uniformly distributed load (q/unit area), it will deform and take a bowl shape (Figure 4.38). Maximum deflection will be at the center; however, the contact stress over the footing will be uniform (q per unit area).

A rigid foundation resting on the same clay will show a uniform settlement (Figure 4.38). The contact stress distribution will take a form such as that shown in Figure 4.38, with only one exception: the stress at the edges of the footing cannot be infinity. Soil is not an infinitely elastic material; beyond a certain limiting stress [$q_{c(max)}$], plastic flow will begin.

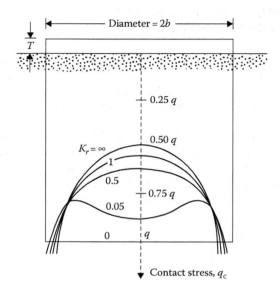

Figure 4.37 Contact stress over a rigid circular foundation resting on an elastic medium.

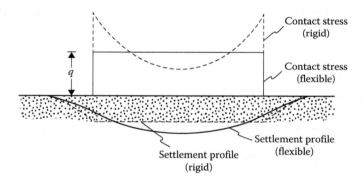

Figure 4.38 Contact pressure and settlement profiles for foundations on clay.

4.17.2 Foundations on sand

For a flexible foundation resting on a cohesionless soil, the distribution of contact pressure will be uniform (Figure 4.39). However, the edges of the foundation will undergo a larger settlement than the center. This occurs because the soil located at the edge of the foundation lacks lateral-confining pressure and hence possesses less strength. The lower strength of the soil at the edge of the foundation will result in larger settlement.

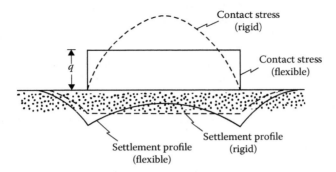

Figure 4.39 Contact pressure and settlement profiles for foundations on sand.

A rigid foundation resting on a sand layer will settle uniformly. The contact pressure on the foundation will increase from zero at the edge to a maximum at the center, as shown in Figure 4.39.

4.18 RELIABILITY OF STRESS CALCULATION USING THE THEORY OF ELASTICITY

Only a limited number of attempts have been made so far to compare theoretical results for stress distribution with the stresses observed under field conditions. The latter, of course, requires elaborate field instrumentation. However, from the results available at present, fairly good agreement is shown between theoretical considerations and field conditions, especially in the case of vertical stress. In any case, a variation of about 20%–30% between the theory and the field conditions may be expected.

REFERENCES

Ahlvin, R. G. and H. H. Ulery, Tabulated values for determining the complete pattern of stresses, strains and deflections beneath a uniform load on a homogeneous half space, *Bull. 342*, Highway Research Record, 1–13, 1962.

Borowicka, H., The distribution of pressure under a uniformly loaded elastic strip resting on elastic-isotropic ground, *2nd Congr. Int. Assoc. Bridge Struct. Eng.*, Berlin, Germany, Final Report, 8(3), 1938.

Borowicka, H., Influence of rigidity of a circular foundation slab on the distribution of pressures over the contact surface, *Proc. 1st Int. Conf. Soil Mech. Found. Eng.*, 2, 144–149, 1936.

Boussinesq, J., *Application des Potentials a L'Etude de L'Equilibre et due Mouvement des Solides Elastiques,* Gauthier-Villars, Paris, France, 1883.

Burmister, D. M., Evaluation of pavement systems of the WASHO road testing layered system methods, *Bull. 177*, Highway Research Board, 1958.

Burmister, D. M., The theory of stresses and displacements in layer systems and application to design of airport runways, *Proc. Highway Res. Board*, 23, 126, 1943.

Deresiewicz, H., The half-space under pressure distributed over an elliptical portion of its plane boundary, *J. Appl. Mech.*, 27(1), 111–119, 1960.

Fox, L., Computation of traffic stresses in a simple road structure, *Proc. 2nd Int. Conf. Soil Mech. Found. Eng.*, 2, 236–246, 1948.

Giroud, J. P., Stresses under linearly loaded rectangular area, *J. Soil Mech. Found. Div.*, ASCE, 98(1), 263–268, 1970.

Harr, M. E., *Foundations of Theoretical Soil Mechanics*, McGraw-Hill, New York, 1966.

Harr, M.D. and C.W. Lovell, Jr., Vertical stresses under certain axisymmetrical loadings, *Highway Res. Rec.*, 39, 68–77, 1963.

Holl, D. L., Stress transmissions on earth, *Proc. Highway Res. Board*, 20, 709–722, 1940.

Jones, A., Tables of stresses in three-layer elastic systems, *Bull. 342*, Highway Research Board, 176–214, 1962.

Newmark, N. M., Simplified computation of vertical pressure in elastic foundations, University of Illinois Engineering Experiment Station, Circular 24, 1935.

Peattie, K. R., Stresses and strain factors for three-layer systems, *Bull. 342*, Highway Research Board, 215–253, 1962.

Schiffman, R. L., Discussions, *Highway Res. Rec.*, 39, 78–81, 1963.

Westergaard, H. M., A problem of elasticity suggested by a problem in soil mechanics: Soft material reinforced by numerous strong horizontal sheets, in *Contributions to the Mechanics of Solids*, Dedicated to S. Timoshenko by His Friends on the Occasion of His Sixtieth Birthday Anniversary, The Macmillan Company, New York, 268–277, 1938.

Chapter 5

Pore water pressure due to undrained loading

5.1 INTRODUCTION

In 1925, Terzaghi suggested the principles of *effective stress* for a saturated soil, according to which the *total vertical stress* σ at a point O (Figure 5.1) can be given as

$$\sigma = \sigma' + u \tag{5.1}$$

where

$$\sigma = h_1\gamma + h_2\gamma_{sat} \tag{5.2}$$

σ' is the effective stress

$$u = \text{pore water pressure} = h_2\gamma_w \tag{5.3}$$

γ_w is the unit weight of water

Combining Equations 5.1 through 5.3 gives

$$\sigma' = \sigma - u = (h_1\gamma + h_2\gamma_{sat}) - h_2\gamma_w = h_1\gamma + h_2\gamma' \tag{5.4}$$

where γ' is the effective unit weight of soil = $\gamma_{sat} - \gamma_w$.

In general, if the normal total stresses at a point in a soil mass are σ_1, σ_2, and σ_3 (Figure 5.2), the effective stresses can be given as follows:

Direction 1: $\sigma_1' = \sigma_1 - u$

Direction 2: $\sigma_2' = \sigma_2 - u$

Direction 3: $\sigma_3' = \sigma_3 - u$

where
  σ_1', σ_2', and σ_3' are the effective stresses
  u is the pore water pressure, $h\gamma_w$

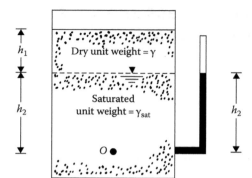

Figure 5.1 Definition of effective stress.

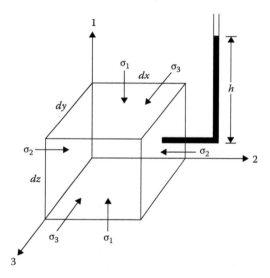

Figure 5.2 Normal total stresses in a soil mass.

A knowledge of the increase of pore water pressure in soils due to various loading conditions without drainage is important in both theoretical and applied soil mechanics. If a load is applied very slowly on a soil such that sufficient time is allowed for pore water to drain out, there will be practically no increase of pore water pressure. However, when a soil is subjected to rapid loading and if the coefficient of permeability is small (e.g., as in the case of clay), there will be insufficient time for drainage of pore water. This will lead to an increase of the excess hydrostatic pressure. In this chapter, mathematical formulations for the excess pore water pressure for various types of undrained loading will be developed.

5.2 PORE WATER PRESSURE DEVELOPED DUE TO ISOTROPIC STRESS APPLICATION

Figure 5.3 shows an isotropic *saturated* soil element subjected to an isotropic stress increase of magnitude $\Delta\sigma$. If drainage from the soil is not allowed, the pore water pressure will increase by Δu.

The increase of pore water pressure will cause a change in volume of the pore fluid by an amount ΔV_p. This can be expressed as

$$\Delta V_p = n V_o C_p \Delta u \tag{5.5}$$

where
    n is the porosity
    C_p is the compressibility of pore water
    V_o is the original volume of soil element

The effective stress increase in all directions of the element is $\Delta\sigma' = \Delta\sigma - \Delta u$. The change in volume of the soil skeleton due to the effective stress increase can be given by

$$\Delta V = 3 C_c V_o \Delta\sigma' = 3 C_c V_o (\Delta\sigma - \Delta u) \tag{5.6}$$

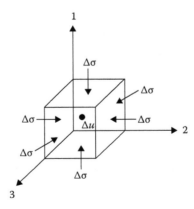

Figure 5.3 Soil element under isotropic stress application.

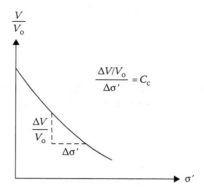

Figure 5.4 Definition of C_c: volume change due to uniaxial stress application with zero excess pore water pressure. (*Note:* V is the volume of the soil element at any given value of σ'.)

In Equation 5.6, C_c is the compressibility of the soil skeleton obtained from laboratory compression results under uniaxial loading with zero excess pore water pressure, as shown in Figure 5.4. It should be noted that compression, that is, a reduction of volume, is taken as positive.

Since the change in volume of the pore fluid, ΔV_p, is equal to the change in the volume of the soil skeleton, ΔV, we obtain from Equations 5.5 and 5.6

$$nV_o C_p \Delta u = 3C_c V_o (\Delta\sigma - \Delta u)$$

and hence

$$\frac{\Delta u}{\Delta\sigma} = B = \frac{1}{1 + n(C_p/3C_c)} \tag{5.7}$$

where B is the pore pressure parameter (Skempton, 1954).

If the pore fluid is water,

$$C_p = C_w = \text{compressibility of water}$$

and

$$3C_c = C_{sk} = \frac{3(1-v)}{E}$$

where E and v are the Young's modulus and Poisson's ratio with respect to changes in effective stress. Hence

$$B = \frac{1}{1 + n(C_w/C_{sk})} \tag{5.8}$$

Table 5.1 Soils considered by Black and Lee (1973) for evaluation of B

Soil type	Description	Void ratio	C_{sk}	B at 100% saturation
Soft soil	Normally consolidated clay	≈2	≈0.145 × 10^{-2} m²/kN	0.9998
Medium soil	Compacted silts and clays and lightly overconsolidated clay	≈0.6	≈0.145 × 10^{-3} m²/kN	0.9988
Stiff soil	Overconsolidated stiff clays, average sand of most densities	≈0.6	≈0.145 × 10^{-4} m²/kN	0.9877
Very stiff soil	Dense sands and stiff clays, particularly at high confining pressure	≈0.4	≈0.145 × 10^{-5} m²/kN	0.9130

5.3 PORE WATER PRESSURE PARAMETER B

Black and Lee (1973) provided the theoretical values of B for various types of soil at complete or near complete saturation. A summary of the soil types and their parameters and the B values at saturation that were considered by Black and Lee is given in Table 5.1.

Figure 5.5 shows the theoretical variation of B parameters for the soils described in Table 5.1 with the degree of saturation. It is obvious from this figure that, for stiffer soils, the B value rapidly decreases with the degree of saturation. This is consistent with the experimental values for several soils shown in Figure 5.6.

As noted in Table 5.1, the B value is also dependent on the effective isotropic consolidation stress (σ') of the soil. An example of such behavior in saturated varved Fort William clay as reported by Eigenbrod and Burak (1990) is shown in Figure 5.7. The decrease in the B value with an increase in σ' is primarily due to the increase in skeletal stiffness (i.e., C_{sk}).

Hence, in general, for soft soils at saturation or near saturation, $B \approx 1$.

5.4 PORE WATER PRESSURE DUE TO UNIAXIAL LOADING

A saturated soil element under a uniaxial stress increment is shown in Figure 5.8. Let the increase of pore water pressure be equal to Δu. As explained in the previous section, the change in the volume of the pore water is

$$\Delta V_p = nV_o C_p \Delta u$$

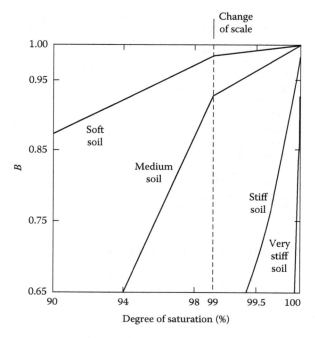

Figure 5.5 Theoretical variation of B with degree of saturation for soils described in Table 5.1. (*Note*: Back pressure = 207 kN/m², Δσ = 138 kN/m².)

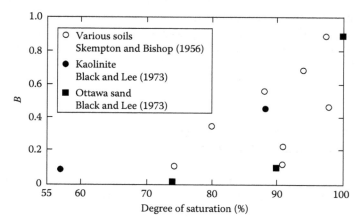

Figure 5.6 Variation of B with degree of saturation.

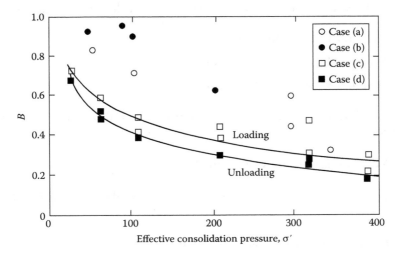

Figure 5.7 Dependence of B values on the level of isotropic consolidation stress (varved clay) for (a) regular triaxial specimens before shearing, (b) regular triaxial specimens after shearing, (c) special series of B tests on one single specimen in loading, and (d) special series of B tests on one single specimen in unloading. [After Eigenbrod, K. D. and Burak, J. P., *Geotech. Test. J.*, 13(4), 370, 1990.]

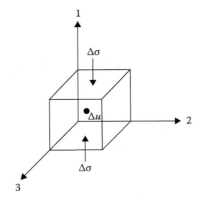

Figure 5.8 Saturated soil element under uniaxial stress increment.

The increases of the effective stresses on the soil element in Figure 5.8 are

Direction 1: $\Delta\sigma' = \Delta\sigma - \Delta u$

Direction 2: $\Delta\sigma' = 0 - \Delta u = -\Delta u$

Direction 3: $\Delta\sigma' = 0 - \Delta u = -\Delta u$

This will result in a change in the volume of the soil skeleton, which may be written as

$$\Delta V = C_c V_o (\Delta\sigma - \Delta u) + C_e V_o (-\Delta u) + C_e V_o (-\Delta u) \qquad (5.9)$$

where C_e is the coefficient of the volume expansibility (Figure 5.9). Since $\Delta V_p = \Delta V$

$$n V_o C_p \Delta u = C_c V_o (\Delta\sigma - \Delta u) - 2C_e V_o \Delta u$$

or

$$\frac{\Delta u}{\Delta\sigma} = A = \frac{C_c}{n C_p + C_c + 2C_e} \qquad (5.10)$$

where A is the pore pressure parameter (Skempton, 1954).

If we assume that the soil element is elastic, then $C_c = C_e$, or

$$A = \frac{1}{n(C_p / C_c) + 3} \qquad (5.11)$$

Again, as pointed out previously, C_p is much smaller than C_e. So $C_p/C_c \approx 0$, which gives $A = 1/3$. However, in reality, this is not the case,

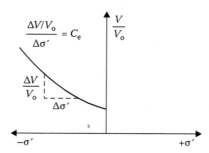

Figure 5.9 Definition of C_e: coefficient of volume expansion under uniaxial loading.

that is, soil is not a perfectly elastic material, and the actual value of A varies widely.

The magnitude of A for a given soil is not a constant and depends on the stress level. If a consolidated drained triaxial test is conducted on a saturated clay soil, the general nature of variation of $\Delta\sigma$, Δu, and $A = \Delta u / \Delta \sigma$ with axial strain will be as shown in Figure 5.10. For highly overconsolidated clay soils, the magnitude of A at failure (i.e., A_f) may be negative. Table 5.2 gives the typical values of A at failure ($=A_f$) for some normally consolidated clay soils. Figure 5.11 shows the variation of A_f with overconsolidation ratio for Weald clay. Table 5.3 gives the typical range of A values at failure for various soils.

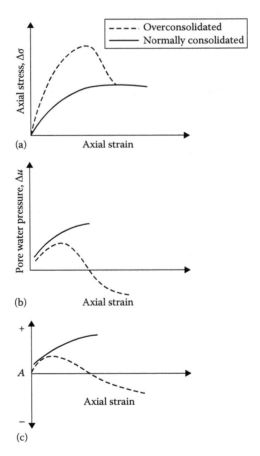

Figure 5.10 Variation of $\Delta\sigma$, Δu, and A for a consolidated drained triaxial test in clay: (a) plot of $\Delta\sigma$ vs. axial strain; (b) plot of Δu vs. axial strain; (c) plot of A vs. axial strain.

Table 5.2 Values of A_f for normally consolidated clays

Clay	Type	Liquid limit	Plasticity index	Sensitivity	A_f
Natural soils					
Toyen	Marine	47	25	8	1.50
		47	25	8	1.48
Drammen	Marine	36	16	4	1.2
		36	16	4	2.4
Saco River	Marine	46	17	10	0.95
Boston	Marine	—	—	—	0.85
Bersimis	Estuarine	39	18	6	0.63
Chew Stoke	Alluvial	28	10	—	0.59
Kapuskasing	Lacustrine	39	23	4	0.46
Decomposed Talus	Residual	50	18	1	0.29
St. Catherines	Till (?)	49	28	3	0.26
Remolded soils					
London	Marine	78	52	1	0.97
Weald	Marine	43	25	1	0.95
Beauharnois	Till (?)	44	24	1	0.73
Boston	Marine	48	24	1	0.69
Beauharnois	Estuarine	70	42	1	0.65
Bersimis	Estuarine	33	13	1	0.38

Source: After Kenney, T. C., *J. Soil Mech. Found. Eng. Div.*, 85(SM3), 67, 1959.

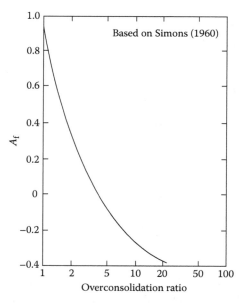

Figure 5.11 Variation of A_f with overconsolidation ratio for Weald clay.

Table 5.3 Typical values of A at failure

Type of soil	A
Clay with high sensitivity	$\frac{3}{4} - 1\frac{1}{2}$
Normally consolidated clay	$\frac{1}{2} - 1$
Overconsolidated clay	$-\frac{1}{2} - 0$
Compacted sandy clay	$\frac{1}{2} - \frac{3}{4}$

5.5 DIRECTIONAL VARIATION OF A_f

Owing to the nature of deposition of cohesive soils and subsequent consolidation, clay particles tend to become oriented perpendicular to the direction of the major principal stress. Parallel orientation of clay particles could cause the strength of clay and thus A_f to vary with direction. Kurukulasuriya et al. (1999) conducted undrained triaxial tests on kaolin clay specimens obtained at various inclinations i as shown in Figure 5.12. Figure 5.13 shows the directional variation of A_f with overconsolidation ratio. It can be seen from this figure that A_f is maximum between $\alpha = 30° - 60°$.

5.6 PORE WATER PRESSURE UNDER TRIAXIAL TEST CONDITIONS

A typical stress application on a soil element under triaxial test conditions is shown in Figure 5.14a ($\Delta\sigma_1 > \Delta\sigma_3$). Δu is the increase in the pore water pressure without drainage. To develop a relation between Δu, $\Delta\sigma_1$, and $\Delta\sigma_3$,

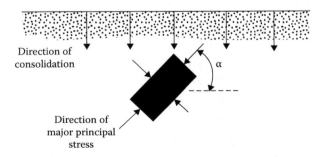

Direction of consolidation

α

Direction of major principal stress

Figure 5.12 Directional variation of major principal stress application.

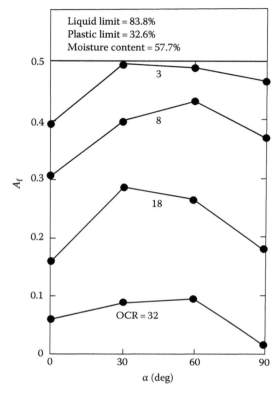

Figure 5.13 Variation of A_f with α and overconsolidation ratio (OCR) for kaolin clay based on the triaxial results of Kurukulasuriya et al. *Soils Found.*, 39(1), 21–29, 1999.

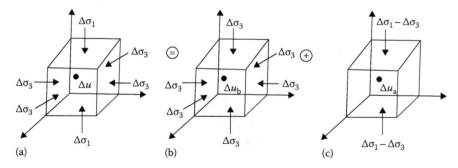

Figure 5.14 Excess pore water pressure under undrained triaxial test conditions: (a) triaxial test condition; (b) application of isotropic stress $\Delta\sigma_3$; (c) application of axial stress $\Delta\sigma_1 - \Delta\sigma_3$.

we can consider that the stress conditions shown in Figure 5.14a are the sum of the stress conditions shown in Figure 5.14b and c.

For the isotropic stress $\Delta\sigma_3$ as applied in Figure 5.14b,

$$\Delta u_b = B\Delta\sigma_3 \tag{5.12}$$

(from Equation 5.7), and for a uniaxial stress $\Delta\sigma_1 - \Delta\sigma_3$ as applied in Figure 5.14c,

$$\Delta u_a = A(\Delta\sigma_1 - \Delta\sigma_3) \tag{5.13}$$

(from Equation 5.10). Now,

$$\Delta u = \Delta u_b + \Delta u_a = B\Delta\sigma_3 + A(\Delta\sigma_1 - \Delta\sigma_3) \tag{5.14}$$

For saturated soil, if $B = 1$; then

$$\Delta u = \Delta\sigma_3 + A(\Delta\sigma_1 - \Delta\sigma_3) \tag{5.15}$$

5.7 HENKEL'S MODIFICATION OF PORE WATER PRESSURE EQUATION

In several practical considerations in soil mechanics, the intermediate and minor principal stresses are not the same. To take the intermediate principal stress into consideration (Figure 5.15), Henkel (1960) suggested a modification of Equation 5.15:

$$\Delta u = \frac{\Delta\sigma_1 + \Delta\sigma_2 + \Delta\sigma_3}{3} + a\sqrt{(\Delta\sigma_1 - \Delta\sigma_2)^2 + (\Delta\sigma_2 - \Delta\sigma_3)^2 + (\Delta\sigma_3 - \Delta\sigma_1)^2} \tag{5.16}$$

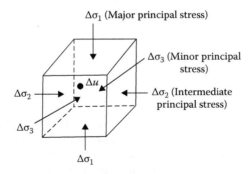

Figure 5.15 Saturated soil element with major, intermediate, and minor principal stresses.

or

$$\Delta u = \Delta\sigma_{oct} + 3a\Delta\tau_{oct} \tag{5.17}$$

where
 a is Henkel's pore pressure parameter
 $\Delta\sigma_{oct}$ and $\Delta\tau_{oct}$ are the increases in the octahedral normal and shear stresses, respectively

In triaxial compression tests, $\Delta\sigma_2 = \Delta\sigma_3$. For that condition,

$$\Delta u = \frac{\Delta\sigma_1 + 2\Delta\sigma_3}{3} + a\sqrt{2}(\Delta\sigma_1 - \Delta\sigma_3) \tag{5.18}$$

For uniaxial tests as in Figure 5.14c, we can substitute $\Delta\sigma_1 - \Delta\sigma_3$ for $\Delta\sigma_1$ and zero for $\Delta\sigma_2$ and $\Delta\sigma_3$ in Equation 5.16, which will yield

$$\Delta u = \frac{\Delta\sigma_1 - \Delta\sigma_3}{3} + a\sqrt{2}(\Delta\sigma_1 - \Delta\sigma_3)$$

or

$$\Delta u = \left(\frac{1}{3} + a\sqrt{2}\right)(\Delta\sigma_1 - \Delta\sigma_3) \tag{5.19}$$

A comparison of Equations 5.13 and 5.19 gives

$$A = \left(\frac{1}{3} + a\sqrt{2}\right)$$

or

$$a = \frac{1}{\sqrt{2}}\left(A - \frac{1}{3}\right) \tag{5.20}$$

The usefulness of this more fundamental definition of pore water pressure is that it enables us to predict the excess pore water pressure associated with loading conditions such as plane strain. This can be illustrated by deriving an expression for the excess pore water pressure developed in a saturated soil (undrained condition) below the centerline of a flexible strip loading of

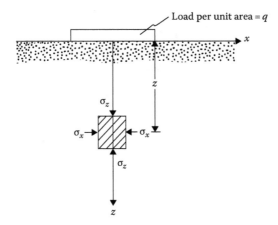

Figure 5.16 Estimation of excess pore water pressure in saturated soil below the center-line of a flexible strip loading (undrained condition).

uniform intensity, q (Figure 5.16). The expressions for σ_x, σ_y, and σ_z for such loading are given in Chapter 3. Note that $\sigma_z > \sigma_y > \sigma_x$, and $\sigma_y = \upsilon(\sigma_x + \sigma_z)$. Substituting σ_z, σ_y, and σ_x for σ_1, σ_2, and σ_3 in Equation 5.16 yields

$$\Delta u = \frac{\sigma_z + \upsilon(\sigma_x + \sigma_z) + \sigma_x}{3} + \frac{1}{\sqrt{2}}\left(A - \frac{1}{3}\right)$$

$$\sqrt{[\sigma_z - \upsilon(\sigma_z + \sigma_x)]^2 + [\upsilon(\sigma_z + \sigma_x) - \sigma_x]^2 + (\sigma_x - \sigma_z)^2}$$

For $\upsilon = 0.5$

$$\Delta u = \sigma_x + \left[\frac{\sqrt{3}}{2}\left(A - \frac{1}{3}\right) + \frac{1}{2}\right](\sigma_z - \sigma_x) \tag{5.21}$$

If a representative value of A can be determined from standard triaxial tests, Δu can be estimated.

Example 5.1

Figure 5.17 shows an inclined line load. Calculate the increase in pore water pressure at M immediately after the application of the load for the following cases with $q = 60$ kN/m and $\alpha = 30°$.

  a. $z = 10$ m, $x = 0$, $v = 0.5$, $A = 0.45$
  b. $z = 10$ m, $x = 2$ m, $v = 0.45$, $A = 0.6$

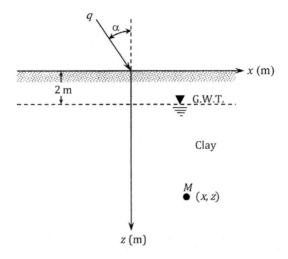

Figure 5.17 Inclined line load, *q*.

Solution

Part a:

Vertical component of q: $q_v = q \cos 30 = (60)(\cos 30) = 51.96$ kN/m
Horizontal component of q: $q_H = q \sin 30 = (60)(\sin 30) = 30$ kN/m
From Equations 3.2 and 3.19,

$$\Delta\sigma_r = \frac{2q_v}{\pi r}\cos\theta + \frac{2q_H}{\pi r}\sin\theta$$

Note: $\theta = 0$

$$\Delta\sigma_r = \frac{(2)(51.96)(\cos 0)}{\pi(10)} = 3.31 \text{ kN/m}^2$$

Equations 3.3 and 3.20:

$$\Delta\sigma_\theta = 0$$

$$\Delta\sigma_y = v(\Delta\sigma_r + \Delta\sigma_\theta) = (0.5)(3.31 + 0) = 1.67 \text{ kN/m}^2$$

Equation 5.20:

$$a = \frac{1}{\sqrt{2}}\left(A - \frac{1}{3}\right) = \frac{1}{\sqrt{2}}\left(0.45 - \frac{1}{3}\right) = 0.0825$$

From Equation 3.16:

$$\Delta u = \frac{3.31 + 1.67 + 0}{3} + 0.0825\left[(3.31-1.67)^2 + (1.67-0)^2 + (0-3.31)^2\right]^{0.5}$$

$$= 1.99 \text{ kN/m}^2$$

Part b:

$x = 2$ m, $z = 10$ m,

$\theta = \tan^{-1}(2/10) = 11.31°$

$r = \sqrt{x^2 + z^2} = \sqrt{(2)^2 + (10)^2} = 10.2$ m

$\cos\theta = 0.98; \sin\theta = 0.196$

From Equations 3.2 and 3.19:

$$\Delta\sigma_r = \frac{(2)(51.96)}{(\pi)(10.2)}(0.98) + \frac{(2)(30)}{(\pi)(10.2)}(0.196) = 3.55 \text{ kN/m}^2$$

From Equations 3.3 and 3.20:

$$\Delta\sigma_\theta = 0$$

$$\Delta\sigma_y = v(\Delta\sigma_r + \Delta\sigma_\theta) = (0.45)(3.55+0) = 1.6 \text{ kN/m}^2$$

Equation 5.20:

$$a = \frac{1}{\sqrt{2}}\left(A - \frac{1}{3}\right) = \frac{1}{\sqrt{2}}\left(0.6 - \frac{1}{3}\right) = 0.189$$

$$\Delta u = \frac{3.55 + 1.6 + 0}{3} + (0.189)\left[(3.55-1.6)^2 + (1.6-0)^2 + (0-3.55)^2\right]^{0.5}$$

$$= 2.54 \text{ kN/m}^2$$

Example 5.2

A uniform vertical load of 145 kN/m² is applied instantaneously over a very long strip, as shown in Figure 5.18. Estimate the excess pore water pressure that will be developed due to the loading at A and B. Assume that $v = 0.45$ and that the representative value of the pore water pressure parameter A determined from standard triaxial tests for such loading is 0.6.

Solution

The values of σ_x, σ_z, and τ_{xz} at A and B can be determined from Tables 3.5 through 3.7.

- At A: $x/b = 0$, $z/b = 2/2 = 1$, and hence
  1. $\sigma_z/q = 0.818$, so $\sigma_z = 0.818 \times 145 = 118.6$ kN/m²

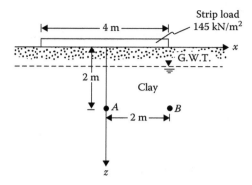

Figure 5.18 Uniform vertical strip load on ground surface.

$\quad$ 2. $\sigma_x/q = 0.182$, so $\sigma_x = 26.39$ kN/m²
$\quad$ 3. $\tau_{xz}/q = 0$, so $\tau_{xz} = 0$

Note that in this case σ_z and σ_x are the major (σ_1) and minor (σ_3) principal stresses, respectively.

This is a plane strain case. So, the intermediate principal stress is

$$\sigma_2 = v(\sigma_1 + \sigma_3) = 0.45(118.6 + 26.39) = 65.25 \text{ kN/m}^2$$

From Equation 5.20

$$a = \frac{1}{\sqrt{2}}\left(A - \frac{1}{3}\right) = \frac{1}{\sqrt{2}}\left(0.6 - \frac{1}{3}\right) = 0.189$$

So

$$\Delta u = \frac{\sigma_1 + \sigma_2 + \sigma_3}{3} + a\sqrt{(\sigma_1 - \sigma_2)^2 + (\sigma_2 - \sigma_3)^2 + (\sigma_3 - \sigma_1)^2}$$

$$= \frac{118.6 + 65.25 + 26.39}{3}$$

$$+ 0.189\sqrt{(118.6 - 65.25)^2 + (65.25 - 26.39)^2 + (26.39 - 118.6)^2}$$

$$= 91.51 \text{ kN/m}^2$$

- At *B*: $x/b = 2/2 = 1$, $z/b = 2/2 = 1$, and hence
  1. $\sigma_z/q = 0.480$, so $\sigma_z = 0.480 \times 145 = 69.6$ kN/m²
  2. $\sigma_x/q = 0.2250$, so $\sigma_x = 0.2250 \times 145 = 32.63$ kN/m²
  3. $\tau_{xz}/q = 0.255$, so $\tau_{xz} = 0.255 \times 145 = 36.98$ kN/m²

Calculation of the major and minor principal stresses is as follows:

$$\sigma_1, \sigma_3 = \frac{\sigma_z + \sigma_x}{2} \pm \sqrt{\left(\frac{\sigma_z - \sigma_x}{2}\right)^2 + \tau_{xz}^2}$$

$$= \frac{69.6 + 32.63}{2} \pm \sqrt{\left(\frac{69.6 - 32.63}{2}\right)^2 + 36.98^2}$$

Hence

$$\sigma_1 = 92.46 \ \text{kN/m}^2 \qquad \sigma_3 = 9.78 \ \text{kN/m}^2$$

$$\sigma_2 = 0.45(92.46 + 9.78) = 46 \ \text{kN/m}^2$$

$$\Delta u = \frac{92.46 + 9.78 + 46}{3}$$

$$+ 0.189\sqrt{(92.46 - 46)^2 + (46 - 9.78)^2 + (9.78 - 92.46)^2}$$

$$= 68.6 \ \text{kN/m}^2$$

Example 5.3

A surcharge of 195 kN/m^2 was applied over a circular area of diameter 3 m, as shown in Figure 5.19. Estimate the height of water h that a piezometer would show immediately after the application of the surcharge. Assume that $A \approx 0.65$ and $v = 0.5$.

Solution

From Chapter 4 (Equations 4.38 to 4.41):

$$\Delta\sigma_z = q(A' + B')$$

$$\Delta\sigma_r = q[2vA' + C + (1 - 2v)F]$$

$$\Delta\sigma_\theta = q[2vA' - D + (1 - 2v)E]$$

$$\Delta\tau_{rz} = \tau_{zr} = qG$$

$q = 195 \ \text{kN/m}^2; \quad s/b = 1.5/1.5 = 1; \quad z/b = 3/1.5 = 2$

From Tables 4.4 through 4.10:

$A' = 0.08219 \qquad C = -0.04144 \qquad E = 0.04675 \qquad G = 0.07738$
$B' = 0.11331 \qquad D = 0.07187 \qquad F = 0.03593$

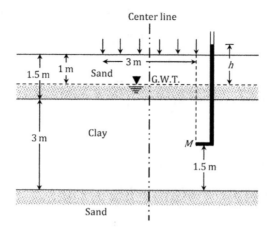

Figure 5.19 Application of surcharge on a circular area.

Substituting the above values in Equations 4.38 to 4.41:

$\Delta\sigma_z = 38.22$ kN/m²; $\Delta\sigma_r = 8.04$ kN/m²; $\Delta\sigma_\theta = 2.11$ kN/m²; $\Delta\tau_{rz} = 15.09$ kN/m²

$$\Delta\sigma_1,\ \Delta\sigma_3 = \frac{\Delta\sigma_z + \Delta\sigma_r}{2} \pm \sqrt{\left(\frac{\Delta\sigma_z - \Delta\sigma_r}{2}\right)^2 + \Delta\tau_{rz}^2}$$

$\Delta\sigma_1 = 44.47$ kN/m²; $\Delta\sigma_3 = 1.79$ kN/m²

$\Delta\sigma_2 = \Delta\sigma_\theta = 2.11$ kN/m²

From Equation 4.20:

$$a = \frac{1}{\sqrt{2}}\left(0.65 - \frac{1}{3}\right) = 0.224$$

Equation 5.16:

$$\Delta u = \frac{\Delta\sigma_1 + \Delta\sigma_2 + \Delta\sigma_3}{3} + a\sqrt{(\Delta\sigma_1 - \Delta\sigma_2)^2 + (\Delta\sigma_2 - \Delta\sigma_3)^2 + (\Delta\sigma_3 - \Delta\sigma_1)^2}$$

$$= \frac{44.47 + 2.11 + 1.79}{3} + (0.224) \times \left[\begin{array}{l}(44.47 - 2.11)^2 + (2.11 - 1.79)^2 \\ + (1.79 - 44.47)^2\end{array}\right]^{0.5}$$

$$= 29.59 \text{ kN/m}^2$$

$$h = \frac{\Delta u}{\gamma_w} = \frac{29.59}{9.81} = 3.02 \text{ m}$$

5.8 PORE WATER PRESSURE DUE TO ONE-DIMENSIONAL STRAIN LOADING (OEDOMETER TEST)

In Section 5.4, the development of pore water pressure due to uniaxial loading (Figure 5.8) is discussed. In that case, the soil specimen was allowed to undergo axial and lateral strains. However, in oedometer tests the soil specimens are confined laterally, thereby allowing only one directional strain, that is, strain in the direction of load application. For such a case, referring to Figure 5.8,

$$\Delta V_p = nV_o C_p \Delta u$$

and

$$\Delta V = C_c V_o (\Delta \sigma - \Delta u)$$

However, $\Delta V_p = \Delta V$. So

$$nV_o C_p \Delta u = C_c V_o (\Delta \sigma - \Delta u)$$

or

$$\frac{\Delta u}{\Delta \sigma} = C = \frac{1}{1 + n(C_p/C_c)} \qquad (5.22)$$

If $C_p < C_c$, the ratio $C_p/C_c \approx 0$; hence $C \approx 1$. Lambe and Whitman (1969) reported the following C values:

Vicksburg buckshot clay slurry	0.99983
Lagunillas soft clay	0.99957
Lagunillas sandy clay	0.99718

Veyera et al. (1992) reported the C values in *reloading* for two poorly graded sands (i.e., Monterrey no. 0/30 sand and Enewetak coral sand) at various relative densities of compaction (D_r). In conducting the tests, the specimens were first consolidated by application of an initial stress (σ_c'), and then the stress was reduced by 69 kN/m^2. Following that, under undrained conditions, the stress was increased by 69 kN/m^2 in increments of 6.9 kN/m^2. The results of those tests for Monterey no. 0/30 sand are given in Table 5.4.

From Table 5.4, it can be seen that the magnitude of the C value can decrease well below 1.0, depending on the soil stiffness. An increase in the initial relative density of compaction as well as an increase in the effective confining pressure does increase the soil stiffness.

Table 5.4 C values in reloading for
Monterrey no. 0/30 sand

Relative density D_r (%)	Effective confining pressure σ'_c (kN/m²)	C
6	86	1.00
6	172	0.85
6	345	0.70
27	86	1.00
27	172	0.83
27	345	0.69
27	690	0.56
46	86	1.00
46	172	0.81
46	345	0.66
46	690	0.55
65	86	1.00
65	172	0.79
65	345	0.62
65	690	0.53
85	86	1.00
85	172	0.74
85	345	0.61
85	690	0.51

Source: Compiled from the results of Veyera, G. E. et al., Geotech. Test. J., 15(3), 223, 1992.

REFERENCES

Black, D. K. and K. L. Lee, Saturating samples by back pressure, *J. Soil Mech. Found. Eng. Div.*, ASCE, 99(1), 75–93, 1973.

Eigenbrod, K. D. and J. P. Burak, Measurement of *B* values less than unity for thinly interbedded varved clay, *Geotech. Test. J.*, ASTM, 13(4), 370–374, 1990.

Henkel, D. J., The shear strength of saturated remolded clays, in *Proc. Res. Conf. Shear Strength of Cohesive Soils*, ASCE, Bourder, CO, USA, pp. 533–554, 1960.

Kenney, T. C., Discussion, *J. Soil Mech. Found. Eng. Div.*, ASCE, 85(3), 67–79, 1959.

Kurukulasuriya, L. C., M. Oda, and H. Kazama, Anisotropy of undrained shear strength of an over-consolidated soil by triaxial and plane strain tests, *Soils Found.*, 39(1), 21–29, 1999.

Lambe, T. W. and R. V. Whitman, *Soil Mechanics*, Wiley, New York, 1969.

Simons, N. E., The effect of overconsolidation on the shear strength characteristics of an undisturbed Oslo clay, in *Proc. Res. Conf. Shear Strength of Cohesive Soils*, ASCE, Bourder, CO, USA, pp. 747–763, 1960.

Skempton, A. W., The pore pressure coefficients *A* and *B*, *Geotechnique*, 4, 143–147, 1954.

Skempton, A. W. and A. W. Bishop, Soils, in *Building Materials, Their Elasticity and Inelasticity*, M. Reiner, Ed., North Holland, Amsterdam, the Netherlands, 1956.

Terzaghi, K., *Erdbaumechanik auf Bodenphysikalisher Grundlage*, Deuticke, Vienna, Austria, 1925.

Veyera, G. E., W. A. Charlie, D. O. Doehring, and M. E. Hubert, Measurement of the pore pressure parameter C less than unity in saturated sands, *Geotech. Test. J.*, ASTM, 15(3), 223–230, 1992.

Chapter 6

Permeability

6.1 INTRODUCTION

Any given mass of soil consists of solid particles of various sizes with interconnected void spaces. The continuous void spaces in a soil permit water to flow from a point of high energy to a point of low energy. Permeability is defined as the property of a soil that allows the seepage of fluids through its interconnected void spaces. This chapter is devoted to the study of the basic parameters involved in the flow of water through soils.

6.2 DARCY'S LAW

In order to obtain a fundamental relation for the quantity of seepage through a soil mass under a given condition, consider the case shown in Figure 6.1. The cross-sectional area of the soil is equal to A and the rate of seepage is q.

According to Bernoulli's theorem, the total head for flow at any section in the soil can be given by

$$\text{Total head} = \text{Elevation head} + \text{pressure head} + \text{velocity head} \qquad (6.1)$$

The velocity head for flow through soil is very small and can be neglected. The total heads at sections A and B can thus be given by

$$\text{Total head at A} = z_A + h_A$$

$$\text{Total head at B} = z_B + h_B$$

where
z_A and z_B are the elevation heads
h_A and h_B are the pressure heads

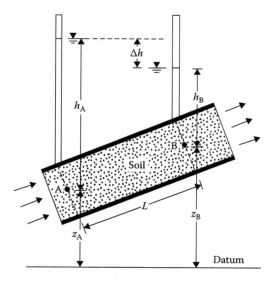

Figure 6.1 Development of Darcy's law.

The loss of head Δh between sections A and B is

$$\Delta h = (z_A + h_A) - (z_A + h_B) \tag{6.2}$$

The hydraulic gradient i can be written as

$$i = \frac{\Delta h}{L} \tag{6.3}$$

where L is the distance between sections A and B.

Darcy (1856) published a simple relation between the discharge velocity and the hydraulic gradient:

$$\upsilon = ki \tag{6.4}$$

where

υ is the discharge velocity
i is the hydraulic gradient
k is the coefficient of permeability

Hence, the rate of seepage q can be given by

$$q = kiA \tag{6.5}$$

Note that A is the cross-section of the soil perpendicular to the direction of flow.

The coefficient of permeability k has the units of velocity, such as cm/s or mm/s, and is a measure of the resistance of the soil to flow of water. When the properties of water affecting the flow are included, we can express k by the relation

$$k\,(\text{cm/s}) = \frac{K\rho g}{\mu} \tag{6.6}$$

where

K is the intrinsic (or absolute) permeability, cm^2
ρ is the mass density of the fluid, g/cm^3
g is the acceleration due to gravity, cm/s^2
μ is the absolute viscosity of the fluid, poise [i.e., g/(cm·s)]

It must be pointed out that the velocity υ given by Equation 6.4 is the discharge velocity calculated on the basis of the gross cross-sectional area. Since water can flow only through the interconnected pore spaces, the actual velocity of seepage through soil, υ_s, can be given by

$$\upsilon_s = \frac{\upsilon}{n} \tag{6.7}$$

where n is the porosity of the soil.

Some typical values of the coefficient of permeability are given in Table 6.1. The coefficient of permeability of soils is generally expressed at a temperature of 20°C. At any other temperature T, the coefficient of permeability can be obtained from Equation 6.6 as

$$\frac{k_{20}}{k_T} = \frac{(\rho_{20})(\mu_T)}{(\rho_T)(\mu_{20})}$$

where

k_T, k_{20} are the coefficient of permeability at T°C and 20°C, respectively
ρ_T, ρ_{20} are the mass density of the fluid at T°C and 20°C, respectively
μ_T, μ_{20} are the coefficient of viscosity at T°C and 20°C, respectively

Table 6.1 Typical values of coefficient of permeability for various soils

Material	Coefficient of permeability (mm/s)
Coarse	$10 - 10^3$
Fine gravel, coarse, and medium sand	$10^{-2} - 10$
Fine sand, loose silt	$10^{-4} - 10^{-2}$
Dense silt, clayey silt	$10^{-5} - 10^{-4}$
Silty clay, clay	$10^{-8} - 10^{-5}$

Table 6.2 Values of μ_T/μ_{20}

Temperature T (°C)	μ_T/μ_{20}	Temperature T (°C)	μ_T/μ_{20}
10	1.298	21	0.975
11	1.263	22	0.952
12	1.228	23	0.930
13	1.195	24	0.908
14	1.165	25	0.887
15	1.135	26	0.867
16	1.106	27	0.847
17	1.078	28	0.829
18	1.051	29	0.811
19	1.025	30	0.793
20	1.000		

Since the value of ρ_{20}/ρ_T is approximately 1, we can write

$$k_{20} = k_T \frac{(\mu_T)}{(\mu_{20})} \tag{6.8}$$

Table 6.2 gives the values of μ_T/μ_{20} for a temperature T varying from 10°C to 30°C.

6.3 VALIDITY OF DARCY'S LAW

Darcy's law given by Equation 6.4, $v = ki$, is true for laminar flow through the void spaces. Several studies have been made to investigate the range over which Darcy's law is valid, and an excellent summary of these works was given by Muskat (1937). A criterion for investigating the range can be furnished by the Reynolds number. For flow through soils, Reynolds number R_n can be given by the relation

$$R_n = \frac{vD\rho}{\mu} \tag{6.9}$$

where
  v is the discharge (superficial) velocity, cm/s
  D is the average diameter of the soil particle, cm
  ρ is the density of the fluid, g/cm^3
  μ is the coefficient of viscosity, g/(cm·s)

For laminar flow conditions in soils, experimental results show that

$$R_n = \frac{\upsilon D\rho}{\mu} \leq 1 \tag{6.10}$$

with coarse sand, assuming $D = 0.45$ mm and $k \approx 100D^2 = 100(0.045)^2 = 0.203$ cm/s. Assuming $i = 1$, then $\upsilon = ki = 0.203$ cm/s. Also, $\rho_{water} \approx 1$ g/cm^3, and $\mu_{20°C} = (10^{-5})(981)$ g/(cm·s). Hence

$$R_n = \frac{(0.203)(0.045)(1)}{(10^{-5})(986)} = 0.931 < 1$$

From the previous calculations, we can conclude that, for flow of water through all types of soil (sand, silt, and clay), the flow is laminar and Darcy's law is valid. With coarse sands, gravels, and boulders, turbulent flow of water can be expected, and the hydraulic gradient can be given by the relation

$$i = a\upsilon + b\upsilon^2 \tag{6.11}$$

where a and b are experimental constants (e.g., see Forchheimer [1902]).

Darcy's law as defined by Equation 6.4 implies that the discharge velocity bears a linear relation with the hydraulic gradient. Hansbo (1960) reported the test results of four undisturbed natural clays. On the basis of his results (Figure 6.2),

$$\upsilon = k(i - i') \quad i \geq i' \tag{6.12}$$

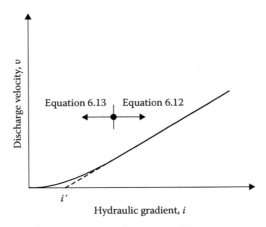

Figure 6.2 Variation of υ with i (Equations 6.12 and 6.13).

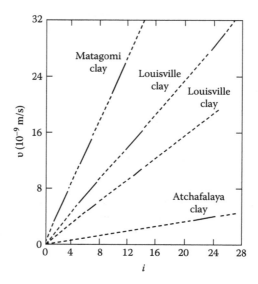

Figure 6.3 Discharge velocity–gradient relationship for four clays. [After Tavenas, F. et al., *Can. Geotech. J.*, 20(4), 629, 1983b.]

and

$$v = ki^n \quad i < i' \tag{6.13}$$

The value of n for the four Swedish clays was about 1.6. There are several studies, however, that refute the preceding conclusion.

Figure 6.3 shows the laboratory test results between v and i for four clays (Tavenas et al., 1983a,b). These tests were conducted using triaxial test equipment, and the results show that Darcy's law is valid.

6.4 DETERMINATION OF THE COEFFICIENT OF PERMEABILITY IN THE LABORATORY

The three most common laboratory methods for determining the coefficient of permeability of soils are the following:

1. Constant-head test
2. Falling-head test
3. Indirect determination from consolidation test

The general principles of these methods are given later.

6.4.1 Constant-head test

The constant-head test is suitable for more permeable granular materials. The basic laboratory test arrangement is shown in Figure 6.4. The soil specimen is placed inside a cylindrical mold, and the constant-head loss h of water flowing through the soil is maintained by adjusting the supply. The outflow water is collected in a measuring cylinder, and the duration of the collection period is noted. From Darcy's law, the total quantity of flow Q in time t can be given by

$$Q = qt = kiAt$$

where A is the area of cross-section of the specimen. However, $i = h/L$, where L is the length of the specimen, and so $Q = k(h/L)At$. Rearranging gives

$$k = \frac{QL}{hAt} \tag{6.14}$$

Once all the quantities on the right-hand side of Equation 6.14 have been determined from the test, the coefficient of permeability of the soil can be calculated.

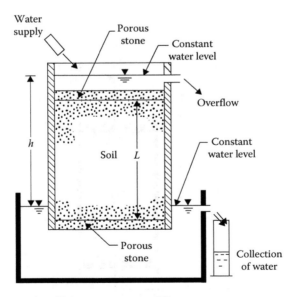

Figure 6.4 Constant-head laboratory permeability test.

6.4.2 Falling-head test

The falling-head permeability test is more suitable for fine-grained soils. Figure 6.5 shows the general laboratory arrangement for the test. The soil specimen is placed inside a tube, and a standpipe is attached to the top of the specimen. Water from the standpipe flows through the specimen. The initial head difference h_1 at time $t = 0$ is recorded, and water is allowed to flow through the soil such that the final head difference at time $t = t$ is h_2.

The rate of flow through the soil is

$$q = kiA = k\frac{h}{L}A = -a\frac{dh}{dt}$$

(6.15)

where

> h is the head difference at any time t
> A is the area of specimen
> a is the area of standpipe
> L is the length of specimen

From Equation 6.15

$$\int_0^t dt = \int_{h_1}^{h_2} \frac{aL}{Ak}\left(-\frac{dh}{h}\right)$$

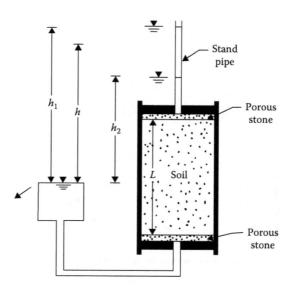

Figure 6.5 Falling-head laboratory permeability test.

or

$$k = 2.303 \frac{aL}{At} \log \frac{h_1}{h_2}$$

(6.16)

The values of a, L, A, t, h_1, and h_2 can be determined from the test, and the coefficient of the permeability k for a soil can then be calculated from Equation 6.16.

6.4.3 Permeability from consolidation test

The coefficient of permeability of clay soils is often determined by the consolidation test, the procedures of which are explained in Section 8.5. From Equation 8.32

$$T_v = \frac{C_v t}{H^2}$$

where

T_v is the time factor
C_v is the coefficient of consolidation
H is the length of average drainage path
t is time

The coefficient of consolidation is (see Equation 8.22)

$$C_v = \frac{k}{\gamma_w m_v}$$

where

γ_w is the unit weight of water
m_v is the volume coefficient of compressibility

Also

$$m_v = \frac{\Delta e}{\Delta \sigma (1 + e)}$$

where

Δe is the change of void ratio for incremental loading
$\Delta \sigma$ is the incremental pressure applied
e is the initial void ratio

Combining these three equations, we have

$$k = \frac{T_v \gamma_w \Delta e H^2}{t \Delta \sigma (1+e)} \tag{6.17}$$

For 50% consolidation, $T_v = 0.197$, and the corresponding t_{50} can be estimated according to the procedure presented in Section 8.10. Hence

$$k = \frac{0.197 \gamma_w \Delta e H^2}{t_{50} \Delta \sigma (1+e)} \tag{6.18}$$

6.5 VARIATION OF THE COEFFICIENT OF PERMEABILITY FOR GRANULAR SOILS

For fairly uniform sand (i.e., small uniformity coefficient), Hazen (1911) proposed an empirical relation for the coefficient of permeability in the form

$$k \,(\text{cm/s}) = c D_{10}^2 \tag{6.19}$$

where
   c is a constant that varies from 1.0 to 1.5
   D_{10} is the effective size, in millimeters, and is defined in Chapter 1

Equation 6.19 is based primarily on observations made by Hazen on loose, clean filter sands. A small quantity of silts and clays, when present in a sandy soil, may substantially change the coefficient of permeability.

Casagrande proposed a simple relation for the coefficient of permeability for fine to medium clean sand in the following form:

$$k = 1.4 e^2 k_{0.85} \tag{6.20}$$

where
   k is the coefficient of permeability at a void ratio e
   $k_{0.85}$ is the corresponding value at a void ratio of 0.85

A theoretical solution for the coefficient of permeability also exists in the literature. This is generally referred to as the Kozeny–Carman equation, which is derived later.

It was pointed out earlier in this chapter that the flow through soils finer than coarse gravel is laminar. The interconnected voids in a given soil mass can be visualized as a number of capillary tubes through which water can flow (Figure 6.6).

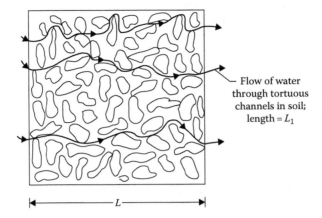

Flow of water through tortuous channels in soil; length = L_1

Figure 6.6 Flow of water through tortuous channels in soil.

According to the Hagen–Poiseuille equation, the quantity of flow of water in unit time, q, through a capillary tube of radius R can be given by

$$q = \frac{\gamma_w S}{8\mu} R^2 a \tag{6.21}$$

where
  γ_w is the unit weight of water
  μ is the absolute coefficient of viscosity
  a is the area cross-section of tube
  S is the hydraulic gradient

The hydraulic radius R_H of the capillary tube can be given by

$$R_H = \frac{\text{area}}{\text{wetted perimeter}} = \frac{\pi R^2}{2\pi R} = \frac{R}{2} \tag{6.22}$$

From Equations 6.21 and 6.22

$$q = \frac{1}{2} \frac{\gamma_w S}{\mu} R_H^2 a \tag{6.23}$$

For flow through two parallel plates, we can also derive

$$q = \frac{1}{3} \frac{\gamma_w S}{\mu} R_H^2 a \tag{6.24}$$

So, for laminar flow conditions, the flow through any cross section can be given by a general equation

$$q = \frac{\gamma_w S}{C_s \mu} R_H^2 a \tag{6.25}$$

where C_s is the shape factor. Also, the average velocity of flow v_a is given by

$$v_a = \frac{q}{a} = \frac{\gamma_w S}{C_s \mu} R_H^2 \tag{6.26}$$

For an actual soil, the interconnected void spaces can be assumed to be a number of tortuous channels (Figure 6.6), and for these, the term S in Equation 6.26 is equal to $\Delta h / \Delta L_1$. Now

$$R_H = \frac{\text{area}}{\text{perimeter}} = \frac{(\text{area})(\text{length})}{(\text{perimeter})(\text{length})} = \frac{\text{volume}}{\text{surface area}}$$

$$= \frac{1}{(\text{surface area})/(\text{volume of pores})} \tag{6.27}$$

If the total volume of soil is V, the volume of voids is $V_v = nV$, where n is porosity. Let S_v be equal to the surface area per unit volume of soil (bulk). From Equation 6.27

$$R_H = \frac{\text{volume}}{\text{surface area}} = \frac{nV}{S_v V} = \frac{n}{S_v} \tag{6.28}$$

Substituting Equation 6.28 into Equation 6.26 and taking $v_a = v_s$ (where v_s is the actual seepage velocity through soil), we get

$$v_s = \frac{\gamma_w}{C_s \mu} S \frac{n^2}{S_v^2} \tag{6.29}$$

It must be pointed out that the hydraulic gradient i used for soils is the macroscopic gradient. The factor S in Equation 6.29 is the microscopic gradient for flow through soils. Referring to Figure 6.6, $i = \Delta h / \Delta L$ and $S = \Delta h / \Delta L_1$. So

$$i = \frac{\Delta h}{\Delta L_1} \frac{\Delta L_1}{\Delta L} = ST \tag{6.30}$$

or

$$S = \frac{i}{T} \tag{6.31}$$

where T is tortuosity, $\Delta L_1 / \Delta L$.

Again, the seepage velocity in soils is

$$\upsilon_s = \frac{\upsilon}{n} \frac{\Delta L_1}{\Delta L} = \frac{\upsilon}{n} T \tag{6.32}$$

where υ is the discharge velocity. Substitution of Equations 6.32 and 6.31 into Equation 6.29 yields

$$\upsilon_s = \frac{\upsilon}{n} T = \frac{\gamma_w}{C_s \mu} \frac{i}{T} \frac{n^2}{S_v^2}$$

or

$$\upsilon = \frac{\gamma_w}{C_s \mu S_v^2} \frac{n^3}{T^2} i \tag{6.33}$$

In Equation 6.33, S_v is the surface area per unit volume of soil. If we define S_s as the surface area per unit volume of soil solids, then

$$S_s V_s = S_v V \tag{6.34}$$

where V_s is the volume of soil solids in a bulk volume V, that is,

$$V_s = (1 - n)V$$

So

$$S_s = \frac{S_v V}{V_s} = \frac{S_v V}{(1 - n)V} = \frac{S_v}{1 - n} \tag{6.35}$$

Combining Equations 6.33 and 6.35, we obtain

$$\upsilon = \frac{\gamma_w}{C_s \mu S_s^2 T^2} \frac{n^3}{(1 - n)^2} i$$

$$= \frac{1}{C_s S_s^2 T^2} \frac{\gamma_w}{\mu} \frac{e^3}{1 + e} i \tag{6.36}$$

where e is the void ratio. This relation is the Kozeny–Carman equation (Kozeny, 1927, 1933; Carman, 1956). Comparing Equations 6.4 and 6.36, we find that the coefficient of permeability is

$$k = \frac{1}{C_s S_s^2 T^2} \frac{\gamma_w}{\mu} \frac{e^3}{1+e} \qquad (6.37)$$

The absolute permeability was defined by Equation 6.6 as

$$K = k \frac{\mu}{\gamma_w}$$

Comparing Equations 6.6 and 6.37,

$$K = \frac{1}{C_s S_s^2 T^2} \frac{e^3}{1+e} \qquad (6.38)$$

The Kozeny–Carman equation works well for describing coarse-grained soils such as sand and some silts. For these cases, the coefficient of permeability bears a linear relation to $e^3/(1 + e)$. However, serious discrepancies are observed when the Kozeny–Carman equation is applied to clayey soils.

For granular soils, the shape factor C_s is approximately 2.5, and the tortuosity factor T is about $\sqrt{2}$.

From Equation 6.20, we write that

$$k \propto e^2 \qquad (6.39)$$

Similarly, from Equation 6.37

$$k \propto \frac{e^3}{1+e} \qquad (6.40)$$

Amer and Awad (1974) used the preceding relation and their experimental results in granular soil to provide

$$k = 3.5 \times 10^{-4} \left(\frac{e^3}{1+e} \right) C_u^{0.6} D_{10}^{2.32} \left(\frac{\rho_w}{\eta} \right) \qquad (6.41a)$$

where k is in cm/s

C_u = uniformity coefficient
D_{10} = effective size (mm)
ρ_w = density of water (g/cm³)
η = viscosity (g·s/cm²)

At 20°C, $\rho_w = 1$ g/cm³ and $\eta \approx 0.1 \times 10^{-4}$ g·s/cm². So

$$k = 3.5 \times 10^{-4} \left(\frac{e^3}{1+e} \right) C_u^{0.6} D_{10}^{2.32} \left(\frac{1}{0.1 \times 10^{-4}} \right) \tag{6.41a}$$

or

$$k \ (\text{cm/s}) = 35 \left(\frac{e^3}{1+e} \right) C_u^{0.6} (D_{10})^{2.32} \tag{6.41b}$$

Another form of relation for the coefficient of permeability and void ratio for granular soils has also been used, namely

$$k \propto \frac{e^2}{1+e} \tag{6.42}$$

For comparison of the validity of the relations given in Equations 6.39 through 6.42, the experimental results (laboratory constant-head test) for a uniform Madison sand are shown in Figure 6.7. From the plot, it appears that all three relations are equally good.

More recently, Chapuis (2004) proposed an empirical relationship for k in conjunction with Equation 6.40 as

$$k \,(\text{cm/s}) = 2.4622 \left[D_{10}^2 \frac{e^3}{1+e} \right]^{0.7825} \tag{6.43}$$

where D_{10} is the effective size (mm).

The preceding equation is valid for natural, uniform sand and gravel to predict k that is in the range of 10^{-1}–10^{-3} cm/s. This can be extended to natural, silty sands without plasticity. It is not valid for crushed materials or silty soils with some plasticity.

Mention was made in Section 6.3 that turbulent flow conditions may exist in very coarse sands and gravels and that Darcy's law may not be valid for these materials. However, under a low hydraulic gradient, laminar flow conditions usually exist. Kenney et al. (1984) conducted

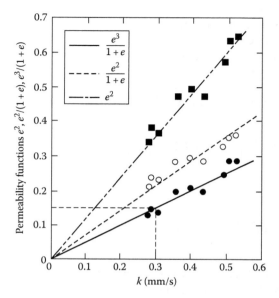

Figure 6.7 Plot of k against permeability function for Madison sand.

laboratory tests on granular soils in which the particle sizes in various specimens ranged from 0.074 to 25.4 mm. The uniformity coefficients of these specimens, C_u, ranged from 1.04 to 12. All permeability tests were conducted at a relative density of 80% or more. These tests showed that, for laminar flow conditions, the absolute permeability can be approximated as

$$K\,(\text{mm}^2) = (0.05 - 1)D_5^2 \tag{6.44}$$

where D_5 is the diameter (mm) through which 5% of soil passes.

6.5.1 Modification of Kozeny–Carman equation for practical application

For practical use, Carrier (2003) modified Equation 6.37 in the following manner. At 20°C, γ_w/μ for water is about 9.33×10^4 (1/cm·s). Also, $(C_s T^2)$ is approximately equal to 5. Substituting these values into (6.37), we obtain

$$k\,(\text{cm/s}) = 1.99 \times 10^4 \left(\frac{1}{S_s}\right)^2 \frac{e^3}{1+e} \tag{6.45}$$

Again

$$S_s = \frac{SF}{D_{eff}}\left(\frac{1}{cm}\right) \qquad (6.46)$$

with

$$D_{eff} = \frac{100\%}{\sum\left(f_i/D_{(av)i}\right)} \qquad (6.47)$$

where f_i is the fraction of particles between two sieve seizes, in percent (*Note:* larger sieve, l; smaller sieve, s)

$$D_{(av)i}(cm) = [D_{li}(cm)]^{0.5} \times [D_{si}(cm)]^{0.5} \qquad (6.48)$$

SF is the shape factor
   Combining Equations 6.45 through 6.48

$$k(cm/s) = 1.99 \times 10^4 \left[\frac{100\%}{\sum f_i/\left(D_{li}^{0.5} \times D_{si}^{0.5}\right)}\right]^2 \left(\frac{1}{SF}\right)^2 \left(\frac{e^3}{1+e}\right) \qquad (6.49)$$

The magnitude of SF may vary between 6 and 8, depending on the angularity of the soil particles.
   Carrier (2003) further suggested a slight modification of Equation 6.49, which can be written as

$$k(cm/s) = 1.99 \times 10^4 \left[\frac{100\%}{\sum f_i/\left(D_{li}^{0.404} \times D_{si}^{0.595}\right)}\right]^5 \left(\frac{1}{SF}\right)^2 \left(\frac{e^3}{1+e}\right) \qquad (6.50)$$

Example 6.1

The results of a sieve analysis on sand are given as follows:

Sieve no	Sieve opening (cm)	Percent passing	Fraction of particles between two consecutive sieves (%)
30	0.06	100	
			4
40	0.0425	96	
			12
60	0.02	84	
			34
100	0.015	50	
			50
200	0.0075	0	

Estimate the hydraulic conductivity using Equation 6.50. Given: the void ratio of the sand is 0.6. Use SF = 7.

Solution

For fraction between Nos. 30 and 40 sieves

$$\frac{f_i}{D_{li}^{0.404} \times D_{si}^{0.595}} = \frac{4}{(0.06)^{0.404} \times (0.0425)^{0.595}} = 81.62$$

For fraction between Nos. 40 and 60 sieves

$$\frac{f_i}{D_{li}^{0.404} \times D_{si}^{0.595}} = \frac{12}{(0.0425)^{0.404} \times (0.02)^{0.595}} = 440.76$$

Similarly, for fraction between Nos. 60 and 100 sieves

$$\frac{f_i}{D_{li}^{0.404} \times D_{si}^{0.595}} = \frac{34}{(0.02)^{0.404} \times (0.015)^{0.595}} = 2009.5$$

And, for between Nos. 100 and 200 sieves

$$\frac{f_i}{D_{li}^{0.404} \times D_{si}^{0.595}} = \frac{50}{(0.015)^{0.404} \times (0.0075)^{0.595}} = 5013.8$$

$$\frac{100\%}{\sum f_i / \left(D_{li}^{0.404} \times D_{si}^{0.595} \right)} = \frac{100}{81.62 + 440.76 + 2009.5 + 5013.8}$$

$$\approx 0.0133$$

From Equation 6.50

$$k = (1.99 \times 10^4)(0.0133)^2 \left(\frac{1}{7} \right)^2 \left(\frac{0.6^3}{1 + 0.6} \right) = 0.0097 \, \text{cm/s}$$

Example 6.2

Refer to Figure 6.7. For the soil, (a) calculate the "composite shape factor," $C_S S_s^2 T^2$, of the Kozeny–Carman equation, given $\mu_{20°C} = 10.09 \times 10^{-3}$ poise, (b) If $C_s = 2.5$ and $T = \sqrt{2}$, determine S_s. Compare this value with the theoretical value for a sphere of diameter $D_{10} = 0.2$ mm.

Solution

Part a:

From Equation 6.37,

$$k = \frac{1}{C_s S_s^2 T^2} \frac{\gamma_w}{\mu} \frac{e^3}{1 + e}$$

$$C_s S_s^2 T^2 = \frac{\gamma_w}{\mu} \frac{e^3/(1+e)}{k}$$

The value of $[e^3/(1 + e)]/k$ is the slope of the straight line for the plot of $e^3/(1 + e)$ against k (Figure 6.7). So

$$\frac{e^3/(1+e)}{k} = \frac{0.15}{0.03\,\text{cm/s}} = 5$$

$$C_s S_s^2 T^2 = \frac{(1\,\text{g/cm}^3)(981\,\text{cm/s}^2)}{10.09\times10^{-3}\,\text{poise}}(5) = 4.86\times10^5\,\text{cm}^{-2}$$

Part b:

(Note the units carefully.)

$$S_s = \sqrt{\frac{4.86\times10^5}{C_s T^2}} = \sqrt{\frac{4.86\times10^5}{2.5\times(\sqrt{2})^2}} = 311.8\,\text{cm}^2/\text{cm}^3$$

For $D_{10} = 0.2$ mm

$$S_s = \frac{\text{Surface area of a sphere of radius 0.01 cm}}{\text{Volume of sphere of radius 0.01 cm}}$$

$$= \frac{4\pi(0.01)^2}{(4/3)\pi(0.01)^3} = \frac{3}{0.01} = 300\,\text{cm}^2/\text{cm}^3$$

This value of $S_s = 300$ cm²/cm³ agrees closely with the estimated value of $S_s = 311.8$ cm²/cm³.

Example 6.3

Solve Example 6.1 using Equation 6.43.

Solution

If a grain-size distribution curve is drawn from the data given in Example 6.1, we find:

$$D_{60} = 0.16\text{ mm}; D_{10} = 0.09\text{ mm}$$

From Equation 6.43:

$$k = 2.4622\left[D_{10}^2\,\frac{e^3}{1+e}\right]^{0.7825} = 2.4622\left[(0.09)^2\,\frac{0.6^3}{1+0.6}\right]^{0.7825} = 0.0119\,\text{cm/s}$$

Example 6.4

Solve Example 6.1 using Equation 6.41b.

Solution

As given in Example 6.3, $D_{60} = 0.16$ mm and $D_{10} = 0.09$ mm. Thus

$$C_u = \frac{D_{60}}{D_{10}} = \frac{0.16}{0.09} = 1.78$$

From Equation 6.41(b):

$$k = 35\left(\frac{e^3}{1+e}\right)C_u^{0.6}(D_{10})^{2.32} = 35\left(\frac{0.6^3}{1+0.6}\right)(1.78)^{0.6}(0.09)^{2.32} = 0.025 \text{ cm/s}$$

6.6 VARIATION OF THE COEFFICIENT OF PERMEABILITY FOR COHESIVE SOILS

The Kozeny–Carman equation does not successfully explain the variation of the coefficient of permeability with void ratio for clayey soils. The discrepancies between the theoretical and experimental values are shown in Figures 6.8

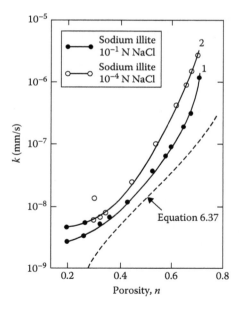

Figure 6.8 Coefficient of permeability for sodium illite. (After Olsen, H. W., Hydraulic flow through saturated clay, ScD thesis, Massachusetts Institute of Technology, Cambridge, MA, 1961.)

and 6.9. These results are based on consolidation–permeability tests (Olsen, 1961, 1962). The marked degrees of variation between the theoretical and the experimental values arise from several factors, including deviations from Darcy's law, high viscosity of the pore water, and unequal pore sizes. Olsen developed a model to account for the variation of permeability due to unequal pore sizes.

Several other empirical relations were proposed from laboratory and field permeability tests on clayey soil. They are summarized in Table 6.3.

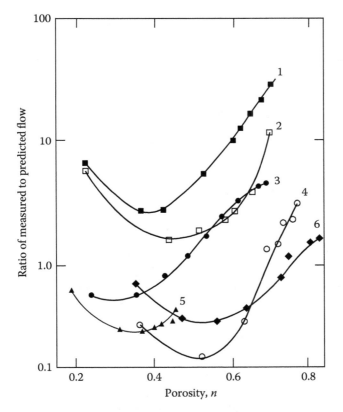

Figure 6.9 Ratio of the measured flow rate to that predicted by the Kozeny–Carman equation for several clays. *Notes:* curve 1: sodium illite, 10^{-1} N NaCl; curve 2: sodium illite, 10^{-4} N NaCl; curve 3: natural kaolinite, distilled water H_2O; curve 4: sodium Boston blue clay, 10^{-1} N NaCl; curve 5: sodium kaolinite and 1% (by weight) sodium tetraphosphate; curve 6: calcium Boston blue clay, 10^{-4} N NaCl. (After Olsen, H. W., Hydraulic flow through saturated clay, ScD thesis, Massachusetts Institute of Technology, Cambridge, MA, 1961.)

Table 6.3 Empirical relations for the coefficient of permeability in clayey soils

Investigator	Relation	Notation	Remarks
Mesri and Olson (1971)	$\log k = C_2 \log e + C_3$	C_2, C_3 = constants	Based on artificial and remolded soils
Taylor (1948)	$\log k = \log k_0 - \dfrac{e_0 - e}{C_k}$	k_0 = coefficient of in situ permeability at void ratio e_0 k = coefficient of permeability at void ratio e C_k = permeability change index	$C_k \approx 0.5 e_0$ (Tavenas et al., 1983a,b)
Samarsinghe et al. (1982)	$k = C_4 \dfrac{e^n}{1+e}$	C_4 = constant $\log [k(1 + e)] = \log C_4 + n \log e$	Applicable only to normally consolidated clays
Raju et al. (1995)	$\dfrac{e}{e_L} = 2.23 + 0.204 \log k$	k is in cm/s e_L = void ratio at liquid limit = $w_{LL} G_s$ w_{LL} = moisture content at liquid limit	Normally consolidated clay
Tavenas et al. (1983a,b)	$k = f$	f = function of void ratio, and PI + CF PI = plasticity index in decimals CF = clay size fraction in decimals	See Figure 6.10

Example 6.5

For a normally consolidated clay soil, the following values are given:

Void ratio	k (cm/s)
1.1	0.302×10^{-7}
0.9	0.12×10^{-7}

Estimate the hydraulic conductivity of the clay at a void ratio of 0.75. Use the equation proposed by Samarsinghe et al. (1982; see Table 6.3; see also Figure 6.10).

Solution

$$k = C_4 \left(\frac{e^n}{1+e} \right)$$

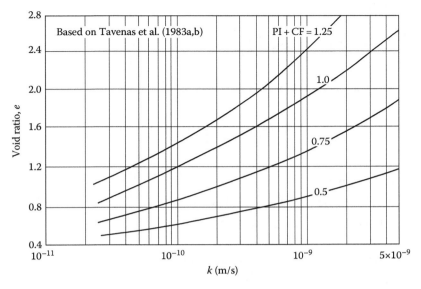

Figure 6.10 Plot of e versus k for various values of PI + CF (See Table 6.3).

$$\frac{k_1}{k_2} = \frac{\left(e_1^n/(1+e_1)\right)}{\left(e_2^n/(1+e_2)\right)}$$

$$\frac{0.302\times10^{-7}}{0.12\times10^{-7}} = \frac{(1.1)^n/(1+1.1)}{(0.9)^n/(1+0.9)}$$

$$2.517 = (1.9/2.1)(1.1/0.9)^n$$

$$2.782 = (1.222)^n$$

$$n = \frac{\log(2.782)}{\log(1.222)} = \frac{0.444}{0.087} = 5.1$$

so

$$k = C_4\left(\frac{e^{5.1}}{1+e}\right)$$

To find C_4

$$0.302\times10^{-7} = C_4\left[\frac{(1.1)^{5.1}}{1+1.1}\right] = \left(\frac{1.626}{2.1}\right)C_4$$

$$C_4 = \frac{(0.302\times10^{-7})(2.1)}{1.626} = 0.39\times10^{-7}$$

Hence

$$k = (0.39 \times 10^{-7} \text{ cm/s}) \left(\frac{e^{5.1}}{1+e} \right)$$

At a void ratio of 0.75

$$k = (0.39 \times 10^{-7} \text{ cm/s}) \left(\frac{0.75^{5.1}}{1+0.75} \right) = 0.514 \times 10^{-8} \text{ cm/s}$$

Example 6.6

A soft saturated clay has the following:

Percent less than 0.02 mm = 32%
Plasticity index = 21
Saturated unit weight, γ_{sat} = 19.4 kN/m³
Specific gravity of soil solids = 2.76

Estimate the hydraulic conductivity of the clay. Use Figure 6.10.

Solution

Given: PI (in fraction) = 0.21; clay-size friction, CF = 0.32

$$CF + PI = 0.32 + 0.21 = 0.53$$

$$\gamma_{sat} = \frac{(G_s + e)\gamma_w}{1+e} = \frac{(2.76 + e)(9.81)}{1+e}; \text{ with } \gamma_{sat} = 19.4 \text{ kN/m}^3, e = 0.8$$

From Figure 6.10, for $e = 0.8$ and $CF + PI = 0.53$, the value of

$$k = 3.59 \times 10^{-10} \text{ m/s} = 3.59 \times 10^{-8} \text{ cm/s}$$

Example 6.7

The void ratio and hydraulic conductivity relation for a normally consolidate clay are given here.

Void ratio	k (cm/s)
1.2	0.6×10^{-7}
1.52	1.519×10^{-7}

Estimate the value of k for the same clay with a void ratio of 1.4. Use the Mesri and Olson (1971) equation (see Table 6.3).

Solution

From Table 6.3:

$$\log k = C_2 \log e + C_3$$

So,

$$\log(0.6 \times 10^{-7}) = C_2 \log(1.2) + C_3 \tag{a}$$

$$\log(1.519 \times 10^{-7}) = C_2 \log(1.52) + C_3 \tag{b}$$

From Equations (a) and (b),

$$\log\left(\frac{0.6 \times 10^{-7}}{1.519 \times 10^{-7}}\right) = C_2 \log\left(\frac{1.2}{1.52}\right)$$

$$C_2 = \frac{-0.4034}{-0.1027} = 3.928 \tag{c}$$

From Equations (a) and (c),

$$C_3 = \log(0.6 \times 10^{-7}) - (3.928)(\log 1.2) = -7.531$$

Thus,

$$\log k = 3.928 \log e - 7.531$$

With $e = 1.4$,

$$\log k = 3.928 \log(1.4) - 7.531 = -6.957$$

Hence,

$$k = 1.1 \times 10^{-7} \text{ cm/s}$$

6.7 DIRECTIONAL VARIATION OF PERMEABILITY IN ANISOTROPIC MEDIUM

Most natural soils are anisotropic with respect to the coefficient of permeability, and the degree of anisotropy depends on the type of soil and the nature of its deposition. In most cases, the anisotropy is more predominant in clayey soils compared to granular soils. In anisotropic soils, the directions of the maximum and minimum permeabilities are generally at right angles to each other, maximum permeability being in the horizontal direction.

Figure 6.11a shows the seepage of water around a sheet pile wall. Consider a point O at which the flow line and the equipotential line are as shown in the figure. The flow line is a line along which a water particle at O will move from left to right. For the definition of an equipotential line, refer to Section 7.2. Note that in anisotropic soil, the flow line and equipotential line are not orthogonal. Figure 6.11b shows the flow line and equipotential line at O. The coefficients of permeability in the x and z directions are k_h and k_v, respectively.

In Figure 6.11, m is the direction of the tangent drawn to the flow line at O, and thus that is the direction of the resultant discharge velocity. Direction n is perpendicular to the equipotential line at O, and so it is the direction of the resultant hydraulic gradient. Using Darcy's law,

$$v_x = -k_h \frac{\partial h}{\partial x} \tag{6.51}$$

$$v_z = -k_v \frac{\partial h}{\partial z} \tag{6.52}$$

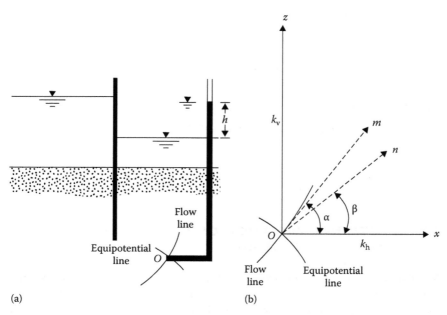

(a)

(b)

Figure 6.11 Directional variation of the coefficient of permeability: (a) seepage of water around a sheet pile wall; (b) flow and equipotential lines at O.

$$v_m = -k_\alpha \frac{\partial h}{\partial m}$$

(6.53)

$$v_n = -k_\beta \frac{\partial h}{\partial n}$$

(6.54)

where

k_h is the maximum coefficient of permeability (in the horizontal x direction)

k_v is the minimum coefficient of permeability (in the vertical z direction)

k_α, k_β are the coefficients of permeability in m, n directions, respectively

Now, we can write

$$\frac{\partial h}{\partial m} = \frac{\partial h}{\partial x}\cos\alpha + \frac{\partial h}{\partial z}\sin\alpha$$

(6.55)

From Equations 6.51 through 6.53, we have

$$\frac{\partial h}{\partial x} = -\frac{v_x}{k_h} \qquad \frac{\partial h}{\partial z} = -\frac{v_z}{k_v} \qquad \frac{\partial h}{\partial m} = -\frac{v_m}{k_\alpha}$$

Also, $v_x = v_m \cos\alpha$ and $v_z = v_m \sin\alpha$.

Substitution of these into Equation 6.55 gives

$$-\frac{v_m}{k_\alpha} = -\frac{v_x}{k_h}\cos\alpha - \frac{v_z}{k_v}\sin\alpha$$

or

$$\frac{v_m}{k_\alpha} = \frac{v_m}{k_h}\cos^2\alpha + \frac{v_m}{k_v}\sin^2\alpha$$

so

$$\frac{1}{k_\alpha} = \frac{\cos^2\alpha}{k_h} + \frac{\sin^2\alpha}{k_v}$$

(6.56)

The nature of the variation of k_α with α as determined by Equation 6.56 is shown in Figure 6.12. Again, we can say that

$$\upsilon_n = \upsilon_x \cos\beta + \upsilon_z \sin\beta \tag{6.57}$$

Combining Equations 6.51, 6.52, and 6.54

$$k_\beta \frac{\partial h}{\partial n} = k_\mathrm{h} \frac{\partial h}{\partial x} \cos\beta + k_\mathrm{v} \frac{\partial h}{\partial z} \sin\beta \tag{6.58}$$

However

$$\frac{\partial h}{\partial x} = \frac{\partial h}{\partial n} \cos\beta \tag{6.59}$$

and

$$\frac{\partial h}{\partial z} = \frac{\partial h}{\partial n} \sin\beta \tag{6.60}$$

Substitution of Equations 6.59 and 6.60 into Equation 6.58 yields

$$k_\beta = k_\mathrm{h} \cos^2\beta + k_\mathrm{v} \sin^2\beta \tag{6.61}$$

The variation of k_β with β is also shown in Figure 6.12. It can be seen that, for given values of k_h and k_v, Equations 6.56 and 6.61 yield slightly different values of the directional permeability. However, the maximum difference will not be more than 25%.

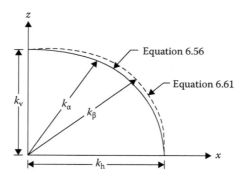

Figure 6.12 Directional variation of permeability.

There are several studies available in the literature providing the experimental values of k_h/k_v. Some are given in the following:

Soil type	k_h/k_v	Reference
Organic silt with peat	1.2–1.7	Tsien (1955)
Plastic marine clay	1.2	Lumb and Holt (1968)
Soft clay	1.5	Basett and Brodie (1961)
Soft marine clay	1.05	Subbaraju (1973)
Boston blue clay	0.7–3.3	Haley and Aldrich (1969)

Figure 6.13 shows the laboratory test results obtained by Fukushima and Ishii (1986) related to k_h and k_v on compacted Maso-do soil (weathered granite). All tests were conducted after full saturation of the compacted soil specimens. The results show that k_h and k_v are functions of molding moisture content and confining pressure. For given molding moisture contents and confining pressures, the ratios of k_h/k_v are in the same general range as shown in the preceding table.

6.8 EFFECTIVE COEFFICIENT OF PERMEABILITY FOR STRATIFIED SOILS

In general, natural soil deposits are stratified. If the stratification is continuous, the effective coefficients of permeability for flow in the horizontal and vertical directions can be readily calculated.

6.8.1 Flow in the horizontal direction

Figure 6.14 shows several layers of soil with horizontal stratification. Owing to fabric anisotropy, the coefficient of permeability of each soil layer may vary depending on the direction of flow. So, let us assume that $k_{h_1}, k_{h_2}, k_{h_3}, \ldots$, are the coefficients of permeability of layers 1, 2, 3, ..., respectively, for flow in the horizontal direction. Similarly, let $k_{v_1}, k_{v_2}, k_{v_3}, \ldots$, be the coefficients of permeability for flow in the vertical direction.

Considering a unit length of the soil layers at right angle to the cross-section as shown in Figure 6.14, the rate of seepage in the horizontal direction can be given by

$$q = q_1 + q_2 + q_3 + \cdots + q_n \tag{6.62}$$

where
   q is the flow rate through the stratified soil layers combined
   $q_1, q_2, q_3, \ldots$, is the rate of flow through soil layers 1, 2, 3, ..., respectively

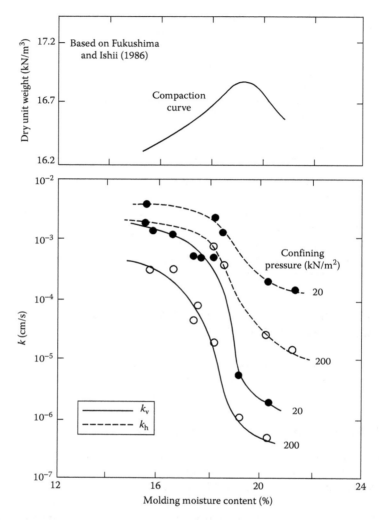

Figure 6.13 Variation of k_v and k_h for Masa-do soil compacted in the laboratory.

Note that for flow in the horizontal direction (which is the direction of stratification of the soil layers), the hydraulic gradient is the same for all layers. So

$$
\left.
\begin{aligned}
q_1 &= k_{h_1} i H_1 \\
q_2 &= k_{h_2} i H_2 \\
q_3 &= k_{h_3} i H_3
\end{aligned}
\right\}
\tag{6.63}
$$

$\vdots$

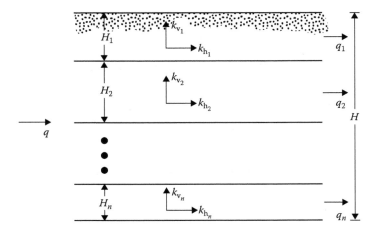

Figure 6.14 Flow in horizontal direction in stratified soil.

and

$$q = k_{e(h)}iH \tag{6.64}$$

where
  i is the hydraulic gradient
  $k_{e(h)}$ is the effective coefficient of permeability for flow in horizontal direction
  H_1, H_2, H_3 are the thicknesses of layers 1, 2, 3, respectively
  $H = H_1 + H_2 + H_3 + \cdots$

Substitution of Equations 6.63 and 6.64 into Equation 6.62 yields

$$k_{e(h)}H = k_{h_1}H_1 + k_{h_2}H_2 + k_{h_3}H_3 + \cdots$$

Hence

$$k_{e(h)} = \frac{1}{H}\left(k_{h_1}H_1 + k_{h_2}H_2 + k_{h_3}H_3 + \cdots\right) \tag{6.65}$$

6.8.2 Flow in the vertical direction

For flow in the vertical direction for the soil layers shown in Figure 6.15,

$$\upsilon = \upsilon_1 = \upsilon_2 = \upsilon_3 = \cdots = \upsilon_n \tag{6.66}$$

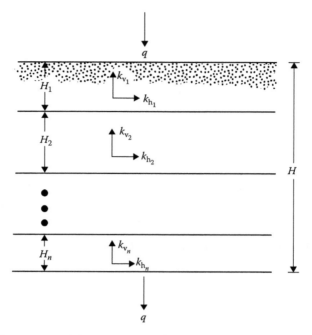

Figure 6.15 Flow in vertical direction in stratified soil.

where v_1, v_2, v_3, ... are the discharge velocities in layers 1, 2, 3, ..., respectively; or

$$v = k_{e(v)}i = k_{v_1}i_1 = k_{v_2}i_2 = k_{v_3}i_3 = \cdots \qquad (6.67)$$

where

$k_{e(v)}$ is the effective coefficient of permeability for flow in the vertical direction

$k_{v_1}, k_{v_2}, k_{v_3}, \ldots$ are the coefficients of permeability of layers 1, 2, 3, ..., respectively, for flow in the vertical direction

i_1, i_2, i_3, ... are the hydraulic gradient in soil layers 1, 2, 3, ..., respectively

For flow at right angles to the direction of stratification

Total head loss = (head loss in layer 1) + (head loss in layer 2) + $\cdots$

or

$$iH = i_1 H_1 + i_2 H_2 + i_3 H_3 + \cdots \qquad (6.68)$$

Combining Equations 6.67 and 6.68 gives

$$\frac{v}{k_{e(v)}}H = \frac{v}{k_{v_1}}H_1 + \frac{v}{k_{v_2}}H_2 + \frac{v}{k_{v_3}}H_3 + \cdots$$

or

$$k_{e(v)} = \frac{H}{H_1/k_{v_1} + H_2/k_{v_2} + H_3/k_{v_3} + \cdots} \tag{6.69}$$

Varved soils are excellent examples of continuously layered soil. Figure 6.16 shows the nature of the layering of New Liskeard varved clay (Chan and Kenny, 1973) along with the variation of moisture content and grain size distribution of various layers. The ratio of $k_{e(h)}/k_{e(v)}$ for this soil varies from about 1.5 to 3.7. Casagrande and Poulos (1969) provided the ratio $k_{e(h)}/k_{e(v)}$ for a varved clay that varies from 4 to 40.

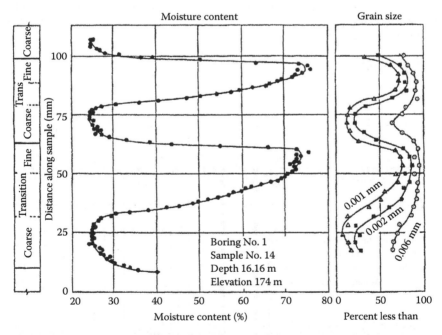

Figure 6.16 Variations of moisture content and grain size across thick-layer varves of New Liskeard varved clay. [After Chan, H. T. and Kenney, T. C., *Can. Geotech. J.*, 10(3), 453, 1973.]

6.9 DETERMINATION OF COEFFICIENT OF PERMEABILITY IN THE FIELD

It is sometimes difficult to obtain undisturbed soil specimens from the field. For large construction projects, it is advisable to conduct permeability tests in situ and compare the results with those obtained in the laboratory. Several techniques are presently available for determination of the coefficient of permeability in the field, such as pumping from wells and borehole tests, and some of these methods will be treated briefly in this section.

6.9.1 Pumping from wells

6.9.1.1 Gravity wells

Figure 6.17 shows a permeable layer underlain by an impermeable stratum. The coefficient of permeability of the top permeable layer can be determined by pumping from a well at a constant rate and observing the steady-state water table in nearby observation wells. The steady state is established when the water levels in the test well and the observation wells become constant. At steady state, the rate of discharge due to pumping can be expressed as

$$q = kiA$$

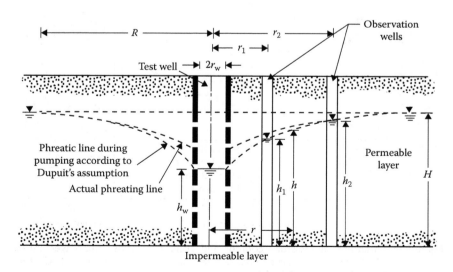

Figure 6.17 Determination of the coefficient of permeability by pumping from wells—gravity well.

From Figure 6.17, $i \approx dh/dr$ (this is referred to as Dupuit's assumption) and $A = 2\pi rh$. Substituting these into the previous equation for rate of discharge gives

$$q = k\frac{dh}{dr}2\pi rh$$

$$\int_{r_1}^{r_2}\frac{dr}{r} = \frac{2\pi k}{q}\int_{h_1}^{h_2}h\,dh$$

So

$$k = \frac{2.303q[\log(r_2/r_1)]}{\pi(h_2^2 - h_1^2)} \tag{6.70}$$

If the values of r_1, r_2, h_1, h_2, and q are known from field measurements, the coefficient of permeability can be calculated from the simple relation given in Equation 6.70. According to Kozeny (1933), the maximum radius of influence, R (Figure 6.17), for drawdown due to pumping can be given by

$$R = \sqrt{\frac{12t}{n}}\sqrt{\frac{qk}{\pi}} \tag{6.71}$$

where
   n is the porosity
   R is the radius of influence
   t is the time during which discharge of water from well has been
      established

Also note that if we substitute $h_1 = h_w$ at $r_1 = r_w$ and $h_2 = H$ at $r_2 = R$, then

$$k = \frac{2.303q[\log(R/r_w)]}{\pi(H^2 - h_w^2)} \tag{6.72}$$

where H is the depth of the original groundwater table from the impermeable layer.
   The depth h at any distance r from the well ($r_w \leq r \leq R$) can be determined from Equation 6.70 by substituting $h_1 = h_w$ at $r_1 = r_w$ and $h_2 = h$ at $r_2 = r$. Thus

$$k = \frac{2.303q[\log(r/r_w)]}{\pi(h^2 - h_w^2)}$$

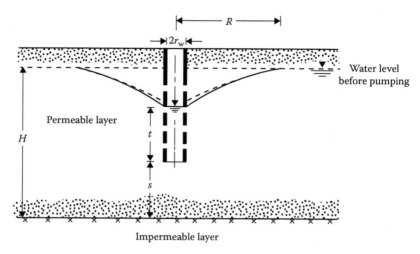

Figure 6.18 Pumping from partially penetrating gravity wells.

or

$$h = \sqrt{\frac{2.303q}{\pi k}\log\frac{r}{r_w} + h_w^2} \tag{6.73}$$

It must be pointed out that Dupuit's assumption (i.e., that $i = dh/dr$) does introduce large errors in regard to the actual phreatic line near the wells during steady-state pumping. This is shown in Figure 6.17. For $r > H - 1.5H$, the phreatic line predicted by Equation 6.73 will coincide with the actual phreatic line.

The relation for the coefficient of permeability given by Equation 6.70 has been developed on the assumption that the well fully penetrates the permeable layer. If the well partially penetrates the permeable layer as shown in Figure 6.18, the coefficient of permeability can be better represented by the following relation (Mansur and Kaufman, 1962):

$$q = \frac{\pi k[(H-s)^2 - t^2]}{2.303\log(R/r_w)}\left[1 + \left(0.30 + \frac{10r_w}{H}\right)\sin\frac{1.8s}{H}\right] \tag{6.74}$$

The notations used on the right-hand side of Equation 6.74 are shown in Figure 6.18.

6.9.1.2 Artesian wells

The coefficient of permeability for a confined aquifer can also be determined from well pumping tests. Figure 6.19 shows an artesian well penetrating the full depth of an aquifer from which water is pumped out at a

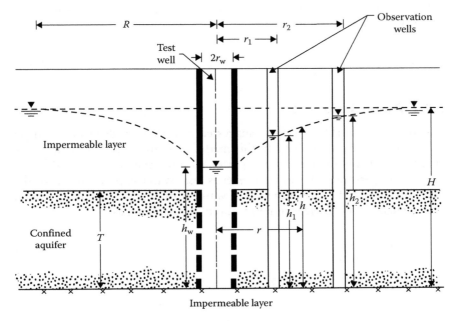

Figure 6.19 Determination of the coefficient of permeability by pumping from wells—confined aquifer.

constant rate. Pumping is continued until a steady state is reached. The rate of water pumped out at steady state is given by

$$q = kiA = k\frac{dh}{dr}2\pi rT \tag{6.75}$$

where T is the thickness of the confined aquifer, or

$$\int_{r_1}^{r_2}\frac{dr}{r} = \int_{h_1}^{h_2}\frac{2\pi kT}{q}dh \tag{6.76}$$

Solution of Equation 6.76 gives

$$k = \frac{q\log(r_2/r_1)}{2.727T(h_2 - h_1)}$$

Hence, the coefficient of permeability k can be determined by observing the drawdown in two observation wells, as shown in Figure 6.19.

If we substitute $h_1 = h_w$ at $r_1 = r_w$ and $h_2 = H$ at $r_2 = R$ in the previous equation, we get

$$k = \frac{q \log(R/r_w)}{2.727 T (H - h_w)} \tag{6.77}$$

6.9.2 Auger hole test

Van Bavel and Kirkham (1948) suggested a method to determine k from an auger hole (Figure 6.20a). In this method, an auger hole is made in the ground that should extend to a depth of 10 times the diameter of the hole or to an impermeable layer, whichever is less. Water is pumped out of the hole, after which the rate of the rise of water with time is observed in several increments. The coefficient of permeability is calculated as

$$k = 0.617 \frac{r_w}{Sd} \frac{dh}{dt} \tag{6.78}$$

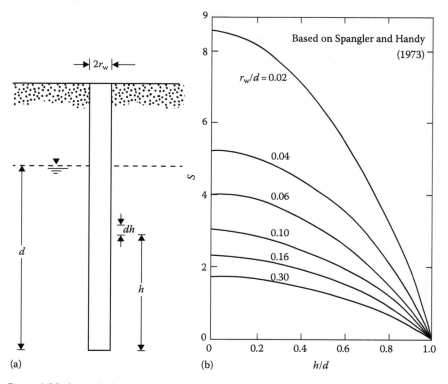

Figure 6.20 Auger hole test: (a) auger hole; (b) plot of S with h/d and r_w/d.

where
  r_w is the radius of the auger hole
  d is the depth of the hole below the water table
  S is the shape factor for auger hole
  dh/dt is the rate of increase of water table at a depth h measured from
    the bottom of the hole

The variation of S with r_w/d and h/d is given in Figure 6.20b (Spangler
and Handy, 1973). There are several other methods of determining the field
coefficient of permeability. For a more detailed description, the readers are
directed to the U.S. Bureau of Reclamation (1961) and the U.S. Department
of the Navy (1971).

Example 6.8

Refer to Figure 6.18. For the steady-state condition, $r_w = 0.4$ m,
$H = 28$ m, $s = 8$ m, and $t = 10$ m. The coefficient of permeability of the
layer is 0.03 mm/s. For the steady-state pumping condition, estimate
the rate of discharge q in m³/min.

Solution

From Equation 6.74

$$q = \frac{\pi k[(H-s)^2 - t^2]}{2.303[\log(R/r_w)]}\left[1 + \left(0.30 + \frac{10r_w}{H}\right)\sin\frac{1.8s}{H}\right]$$

$k = 0.03$ mm/s $= 0.0018$ m/min

So

$$q = \frac{\pi(0.0018)[(28-8)^2 - 10^2]}{2.303[\log(R/0.4)]}\left\{1 + \left[0.30 + \frac{(10)(0.4)}{28}\right]\sin\frac{1.8(8)}{28}\right\}$$
$$\qquad\qquad\qquad\qquad\qquad\qquad\qquad\qquad\qquad\qquad\uparrow$$

$$= \frac{0.8976}{\log(R/0.4)} \qquad\qquad\qquad\qquad \text{radian}$$

From the equation for q, we can construct the following table:

R (m)	q (m³)
25	0.5
30	0.48
40	0.45
50	0.43
100	0.37

From the aforementioned table, the rate of discharge is approximately
0.45 m³/min.

6.10 FACTORS AFFECTING THE COEFFICIENT OF PERMEABILITY

The coefficient of permeability depends on several factors, most of which are listed in the following:

1. Shape and size of the soil particles.
2. Void ratio. Permeability increases with increase in void ratio.
3. Degree of saturation. Permeability increases with increase in degree of saturation.
4. Composition of soil particles. For sands and silts, this is not important; however, for soils with clay minerals, this is one of the most important factors. Permeability depends on the thickness of water held to the soil particles, which is a function of the cation exchange capacity, valence of the cations, and so forth. Other factors remaining the same, the coefficient of permeability decreases with increasing thickness of the diffuse double layer.
5. Soil structure. Fine-grained soils with a flocculated structure have a higher coefficient of permeability than those with a dispersed structure.
6. Viscosity of the permeant.
7. Density and concentration of the permeant.

6.11 ELECTROOSMOSIS

The coefficient of permeability—and hence the rate of seepage—through clay soils is very small compared to that in granular soils, but the drainage can be increased by the application of an external electric current. This phenomenon is a result of the exchangeable nature of the adsorbed cations in clay particles and the dipolar nature of the water molecules. The principle can be explained with the help of Figure 6.21. When dc electricity is applied to the soil, the cations start to migrate to the cathode, which consists of a perforated metallic pipe. Since water is adsorbed on the cations, it is also dragged along. When the cations reach the cathode, they release the water, and the subsequent build up of pressure causes the water to drain out. This process is called *electroosmosis* and was first used by L. Casagrande in 1939 for soil stabilization in Germany (See Casagrande, 1952).

6.11.1 Rate of drainage by electroosmosis

Figure 6.22 shows a capillary tube formed by clay particles. The surface of the clay particles has negative charges, and the cations are concentrated in a layer of liquid. According to the Helmholtz–Smoluchowski theory (Helmholtz, 1879; Smoluchowski, 1914; see also Mitchell, 1970, 1976), the flow velocity due to an applied dc voltage E can be given by

$$v_e = \frac{D\zeta}{4\pi\eta} \frac{E}{L} \tag{6.79}$$

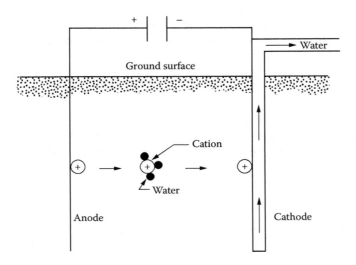

Figure 6.21 Principles of electroosmosis.

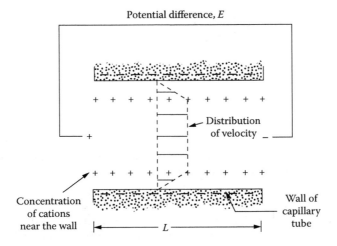

Figure 6.22 Helmholtz–Smoluchowski theory for electroosmosis.

where

υ_e is the flow velocity due to applied voltage
D is the dielectric constant
ζ is the zeta potential
η is the viscosity
L is the electrode spacing

Equation 6.79 is based on the assumptions that the radius of the capillary tube is large compared to the thickness of the diffuse double layer surrounding the clay particles and that all the mobile charge is concentrated near the wall. The rate of flow of water through the capillary tube can be given by

$$q_c = a\upsilon_e \tag{6.80}$$

where a is the area of cross section of the capillary tube.

If a soil mass is assumed to have a number of capillary tubes as a result of interconnected voids, the cross-sectional area A_υ of the voids is

$$A_\upsilon = nA$$

where

A is the gross cross-sectional area of the soil
n is the porosity

The rate of discharge q through a soil mass of gross cross-sectional area A can be expressed by the relation

$$q = A_\upsilon \upsilon_e = nA\upsilon_e = n\frac{D\zeta}{4\pi\eta}\frac{E}{L}A \tag{6.81}$$

or,

$$q = k_e i_e A \tag{6.82}$$

where

$k_e = n(D\zeta/4\pi\eta)$ is the electroosmotic coefficient of permeability
i_e is the electrical potential gradient

The units of k_e can be cm²/(s·V) and the units of i_e can be V/cm.

In contrast to the Helmholtz–Smoluchowski theory (Equation 6.79), which is based on flow through large capillary tubes, Schmid (1951) proposed a theory in which it was assumed that the capillary tubes formed by

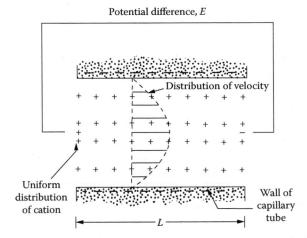

Figure 6.23 Schmid theory for electroosmosis.

the pores between clay particles are small in diameter and that the excess cations are uniformly distributed across the pore cross-sectional area (Figure 6.23). According to this theory

$$v_e = \frac{r^2 A_o F}{8\eta} \frac{E}{L} \tag{6.83}$$

where
   r is the pore radius
   A_o is the volume charge density in pore
   F is the Faraday constant

Based on Equation 6.83, the rate of discharge q through a soil mass of gross cross-sectional area A can be written as

$$q = n\frac{r^2 A_o F}{8\eta} \frac{E}{L} A = k_e i_e A \tag{6.84}$$

where
   n is porosity
   $k_e = n(r^2 A_o F/8\eta)$ is the electroosmotic coefficient of permeability

Without arguing over the shortcomings of the two theories proposed, our purpose will be adequately served by using the flow-rate relation as $q = k_e i_e A$. Some typical values of k_e for several soils are as follows (Mitchell, 1976):

Material	Water content (%)	k_e (cm²/(s · V))
London clay	52.3	5.8×10^{-5}
Boston blue clay	50.8	5.1×10^{-5}
Kaolin	67.7	5.7×10^{-5}
Clayey silt	31.7	5.0×10^{-5}
Rock flour	27.2	4.5×10^{-5}
Na-Montmorillonite	170	2.0×10^{-5}
Na-Montmorillonite	2000	12.0×10^{-5}

These values are of the same order of magnitude and range from 1.5×10^{-5} to 12×10^{-5} cm²/(s · V) with an average of about 6×10^{-5} cm²/(s · V).

Electroosmosis is costly and is not generally used unless drainage by conventional means cannot be achieved. Gray and Mitchell (1967) have studied the factors that affect the amount of water transferred per unit charge passed, such as water content, cation exchange capacity, and free electrolyte content of the soil.

6.12 COMPACTION OF CLAY FOR CLAY LINERS IN WASTE DISPOSAL SITES

When a clay soil is compacted at a lower moisture content, it possesses a flocculent structure. Approximately at the optimum moisture content of compaction, the clay particles have a lower degree of flocculation. A further increase in the moisture content at compaction provides a greater degree of particle orientation; however, the dry unit weight decreases because the added water dilutes the concentration of soil solids per unit volume.

Figure 6.24 shows the results of laboratory compaction tests on a clay soil as well as the variation of the coefficient of permeability of the compacted clay specimens. From the laboratory test results shown, the following observations can be made:

1. For a given compaction effort, the coefficient of permeability k decreases with the increase in molding moisture content, reaching a minimum value at about the optimum moisture content (i.e., approximately where the soil has a higher dry unit weight with the clay particles having a lower degree of flocculation). Beyond the optimum moisture content, the coefficient of permeability increases slightly.

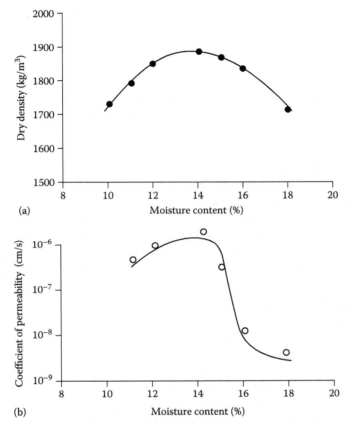

Figure 6.24 Test on clay soil: (a) modified Proctor compaction curve; (b) variation of *k* with molding moisture content.

2. For similar compaction effort and dry unit weight, a soil will have a lower coefficient of permeability when it is compacted on the wet side of the optimum moisture content.

Benson and Daniel (1990) conducted laboratory compaction tests by varying the size of clods of moist clayey soil. These tests show that, for similar compaction effort and molding moisture content, the magnitude of *k* decreases with the decrease in clod size.

In some compaction work in clayey soils, the compaction must be done in a manner so that a certain specified upper level of coefficient of permeability of the soil is achieved. Examples of such works are compaction of the core of an earth dam and installation of clay liners in solid-waste disposal sites.

To prevent groundwater pollution from leachates generated from solid-waste disposal sites, the U.S. Environmental Protection Agency (EPA) requires that clay liners have a hydraulic conductivity of 10^{-7} cm/s or less. To achieve this value, the contractor must ensure that the soil meets the following criteria (U.S. Environmental Protection Agency, 1989):

1. The soil should have at least 20% fines (fine silt and clay-sized particles).
2. The plasticity index (PI) should be greater than 10. Soils that have a PI greater than about 30 are difficult to work with in the field.
3. The soil should not include more than 10% gravel-sized particles.
4. The soil should not contain any particles or chunks of rock that are larger than 25–30 mm.

In many instances, the soil found at the construction site may be somewhat nonplastic. Such soil may be blended with imported clay minerals (like sodium bentonite) to achieve the desired range of coefficient of permeability. In addition, during field compaction, a heavy sheepsfoot roller can introduce larger shear strains during compaction that create a more dispersed structure in the soil. This type of compacted soil will have an even lower coefficient of permeability. Small lifts should be used during compaction so that the feet of the compactor can penetrate the full depth of the lift.

The size of the clay clods has a strong influence on the coefficient of permeability of a compacted clay. Hence, during compaction, the clods must be broken down mechanically to as small as possible. A very heavy roller used for compaction helps to break them down.

Bonding between successive lifts is also an important factor; otherwise, permeant can move through a vertical crack in the compacted clay and then travel along the interface between two lifts until it finds another crack to travel down.

In the construction of clay liners for solid-waste disposal sites where it is required that $k \leq 10^{-7}$ cm/s, it is important to establish the moisture content–unit weight criteria in the laboratory for the soil to be used in field construction. This helps in the development of proper specifications.

Daniel and Benson (1990) developed a procedure to establish the moisture content–unit weight criteria for clayey soils to meet the coefficient of permeability requirement. The following is a step-by-step procedure to develop the criteria.

Step 1: Conduct Proctor tests to establish the dry unit weight versus molding moisture content relationships (Figure 6.25a).

Step 2: Conduct permeability tests on the compacted soil specimens (from Step 1) and plot the results as shown in Figure 6.25b. In this figure, also plot the maximum allowable value of k (i.e., k_{all}).

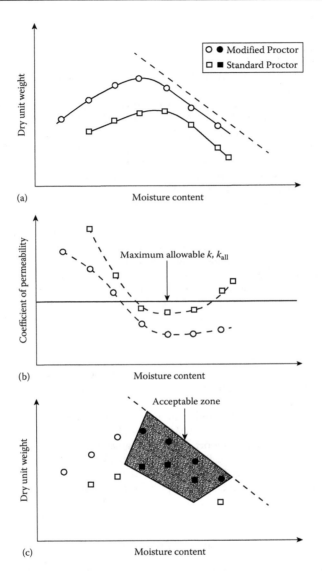

Figure 6.25 (a) Proctor curves; (b) variation of k of compacted specimens; (c) determination of acceptable zone.

Step 3: Replot the dry unit weight–moisture content points (Figure 6.25c) with different symbols to represent the compacted specimens with $k > k_{all}$ and $k \leq k_{all}$.

Step 4: Plot the acceptable zone for which k is less than or equal to k_{all} (Figure 6.25c).

REFERENCES

Amer, A. M. and A. A. Awad, Permeability of cohesionless soils, *J. Geotech. Eng. Div.*, ASCE, 100(12), 1309–1316, 1974.

Basett, D. J. and A. F. Brodie, A study of matabitchuan varved clay, *Ont. Hydro Res. News*, 13(4), 1–6, 1961.

Benson, C. H. and D. E. Daniel, Influence of clods on hydraulic conductivity of compacted clay, *J. Geotech. Eng.*, ASCE, 116(8), 1231–1248, 1990.

Carman, P. E., *Flow of Gases through Porous Media*, Academic, New York, 1956.

Carrier III, W. D., Goodbye Hazen; Hello Kozeny–Carman, *J. Geotech. Geoenviron. Eng.*, ASCE, 129(11), 1054–1056, 2003.

Casagrande, L., Electro-osmotic stabilization of soils, *Journal of The Boston Society of Civil Engineers*, 39(1), 51–83, 1952.

Casagrande, L. and S. J. Poulos, On effectiveness of sand drains, *Can. Geotech. J.*, 6(3), 287–326, 1969.

Chan, H. T. and T. C. Kenney, Laboratory investigation of permeability ratio of new liskeard varved clay, *Can. Geotech. J.*, 10(3), 453–472, 1973.

Chapuis, R. P., Predicting the saturated hydraulic conductivity of sand and gravel using effective diameter and void ratio, *Can. Geotech. J.*, 41(5), 785–795, 2004.

Daniel, D. E. and C. H. Benson, *Water content*–density criteria for compacted clay liners, *J. Geotech. Eng.*, ASCE, 116(12), 1811–1830, 1990.

Darcy, H., *Les Fontaines Publiques de la Ville de Dijon*, Dalmont, Paris, France, 1856.

Forchheimer, P., Discussion of "The Bohio Dam" by George S. Morrison, *Trans.* ASCE, 48, 302, 1902.

Fukushima, S. and T. Ishii, An experimental study of the influence of confining pressure on permeability coefficients of filldam core materials, *Soils Found.*, 26(4), 32–46, 1986.

Gray, D. H. and J. K. Mitchell, Fundamental aspects of electro-osmosis in soils, *J. Soil Mech. Found. Div.*, ASCE, 93(SM6), 209–236, 1967.

Haley, X. and X. Aldrich, Report No. 1—Engineering properties of foundation soils at Long Creek–Fore River areas and Back Cove, Report to Maine State Highway Commission, 1969.

Hansbo, S., Consolidation of clay with special reference to influence of vertical sand drain, *Proc. Swedish Geotech. Inst.*, (18), 41–61, 1960.

Hazen, A., Discussion of "Dams on Sand Foundation" by A. C. Koenig, *Trans.* ASCE, 73, 199, 1911.

Helmholtz, H., Studien über electrische Grenzschichten, *Wiedemanns Ann. Phys.*, 7, 137, 1879.

Kenney, T. C., D. Lau, and G. I. Ofoegbu, Permeability of compacted granular materials, *Can. Geotech. J.*, 21(4), 726–729, 1984.

Kozeny, J., Theorie und Berechnung der Brunnen, *Wasserkr. Wasserwritsch.*, 28, 104, 1933.

Kozeny, J., Ueber kapillare Leitung des Wassers in Boden, *Wien Akad. Wiss.*, 136, part 2a, 271, 1927.

Lumb, P. and J. K. Holt, The undrained shear strength of a soft marine clay from Hong Kong, *Geotechnique*, 18, 25–36, 1968.

Mansur, C. I. and R. I. Kaufman, Dewatering, in *Foundation Engineering*, McGraw-Hill, New York, 241–350, 1962.

Mesri, G. and R. E. Olson, Mechanism controlling the permeability of clays, *Clay Clay Miner.*, 19, 151–158, 1971.

Mitchell, J. K., *Fundamentals of Soil Behavior*, Wiley, New York, 1976.

Mitchell, J. K., In-place treatment of foundation soils, *J. Soil Mech. Found. Div.*, ASCE, 96(SM1), 73–110, 1970.

Muskat, M., *The Flow of Homogeneous Fluids through Porous Media*, McGraw-Hill, New York, 1937.

Olsen, H. W., Hydraulic flow through saturated clay, ScD thesis, Massachusetts Institute of Technology, Cambridge, MA, 1961.

Olsen, H. W., Hydraulic flow through saturated clays, *Proc. 9th Natl. Conf. Clay Clay Miner.*, Lafayette, IN, USA, pp. 131–161, 1962.

Raju, P. S. R. N., N. S. Pandian, and T. S. Nagaraj, Analysis and estimation of coefficient of consolidation, *Geotech. Test. J.*, ASTM, 18(2), 252–258, 1995.

Samarsinghe, A. M., Y. H. Huang, and V. P. Drnevich, Permeability and consolidation of normally consolidated soils, *J. Geotech. Eng. Div.*, ASCE, 108(6), 835–850, 1982.

Schmid, G., Zur Elektrochemie Feinporiger Kapillarsystems, *Zh. Elektrochem.*, 54, 425, 1950; 55, 684, 1951.

Smoluchowski, M., in L. Graetz (Ed.), Elektrische Endosmose and Stromungsstrome, *Handbuch der Elektrizital und Magnetismus*, vol. 2, Barth, Leipzig, Germany, 1914.

Spangler, M. G. and R. L. Handy, *Soil Engineering*, 3rd edn., Intext Educational, New York, 1973.

Subbaraju, B. H., Field performance of drain wells designed expressly for strength gain in soft marine clays, *Proc. 8th Int. Conf. Soil Mech. Found. Eng.*, Moscow, vol. 2.2, pp. 217–220, 1973.

Tavenas, F., P. Jean, P. Leblong, and S. Leroueil, The permeability of natural soft clays. Part II: Permeability characteristics, *Can. Geotech. J.*, 20(4), 645–660, 1983a.

Tavenas, F., P. Leblong, P. Jean, and S. Leroueil, The permeability of natural soft clays. Part I: Methods of laboratory measurement, *Can. Geotech. J.*, 20(4), 629–644, 1983b.

Taylor, D. W., *Fundamentals of Soil Mechanics*, Wiley, New York, 1948.

Tsien, S. I., Stability of marsh deposits, *Bull. 15*, Highway Research Board, 15–43, 1955.

U.S. Bureau of Reclamation, Department of the Interior, *Design of Small Dams*, U.S. Govt. Print office, Washington, D.C., 1961.

U.S. Department of the Navy, Naval Facilities Engineering Command, *Design Manual—Soil Mechanics, Foundations and Earth Structures*, NAVFAC DM-7, Washington, DC, 1971.

U.S. Environmental Protection Agency, *Requirements for Hazard Waste Landfill Design, Construction and Closure*, Publication No. EPA-625/4-89-022, Cincinnati, OH, 1989.

Van Bavel, C. H. M. and D. Kirkham, Field measurement of soil permeability using auger holes, *Soil Sci. Soc. Am., Proc.*, 13, 90–96, 1948.

Chapter 7

Seepage

7.1 INTRODUCTION

In many practical cases, the nature of the flow of water through soil is such that the velocity and gradient vary throughout the medium. For these problems, calculation of flow is generally made by use of graphs referred to as *flow nets*. The concept of the flow net is based on Laplace's equation of continuity, which describes the steady flow condition for a given point in the soil mass. In this chapter, we will derive Laplace's equation of continuity and study its applications as related to problems such as the flow under hydraulic structures and seepage through earth dams.

7.2 EQUATION OF CONTINUITY

To derive the equation of continuity of flow, consider an elementary soil prism at point A (Figure 7.1b) for the hydraulic structure shown in Figure 7.1a. The flows entering the soil prism in the x, y, and z directions can be given from Darcy's law as

$$q_x = k_x i_x A_x = k_x \frac{\partial h}{\partial x} dy \, dz \tag{7.1}$$

$$q_y = k_y i_y A_y = k_y \frac{\partial h}{\partial y} dx \, dz \tag{7.2}$$

$$q_z = k_z i_z A_z = k_z \frac{\partial h}{\partial z} dx \, dy \tag{7.3}$$

where
   q_x, q_y, q_z are the flow entering in directions x, y, and z, respectively
   k_x, k_y, k_z are the coefficients of permeability in directions x, y, and z, respectively
   h is the hydraulic head at point A

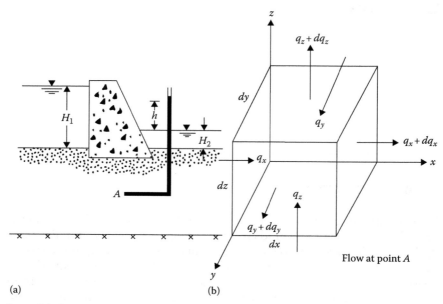

Figure 7.1 Derivation of continuity equation: (a) hydraulic structure; (b) flow in an elementary soil prism at A.

The respective flows leaving the prism in the x, y, and z directions are

$$q_x + dq_x = k_x(i_x + di_x)A_x$$

$$= k_x \left(\frac{\partial h}{\partial x} + \frac{\partial^2 h}{\partial x^2} dx \right) dy\, dz \tag{7.4}$$

$$q_y + dq_y = k_y \left(\frac{\partial h}{\partial y} + \frac{\partial^2 h}{\partial y^2} dy \right) dx\, dz \tag{7.5}$$

$$q_z + dq_z = k_z \left(\frac{\partial h}{\partial z} + \frac{\partial^2 h}{\partial z^2} dz \right) dx\, dy \tag{7.6}$$

For steady flow through an incompressible medium, the flow entering the elementary prism is equal to the flow leaving the elementary prism. So,

$$q_x + q_y + q_z = (q_x + dq_x) + (q_y + dq_y) + (q_z + dq_z)$$

Substituting Equations 7.1 through 7.6 in the preceeding equation we obtain

$$k_x \frac{\partial^2 h}{\partial x^2} + k_y \frac{\partial^2 h}{\partial y^2} + k_z \frac{\partial^2 h}{\partial z^2} = 0 \tag{7.7}$$

For two-dimensional flow in the xz plane, Equation 7.7 becomes

$$k_x \frac{\partial^2 h}{\partial x^2} + k_z \frac{\partial^2 h}{\partial z^2} = 0 \tag{7.8}$$

If the soil is isotropic with respect to permeability, $k_x = k_z = k$, and the continuity equation simplifies to

$$\frac{\partial^2 h}{\partial x^2} + \frac{\partial^2 h}{\partial z^2} = 0 \tag{7.9}$$

This is generally referred to as Laplace's equation.

7.2.1 Potential and stream functions

Consider a function $\phi(x, z)$ such that

$$\frac{\partial \phi}{\partial x} = v_x = -k \frac{\partial h}{\partial x} \tag{7.10}$$

and

$$\frac{\partial \phi}{\partial z} = v_z = -k \frac{\partial h}{\partial z} \tag{7.11}$$

If we differentiate Equation 7.10 with respect to x and Equation 7.11 with respect to z and substitute in Equation 7.9, we get

$$\frac{\partial^2 \phi}{\partial x^2} + \frac{\partial^2 \phi}{\partial z^2} = 0 \tag{7.12}$$

Therefore, $\phi(x, z)$ satisfies the Laplace equation. From Equations 7.10 and 7.11

$$\phi(x, z) = -kh(x, z) + f(z) \tag{7.13}$$

and

$$\phi(x, z) = -kh(x, z) + g(x) \tag{7.14}$$

Since x and z can be varied independently, $f(z) = g(x) = C$, a constant. So

$$\phi(x, z) = -kh(x, z) + C$$

and

$$h(x, z) = \frac{1}{k}[C - \phi(x,\ z)] \tag{7.15}$$

If $h(x,\ z)$ is a constant equal to h_1, Equation 7.15 represents a curve in the xz plane. For this curve, ϕ will have a constant value ϕ_1. This is an *equipotential line*. So, by assigning to ϕ a number of values such as ϕ_1, ϕ_2, ϕ_3, ..., we can get a number of equipotential lines along which $h = h_1$, h_2, h_3, ..., respectively. The slope along an equipotential line ϕ can now be derived:

$$d\phi = \frac{\partial \phi}{\partial x} dx + \frac{\partial \phi}{\partial z} dz \tag{7.16}$$

If ϕ is a constant along a curve, $d\phi = 0$. Hence

$$\left(\frac{dz}{dx}\right)_\phi = -\frac{\partial \phi / \partial x}{\partial \phi / \partial z} = -\frac{\upsilon_x}{\upsilon_z} \tag{7.17}$$

Again, let $\psi(x,\ z)$ be a function such that

$$\frac{\partial \psi}{\partial z} = \upsilon_x = -k \frac{\partial h}{\partial z} \tag{7.18}$$

and

$$-\frac{\partial \psi}{\partial x} = \upsilon_z = -k \frac{\partial h}{\partial z} \tag{7.19}$$

Combining Equations 7.10 and 7.18, we obtain

$$\frac{\partial \phi}{\partial x} = \frac{\partial \psi}{\partial z}$$

$$\frac{\partial^2 \psi}{\partial z^2} = \frac{\partial^2 \phi}{\partial x \, \partial z} \qquad (7.20)$$

Again, combining Equations 7.11 and 7.19

$$-\frac{\partial \phi}{\partial z} = \frac{\partial \psi}{\partial x}$$

$$-\frac{\partial^2 \phi}{\partial x \, \partial z} = \frac{\partial^2 \psi}{\partial x^2} \qquad (7.21)$$

From Equations 7.20 and 7.21

$$\frac{\partial^2 \psi}{\partial x^2} + \frac{\partial^2 \psi}{\partial z^2} = -\frac{\partial^2 \phi}{\partial x \partial z} + \frac{\partial^2 \phi}{\partial x \partial y} = 0$$

So $\psi(x, z)$ also satisfies Laplace's equation. If we assign to $\psi(x, z)$ various values ψ_1, ψ_2, ψ_3, ..., we get a family of curves in the xz plane. Now

$$d\psi = \frac{\partial \psi}{\partial x} dx + \frac{\partial \psi}{\partial z} dz \qquad (7.22)$$

For a given curve, if ψ is constant, then $d\psi = 0$. Thus, from Equation 7.22

$$\left(\frac{dz}{dx} \right)_\psi = \frac{\partial \psi / \partial x}{\partial \psi / \partial z} = \frac{\upsilon_z}{\upsilon_x} \qquad (7.23)$$

Note that the slope $(dz/dx)_\psi$ is in the same direction as the resultant velocity. Hence, the curves $\psi = \psi_1$, ψ_2, ψ_3, ... are the flow lines.

From Equations 7.17 and 7.23, we can see that at a given point (x, z) the equipotential line and the flow line are orthogonal.

The functions $\phi(x, z)$ and $\psi(x, z)$ are called the potential function and the stream function, respectively.

7.3 USE OF CONTINUITY EQUATION FOR SOLUTION OF SIMPLE FLOW PROBLEM

To understand the role of the continuity equation (Equation 7.9), consider a simple case flow of water through two layers of soil as shown in Figure 7.2. The flow is in one direction only, that is, in the direction of the x axis. The lengths of the two soil layers (L_A and L_B) and their coefficients of permeability in the direction of the x axis (k_A and k_B) are known. The total heads

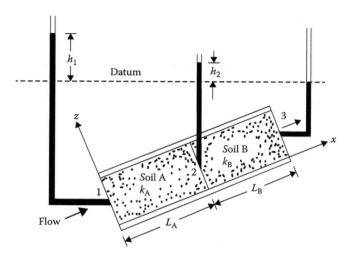

Figure 7.2 One-directional flow through two layers of soil.

at sections 1 and 3 are known. We are required to plot the total head at any other section for $0 < x < L_A + L_B$.

For one-dimensional flow, Equation 7.9 becomes

$$\frac{\partial^2 h}{\partial x^2} = 0 \tag{7.24}$$

Integration of Equation 7.24 twice gives

$$h = C_2 x + C_1 \tag{7.25}$$

where C_1 and C_2 are constants.

For flow through soil A, the boundary conditions are

1. At $x = 0$, $h = h_1$
2. At $x = L_A$, $h = h_2$

However, h_2 is unknown ($h_1 > h_2$). From the first boundary condition and Equation 7.25, $C_1 = h_1$. So

$$h = C_2 x + h_1 \tag{7.26}$$

From the second boundary condition and Equation 7.25,

$$h_2 = C_2 L_A + h_1 \quad \text{or} \quad C_2 = \frac{(h_2 - h_1)}{L_A}$$

So

$$h = -\frac{h_1 - h_2}{L_A}x + h_1 \quad 0 \le x \le L_A \tag{7.27}$$

For flow through soil B, the boundary conditions for solution of C_1 and C_2 in Equation 7.25 are

1. At $x = L_A$, $h = h_2$
2. At $x = L_A + L_B$, $h = 0$

From the first boundary condition and Equation 7.25, $h_2 = C_2L_A + C_1$, or

$$C_1 = h_2 - C_2L_A \tag{7.28}$$

Again, from the secondary boundary condition and Equation 7.25, $0 = C_2(L_A + L_B) + C_1$, or

$$C_1 = -C_2(L_A + L_B) \tag{7.29}$$

Equating the right-hand sides of Equations 7.28 and 7.29,

$$h_2 - C_2L_A = -C_2(L_A + L_B)$$

$$C_2 = -\frac{h_2}{L_B} \tag{7.30}$$

and then substituting Equation 7.30 into Equation 7.28 gives

$$C_1 = h_2 + \frac{h_2}{L_B}L_A = h_2\left(1 + \frac{L_A}{L_B}\right) \tag{7.31}$$

Thus, for flow through soil B,

$$h = -\frac{h_2}{L_B}x + h_2\left(1 + \frac{L_A}{L_B}\right) \quad L_A \le x \le L_A + L_B \tag{7.32}$$

With Equations 7.27 and 7.32, we can solve for h for any value of x from 0 to $L_A + L_B$, provided that h_2 is known. However

q = rate of flow through soil A = rate of flow through soil B

So

$$q = k_A \left(\frac{h_1 - h_2}{L_A} \right) A = k_B \left(\frac{h_2}{L_B} \right) A \tag{7.33}$$

where

k_A and k_B are the coefficients of permeability of soils A and B, respectively

A is the area of cross section of soil perpendicular to the direction of flow

From Equation 7.33

$$h_2 = \frac{k_A h_1}{L_A \left(k_A / L_A + k_B / L_B \right)} \tag{7.34}$$

Substitution of Equation 7.34 into Equations 7.27 and 7.32 yields, after simplification,

$$h = h_1 \left(1 - \frac{k_B x}{k_A L_B + k_B L_A} \right) (\text{for } x = 0 \text{ to } L_A) \tag{7.35}$$

$$h = h_1 \left[\frac{k_A}{k_A L_B + k_B L_A} (L_A + L_B - x) \right] [\text{for } x = L_A \text{ to } (L_A + L_B)] \tag{7.36}$$

7.4 FLOW NETS

7.4.1 Definition

A set of flow lines and equipotential lines is called a flow net. As discussed in Section 7.2, a flow line is a line along which a water particle will travel. An equipotential line is a line joining the points that show the same piezometric elevation (i.e., hydraulic head = $h(x, z)$ = constant). Figure 7.3 shows an example of a flow net for a single row of sheet piles. The permeable layer is isotropic with respect to the coefficient of permeability, that is, $k_x = k_z = k$. Note that the solid lines in Figure 7.3 are the flow lines and the broken lines are the equipotential lines. In drawing a flow net, the boundary conditions must be kept in mind. For example, in Figure 7.3,

1. AB is an equipotential line.
2. EF is an equipotential line.

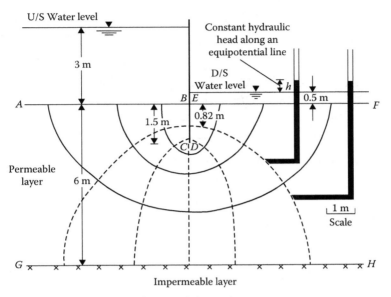

Figure 7.3 Flow net around a single row of sheet pile structures.

3. *BCDE* (i.e., the sides of the sheet pile) is a flow line.
4. *GH* is a flow line.

The flow lines and the equipotential lines are drawn by trial and error. It must be remembered that the flow lines intersect the equipotential lines at right angles. The flow and equipotential lines are usually drawn in such a way that the flow elements are approximately squares. Drawing a flow net is time consuming and tedious because of the trial-and-error process involved. Once a satisfactory flow net has been drawn, it can be traced out.

Some other examples of flow nets are shown in Figures 7.4 and 7.5 for flow under dams.

7.4.2 Calculation of seepage from a flow net under a hydraulic structure

A *flow channel* is the strip located between two adjacent flow lines. To calculate the seepage under a hydraulic structure, consider a flow channel as shown in Figure 7.6.

The equipotential lines crossing the flow channel are also shown, along with their corresponding hydraulic heads. Let Δq be the flow through the

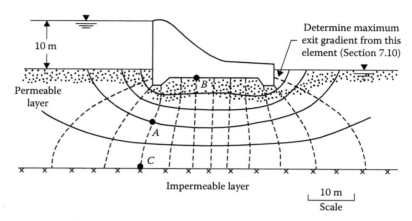

Figure 7.4 Flow net under a dam.

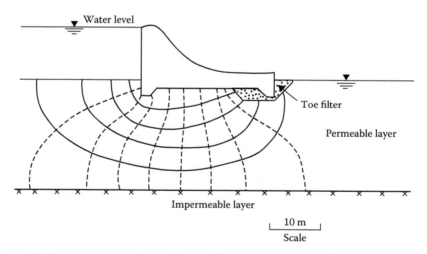

Figure 7.5 Flow net under a dam with a toe filter.

flow channel per unit length of the hydraulic structure (i.e., perpendicular to the section shown). According to Darcy's law

$$\Delta q = kiA = k\left(\frac{h_1 - h_2}{l_i}\right)(b_1 \times 1) = k\left(\frac{h_2 - h_3}{l_2}\right)(b_2 \times 1)$$

$$= k\left(\frac{h_3 - h_4}{l_3}\right)(b_3 \times 1) = \cdots \tag{7.37}$$

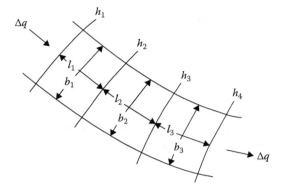

Figure 7.6 Flow through a flow channel.

If the flow elements are drawn as squares, then

$$l_1 = b_1$$

$$l_2 = b_2$$

$$l_3 = b_3$$

$$\vdots$$

So, from Equation 7.37, we get

$$h_1 - h_2 = h_2 - h_3 = h_3 - h_4 = \cdots = \Delta h = \frac{h}{N_d} \tag{7.38}$$

where
    Δh is the potential drop (= drop in piezometric elevation between two consecutive equipotential lines)
    h is the total hydraulic head (= difference in elevation of water between the upstream and downstream side)
    N_d is the number of potential drops

Equation 7.38 demonstrates that the loss of head between any two consecutive equipotential lines is the same. Combining Equations 7.37 and 7.38 gives

$$\Delta q = k \frac{h}{N_d} \tag{7.39}$$

If there are N_f flow channels in a flow net, the rate of seepage per unit length of the hydraulic structure is

$$q = N_f \Delta q = kh \frac{N_f}{N_d} \tag{7.40}$$

Although flow nets are usually constructed in such a way that all flow elements are approximately squares, that need not always be the case. We could construct flow nets with all the flow elements drawn as rectangles. In that case, the width-to-length ratio of the flow nets must be a constant, that is

$$\frac{b_1}{l_1} = \frac{b_2}{l_2} = \frac{b_3}{l_3} = \cdots = n \tag{7.41}$$

For such flow nets, the rate of seepage per unit length of hydraulic structure can be given by

$$q = kh \frac{N_f}{N_d} n \tag{7.42}$$

Example 7.1

For the flow net shown in Figure 7.4

   a. How high would water rise if a piezometer is placed at (i) A (ii) B (iii) C?
   b. If $k = 0.01$ mm/s, determine the seepage loss of the dam in m³/(day · m).

Solution

The maximum hydraulic head h is 10 m. In Figure 7.4, $N_d = 12$, $\Delta h = h/N_d = 10/12 = 0.833$.

Part a(i):

To reach A, water must go through three potential drops. So head lost is equal to $3 \times 0.833 = 2.5$ m. Hence, the elevation of the water level in the piezometer at A will be $10 - 2.5 = 7.5$ m above the ground surface.

Part a(ii):

The water level in the piezometer above the ground level is $10 - 5(0.833) = 5.84$ m.

Part a(iii):

Points A and C are located on the same equipotential line. So water in a piezometer at C will rise to the same elevation as at A, that is, 7.5 m above the ground surface.

Part b:

The seepage loss is given by $q = kh(N_f/N_d)$. From Figure 7.4, $N_f = 5$ and $N_d = 12$. Since

$$k = 0.01\,\text{mm/s} = \left(\frac{0.01}{1000}\right)(60\times60\times24) = 0.864\,\text{m/day}$$

$$q = 0.864(10)\left(\frac{5}{12}\right) = 3.6\,\text{m}^3/(\text{day}\cdot\text{m})$$

Example 7.2

Seepage takes place around a retaining wall shown in Figure 7.7. The coefficient of permeability of the sand is 2×10^{-3} cm/s. The retaining wall is 30 m long. Determine the quantity of seepage across the entire wall per day.

Solution

For the flow net shown in Figure 7.7, $N_f = 3$ and $N_d = 10$. The total head loss from right to left, $H = 5.0$ m. The flow rate is given by (Equation 7.40),

$$q = kh\frac{N_f}{N_d} = (2\times10^{-5}\,\text{m/s})(5.0)\left(\frac{3}{10}\right) = 3.0\times10^{-5}\,\text{m}^3/\text{s/m}$$

Seepage across the entire wall,

$$Q = 3.0\times10^{-5} \times 30.0\times24\times3600 \text{ m}^3/\text{day} = \textbf{77.76 m}^3/\textbf{day}$$

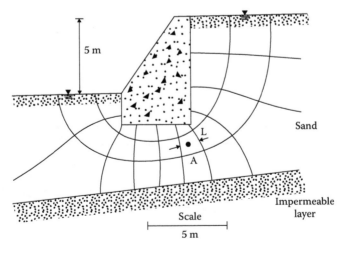

Figure 7.7 Flow net around a retaining wall.

7.5 HYDRAULIC UPLIFT FORCE UNDER A STRUCTURE

Flow nets can be used to determine the hydraulic uplifting force under a structure. The procedure can best be explained through a numerical example. Consider the dam section shown in Figure 7.4, the cross section of which has been replotted in Figure 7.8. To find the pressure head at point D (Figure 7.8), we refer to the flow net shown in Figure 7.4; the pressure head is equal to $(10 + 3.34 \text{ m})$ minus the hydraulic head loss. Point D coincides with the third equipotential line beginning with the upstream side, which means that the hydraulic head loss at that point is $2(h/N_d) = 2(10/12) = 1.67$ m. So

Pressure head at $D = 13.34 - 1.67 = 11.67$ m

Similarly

Pressure head at $E = (10 + 3.34) - 3(10/12) = 10.84$ m

Pressure head at $F = (10 + 1.67) - 3.5(10/12) = 8.75$ m

(Note that point F is approximately midway between the fourth and the fifth equipotential lines starting from the upstream side.)

Pressure head at $G = (10 + 1.67) - 8.5(10/12) = 4.56$ m

Pressure head at $H = (10 + 3.34) - 9(10/12) = 5.84$ m

Pressure head at $I = (10 + 3.34) - 10(10/12) = 5$ m

The pressure heads calculated earlier are plotted in Figure 7.8. Between points F and G, the variation of pressure heads will be approximately

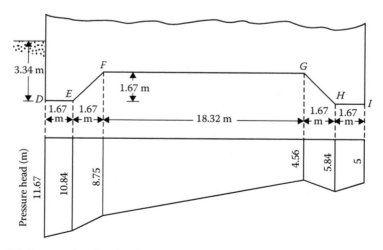

Figure 7.8 Pressure head under the dam section shown in Figure 7.4.

linear. The hydraulic uplift force per unit length of the dam, U, can now be calculated as

$$U = \gamma_w(\text{area of the pressure head diagram}) \quad (1)$$

$$= 9.81\left[\left(\frac{11.67 + 10.84}{2}\right)(1.67) + \left(\frac{10.84 + 8.75}{2}\right)(1.67)\right.$$

$$+ \left(\frac{8.75 + 4.56}{2}\right)(18.32) + \left(\frac{4.56 + 5.84}{2}\right)(1.67)$$

$$+ \left.\left(\frac{5.84 + 5}{2}\right)(1.67)\right]$$

$$= 9.81(18.8 + 16.36 + 121.92 + 8.68 + 9.05)$$

$$= 1714.9\,\text{kN/m}$$

7.6 FLOW NETS IN ANISOTROPIC MATERIAL

In developing the procedure described in Section 7.4 for plotting flow nets, we assumed that the permeable layer is isotropic, that is, $k_{\text{horizontal}} = k_{\text{vertical}} = k$. Let us now consider the case of constructing flow nets for seepage through soils that show anisotropy with respect to permeability. For two-dimensional flow problems, we refer to Equation 7.8:

$$k_x\frac{\partial^2 h}{\partial x^2} + k_z\frac{\partial^2 h}{\partial z^2} = 0$$

where
$k_x = k_{\text{horizontal}}$
$k_z = k_{\text{vertical}}$

This equation can be rewritten as

$$\frac{\partial^2 h}{(k_z/k_x)\partial x^2} + \frac{\partial^2 h}{\partial z^2} = 0 \quad (7.43)$$

Let $x' = \sqrt{k_z/k_x}\,x$, then

$$\frac{\partial^2 h}{(k_z/k_x)\partial x^2} = \frac{\partial^2 h}{\partial x'^2} \quad (7.44)$$

Substituting Equation 7.44 into Equation 7.43, we obtain

$$\frac{\partial^2 h}{\partial x'^2} + \frac{\partial^2 h}{\partial z^2} = 0 \tag{7.45}$$

Equation 7.45 is of the same form as Equation 7.9, which governs the flow in isotropic soils and should represent two sets of orthogonal lines in the $x'z$ plane. The steps for construction of a flow net in an anisotropic medium are as follows:

1. To plot the section of the hydraulic structure, adopt a *vertical scale*.
2. Determine $\sqrt{k_z/k_x} = \sqrt{k_{vertical}/k_{horizontal}}$.
3. Adopt a horizontal scale such that $scale_{horizontal} = \sqrt{k_z/k_x}\,(scale_{vertical})$.
4. With the scales adopted in steps 1 and 3, plot the cross section of the structure.
5. Draw the flow net for the transformed section plotted in step 4 in the same manner as is done for seepage through isotropic soils.
6. Calculate the rate of seepage as

$$q = \sqrt{k_x k_z}\, h \frac{N_f}{N_d} \tag{7.46}$$

Compare Equations 7.39 and 7.46. Both equations are similar except for the fact that k in Equation 7.39 is replaced by $\sqrt{k_x k_z}$ in Equation 7.46.

Example 7.3

A dam section is shown in Figure 7.9a. The coefficients of permeability of the permeable layer in the vertical and horizontal directions are 2×10^{-2} and 4×10^{-2} mm/s, respectively. Draw a flow net and calculate the seepage loss of the dam in $m^3/(day \cdot m)$.

Solution

From the given data

$$k_z = 2 \times 10^{-2}\,mm/s = 1.728\ m/day$$

$$k_x = 4 \times 10^{-2}\,mm/s = 3.456\ m/day$$

and $h = 10$ m. For drawing the flow net,

$$\text{Horizontal scale} = \sqrt{\frac{2 \times 10^{-2}}{4 \times 10^{-2}}}\,(\text{vertical scale})$$

$$= \frac{1}{\sqrt{2}}\,(\text{vertical scale})$$

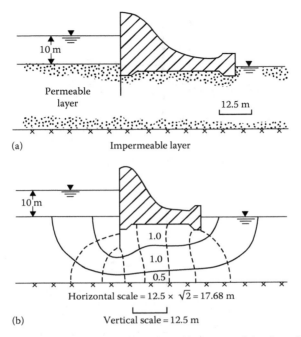

10 m

Permeable
layer

12.5 m

(a)

Impermeable layer

10 m

1.0

1.0

0.5

Horizontal scale = 12.5 × √2 = 17.68 m

(b)

Vertical scale = 12.5 m

Figure 7.9 Construction of flow net under a dam: (a) section of the dam; (b) flow net.

On the basis of this, the dam section is replotted, and the flow net drawn as in Figure 7.9b. The rate of seepage is given by $q = \sqrt{k_x k_z}\, h(N_f/N_d)$. From Figure 7.9b, $N_d = 8$ and $N_f = 2.5$ (the lowermost flow channel has a width-to-length ratio of 0.5). So

$$q = \sqrt{(1.728)(3.456)}(10)(2.5/8) = 7.637\ \text{m}^3/(\text{day}\cdot\text{m})$$

Example 7.4

A single row of sheet pile structure is shown in Figure 7.10a. Draw a flow net for the transformed section. Replot this flow net in the natural scale also. The relationship between the permeabilities is given as $k_x = 6k_z$.

Solution

For the transformed section

$$\text{Horizontal scale} = \sqrt{\frac{k_z}{k_x}}\,(\text{vertical scale})$$

$$= \frac{1}{\sqrt{6}}\,(\text{vertical scale})$$

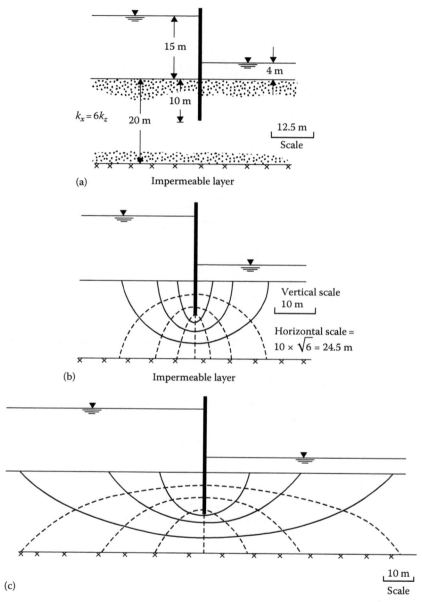

Figure 7.10 Flow net construction in anisotropic soil: (a) sheet pile structure; (b) flow net in transformed scale; (c) flow net in natural scale.

The transformed section and the corresponding flow net are shown in Figure 7.10b.

Figure 7.10c shows the flow net constructed to the natural scale. One important fact to be noticed from this is that when the soil is anisotropic with respect to permeability, *the flow and equipotential lines are not necessarily orthogonal.*

7.7 CONSTRUCTION OF FLOW NETS FOR HYDRAULIC STRUCTURES ON NONHOMOGENEOUS SUBSOILS

The flow net construction technique described in Section 7.4 is for the condition where the subsoil is homogeneous. Rarely in nature do such ideal conditions occur; in most cases, we encounter stratified soil deposits (such as those shown in Figure 7.13 later in the chapter). When a flow net is constructed across the boundary of two soils with different permeabilities, the flow net deflects at the boundary. This is called a *transfer condition*. Figure 7.11 shows a general condition where a flow channel crosses the boundary of two soils. Soil layers 1 and 2 have permeabilities of k_1 and k_2, respectively. The dashed lines drawn across the flow channel are the equipotential lines.

Let Δh be the loss of hydraulic head between two consecutive equipotential lines. Considering a unit length perpendicular to the section shown, the rate of seepage through the flow channel is

$$\Delta q = k_1 \frac{\Delta h}{l_1}(b_1 \times 1) = k_2 \frac{\Delta h}{l_2}(b_2 \times 1)$$

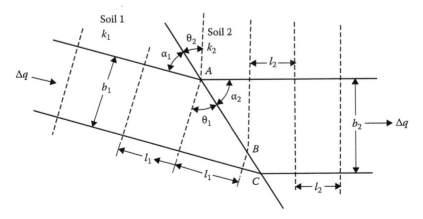

Figure 7.11 Transfer condition.

or

$$\frac{k_1}{k_2} = \frac{b_2/l_2}{b_1/l_1} \tag{7.47}$$

where
l_1 and b_1 are the length and width of the flow elements in soil layer 1
l_2 and b_2 are the length and width of the flow elements in soil layer 2

Referring again to Figure 7.11

$$l_1 = AB \sin\theta_1 = AB \cos\alpha_1 \tag{7.48a}$$

$$l_2 = AB \sin\theta_2 = AB \cos\alpha_2 \tag{7.48b}$$

$$b_1 = AC \cos\theta_1 = AC \sin\alpha_1 \tag{7.48c}$$

$$b_2 = AC \cos\theta_2 = AC \sin\alpha_2 \tag{7.48d}$$

From Equations 7.48a,c

$$\frac{b_1}{l_1} = \frac{\cos\theta_1}{\sin\theta_1} = \frac{\sin\alpha_1}{\cos\alpha_1}$$

or

$$\frac{b_1}{l_1} = \frac{1}{\tan\theta_1} = \tan\alpha_1 \tag{7.49}$$

Also, from Equations 7.48b,d

$$\frac{b_2}{l_2} = \frac{\cos\theta_2}{\sin\theta_2} = \frac{\sin\alpha_2}{\cos\alpha_2}$$

or

$$\frac{b_2}{l_2} = \frac{1}{\tan\theta_2} = \tan\alpha_2 \tag{7.50}$$

Combining Equations 7.47, 7.49, and 7.50

$$\frac{k_1}{k_2} = \frac{\tan\theta_1}{\tan\theta_2} = \frac{\tan\alpha_2}{\tan\alpha_1} \tag{7.51}$$

Flow nets in nonhomogeneous subsoils can be constructed using the relations given by Equation 7.51 and other general principles outlined in Section 7.4. It is useful to keep the following points in mind while constructing the flow nets:

1. If $k_1 > k_2$, we may plot square flow elements in layer 1. This means that $l_1 = b_1$ in Equation 7.47. So, $k_1/k_2 = b_2/l_2$. Thus, the flow elements in layer 2 will be rectangles and their width-to-length ratios will be equal to k_1/k_2. This is shown in Figure 7.12a.
2. If $k_1 < k_2$, we may plot square flow elements in layer 1 (i.e., $l_1 = b_1$). From Equation 7.47, $k_1/k_2 = b_2/l_2$. So, the flow elements in layer 2 will be rectangles. This is shown in Figure 7.12b.

An example of the construction of a flow net for a dam section resting on a two-layered soil deposit is given in Figure 7.13. Note that $k_1 = 5 \times 10^{-2}$ mm/s and $k_2 = 2.5 \times 10^{-2}$ mm/s. So

$$\frac{k_1}{k_2} = \frac{5.0 \times 10^{-2}}{2.5 \times 10^{-2}} = 2 = \frac{\tan\alpha_2}{\tan\alpha_1} = \frac{\tan\theta_1}{\tan\theta_2}$$

In soil layer 1, the flow elements are plotted as squares, and since $k_1/k_2 = 2$, the length-to-width ratio of the flow elements in soil layer 2 is 1/2.

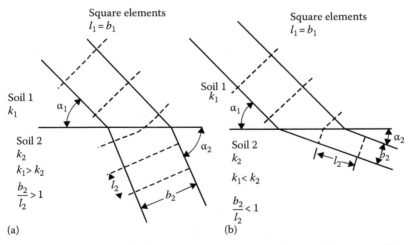

Figure 7.12 Flow channel at the boundary between two soils with different coefficients of permeability: (a) $k_1 > k_2$; (b) $k_1 < k_2$.

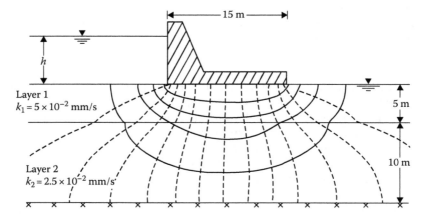

Figure 7.13 Flow net under a dam resting on a two-layered soil deposit.

7.8 NUMERICAL ANALYSIS OF SEEPAGE

7.8.1 General seepage problems

In this section, we develop some approximate finite difference equations for solving seepage problems. We start from Laplace's equation, which was derived in Section 7.2. For two-dimensional seepage

$$k_x \frac{\partial^2 h}{\partial x^2} + k_z \frac{\partial^2 h}{\partial z^2} = 0 \tag{7.52}$$

Figure 7.14 shows a part of a region in which flow is taking place. For flow in the horizontal direction, using Taylor's series, we can write

$$h_1 = h_0 + \Delta x \left(\frac{\partial h}{\partial x} \right)_0 + \frac{(\Delta x)^2}{2!} \left(\frac{\partial^2 h}{\partial x^2} \right)_0 + \frac{(\Delta x)^3}{3!} \left(\frac{\partial^3 h}{\partial x^3} \right)_0 + \cdots \tag{7.53}$$

and

$$h_3 = h_0 - \Delta x \left(\frac{\partial h}{\partial x} \right)_0 + \frac{(\Delta x)^2}{2!} \left(\frac{\partial^2 h}{\partial x^2} \right)_0 - \frac{(\Delta x)^3}{3!} \left(\frac{\partial^3 h}{\partial x^3} \right)_0 + \cdots \tag{7.54}$$

Adding Equations 7.53 and 7.54, we obtain

$$h_1 + h_3 = 2h_0 + \frac{2(\Delta x)^2}{2!} \left(\frac{\partial^2 h}{\partial x^2} \right)_0 + \frac{2(\Delta x)^4}{4!} \left(\frac{\partial^4 h}{\partial x^4} \right)_0 + \cdots \tag{7.55}$$

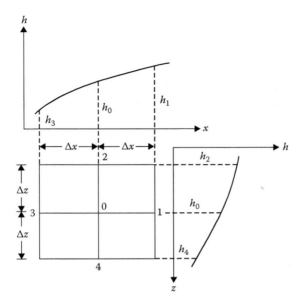

Figure 7.14 Hydraulic heads for flow in a region.

Assuming Δx to be small, we can neglect the third and subsequent terms on the right-hand side of Equation 7.55. Thus

$$\left(\frac{\partial^2 h}{\partial x^2}\right)_0 = \frac{h_1 + h_3 - 2h_0}{(\Delta x)^2} \tag{7.56}$$

Similarly, for flow in the z direction we can obtain

$$\left(\frac{\partial^2 h}{\partial z^2}\right)_0 = \frac{h_2 + h_4 - 2h_0}{(\Delta z)^2} \tag{7.57}$$

Substitution of Equations 7.56 and 7.57 into Equation 7.52 gives

$$k_x \frac{h_1 + h_3 - 2h_0}{(\Delta x)^2} + k_z \frac{h_2 + h_4 - 2h_0}{(\Delta z)^2} = 0 \tag{7.58}$$

If $k_x = k_y = k$ and $\Delta x = \Delta z$, Equation 7.58 simplifies to

$$h_1 + h_2 + h_3 + h_4 - 4h_0 = 0$$

or

$$h_0 = \frac{1}{4}(h_1 + h_2 + h_3 + h_4) \tag{7.59}$$

Equation 7.59 can also be derived by considering Darcy's law, $q = kiA$. For the rate of flow from point 1 to point 0 through the channel shown in Figure 7.15a, we have

$$q_{1-0} = k\frac{h_1 - h_0}{\Delta x}\Delta z \tag{7.60}$$

Similarly

$$q_{0-3} = k\frac{h_0 - h_3}{\Delta x}\Delta z \tag{7.61}$$

$$q_{2-0} = k\frac{h_2 - h_0}{\Delta z}\Delta x \tag{7.62}$$

$$q_{0-4} = k\frac{h_0 - h_4}{\Delta z}\Delta x \tag{7.63}$$

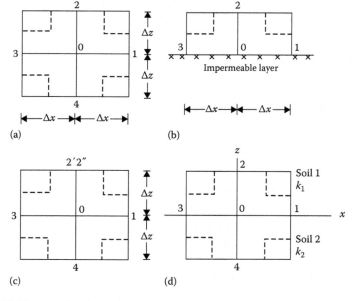

Figure 7.15 Numerical analysis of seepage: (a) derivation of Equation 7.59; (b) derivation of Equation 7.69; (c) derivation of Equation 7.71; (d) derivation of Equation 7.77.

Since the total rate of flow into point 0 is equal to the total rate of flow out of point 0, $q_{in} - q_{out} = 0$. Hence

$$(q_{1-0} + q_{2-0}) - (q_{0-3} + q_{0-4}) = 0 \qquad (7.64)$$

Taking $\Delta x = \Delta z$ and substituting Equations 7.60 through 7.63 into Equation 7.64, we get

$$h_0 = \frac{1}{4}(h_1 + h_2 + h_3 + h_4)$$

If the point 0 is located on the boundary of a pervious and an impervious layer, as shown in Figure 7.15b, Equation 7.59 must be modified as follows:

$$q_{1-0} = k \frac{h_1 - h_0}{\Delta x} \frac{\Delta z}{2} \qquad (7.65)$$

$$q_{0-3} = k \frac{h_0 - h_2}{\Delta x} \frac{\Delta z}{2} \qquad (7.66)$$

$$q_{0-2} = k \frac{h_0 - h_2}{\Delta z} \Delta x \qquad (7.67)$$

For continuity of flow

$$q_{1-0} - q_{0-3} - q_{0-2} = 0 \qquad (7.68)$$

With $\Delta x = \Delta z$, combining Equations 7.65 through 7.68 gives

$$\frac{h_1 - h_0}{2} - \frac{h_0 - h_3}{2} - (h_0 - h_2) = 0$$

$$\frac{h_1}{2} + \frac{h_3}{2} + h_2 - 2h_0 = 0$$

or

$$h_0 = \frac{1}{4}(h_1 + 2h_2 + h_3) \qquad (7.69)$$

When point 0 is located at the bottom of a piling (Figure 7.15c), the equation for the hydraulic head for flow continuity can be given by

$$q_{1-0} + q_{4-0} - q_{0-3} - q_{0-2'} - q_{0-2''} = 0 \tag{7.70}$$

Note that $2'$ and $2''$ are two points at the same elevation on the opposite sides of the sheet pile with hydraulic heads of $h_{2'}$ and $h_{2''}$, respectively. For this condition, we can obtain (for $\Delta x = \Delta z$), through a similar procedure to that mentioned earlier,

$$h_0 = \frac{1}{4}\left[h_1 + \frac{1}{2}(h_{2'} + h_{2''}) + h_3 + h_4 \right] \tag{7.71}$$

7.8.2 Seepage in layered soils

Equation 7.59, which we derived earlier, is valid for seepage in homogeneous soils. However, for the case of flow across the boundary of one homogeneous soil layer to another, Equation 7.59 must be modified. Referring to Figure 7.15d, since the flow region is located half in soil 1 with a coefficient of permeability k_1 and half in soil 2 with a coefficient of permeability k_2, we can say that

$$k_{av} = \frac{1}{2}(k_1 + k_2) \tag{7.72}$$

Now, if we replace soil 2 by soil 1, the replaced soil (i.e., soil 1) will have a hydraulic head of $h_{4'}$ in place of h_4. For the velocity to remain the same

$$k_1 \frac{h_{4'} - h_0}{\Delta z} = k_2 \frac{h_4 - h_0}{\Delta z} \tag{7.73}$$

or

$$h_{4'} = \frac{k_2}{k_1}(h_4 - h_0) + h_0 \tag{7.74}$$

Thus, based on Equation 7.52, we can write

$$\frac{k_1 + k_2}{2} \frac{h_1 + h_3 - 2h_0}{(\Delta x)^2} + k_1 \frac{h_2 + h_{4'} - 2h_0}{(\Delta z)^2} = 0 \tag{7.75}$$

Taking $\Delta x = \Delta z$ and substituting Equation 7.74 into Equation 7.75

$$\frac{1}{2}(k_1 + k_2)\left[\frac{h_1 + h_3 - 2h_0}{(\Delta x)^2}\right] + \frac{k_1}{(\Delta x)^2}\left\{h_2 + \left[\frac{k_2}{k_1}(h_4 - h_0) + h_0\right] - 2h_0\right\} = 0$$

(7.76)

or

$$h_0 = \frac{1}{4}\left(h_1 + \frac{2k_1}{k_1 + k_2}h_2 + h_3 + \frac{2k_2}{k_1 + k_2}h_4\right)$$

(7.77)

The application of the equations developed in this section can best be demonstrated by the use of a numerical example. Consider the problem of determining the hydraulic heads at various points below the dam as shown in Figure 7.13. Let $\Delta x = \Delta z = 1.25$ m. Since the flow net below the dam will be symmetrical, we will consider only the left half. The steps for determining the values of h at various points in the permeable soil layers are as follows:

1. Roughly sketch out a flow net.
2. Based on the rough flow net (step 1), assign some values for the hydraulic heads at various grid points. These are shown in Figure 7.16a. Note that the values of h assigned here are in percent.
3. Consider the heads for row 1 (i.e., $i = 1$). The $h_{(i,j)}$ for $i = 1$ and $j = 1, 2, \ldots, 22$ are 100 in Figure 7.16a; these are correct values based on the boundary conditions. The $h_{(i,j)}$ for $i = 1$ and $j = 23, 24, \ldots, 28$ are estimated values. The flow condition for these grid points is similar to that shown in Figure 7.14b, and according to Equation 7.69, $(h_1 + 2h_2 + h_3) - 4h_0 = 0$, or

$$(h_{(i,j+1)} + 2h_{(i+1,j)} + h_{(i,j-1)}) - 4h_{(i,j)} = 0$$

(7.78)

Since the hydraulic heads in Figure 7.16 are assumed values, Equation 7.78 will not be satisfied. For example, for the grid point $i = 1$ and $j = 23$, $h_{(i,j-1)} = 100$, $h_{(i,j)} = 84$, $h_{(i,j+1)} = 68$, and $h_{(i+1,j)} = 78$. If these values are substituted into Equation 7.78, we get $[68 + 2(78) + 100] - 4(84) = -12$, instead of zero. If we set -12 equal to R (where R stands for *residual*) and add $R/4$ to $h_{(i,j)}$, Equation 7.78 will be satisfied. So the new, corrected value of $h_{(i,j)}$ is equal to $84 + (-3) = 81$, as shown in Figure 7.16b. This is called the relaxation process. Similarly, the corrected head for the grid point $i = 1$ and $j = 24$ can be found as follows:

$$[84 + 2(67) + 61] - 4(68) = 7 = R$$

k_1 k_2

7.5 m

$j = 1$

$i = 1$

i

50	50	50	50	50	50	50	50	50	50	50	50	50	50	50	50	50	50	50	50	50	50	50	50	50	50
50																									
54	54	54	54	53	53	53	53	53	53	53															
54																									
57	58	58	58	57	57	57	57	57	57	57															
57																									
61	62	62	62	61	60	60	60	60	60	60															
61																									
68	67	66	60	65	64	64	64	64	64	64															
67																									
84	73	72	70	68	66	66	66	66	66	66															
78																									
100	80	77	74	72	70	69	68	68	68	68															
90																									

(The body of Figure 7.16(a) is a full grid of hydraulic-head values arranged on an i–j mesh, with constant-head boundaries of 100 (left) and 50 (right). The complete numerical field is shown in the original figure.)

(continued)

Figure 7.16 Hydraulic head calculation by numerical method: (a) initial assumption.

(a)

k_1 ← → k_2

↕ -7.5 m

$j=1$ →→ j

$i=1$

i ↓

100	100	100	100	100	100	100	100	100	100	100	100	100	100	100	100	100	100	100	100	100	81	69.8	61.8	57.3	53.8	50
99	99.3	99.3	99.3	99.3	99.0	99.3	99.3	98.8	98.5	98.0	97.8	97.5	97.0	96.5	95.8	95.3	94.3	92.5	91.8	87.0	78.5	68.5	61.8	57.3	53.8	50
99	99.0	98.8	98.8	98.5	98.3	98.3	98.0	97.8	97.3	96.8	96.5	95.3	95.3	94.0	92.8	91.3	89.5	87.3	84.0	80.8	74.3	67.0	62.0	67.8	54.0	50
99	98.8	98.5	98.3	98.0	96.8	97.3	97.5	96.5	96.0	95.8	94.8	94.8	93.3	92.0	90.5	88.5	86.0	83.3	80.0	76.5	71.5	66.8	62.0	58.0	54.0	50
99	98.6	98.3	97.6	97.4	96.9	96.4	95.9	95.2	94.3	94.3	93.7	92.6	92.1	90.2	88.2	86.2	83.7	81.0	77.9	74.4	70.3	65.8	61.8	57.8	53.8	50
99	98.5	97.5	97.3	96.5	96.0	95.5	95.0	94.0	93.5	92.3	91.3	90.5	88.8	87.3	85.0	82.8	81.0	78.3	75.5	71.8	68.3	65.8	61.0	57.3	53.5	50
99	97.8	97.3	96.5	96.0	95.5	95.0	94.5	93.3	92.8	91.3	89.8	88.0	86.8	84.8	82.5	80.5	78.5	76.3	73.5	70.0	67.0	63.8	60.5	56.8	53.3	50
97	96.5	96.0	95.5	95.0	94.5	94.0	93.3	92.3	91.0	90.0	88.0	86.3	84.8	82.5	81.0	78.8	76.5	74.8	71.5	69.0	66.5	63.5	60.3	56.8	53.3	50
96	95.3	94.5	94.6	93.5	93.0	92.5	92.0	90.5	89.8	88.0	86.3	84.8	82.8	81.5	79.0	77.5	75.3	73.0	70.8	68.3	66.0	63.5	60.3	56.8	53.3	50
95	94.3	93.5	93.0	92.5	92.0	92.0	91.0	89.8	88.5	86.8	85.3	83.5	82.0	80.5	78.5	76.5	74.5	72.5	70.3	68.0	66.0	63.5	60.3	56.8	53.3	50
95	93.8	93.0	92.5	92.0	91.5	91.5	90.5	89.3	88.0	86.5	84.5	83.0	81.5	80.8	78.3	76.0	74.3	72.3	70.3	68.0	66.0	63.5	60.3	56.8	53.3	50
94	93.3	92.5	92.0	91.5	91.0	90.5	90.0	89.0	87.8	86.0	84.3	82.5	81.0	79.5	78.0	76.0	74.0	72.3	70.0	68.0	66.0	63.5	60.3	56.8	53.3	50
94	93.0	92.3	91.8	91.3	90.8	90.3	89.5	88.8	87.8	86.0	84.0	82.0	80.8	79.5	77.8	76.0	74.0	72.0	70.0	68.0	66.0	63.5	60.3	56.8	53.3	50

Figure 7.16 (continued) Hydraulic head calculation by numerical method: (b) at the end of the first interaction.

(b)

Figure 7.16 (continued) Hydraulic head calculation by numerical method: (c) at the end of the tenth iteration.

The figure shows a finite-difference grid of hydraulic-head values (at the end of the tenth iteration). The soil has two permeability zones, k_1 (upper) and k_2 (lower), with a total depth of 7.5 m. Grid indices run $i = 1$ (rows, with arrow i) and $j = 1$ (columns, with arrow j). The computed nodal head values are given below (boundary/upstream nodes = 100, downstream nodes = 50).

100	100	100	100	100	100	100	100	100	100	100	100	100	100	100	100	100	100	100	100	92.0	88.3	82.7	72.4	65.2	59.4	54.5	50
99.0	99.4	99.4	99.3	99.4	99.3	99.2	99.1	99.0	98.8	98.7	98.4	97.9	97.6	97.9	98.3	98.4	98.7	98.8	99.0	86.1	81.9	79.4	71.4	64.7	59.2	54.4	50
99.0	99.0																			82.0	78.1	75.9	69.6	63.9	58.9	54.3	50
99.0	98.8																			79.6	75.8	73.2	68.1	63.1	58.4	54.1	50
99.0	98.6																			76.4	73.0	72.5	67.0	62.4	58.1	54.0	50
99.0	98.2																			74.1	71.1	69.4	65.4	61.5	57.6	53.8	50
97.0	97.7																			72.5	69.7	67.8	64.4	60.8	57.2	53.6	50
96.0	96.5																			71.4	68.9	66.8	63.6	60.3	56.9	53.5	50
95.0	95.6																			70.8	68.3	66.1	63.1	60.0	56.7	53.4	50
95.0	94.7																			70.3	68.0	65.7	62.8	59.8	56.6	53.3	50
94.0	93.8																			70.1	67.8	65.4	62.7	59.7	56.6	53.3	50
94.0	93.6																			70.0	67.8	65.3	62.6	59.7	56.6	53.3	50
																						65.3	62.6	59.7	56.6	53.3	50

(c)

So, $h_{(1,24)} = 68 + 7/4 = 69.75 \approx 69.8$. The corrected values of $h_{(1,25)}$, $h_{(1,26)}$, and $h_{(1,27)}$ can be determined in a similar manner. Note that $h_{(1,28)} = 50$ is correct, based on the boundary condition. These are shown in Figure 7.16b.

4. Consider the rows $i = 2$, 3, and 4. The $h_{(i,j)}$ for $i = 2, ..., 4$ and $j = 2$, 3, ..., 27 should follow Equation 7.59; $(h_1 + h_2 + h_3 + h_4) - 4h_0 = 0$; or

$$(h_{(i,j+1)} + h_{(i-1,j)} + h_{(i,j-1)} + h_{(i+1,j)}) - 4h_{(i,j)} = 0 \qquad (7.79)$$

To find the corrected heads $h_{(i,j)}$, we proceed as in Step 3. The residual R is calculated by substituting values into Equation 7.79, and the corrected head is then given by $h_{(i,j)} + R/4$. Owing to symmetry, the corrected values of $h_{(1,28)}$ for $i = 2$, 3, and 4 are all 50, as originally assumed. The corrected heads are shown in Figure 7.16b.

5. Consider row $i = 5$ (for $j = 2$, 3, ..., 27). According to Equation 7.77

$$h_1 + \frac{2k_1}{k_1 + k_2} h_2 + h_3 + \frac{2k_2}{k_1 + k_2} h_4 - 4h_0 = 0 \qquad (7.80)$$

Since $k_1 = 5 \times 10^{-2}$ mm/s and $k_2 = 2.5 \times 10^{-2}$ mm/s

$$\frac{2k_1}{k_1 + k_2} = \frac{2(5) \times 10^{-2}}{(5 + 2.5) \times 10^{-2}} = 1.333$$

$$\frac{2k_2}{k_1 + k_2} = \frac{2(2.5) \times 10^{-2}}{(5 + 2.5) \times 10^{-2}} = 0.667$$

Using the aforementioned values, Equation 7.80 can be rewritten as

$$h_{(i,j+1)} + 1.333h_{(i-1,j)} + h_{(i,j-1)} + 0.667h_{(i+1,j)} - 4h_{(i,j)} = 0$$

As in step 4, calculate the residual R by using the heads in Figure 7.16a. The corrected values of the heads are given by $h_{(i,j)} + R/4$. These are shown in Figure 7.16b. Note that, owing to symmetry, the head at the grid point $i = 5$ and $j = 28$ is 50, as assumed initially.

6. Consider the rows $i = 6$, 7, ..., 12. The $h_{(i,j)}$ for $i = 6$, 7, ..., 12 and $j = 2$, 3, ..., 27 can be found by using Equation 7.79. Find the corrected head in a manner similar to that in step 4. The heads at $j = 28$ are all 50, as assumed. These values are shown in Figure 7.16b.

7. Consider row $i = 13$. The $h_{(i,j)}$ for $i = 13$ and $j = 2$, 3, ..., 27 can be found from Equation 7.69, $(h_1 + 2h_2 + h_3) - 4h_0 = 0$, or

$$h_{(i,j+1)} + 2h_{(i-1,j)} + h_{(i,j-1)} - 4h_{(i,j)} = 0$$

With proper values of the head given in Figure 7.16a, find the residual and the corrected heads as in step 3. Note that $h_{(13,28)} = 50$ owing to symmetry. These values are given in Figure 7.16b.

8. With the new heads, repeat steps 3 through 7. This iteration must be carried out several times until the residuals are negligible.

Figure 7.16c shows the corrected hydraulic heads after 10 iterations. With these values of h, the equipotential lines can now easily be drawn.

7.9 SEEPAGE FORCE PER UNIT VOLUME OF SOIL MASS

Flow of water through a soil mass results in some force being exerted on the soil itself. To evaluate the *seepage force* per unit volume of soil, consider a soil mass bounded by two flow lines ab and cd and two equipotential lines ef and gh, as shown in Figure 7.17. The soil mass has unit thickness at right angles to the section shown. The self-weight of the soil mass is $(\text{length})(\text{width})(\text{thickness})(\gamma_{sat}) = (L)(L)(1)(\gamma_{sat}) = L^2\gamma_{sat}$. The hydrostatic force on the side ef of the soil mass is $(\text{pressure head})(L)(1) = h_1\gamma_w L$. The hydrostatic force on the side gh of the soil mass is $h_2 L\gamma_w$. For equilibrium

$$\Delta F = h_1\gamma_w L + L^2\gamma_{sat} \sin\alpha - h_2\gamma_w L \tag{7.81}$$

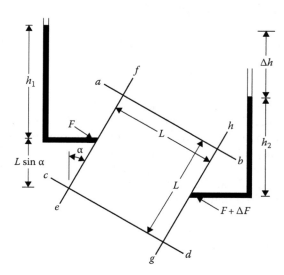

Figure 7.17 Seepage force determination.

However, $h_1 + L \sin \alpha = h_2 + \Delta h$, so

$$h_2 = h_1 + L \sin \alpha - \Delta h \tag{7.82}$$

Combining Equations 7.81 and 7.82

$$\Delta F = h_1 \gamma_w L + L^2 \gamma_{sat} \sin \alpha - (h_1 + L \sin \alpha - \Delta h) \gamma_w L$$

or

$$\Delta F = \underbrace{L^2(\gamma_{sat} - \gamma_w) \sin \alpha}_{} + \underbrace{\Delta h \gamma_w L}_{} = \underbrace{L^2 \gamma' \sin \alpha}_{\substack{\text{effective weight} \\ \text{of soil in the} \\ \text{direction of flow}}} + \underbrace{\Delta h \gamma_w L}_{\substack{\text{seepage} \\ \text{force}}} \tag{7.83}$$

where $\gamma' = \gamma_{sat} - \gamma_w$. From Equation 7.83, we can see that the seepage force on the soil mass considered is equal to $\Delta h \gamma_w L$. Therefore

$$\text{Seepage force per unit volume of soil mass} = \frac{\Delta h \gamma_w L}{L^2}$$

$$= \gamma_w \frac{\Delta h}{L} = \gamma_w i \tag{7.84}$$

where i is the hydraulic gradient.

Example 7.5

Refer to Example 7.2 and Figure 7.7. Estimate the force per unit volume of the sand at A.

Solution

From Equation 7.84, the force per unit volume of the soil mass $= \gamma_w i$.

The hydraulic gradient at A, $i \approx \dfrac{\Delta h}{L}$.

$$\Delta h = \frac{5 \text{ m}}{10} = 0.5 \text{ m}$$

$$L \approx 1.38 \text{ m}$$

$$i \approx \frac{\Delta h}{L} = \frac{0.5}{1.38} = 0.362$$

So, force per unit volume,

$$\gamma_w i = (9.81)(0.362) = \mathbf{3.55 \, kN/m^3}$$

7.10 SAFETY OF HYDRAULIC STRUCTURES AGAINST PIPING

When upward seepage occurs and the hydraulic gradient i is equal to i_{cr}, *piping* or *heaving* originates in the soil mass:

$$i_{cr} = \frac{\gamma'}{\gamma_w}$$

$$\gamma' = \gamma_{sat} - \gamma_w = \frac{G_s\gamma_w + e\gamma_w}{1+e} - \gamma_w = \frac{(G_s - 1)\gamma_w}{1+e}$$

So

$$i_{cr} = \frac{\gamma'}{\gamma_w} = \frac{G_s - 1}{1+e} \qquad (7.85)$$

For the combinations of G_s and e generally encountered in soils, i_{cr} varies within a range of about 0.85–1.1.

Harza (1935) investigated the safety of hydraulic structures against piping. According to his work, the factor of safety against piping, F_S, can be defined as

$$F_S = \frac{i_{cr}}{i_{exit}} \qquad (7.86)$$

where i_{exit} is the maximum exit gradient. The maximum exit gradient can be determined from the flow net. Referring to Figure 7.4, the maximum exit gradient can be given by $\Delta h/l$ (Δh is the head lost between the last two equipotential lines, and l the length of the flow element). A factor of safety of 3–4 is considered adequate for the safe performance of the structure. Harza also presented charts for the maximum exit gradient of dams constructed over deep homogeneous deposits (see Figure 7.18). Using the notations shown in Figure 7.18, the maximum exit gradient can be given by

$$i_{exit} = C\frac{h}{B} \qquad (7.87)$$

A theoretical solution for the determination of the maximum exit gradient for a single row of sheet pile structures as shown in Figure 7.3 is available (see Harr, 1962) and is of the form

$$i_{exit} = \frac{1}{\pi}\frac{\text{maximum hydraulic head}}{\text{depth of penetration of sheet pile}} \qquad (7.88)$$

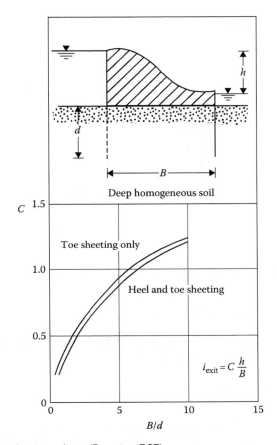

Figure 7.18 Critical exit gradient (Equation 7.87).

Lane (1935) also investigated the safety of dams against piping and sug-gested an empirical approach to the problem. He introduced a term called *weighted creep distance*, which is determined from the shortest flow path:

$$L_w = \frac{\sum L_h}{3} + \sum L_v \qquad (7.89)$$

where
  L_w is the weighted creep distance
  $\sum L_h = L_{h_1} + L_{h_2} + \cdots$ is the sum of horizontal distance along shortest flow path (see Figure 7.19)
  $\sum L_v = L_{v_1} + L_{v_2} + \cdots$ is the sum of vertical distances along shortest flow path (see Figure 7.19)

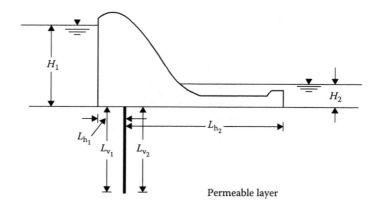

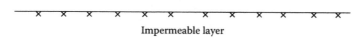

Permeable layer

Impermeable layer

Figure 7.19 Calculation of weighted creep distance.

Once the weighted creep length has been calculated, the weighted creep ratio can be determined as (Figure 7.19)

$$\text{Weighted creep ratio} = \frac{L_w}{H_1 - H_2} \qquad (7.90)$$

For a structure to be safe against piping, Lane (1935) suggested that the weighted creep ratio should be equal to or greater than the safe values shown in Table 7.1.

If the cross section of a given structure is such that the shortest flow path has a slope steeper than 45°, it should be taken as a vertical path. If the slope of the shortest flow path is less than 45°, it should be considered as a horizontal path.

Terzaghi (1922) conducted some model tests with a single row of sheet piles as shown in Figure 7.20 and found that the failure due to piping takes place within a distance of $D/2$ from the sheet piles (D is the depth of penetration of the sheet pile). Therefore, the stability of this type of structure can be determined by considering a soil prism on the downstream side of unit thickness and of section $D \times D/2$. Using the flow net, the hydraulic uplifting pressure can be determined as

$$U = \frac{1}{2}\gamma_w D h_a \qquad (7.91)$$

Table 7.1 Safe values for the weighted creep ratio

Material	Safe weighted creep ratio
Very fine sand or silt	8.5
Fine sand	7.0
Medium sand	6.0
Coarse sand	5.0
Fine gravel	4.0
Coarse gravel	3.0
Soft to medium clay	2.0–3.0
Hard clay	1.8
Hard pan	1.6

where h_a is the average hydraulic head at the base of the soil prism. The submerged weight of the soil prism acting vertically downward can be given by

$$W' = \frac{1}{2}\gamma'D^2 \tag{7.92}$$

Hence, the factor of safety against heave is

$$F_s = \frac{W'}{U} = \frac{\frac{1}{2}\gamma'D^2}{\frac{1}{2}\gamma_w Dh_a} = \frac{D\gamma'}{h_a\gamma_w} \tag{7.93}$$

A factor of safety of about 4 is generally considered adequate.

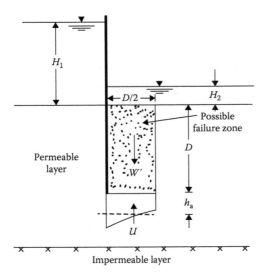

Figure 7.20 Failure due to piping for a single-row sheet pile structure.

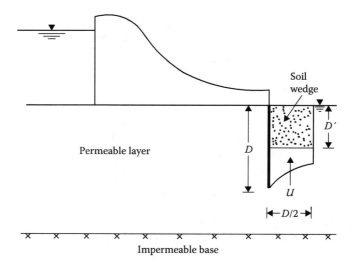

Figure 7.21 Safety against piping under a dam.

For structures other than a single row of sheet piles, such as that shown in Figure 7.21, Terzaghi (1943) recommended that the stability of several soil prisms of size $D/2 \times D' \times 1$ be investigated to find the minimum factor of safety. Note that $0 < D' \leq D$. However, Harr (1962, p. 125) suggested that a factor of safety of 4–5 with $D' = D$ should be sufficient for safe performance of the structure.

Example 7.6

A flow net for a single row of sheet piles is given in Figure 7.3.

    a. Determine the factor of safety against piping by Harza's method.
    b. Determine the factor of safety against piping by Terzaghi's method (Equation 7.93).

Assume $\gamma' = 10.2$ kN/m^3.

Solution

Part a:

$$i_{\text{exit}} = \frac{\Delta h}{L} \quad \Delta h = \frac{3-0.5}{N_d} = \frac{3-0.5}{6} = 0.417\,\text{m}$$

The length of the last flow element can be scaled out of Figure 7.3 and is approximately 0.82 m. So

$$i_{\text{exit}} = \frac{0.417}{0.82} = 0.509$$

(We can check this with the theoretical equation given in Equation 7.88:

$$i_{exit} = \left(\frac{1}{\pi}\right)\left[\frac{(3-0.5)}{1.5}\right] = 0.53$$

which is close to the value obtained earlier.)

$$i_{cr} = \frac{\gamma'}{\gamma_w} = \frac{10.2\,kN/m^3}{9.81\,kN/m^3} = 1.04$$

So, the factor of safety against piping is

$$\frac{i_{cr}}{i_{exit}} = \frac{1.04}{0.509} = 2.04$$

Part b:

A soil prism of cross section $D \times D/2$, where $D = 1.5$ m, on the down-stream side adjacent to the sheet pile is plotted in Figure 7.22a. The approximate hydraulic heads at the bottom of the prism can be evaluated by using the flow net. Referring to Figure 7.3 (note that $N_d = 6$)

$$h_A = \frac{3}{6}(3-0.5) = 1.25\,m$$

$$h_B = \frac{2}{6}(3-0.5) = 0.833\,m$$

$$h_C = \frac{1.8}{6}(3-0.5) = 0.75\,m$$

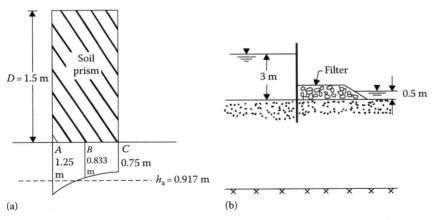

(a)
(b)

Figure 7.22 Factor of safety calculation by Terzaghi's method: (a) hydraulic head at the bottom prism measuring $D \times D/2$; (b) use of filter in the downstream side.

$$h_a = \frac{0.375}{0.75}\left(\frac{1.25+0.75}{2}+0.833\right)=0.917\,m$$

$$F_s = \frac{D\gamma'}{h_a\gamma_w} = \frac{1.5\times10.2}{0.917\times9.81} = 1.7$$

The factor of safety calculated here is rather low. However, it can be increased by placing some filter material on the downstream side above the ground surface, as shown in Figure 7.22b. This will increase the weight of the soil prism (W'; see Equation 7.92).

Example 7.7

A dam section is shown in Figure 7.23. The subsoil is fine sand. Using Lane's method, determine whether the structure is safe against piping.

Solution

From Equation 7.89

$$L_w = \frac{\sum L_h}{3} + \sum L_v$$

$$\sum L_h = 6+10 = 16\,m$$

$$\sum L_v = 1+(8+8)+1+2 = 20\,m$$

$$L_w = \frac{16}{3}+20 = 25.33\,m$$

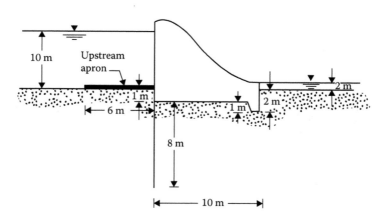

Figure 7.23 Safety against piping under a dam by using Lane's method.

From Equation 7.90

$$\text{Weighted creep ratio} = \frac{L_w}{H_1 - H_2} = \frac{25.33}{10 - 2} = 3.17$$

From Table 7.1, the safe weighted creep ratio for fine sand is about 7. Since the calculated weighted creep ratio is 3.17, the structure is *unsafe*.

Example 7.8

Refer to Figure 7.24.

   a. Draw a flow net for the seepage in the permeable layer.
   b. Determine the exit gradient.

Solution

Part a:
From Equation 7.51,

$$\frac{k_1}{k_2} = \frac{\tan\theta_1}{\tan\theta_2} = \frac{\tan\alpha_2}{\tan\alpha_1} = \frac{4\times10^{-3}}{2\times10^{-2}} = 2$$

The flow net is shown in Figure 7.25.

Part b:

$$i_{\text{exit}} = \frac{\Delta h}{L} = \frac{(3-1)}{L} = \frac{(3-1)/10}{1.1} = 0.18$$

(*Note:* L is the length of the flow element.)

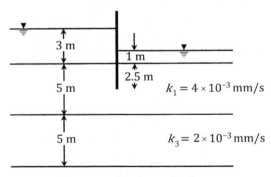

$k_1 = 4\times10^{-3}\,\text{mm/s}$

$k_3 = 2\times10^{-3}\,\text{mm/s}$

Impermeable layer

Figure 7.24 Flow net problem in a two-layered subsoil.

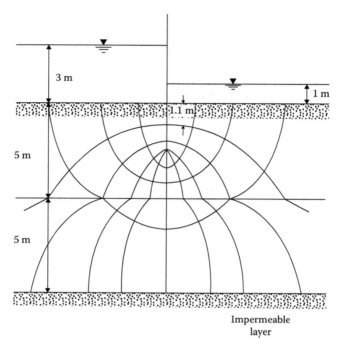

Figure 7.25 Flow net for the case shown in Figure 7.24.

7.11 FILTER DESIGN

When seepage water flows from a soil with relatively fine grains into a coarser material (e.g., Figure 7.22b), there is a danger that the fine soil particles may wash away into the coarse material. Over a period of time, this process may clog the void spaces in the coarser material. Such a situation can be prevented by the use of a filter or protective filter between the two soils. For example, consider the earth dam section shown in Figure 7.26.

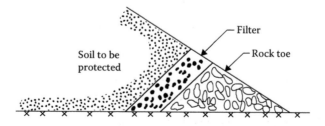

Figure 7.26 Use of filter at the toe of an earth dam.

If rockfills were only used at the toe of the dam, the seepage water would wash the fine soil grains into the toe and undermine the structure. Hence, for the safety of the structure, a filter should be placed between the fine soil and the rock toe (Figure 7.26). For the proper selection of the filter material, two conditions should be kept in mind:

1. The size of the voids in the filter material should be small enough to hold the larger particles of the protected material in place.
2. The filter material should have a high permeability to prevent build up of large seepage forces and hydrostatic pressures.

Based on the experimental investigation of protective filters, Terzaghi and Peck (1948) provided the following criteria to satisfy the above conditions:

$$\frac{D_{15(F)}}{D_{85(B)}} \leq 4-5 \quad \text{(to satisfy condition 1)} \tag{7.94}$$

$$\frac{D_{15(F)}}{D_{15(B)}} \geq 4-5 \quad \text{(to satisfy condition 2)} \tag{7.95}$$

where
$D_{15(F)}$ is the diameter through which 15% of filter material will pass
$D_{15(B)}$ is the diameter through which 15% of soil to be protected will pass
$D_{85(B)}$ is the diameter through which 85% of soil to be protected will pass

The proper use of Equations 7.94 and 7.95 to determine the grain-size distribution of soils used as filters is shown in Figure 7.27. Consider the soil

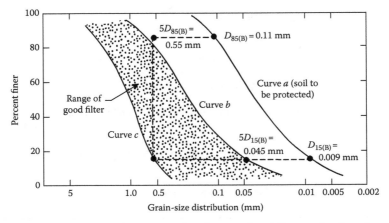

Figure 7.27 Determination of grain-size distribution of soil filters using Equations 7.94 and 7.95.

used for the construction of the earth dam shown in Figure 7.26. Let the grain-size distribution of this soil be given by curve a in Figure 7.27. We can now determine $5D_{85(B)}$ and $5D_{15(B)}$ and plot them as shown in Figure 7.27. The acceptable grain-size distribution of the filter material will have to lie in the shaded zone.

Based on laboratory experimental results, several other filter design criteria have been suggested in the past. These are summarized in Table 7.2.

Table 7.2 Filter criteria developed from laboratory testing

Investigator	Year	Criteria developed
Bertram	1940	$\dfrac{D_{15(F)}}{D_{85(B)}} < 6;\quad \dfrac{D_{15(F)}}{D_{85(B)}} < 9$
U.S. Corps of Engineers	1948	$\dfrac{D_{15(F)}}{D_{85(B)}} < 5;\quad \dfrac{D_{50(F)}}{D_{50(B)}} < 25;\quad \dfrac{D_{15(F)}}{D_{15(B)}} < 20$
Sherman	1953	For $C_{u(base)} < 1.5 : \dfrac{D_{15(F)}}{D_{85(B)}} < 6;\quad \dfrac{D_{15(F)}}{D_{15(B)}} < 20;\quad \dfrac{D_{50(F)}}{D_{50(B)}} < 25$
		For $1.5 < C_{u(base)} < 4.0 : \dfrac{D_{15(F)}}{D_{85(B)}} < 5;\quad \dfrac{D_{15(F)}}{D_{15(B)}} < 20;\quad \dfrac{D_{50(F)}}{D_{50(B)}} < 20$
		For $C_{u(base)} > 4.0 : \dfrac{D_{15(F)}}{D_{85(B)}} < 5;\quad \dfrac{D_{15(F)}}{D_{15(B)}} < 40;\quad \dfrac{D_{50(F)}}{D_{50(B)}} < 25$
Leatherwood and Peterson	1954	$\dfrac{D_{15(F)}}{D_{85(B)}} < 4.1;\quad \dfrac{D_{50(F)}}{D_{50(B)}} < 5.3$
Karpoff	1955	Uniform filter: $5 < \dfrac{D_{50(F)}}{D_{50(B)}} < 10$
		Well-graded filter: $12 < \dfrac{D_{50(F)}}{D_{50(B)}} < 58;\quad 12 < \dfrac{D_{15(F)}}{D_{15(B)}} < 40;$ and parallel grain-size curves
Zweck and Davidenkoff	1957	Base of medium and coarse uniform sand: $5 < \dfrac{D_{50(F)}}{D_{50(B)}} < 10$
		Base of fine uniform sand: $5 < \dfrac{D_{50(F)}}{D_{50(B)}} < 15$
		Base of well-graded fine sand: $5 < \dfrac{D_{50(F)}}{D_{50(B)}} < 25$

Note: $D_{50(F)}$, diameter through which 50% of the filter passes; $D_{50(B)}$, diameter through which 50% of the soil to be protected passes; C_u, uniformity coefficient.

7.12 CALCULATION OF SEEPAGE THROUGH AN EARTH DAM RESTING ON AN IMPERVIOUS BASE

Several solutions have been proposed for determination of the quantity of seepage through a homogeneous earth dam. In this section, some of these solutions will be considered.

7.12.1 Dupuit's solution

Figure 7.28 shows the section of an earth dam in which ab is the *phreatic surface*, that is, the uppermost line of seepage. The quantity of seepage through a unit length at right angles to the cross-section can be given by Darcy's law as $q = kiA$.

Dupuit (1863) assumed that the hydraulic gradient i is equal to the slope of the free surface and is constant with depth, that is, $i = dz/dx$. So

$$q = k\frac{dz}{dx}[(z)(1)] = k\frac{dz}{dx}z$$

$$\int_0^d q\,dx = \int_{H_2}^{H_1} kz\,dz$$

$$qd = \frac{k}{2}\left(H_1^2 - H_2^2\right)$$

or

$$q = \frac{k}{2d}\left(H_1^2 - H_2^2\right) \tag{7.96}$$

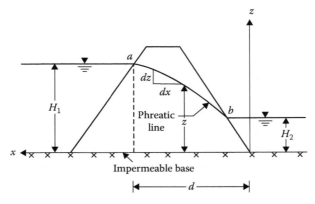

Figure 7.28 Dupuit's solution for flow through an earth dam.

Equation 7.96 represents a parabolic free surface. However, in the derivation of the equation, no attention has been paid to the entrance or exit conditions. Also note that if $H_2 = 0$, the phreatic line would intersect the impervious surface.

7.12.2 Schaffernak's solution

For calculation of seepage through a homogeneous earth dam. Schaffernak (1917) proposed that the phreatic surface will be like line ab in Figure 7.29, that is, it will intersect the downstream slope at a distance l from the impervious base. The seepage per unit length of the dam can now be determined by considering the triangle bcd in Figure 7.29:

$$q = kiA; \quad A = (\overline{bd})(1) = l\sin\beta$$

From Dupuit's assumption, the hydraulic gradient is given by $i = dz/dx = \tan\beta$. So

$$q = kz\frac{dz}{dx} = (k)(l\sin\beta)(\tan\beta) \tag{7.97}$$

or

$$\int_{l\sin\beta}^{H} z\,dz = \int_{l\cos\beta}^{d} (l\sin\beta)(\tan\beta)\,dx$$

$$\frac{1}{2}(H^2 - l^2\sin^2\beta) = (l\sin\beta)(\tan\beta)(d - l\cos\beta)$$

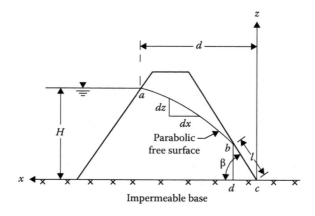

Figure 7.29 Schaffernak's solution for flow through an earth dam.

$$\frac{1}{2}(H^2 - l^2\sin^2\beta) = l\frac{\sin^2\beta}{\cos\beta}(d - l\cos\beta)$$

$$\frac{H^2\cos\beta}{2\sin^2\beta} - \frac{l^2\cos\beta}{2} = ld - l^2\cos\beta$$

$$l^2\cos\beta - 2ld + \frac{H^2\cos\beta}{\sin^2\beta} = 0$$

$$l = \frac{2d \pm \sqrt{4d^2 - 4[(H^2\cos^2\beta)/\sin^2\beta]}}{2\cos\beta} \tag{7.98}$$

so

$$l = \frac{d}{\cos\beta} - \sqrt{\frac{d^2}{\cos^2\beta} - \frac{H^2}{\sin^2\beta}} \tag{7.99}$$

Once the value of l is known, the rate of seepage can be calculated from the equation $q = kl\sin\beta\tan\beta$.

Schaffernak suggested a graphical procedure to determine the value of l. This procedure can be explained with the aid of Figure 7.30:

1. Extend the downstream slope line bc upward.
2. Draw a vertical line ae through the point a. This will intersect the projection of line bc (step 1) at point f.

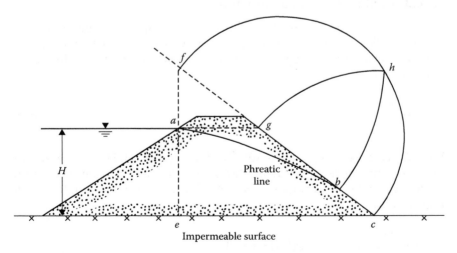

Figure 7.30 Graphical construction for Schaffernak's solution.

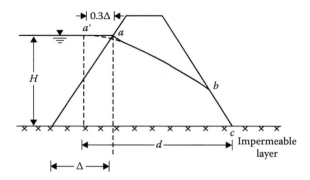

Figure 7.31 Modified distance *d* for use in Equation 7.99.

   3. With *fc* as diameter, draw a semicircle *fhc*.
   4. Draw a horizontal line *ag*.
   5. With *c* as the center and *cg* as the radius, draw an arc of a circle, *gh*.
   6. With *f* as the center and *fh* as the radius, draw an arc of a circle, *hb*.
   7. Measure *bc* = *l*.

   Casagrande (1937) showed experimentally that the parabola *ab* shown in Figure 7.29 should actually start from the point *a'* as shown in Figure 7.31. Note that *aa'* = 0.3Δ. So, with this modification, the value of *d* for use in Equation 7.99 will be the horizontal distance between points *a'* and *c*.

7.12.3 L. Casagrande's solution

Equation 7.99 was obtained on the basis of Dupuit's assumption that the hydraulic gradient *i* is equal to *dz/dx*. Casagrande (1932) suggested that this relation is an approximation to the actual condition. In reality (see Figure 7.32)

$$i = \frac{dz}{ds} \qquad (7.100)$$

For a downstream slope of β > 30°, the deviations from Dupuit's assumption become more noticeable. Based on this assumption (Equation 7.100), the rate of seepage is *q* = *kiA*. Considering the triangle *bcd* in Figure 7.32,

$$i = \frac{dz}{ds} = \sin\beta \quad A = (bd)(1) = l\sin\beta$$

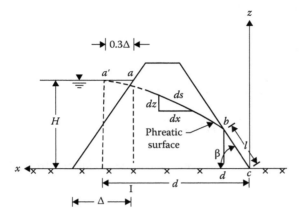

Figure 7.32 L. Casagrande's solution for flow through an earth dam. (*Note:* length of the curve a'bc = S.)

So

$$q = k \frac{dz}{ds} z = kl \sin^2 \beta \tag{7.101}$$

or

$$\int_{l\sin\beta}^{H} z \, dz \int_{l}^{s} (l \sin^2 \beta) \, ds$$

where s is the length of the curve a'bc. Hence

$$\frac{1}{2}(H^2 - l^2 \sin^2 \beta) = l \sin^2 \beta (s - l)$$

$$H^2 - l^2 \sin^2 \beta = 2l s \sin^2 \beta - 2l^2 \sin^2 \beta$$

$$l^2 - 2ls + \frac{H^2}{\sin^2 \beta} = 0 \tag{7.102}$$

The solution to Equation 7.102 is

$$l = s - \sqrt{s^2 - \frac{H^2}{\sin^2 \beta}} \tag{7.103}$$

With about a 4%–5% error, we can approximate s as the length of the straight line $a'c$. So

$$s = \sqrt{d^2 + H^2} \qquad (7.104)$$

Combining Equations 7.103 and 7.104

$$l = \sqrt{d^2 + H^2} - \sqrt{d^2 - H^2 \cot^2 \beta} \qquad (7.105)$$

Once l is known, the rate of seepage can be calculated from the equation

$$q = kl \sin^2 \beta$$

A solution that avoids the approximation introduced in Equation 7.105 was given by Gilboy (1934) and put into graphical form by Taylor (1948), as shown in Figure 7.33. To use the graph

1. Determine d/H
2. For given values of d/H and β, determine m
3. Calculate $l = mH/\sin \beta$
4. Calculate $q = kl \sin^2 \beta$

7.12.4 Pavlovsky's solution

Pavlovsky (1931; also see Harr, 1962) also gave a solution for calculation of seepage through an earth dam. This can be explained with reference to Figure 7.34. The dam section can be divided into three zones, and the rate of seepage through each zone can be calculated as follows.

7.12.4.1 Zone I (area agOf)

In zone I the seepage lines are actually curved, but Pavlovsky assumed that they can be replaced by horizontal lines. The rate of seepage through an elementary strip dz can then be given by

$$dq = ki\,dA$$

$$dA = (dz)(1) = dz$$

$$i = \frac{\text{Loss of head, } l_1}{\text{Length of flow}} = \frac{l_1}{(H_d - z) \cot \beta_1}$$

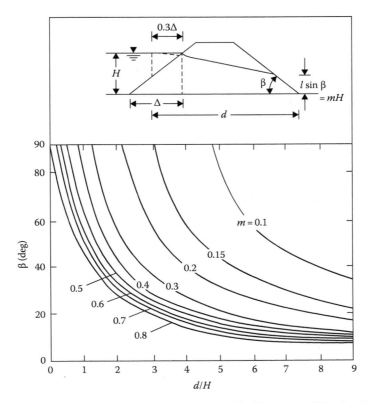

Figure 7.33 Chart for solution by L. Casagrande's method based on Gilboy's solution.

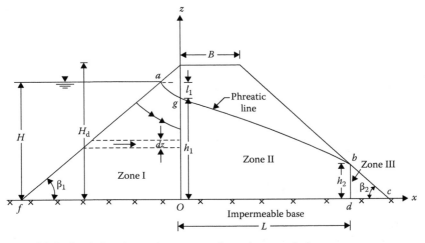

Figure 7.34 Pavlovsky's solution for seepage through an earth dam.

So

$$q = \int dq = \int_0^{h_1} \frac{kl_1}{(H_d - z)\cot\beta_1} dz = \frac{kl_1}{\cot\beta_1} \ln \frac{H_d}{H_d - h_1}$$

However, $l_1 = H - h_1$. So

$$q = \frac{k(H - h_1)}{\cot\beta_1} \ln \frac{H_d}{H_d - h_1} \qquad (7.106)$$

7.12.4.2 Zone II (area Ogbd)

The flow in zone II can be given by the equation derived by Dupuit (Equation 7.96). Substituting h_1 for H_1, h_2 for H_2, and L for d in Equation 7.96, we get

$$q = \frac{k}{2L}\left(h_1^2 - h_2^2\right) \qquad (7.107)$$

where

$$L = B + (H_d - h_2)\cot\beta_2 \qquad (7.108)$$

7.12.4.3 Zone III (area bcd)

As in zone I, the stream lines in zone III are also assumed to be horizontal:

$$q = k \int_0^{h_2} \frac{dz}{\cot\beta_2} = \frac{kh_2}{\cot\beta_2} \qquad (7.109)$$

Combining Equations 7.106 through 7.108

$$h_2 = \frac{B}{\cot\beta_2} + H_d - \sqrt{\left(\frac{B}{\cot\beta_2} + H_d\right)^2 - h_1^2} \qquad (7.110)$$

From Equations 7.106 and 7.109

$$\frac{H - h_1}{\cot\beta_1} \ln \frac{H_d}{H_d - h_1} = \frac{h_2}{\cot\beta_2} \qquad (7.111)$$

Equations 7.110 and 7.111 contain two unknowns, h_1 and h_2, which can be solved graphically (see Example 7.6). Once these are known, the rate of seepage per unit length of the dam can be obtained from any one of the Equations 7.106, 7.107, and 7.109.

7.12.5 Seepage through earth dams with $k_x \neq k_z$

If the soil in a dam section shows anisotropic behavior with respect to permeability, the dam section should first be plotted according to the transformed scale (as explained in Section 7.6):

$$x' = \sqrt{\frac{k_z}{k_x}}x$$

All calculations should be based on this transformed section. Also, for calculating the rate of seepage, the term k in the corresponding equations should be equal to $\sqrt{k_x k_z}$.

Example 7.9

The cross section of an earth dam is shown in Figure 7.35. Calculate the rate of seepage through the dam [q in m³/(min·m)] by (a) Dupuit's method; (b) Schaffernak's method; (c) L. Casagrande's method; and (d) Pavlovsky's method.

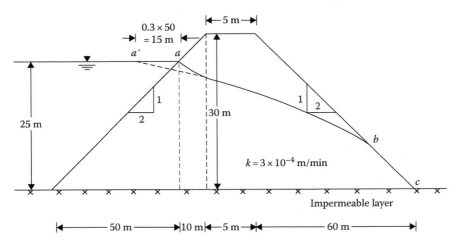

Figure 7.35 Seepage through an earth dam.

Solution

Part a: Dupuit's method.

From Equation 7.96

$$q = \frac{k}{2d}\left(H_1^2 - H_2^2\right)$$

From Figure 7.35, $H_1 = 25$ m and $H_2 = 0$; also, d (the horizontal distance between points a and c) is equal to $60 + 5 + 10 = 75$ m. Hence

$$q = \frac{3 \times 10^{-4}}{2 \times 75}(25)^2 = 12.5 \times 10^{-4} \text{ m}^3/(\text{min·m})$$

Part b: Schaffernak's method.

From Equations 7.97 and 7.99

$$q = (k)(l\sin\beta)(\tan\beta); \quad l = \frac{d}{\cos\beta} - \sqrt{\frac{d^2}{\cos^2\beta} - \frac{H^2}{\sin^2\beta}}$$

Using Casagrande's correction (Figure 7.31), d (the horizontal distance between a' and c) is equal to $60 + 5 + 10 + 15 = 90$ m. Also

$$\beta = \tan^{-1}\frac{1}{2} = 26.57° \quad H = 25\,\text{m}$$

So

$$l = \frac{90}{\cos 26.57°} - \sqrt{\left(\frac{90}{\cos 26.57°}\right)^2 - \left(\frac{25}{\sin 26.57°}\right)^2}$$

$$= 100.63 - \sqrt{(100.63)^2 - (55.89)^2} = 16.95 \text{ m}$$

$$q = (3 \times 10^{-4})(16.95)(\sin 26.57°)(\tan 26.57°) = 11.37 \times 10^{-4} \text{ m}^3/(\text{min} \cdot \text{m})$$

Part c: L. Casagrande's method.

We will use the graph given in Figure 7.33.

$$d = 90\,\text{m} \quad H = 25\,\text{m} \quad \frac{d}{H} = \frac{90}{25} = 3.6 \quad \beta = 26.57°$$

From Figure 7.33 for $\beta = 26.57°$ and $d/H = 3.6$, $m = 0.34$, and

$$l = \frac{mH}{\sin\beta} = \frac{0.34(25)}{\sin 26.57°} = 19.0 \text{ m}$$

$$q = kl \sin^2\beta = (3 \times 10^{-4})\,(19.0)\,(\sin 26.57°)^2 = 11.4 \times 10^{-4} \text{ m}^3/(\text{min} \cdot \text{m})$$

Part d: Pavlovsky's method.

From Equations 7.110 and 7.111

$$h_2 = \frac{B}{\cot \beta_2} + H_d - \sqrt{\left(\frac{B}{\cot \beta_2} + H_d\right)^2 - h_1^2}$$

$$\frac{H - h_1}{\cot \beta_1} \ln \frac{H_d}{H_d - h_1} = \frac{h_2}{\cot \beta_2}$$

From Figure 7.35, $B = 5$ m, $\cot \beta_2 = \cot 26.57° = 2$, $H_d = 30$ m, and $H = 25$ m. Substituting these values in Equation 7.110, we get

$$h_2 = \frac{5}{2} + 30 - \sqrt{\left(\frac{5}{2} + 30\right)^2 - h_1^2}$$

or

$$h_2 = 32.5 - \sqrt{1056.25 - h_1^2} \tag{E7.1}$$

Similarly, from Equation 7.111

$$\frac{25 - h_1}{2} \ln \frac{30}{30 - h_1} = \frac{h_2}{2}$$

or

$$h_2 = (25 - h_1) \ln \frac{30}{30 - h_1} \tag{E7.2}$$

Equations E7.1 and E7.2 must be solved by trial and error:

h_1 (m)	h_2 from Equation E7.1 (m)	h_2 from Equation E7.2 (m)
2	0.062	1.587
4	0.247	3.005
6	0.559	4.240
8	1.0	5.273
10	1.577	6.082
12	2.297	6.641
14	3.170	6.915
16	4.211	6.859
18	5.400	6.414
20	6.882	5.493

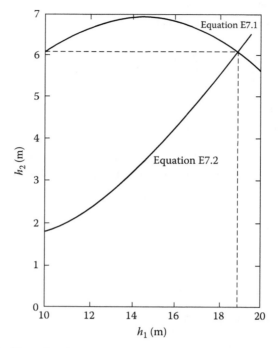

Figure 7.36 Plot of h_2 against h_1.

Using the values of h_1 and h_2 calculated in the preceding table, we can plot the graph as shown in Figure 7.36 and from that, $h_1 = 18.9$ m and $h_2 = 6.06$ m. From Equation 7.109

$$q = \frac{kh_2}{\cot\beta_2} = \frac{(3\times10^{-4})(6.06)}{2} = 9.09\times10^{-4} \text{ m}^3/(\text{min}\cdot\text{m})$$

Example 7.10

Figure 7.37 shows an earth dam section. Given: $k_x = 4\times10^{-5}$ m/min and $k_z = 1\times10^{-5}$ m/min. Determine the rate of seepage through the earth dam using

a. Dupuit's method
b. Schaffernak's method with Casagrande's modification
c. L. Casagrande's method

Note: The upstream and downstream slopes of the earth dam are 1V:2H.

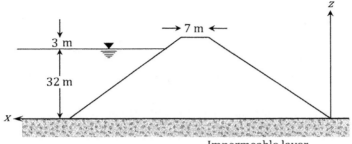

Figure 7.37 An earth dam section on an impermeable layer.

Solution

For this case, $\sqrt{\dfrac{k_z}{k_x}} = \sqrt{\dfrac{1}{4}} = 0.5$

The transformed section of the earth dam is shown in Figure 7.38.

Part a:

$$q = \sqrt{\frac{k_x k_z}{2d}}(H_1^2 - H_2^2)$$

$$d = 35 + 3.5 + 3 = 41.5 \text{ m}$$

$$q = \sqrt{\frac{(4\times10^{-5})(1\times10^{-5})}{(2)(41.5)}} = (32^2 - 0^2) = 24.67\times10^{-5} \text{ m}^3/\text{min}/\text{m}$$

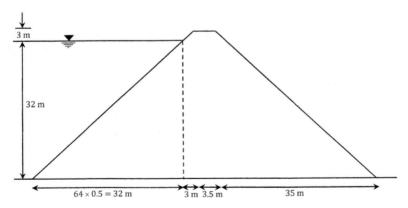

Figure 7.38 Transformed section of the earth dam.

Part b: Given: $\beta = 45°$

$$\Delta = (0.3)(32) = 9.6 \text{ m}$$
$$d = 9.6 + 3 + 3.5 + 35 = 51.1 \text{ m}$$

$$l = \frac{d}{\cos\beta} - \sqrt{\frac{d^2}{\cos^2\beta} - \frac{H^2}{\sin^2\beta}} = \frac{51.1}{\cos 45} - \sqrt{\left(\frac{51.1}{\cos 45}\right)^2 - \left(\frac{32}{\sin 45}\right)^2} = 15.9 \text{ m}$$

$$q = \sqrt{k_x k_z}\, l \sin\beta \tan\beta = \sqrt{(4\times 10^{-5})(10^{-5})}(15.9)\sin 45 \tan 45$$
$$= 22.49 \times 10^{-5} \text{ m}^3/\text{min/m}$$

Part c:

$d = 51.1$ m; $H = 32$ m; $d/H = 51.1/32 = 1.6$ m. From Figure 7.33, $m = 0.4$.

$$l = \frac{(0.4)(32)}{\sin 45} = 18.1 \text{ m}$$

$$q = \sqrt{k_x k_z}\, l \sin^2\beta = \sqrt{(4\times 10^{-5})(10^{-5})}(18.1)(\sin^2 45)$$
$$= 18.1 \times 10^{-5} \text{ m}^3/\text{min/m}$$

7.13 PLOTTING OF PHREATIC LINE FOR SEEPAGE THROUGH EARTH DAMS

For construction of flow nets for seepage through earth dams, the phreatic line needs to be established first. This is usually done by the method proposed by Casagrande (1937) and is shown in Figure 7.39a. Note that *aefb* in Figure 7.39a is the actual phreatic line. The curve *a'efb'c'* is a parabola with its focus at *c'*. The phreatic line coincides with this parabola, but with some deviations at the upstream and the downstream faces. At a point *a*, the phreatic line starts at an angle of 90° to the upstream face of the dam and $aa' = 0.3\Delta$.

The parabola *a'efb'c'* can be constructed as follows:

1. Let the distance *cc'* be equal to *p*. Now, referring to Figure 7.39b, $Ac = AD$ (based on the properties of a parabola), $Ac = \sqrt{x^2 + z^2}$, and $AD = 2p + x$. Thus

$$\sqrt{x^2 + z^2} = 2p + x \tag{7.112}$$

At $x = d$, $z = H$. Substituting these conditions into Equation 7.112 and rearranging, we obtain

$$p = \frac{1}{2}\left(\sqrt{d^2 + H^2} - d\right) \tag{7.113}$$

Since d and H are known, the value of p can be calculated.
2. From Equation 7.112

$$x^2 + z^2 = 4p^2 + x^2 + 4px$$

$$x = \frac{z^2 - 4p^2}{4p} \tag{7.114}$$

With p known, the values of x for various values of z can be calculated from Equation 7.114, and the parabola can be constructed.

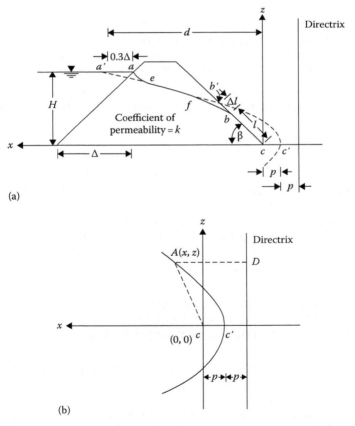

(a)

(b)

Figure 7.39 Determination of phreatic line for seepage through an earth dam: (a) phreatic line; (b) parabola with the focus at c'.

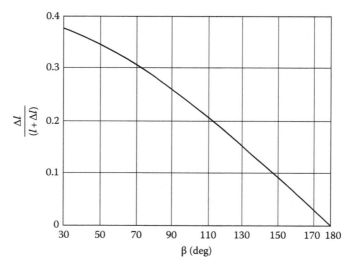

Figure 7.40 Plot of $\Delta l/(l + \Delta l)$ against downstream slope angle. (After Casagrande, A., Seepage through dams, in *Contribution to Soil Mechanics 1925–1940*, Boston Society of Civil Engineering, Boston, MA, p. 295, 1937.)

To complete the phreatic line, the portion *ae* must be approximated and drawn by hand. When $\beta < 30°$, the value of l can be calculated from Equation 7.99 as

$$l = \frac{d}{\cos\beta} - \sqrt{\frac{d^2}{\cos^2\beta} - \frac{H^2}{\sin^2\beta}}$$

Note that $l = bc$ in Figure 7.39a. Once point *b* has been located, the curve *fb* can be approximately drawn by hand.

If $\beta \geq 30°$, Casagrande proposed that the value of Δl can be determined by using the graph given in Figure 7.40. In Figure 7.39a, $b'b = \Delta l$ and $bc = l$. After locating the point *b* on the downstream face, the curve *fb* can be approximately drawn by hand.

7.14 ENTRANCE, DISCHARGE, AND TRANSFER CONDITIONS OF LINE OF SEEPAGE THROUGH EARTH DAMS

A. Casagrande (1937) analyzed the entrance, discharge, and transfer conditions for the line of seepage through earth dams. When we consider the flow from a free-draining material (coefficient of permeability very large; $k_1 \approx \infty$ into a material of permeability k_2, it is called an *entrance*). Similarly, when the flow is from a material of permeability k_1 into a free-draining material ($k_2 \approx \infty$), it is referred to as *discharge*. Figure 7.41 shows various

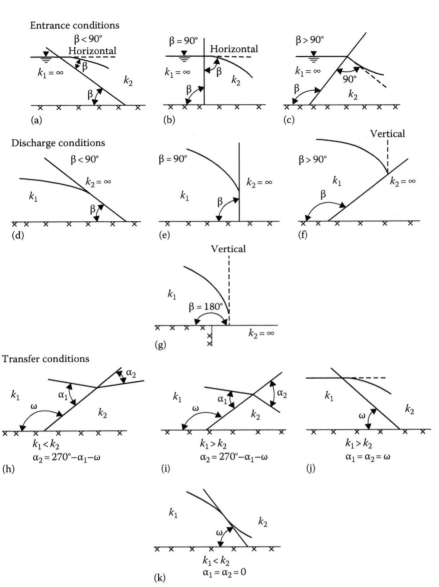

Figure 7.41 Entrance, discharge, and transfer conditions: (a) entrance, $\beta < 90°$;
(b) entrance, $\beta = 90°$; (c) entrance, $\beta > 90°$; (d) discharge, $\beta < 90°$; (e) discharge, $\beta = 90°$; (f) discharge, $\beta > 90°$; (g) discharge, $\beta = 180°$; (h) transfer,
$k_1 < k_2$, $\alpha_2 = 270° - \alpha_1 - \omega$; (i) transfer, $k_1 > k_2$, $\alpha_2 = 270° - \alpha_1 - \omega$; (j) transfer,
$k_1 > k_2$, $\alpha_1 = \alpha_2 = \omega$; (k) transfer, $k_1 < k_2$, $\alpha_1 = \alpha_2 = 0$. (After Casagrande, A.,
Seepage through dams, in *Contribution to Soil Mechanics 1925–1940*, Boston
Society of Civil Engineering, Boston, MA, p. 295, 1937.)

entrance, discharge, and transfer conditions. The transfer conditions show the nature of deflection of the line of seepage when passing from a material of permeability k_1 to a material of permeability k_2.

Using the conditions given in Figure 7.41, we can determine the nature of the phreatic lines for various types of earth dam sections.

7.15 FLOW NET CONSTRUCTION FOR EARTH DAMS

With a knowledge of the nature of the phreatic line and the entrance, discharge, and transfer conditions, we can now proceed to draw flow nets for earth dam sections. Figure 7.42 shows an earth dam section that is homogeneous with respect to permeability. To draw the flow net, the following steps must be taken:

1. Draw the phreatic line, since this is known.
2. Note that ag is an equipotential line and that gc is a flow line.
3. It is important to realize that the pressure head at any point on the phreatic line is zero; hence, the difference of total head between any two equipotential lines should be equal to the difference in elevation between the points where these equipotential lines intersect the phreatic line.

   Since loss of hydraulic head between any two consecutive equipotential lines is the same, determine the number of equipotential drops, N_d, the flow net needs to have and calculate $\Delta h = h/N_d$.
4. Draw the *head lines* for the cross section of the dam. The points of intersection of the head lines and the phreatic lines are the points from which the equipotential lines should start.

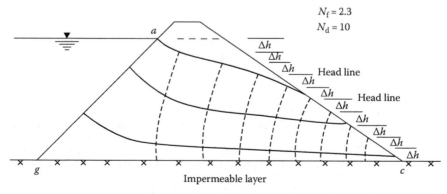

Figure 7.42 Flow net construction for an earth dam.

5. Draw the flow net, keeping in mind that the equipotential lines and flow lines must intersect at right angles.
6. The rate of seepage through the earth dam can be calculated from the relation given in Equation 7.40, $q = kh(N_f/N_d)$.

In Figure 7.42, the number of flow channels, N_f, is equal to 2.3. The top two flow channels have square flow elements, and the bottom flow channel has elements with a width-to-length ratio of 0.3. Also, N_d in Figure 7.42 is equal to 10.

If the dam section is anisotropic with respect to permeability, a transformed section should first be prepared in the manner outlined in Section 7.6. The flow net can then be drawn on the transformed section, and the rate of seepage obtained from Equation 7.46.

Figures 7.43 through 7.45 show some typical flow nets through earth dam sections.

A flow net for seepage through a zoned earth dam section is shown in Figure 7.46. The soil for the upstream half of the dam has a permeability k_1, and the soil for the downstream half of the dam has a permeability $k_2 = 5k_1$. The phreatic line must be plotted by trial and error. As shown in Figure 7.12b, here the seepage is from a soil of low permeability

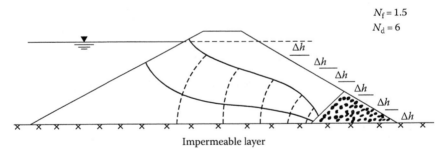

Figure 7.43 Typical flow net for an earth dam with rock toe filter.

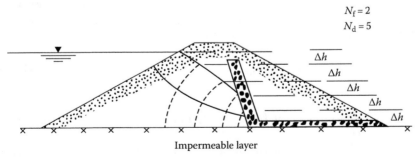

Figure 7.44 Typical flow net for an earth dam with chimney drain.

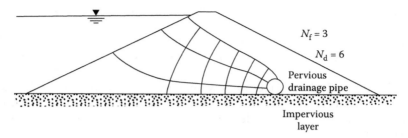

$N_f = 3$

$N_d = 6$

Pervious
drainage pipe

Impervious
layer

Figure 7.45 Typical flow net for an earth dam with a pervious drainage pipe.

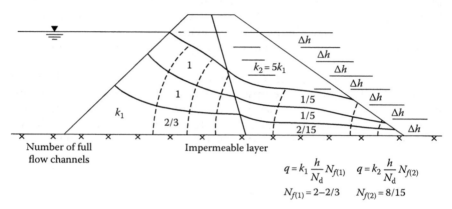

Figure 7.46 Flow net for seepage through a zoned earth dam.

(upstream half) to a soil of high permeability (downstream half). From Equation 7.47

$$\frac{k_1}{k_2} = \frac{b_2/l_2}{b_1/l_1}$$

If $b_1 = l_1$ and $k_2 = 5k_1$, $b_2/l_2 = 1/5$. For that reason, square flow elements have been plotted in the upstream half of the dam, and the flow elements in the downstream half have a width-to-length ratio of 1/5. The rate of seepage can be calculated by using the following equation:

$$q = k_1 \frac{h}{N_d} N_{f(1)} = k_2 \frac{h}{N_d} N_{f(2)} \qquad (7.115)$$

where
   $N_{f(1)}$ is the number of full flow channels in the soil having a permeability k_1
   $N_{f(2)}$ is the number of full flow channels in the soil having a permeability k_2

REFERENCES

Bertram, G. E., An experimental investigation of protective filters, Publication No. 267, Harvard University Graduate School, Cambridge, MA, 1940.

Casagrande, A., Seepage through dams, in *Contribution to Soil Mechanics 1925–1940*, Boston Society of Civil Engineering, Boston, MA, p. 295, 1937.

Casagrande, L., Naeherungsmethoden zur Bestimmurg von Art und Menge der Sickerung durch geschuettete Daemme, Thesis, Technische Hochschule, Vienna, Austria, 1932.

Dupuit, J., *Etudes theoriques et Practiques sur le Mouvement des eaux dans les Canaux Decouverts et a travers les Terrains Permeables*, Dunot, Paris, France, 1863.

Gilboy, G., Mechanics of hydraulic fill dams, in *Contribution to Soil Mechanics 1925–1940*, Boston Society of Civil Engineering, Boston, MA, 1934.

Harr, M. E., *Groundwater and Seepage*, McGraw-Hill, New York, 1962.

Harza, L. F., Uplift and seepage under dams in sand, *Trans.* ASCE, 100(1), 1352–1385, 1935.

Karpoff, K. P., The use of laboratory tests to develop design criteria for protective filters, *Proc.* ASTM, 55, 1183–1193, 1955.

Lane, E. W., Security from under-seepage: Masonry Dams on Earth Foundation, *Trans.* ASCE, 100, 1235, 1935.

Leatherwood, F. N. and D. G. Peterson, Hydraulic head loss at the interface between uniform sands of different sizes, *Eos Trans. AGU*, 35(4), 588–594, 1954.

Pavlovsky, N. N., *Seepage through Earth Dams* (in Russian), Inst. Gidrotekhniki i Melioratsii, Leningrad, Russia, 1931.

Schaffernak, F., Über die Standicherheit durchlaessiger geschuetteter Dämme, *Allgem Bauzeitung*, 1917.

Sherman, W. C., Filter equipment and design criteria, NTIS/AD 771076, U.S. Army Waterways Experiment Station, Vicksburg, MS, 1953.

Taylor, D. W., *Fundamentals of Soil Mechanics*, Wiley, New York, 1948.

Terzaghi, K., Der Grundbrunch on Stauwerken und Seine Verhutung, *Wasserkraft*, 17, 445–449, 1922. [Reprinted in *From Theory to Practice in Soil Mechanics*, Wiley, New York, pp. 146–148, 1960.]

Terzaghi, K., *Theoretical Soil Mechanics*, Wiley, New York, 1943.

Terzaghi, K. and R. B. Peck, *Soil Mechanics in Engineering Practice*, Wiley, New York, 1948.

U.S. Corps of Engineers, Laboratory Investigation of Filters for Enid and Grenada Dams, Tech. Memo. 3–245, U.S. Waterways Experiment Station, Vicksburg, MS, 1948.

Zweck, H. and R. Davidenkoff, Etude Experimentale des Filtres de Granulometrie Uniforme, *Proc. 4th Int. Conf. Soil Mech. Found. Eng., London*, August 12–24, 2, 410–413, 1957.

Chapter 8

Consolidation

8.1 INTRODUCTION

When a saturated soil layer is subjected to a stress increase, the pore water pressure is suddenly increased. In sandy soils that are highly permeable, the drainage caused by the increase of pore water pressure is completed immediately. Pore water drainage is accompanied by a volume reduction in the soil mass, resulting in settlement. Because of rapid drainage of pore water in sandy soils, immediate (or elastic) settlement and consolidation take place simultaneously.

When a saturated compressible clay layer is subjected to a stress increase, elastic settlement occurs immediately. Since the coefficient of permeability of clay is significantly smaller than that of sand, the excess pore water pressure generated due to loading gradually dissipates over a long period of time. Thus the associated volume change (that is, the consolidation) in soft clay may continue long after the immediate (or elastic) settlement. The settlement due to consolidation in soft clay may be several times larger than the immediate settlement.

The time-dependent deformation of saturated clay soils can best be understood by first considering a simple rheological model consisting of a linear elastic spring and dashpot connected in parallel (Kelvin model, Figure 8.1). The stress-strain relationship for the spring and the dashpot can be given by

$$\text{Spring: } \sigma_s = \bar{k}\varepsilon \tag{8.1}$$

$$\text{Dashpot: } \sigma_d = \eta \frac{d\varepsilon}{dt} \tag{8.2}$$

where σ = stress (subscripts s and d refer to spring and dashpot, respectively)
  ε = strain
  $\bar{k}$ = spring constant

349

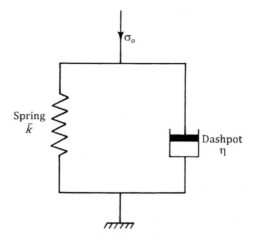

Figure 8.1 Kelvin model.

η = dashpot constant
t = time

The viscoelastic response for the stress σ_o in Figure 8.1 can be written

$$\sigma_o = \bar{k}\varepsilon + \eta\frac{d\varepsilon}{dt} \tag{8.3}$$

If stress σ_o is applied at time $t = 0$ and remains constant thereafter, the equation for strain at any time t can be found by solving the preceding differential equation. Thus,

$$\varepsilon = \frac{\sigma_o}{\bar{k}}\left[1 - e^{-(\bar{k}/\eta)t}\right] + \varepsilon_o e^{-(\bar{k}/\eta)t}$$

where ε_o = strain at time $t = 0$
 If ε_o is taken to be zero

$$\varepsilon = \frac{\sigma_o}{\bar{k}}\left[1 - e^{-(\bar{k}/\eta)t}\right] \tag{8.4}$$

The nature of variation of strain with time represented by Equation 8.4 is shown in Figure 8.2. At time $t = \infty$, the strain will approach the maximum value of $\sigma_o/\bar{k}$. This is the strain that the spring alone would have immediately undergone with the application of the same stress, σ_o, without

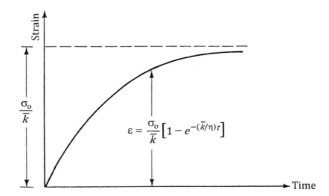

Figure 8.2 Strain-time diagram for the Kelvin model.

the dashpot having been attached to it. The distribution of stress at any time t between the spring and dashpot can be evaluated from Equations 8.3 and 8.4.

Portion of stress carried by the spring:

$$\sigma_s = \bar{k}\varepsilon = \sigma_o\left[1 - e^{-(\bar{k}/\eta)t}\right] \tag{8.5}$$

Portion of stress carried by the dashpot:

$$\sigma_d = \eta\frac{d\varepsilon}{dt} = \sigma_o e^{-(\bar{k}/\eta)t} \tag{8.6}$$

(*Note:* $\sigma_o = \sigma_s + \sigma_d$.)

Figure 8.3 shows the variation of σ_s and σ_d with time. At time $t = 0$, the stress σ_o is totally carried by the dashpot. The share of stress carried by the spring gradually increases with time. The stress carried by the dashpot decreases at an equal rate. At time $t = \infty$, the stress σ_o is carried entirely by the spring.

With this in mind, we can now analyze the strain of a saturated clay layer subjected to a stress increase (Figure 8.4a). Consider the case where a layer of saturated clay of thickness of H_t that is confined between two layers of sand is being subjected to an instantaneous increase of total stress $\Delta\sigma$. This incremental total stress will be transmitted to the pore water and the soil solids. This means that the total stress $\Delta\sigma$ will be divided in some proportion between the effective stress and pore water pressure. The behavior of the effective stress change will be similar to that of the spring in the Kelvin

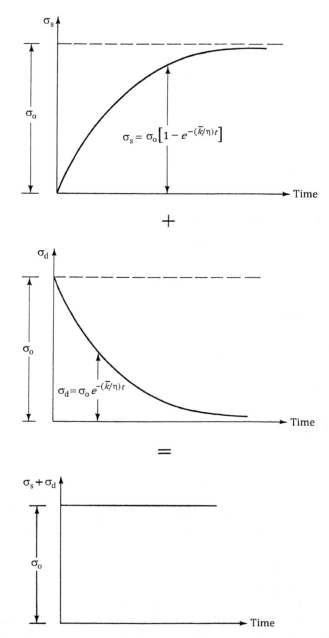

Figure 8.3 Stress–time diagram for the spring and dashpot in Kelvin model.

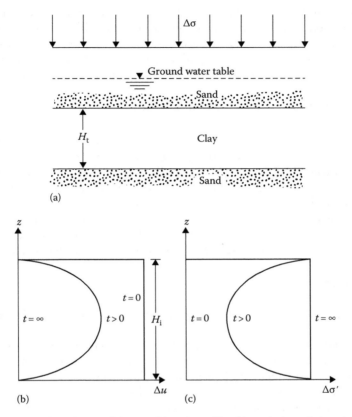

Figure 8.4 Principles of consolidation: (a) soil profile; (b) variation of Δu with depth; (c) variation of $\Delta \sigma'$ with depth.

model, and the behavior of the pore water pressure change will be similar to that of the dashpot. From the principle of effective stress,

$$\Delta \sigma = \Delta \sigma' + \Delta u \tag{8.7}$$

where
 $\Delta \sigma'$ = increase in the effective stress
 Δu = increase in the pore water pressure

Since clay has very low permeability and water is incompressible as compared to the soil skeleton, at time $t = 0$ the entire incremental stress $\Delta \sigma$ will be carried by water ($\Delta \sigma = \Delta u$) at all depths (Figure 8.4b). None will be carried by the soil skeleton (that is, incremental effective stress $\Delta \sigma' = 0$). This is similar to the behavior of the Kelvin model where at time $t = 0$, $\sigma_o = \sigma_d$ and $\sigma_s = 0$.

After the application of incremental stress $\Delta\sigma$ to the clay layer, the water in the void spaces will start to be squeezed out and will drain in both directions into the sand layers. By this process, the excess pore water pressure at any depth in the clay layer will gradually reduce, and the stress carried by the soil solids (effective stress) will increase. Thus, at time $0 < t < \infty$,

$$\Delta\sigma = \Delta\sigma' + \Delta u \qquad (\Delta\sigma' > 0 \text{ and } \Delta u < \Delta\sigma)$$

However, the magnitudes of $\Delta\sigma'$ and Δu at various depths will change, depending on the minimum distance of the drainage path either to the top or bottom sand layer. This is similar to the Kelvin model behavior for $0 < t < \infty$, where the stress carried by the spring increases with similar reduction in the stress carried by the dashpot.

Theoretically, at time $t = \infty$, the entire excess pore water pressure would be dissipated by drainage from all points of the clay layer, thus giving $\Delta u = 0$. Now the total stress increase, $\Delta\sigma$, will be carried by the soil structure. So,

$$\Delta\sigma = \Delta\sigma'$$

Again, this is similar to the spring-dashpot behavior, for which at time $t = \infty$, $\sigma_o = \sigma_s$ and $\sigma_d = 0$.

This gradual process of drainage under additional load applications and the associated transfer of excess pore water pressure to effective stress is the cause of time-dependent settlement in the clay layer.

Several types of rheological models have been used by various investigators for better representation of stress-strain-time behavior of soils. The Kelvin model only has been treated in this section to explain the fundamental concept of consolidation.

8.2 THEORY OF ONE-DIMENSIONAL CONSOLIDATION

The theory for the time rate of one-dimensional consolidation was first proposed by Terzaghi (1925). The underlying assumptions in the derivation of the mathematical equations are as follows:

1. The clay layer is homogeneous.
2. The clay layer is saturated.
3. The compression of the soil layer is due to the change in volume only, which in turn is due to the squeezing out of water from the void spaces.
4. Darcy's law is valid.
5. Deformation of soil occurs only in the direction of the load application.

6. The coefficient of consolidation C_v (Equation 8.22) is constant during the consolidation.

With the assumptions described earlier, let us consider a clay layer of thickness H_t as shown in Figure 8.5. The layer is located between two highly permeable sand layers. When the clay is subjected to an increase of vertical pressure, $\Delta\sigma$, the pore water pressure at any point A will increase by u. Consider an elemental soil mass with a volume of $dx \cdot dy \cdot dz$ at A; this is similar to the one shown in Figure 7.1b. In the case of one-dimensional consolidation, the flow of water into and out of the soil element is in one direction only, that is, in the z direction. This means that q_x, q_y, dq_x, and dq_y in Figure 7.1b are equal to zero, and thus the rate of flow into and out of the soil element can be given by Equations 7.3 and 7.6, respectively. So

$$(q_z + dq_z) - q_z = \text{rate of change of volume of soil element} = \frac{\partial V}{\partial t} \quad (8.8)$$

where

$$V = dx\,dy\,dz \quad (8.9)$$

Substituting the right-hand sides of Equations 7.3 and 7.6 into the left-hand side of Equation 8.8, we obtain

$$k\frac{\partial^2 h}{\partial z^2}\,dx\,dy\,dz = \frac{\partial V}{\partial t} \quad (8.10)$$

where k is the coefficient of permeability (k_z in Equations 7.3 and 7.6).

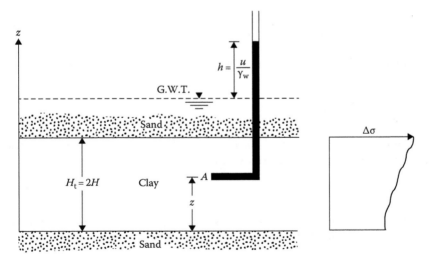

Figure 8.5 Clay layer undergoing consolidation.

However

$$h = \frac{u}{\gamma_w} \tag{8.11}$$

where γ_w is the unit weight of water. Substitution of Equation 8.11 into 8.10 and rearranging gives

$$\frac{k}{\gamma_w} \frac{\partial^2 u}{\partial z^2} = \frac{1}{dx\, dy\, dz} \frac{\partial V}{\partial t} \tag{8.12}$$

During consolidation, the rate of change of volume is equal to the rate of change of the void volume. So

$$\frac{\partial V}{\partial t} = \frac{\partial V_v}{\partial t} \tag{8.13}$$

where V_v is the volume of voids in the soil element. However

$$V_v = eV_s \tag{8.14}$$

where
   V_s is the volume of soil solids in the element, which is constant
   e is the void ratio

So

$$\frac{\partial V}{\partial t} = V_s \frac{\partial e}{\partial t} = \frac{V}{1+e} \frac{\partial e}{\partial t} = \frac{dx\, dy\, dz}{1+e} \frac{\partial e}{\partial t} \tag{8.15}$$

Substituting the aforementioned relation into Equation 8.12, we get

$$\frac{k}{\gamma_w} \frac{\partial^2 u}{\partial z^2} = \frac{1}{1+e} \frac{\partial e}{\partial t} \tag{8.16}$$

The change in void ratio, ∂e, is due to the increase of effective stress; assuming that these are linearly related, then

$$\partial e = -a_v \partial(\Delta\sigma') \tag{8.17}$$

where a_v is the coefficient of compressibility. Again, the increase of effective stress is due to the decrease of excess pore water pressure, ∂u. Hence

$$\partial e = a_v \partial u \tag{8.18}$$

Combining Equations 8.16 and 8.18 gives

$$\frac{k}{\gamma_w} \frac{\partial^2 u}{\partial z^2} = \frac{a_v}{1+e} \frac{\partial u}{\partial t} = m_v \frac{\partial u}{\partial t} \tag{8.19}$$

where

$$m_v = \text{coefficient of volume compressibility} = \frac{a_v}{1+e} \tag{8.20}$$

or

$$\frac{\partial u}{\partial t} = \frac{k}{\gamma_w m_v} \frac{\partial^2 u}{\partial z^2} = C_v \frac{\partial^2 u}{\partial z^2} \tag{8.21}$$

where

$$C_v = \text{coefficient of consolidation} = \frac{k}{\gamma_w m_v} \tag{8.22}$$

Equation 8.21 is the basic differential equation of Terzaghi's consolidation theory and can be solved with proper boundary conditions. To solve the equation, we assume u to be the product of two functions, that is, the product of a function of z and a function of t, or

$$u = F(z)G(t) \tag{8.23}$$

So

$$\frac{\partial u}{\partial t} = F(z)\frac{\partial}{\partial t}G(t) = F(z)G'(t) \tag{8.24}$$

and

$$\frac{\partial^2 u}{\partial z^2} = \frac{\partial^2}{\partial z^2}F(z)G(t) = F''(z)G(t) \tag{8.25}$$

From Equations 8.21, 8.24, and 8.25

$$F(z)G'(t) = C_v F''(z)G(t)$$

or

$$\frac{F''(z)}{F(z)} = \frac{G'(t)}{C_v G(t)} \tag{8.26}$$

The right-hand side of Equation 8.26 is a function of z only and is independent of t; the left-hand side of the equation is a function of t only and is independent of z. Therefore, they must be equal to a constant, say, $-B^2$. So

$$F''(z) = -B^2 F(z) \tag{8.27}$$

A solution to Equation 8.27 can be given by

$$F(z) = A_1 \cos Bz + A_2 \sin Bz \tag{8.28}$$

where A_1 and A_2 are constants.

Again, the right-hand side of Equation 8.26 may be written as

$$G'(t) = -B^2 C_v G(t) \tag{8.29}$$

The solution to Equation 8.29 is given by

$$G(t) = A_3 \exp(-B^2 C_v t) \tag{8.30}$$

where A_3 is a constant. Combining Equations 8.23, 8.28, and 8.30

$$u = (A_1 \cos Bz + A_2 \sin Bz)A_3 \exp(-B^2 C_v t)$$

$$= (A_4 \cos Bz + A_5 \sin Bz)\exp(-B^2 C_v t) \tag{8.31}$$

where
$$A_4 = A_1 A_3$$
$$A_5 = A_2 A_3$$

The constants in Equation 8.31 can be evaluated from the boundary conditions, which are as follows:

1. At time $t = 0$, $u = u_i$ (initial excess pore water pressure at any depth)
2. $u = 0$ at $z = 0$
3. $u = 0$ at $z = H_t = 2H$

Note that H is the length of the longest drainage path. In this case, which is a two-way drainage condition (top *and* bottom of the clay layer), H is equal to half the total thickness of the clay layer, H_t.

The second boundary condition dictates that $A_4 = 0$, and from the third boundary condition we get

$$A_5 \sin 2BH = 0 \quad \text{or} \quad 2BH = n\pi$$

where n is an integer. From the previous equation, a general solution of Equation 8.31 can be given in the form

$$u = \sum_{n=1}^{n=\infty} A_n \sin \frac{n\pi z}{2H} \exp\left(\frac{-n^2\pi^2 T_v}{4}\right) \tag{8.32}$$

where T_v is the nondimensional time factor and is equal to $C_v t/H^2$.

To satisfy the first boundary condition, we must have the coefficients of A_n such that

$$u_i = \sum_{n=1}^{n=\infty} A_n \sin \frac{n\pi z}{2H} \tag{8.33}$$

Equation 8.33 is a Fourier sine series, and A_n can be given by

$$A_n = \frac{1}{H} \int_0^{2H} u_i \sin \frac{n\pi z}{2H} dz \tag{8.34}$$

Combining Equations 8.32 and 8.34

$$u = \sum_{n=1}^{n=\infty} \left(\frac{1}{H} \int_0^{2H} u_i \sin \frac{n\pi z}{2H} dz\right) \sin \frac{n\pi z}{2H} \exp\left(\frac{-n^2\pi^2 T_v}{4}\right) \tag{8.35}$$

So far, no assumptions have been made regarding the variation of u_i with the depth of the clay layer. Several possible types of variation for u_i are shown in Figure 8.6. Each case is considered later.

8.2.1 Constant u_i with depth

If u_i is constant with depth—that is, if $u_i = u_0$ (Figure 8.6a)—then, referring to Equation 8.35

$$\frac{1}{H} \int_0^{2H} \underset{\uparrow}{u_i} \sin \frac{n\pi z}{2H} dz = \frac{2u_0}{n\pi}(1 - \cos n\pi)$$

$$= u_0$$

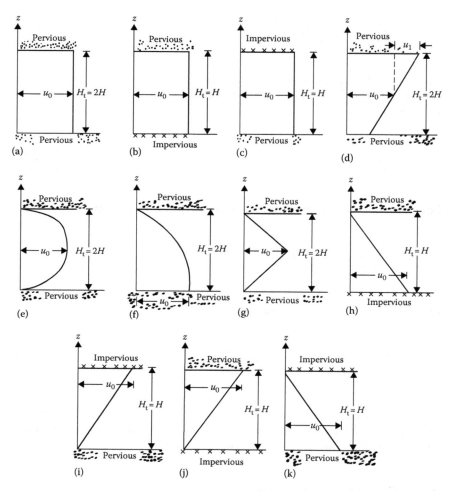

Figure 8.6 Variation of u_i with depth: (a) u_i constant with depth (two-way drainage); (b) u_i constant with depth (drainage at top); (c) u_i constant with depth (drainage at bottom); (d) linear variation of u_i (two-way drainage); (e) sinusoidal variation of u_i (two-way drainage); (f) half sinusoidal variation of u_i (two-way drainage); (g) triangular variation of u_i (two-way drainage); (h) triangular variation of u_i (drainage at top)-base at bottom; (i) triangular variation of u_i (drainage at bottom)-base at top; (j) triangular variation of u_i (drainage at top)-base at top; (k) triangular variation of u_i (drainage at bottom)-base at bottom.

So

$$u = \sum_{n=1}^{n=\infty} \frac{2u_0}{n\pi}(1-\cos n\pi)\sin\frac{n\pi z}{2H}\exp\left(\frac{-n^2\pi^2 T_v}{4}\right) \tag{8.36}$$

Note that the term $1 - \cos n\pi$ in the previous equation is zero for cases when n is even; therefore, u is also zero. For the nonzero terms, it is convenient to substitute $n = 2m + 1$, where m is an integer. So, Equation 8.36 will now read

$$u = \sum_{m=0}^{m=\infty} \frac{2u_0}{(2m+1)\pi}[1-\cos(2m+1)\pi]\sin\frac{(2m+1)\pi z}{2H}$$

$$\times \exp\left[\frac{-(2m+1)^2\pi^2 T_v}{4}\right]$$

or

$$u = \sum_{m=0}^{m=\infty} \frac{2u_0}{M}\sin\frac{Mz}{H}\exp(-M^2 T_v) \tag{8.37}$$

where $M = (2m + 1)\pi/2$. At a given time, the degree of consolidation at any depth z is defined as

$$U_z = \frac{\text{Excess pore water pressure dissipated}}{\text{Initial excess pore water pressure}}$$

$$= \frac{u_i - u}{u_i} = 1 - \frac{u}{u_i} = \frac{\Delta\sigma'}{u_i} = \frac{\Delta\sigma'}{u_0} \tag{8.38}$$

where $\Delta\sigma'$ is the increase of effective stress at a depth z due to consolidation. From Equations 8.37 and 8.38.

$$U_z = 1 - \sum_{m=0}^{m=\infty} \frac{2}{M}\sin\frac{Mz}{H}\exp(-M^2 T_v) \tag{8.39}$$

Figure 8.7 shows the variation of U_z with depth for various values of the nondimensional time factor T_v; these curves are called isochrones. Example 8.1 demonstrates the procedure for calculation of U_z using Equation 8.39.

Example 8.1

Consider the case of an initial excess hydrostatic pore water that is constant with depth, that is, $u_i = u_0$ (Figure 8.6c). For $T_v = 0.3$, determine the degree of consolidation at a depth $H/3$ measured from the top of the layer.

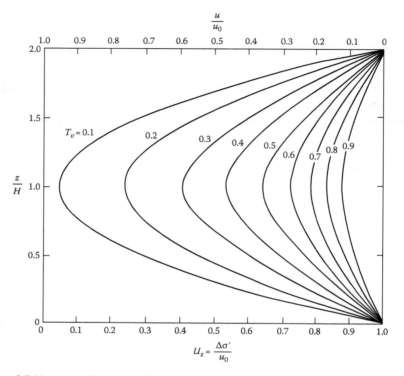

Figure 8.7 Variation of U_z with z/H and T_v.

Solution

From Equation 8.39, for constant pore water pressure increase

$$U_z = 1 - \sum_{m=0}^{m=\infty} \frac{2}{M} \sin \frac{Mz}{H} \exp(-M^2 T_v)$$

Here, $z = H/3$, or $z/H = 1/3$, and $M = (2m + 1)\pi/2$. We can now make a table to calculate U_z.

1. z/H	1/3	1/3	1/3	
2. T_v	0.3	0.3	0.3	
3. m	0	1	2	
4. M	$\pi/2$	$3\pi/2$	$5\pi/2$	
5. Mz/H	$\pi/6$	$\pi/2$	$5\pi/6$	
6. $2/M$	1.273	0.4244	0.2546	
7. $\exp(-M^2 T_v)$	0.4770	0.00128	≈ 0	
8. $\sin(Mz/H)$	0.5	1.0	0.5	
9. $(2/M)[\exp(-M^2 T_v)\sin(Mz/H)]$	0.3036	0.0005	≈ 0	$\Sigma = 0.3041$

Using the value of 0.3041 calculated in step 9, the degree of consolidation at depth $H/3$ is

$$U_{(H/3)} = 1 - 0.3041 = 0.6959 = 69.59\%$$

Note that in the previous table we need not go beyond $m = 2$, since the expression in step 9 is negligible for $m \geq 3$.

In most cases, however, we need to obtain the average degree of consolidation for the entire layer. This is given by

$$U_{av} = \frac{(1/H_t)\int_0^{H_t} u_i dz - (1/H_t)\int_0^{H_t} u\, dz}{(1/H_t)\int_0^{H_t} u_i dz} \tag{8.40}$$

The average degree of consolidation is also the ratio of consolidation settlement at any time to maximum consolidation settlement. Note, in this case, that $H_t = 2H$ and $u_i = u_0$.

Combining Equations 8.37 and 8.40

$$U_{av} = 1 - \sum_{m=0}^{m=\infty} \frac{2}{M^2} \exp(-M^2 T_v) \tag{8.41}$$

Terzaghi suggested the following equations for U_{av} to approximate the values obtained from Equation 8.41:

$$\text{For } U_{av} = 0\% - 53\%: \quad T_v = \frac{\pi}{4}\left(\frac{U_{av}\%}{100}\right)^2 \tag{8.42}$$

$$\text{For } U_{av} = 53\% - 100\%: \quad T_v = 1.781 - 0.933\,[\log(100 - U_{av}\%)] \tag{8.43}$$

Sivaram and Swamee (1977) gave the following equation for U_{av} varying from 0% to 100%:

$$\frac{U_{av}\%}{100} = \frac{(4T_v/\pi)^{0.5}}{[1 + (4T_v/\pi)^{2.8}]^{0.179}} \tag{8.44}$$

or

$$T_v = \frac{(\pi/4)(U_{av}\%/100)^2}{[1-(U_{av}\%/100)^{5.6}]^{0.357}} \tag{8.45}$$

Equations 8.44 and 8.45 give an error in T_v of less than 1% for 0% < U_{av} < 90% and less than 3% for 90% < U_{av} < 100%. Table 8.1 gives the variation of T_v with U_{av} based on Equation 8.41.

It must be pointed out that, if we have a situation of one-way drainage as shown in Figure 8.6b and c, Equation 8.41 would still be valid. Note, however, that the length of the drainage path is equal to the total thickness of the clay layer.

8.2.2 Linear variation of u_i

The linear variation of the initial excess pore water pressure, as shown in Figure 8.6d, may be written as

$$u_i = u_0 - u_1 \frac{H-z}{H} \tag{8.46}$$

Substitution of the earlier relation for u_i into Equation 8.35 yields

$$u = \sum_{n=1}^{n=\infty} \left[\frac{1}{H} \int_0^{2H} \left(u_0 - u_1 \frac{H-z}{H} \right) \sin \frac{n\pi z}{2H} dz \right] \sin \frac{n\pi z}{2H} \exp\left(\frac{-n^2\pi^2 T_v}{4} \right) \tag{8.47}$$

The average degree of consolidation can be obtained by solving Equations 8.40 and 8.47:

$$U_{av} = 1 - \sum_{m=0}^{m=\infty} \frac{2}{M^2} \exp(-M^2 T_v) \tag{8.47a}$$

This is identical to Equation 8.41, which was for the case where the excess pore water pressure is constant with depth, and so the same values as given in Table 8.1 can be used.

8.2.3 Sinusoidal variation of u_i

Sinusoidal variation (Figure 8.6e) can be represented by the equation

$$u_i = u_0 \sin \frac{\pi z}{2H} \tag{8.48}$$

Table 8.1 Variation of T_v with U_{av}

	Value of T_v	
	$u_i = u_0 = const$ (Figure 8.6a through c)	
U_{av}(%)	$u_i = u_0 - u_1\left(\dfrac{H-z}{H}\right)$ (Figure 8.6d)	$u_i = u_0 \sin\dfrac{\pi z}{2H}$ (Figure 8.6e)
0	0	0
1	0.00008	0.0041
2	0.0003	0.0082
3	0.00071	0.0123
4	0.00126	0.0165
5	0.00196	0.0208
6	0.00283	0.0251
7	0.00385	0.0294
8	0.00502	0.0338
9	0.00636	0.0382
10	0.00785	0.0427
11	0.0095	0.0472
12	0.0113	0.0518
13	0.0133	0.0564
14	0.0154	0.0611
15	0.0177	0.0659
16	0.0201	0.0707
17	0.0227	0.0755
18	0.0254	0.0804
19	0.0283	0.0854
20	0.0314	0.0904
21	0.0346	0.0955
22	0.0380	0.101
23	0.0415	0.106
24	0.0452	0.111
25	0.0491	0.117
26	0.0531	0.122
27	0.0572	0.128
28	0.0615	0.133
29	0.0660	0.139
30	0.0707	0.145
31	0.0754	0.150
32	0.0803	0.156
33	0.0855	0.162
34	0.0907	0.168
35	0.0962	0.175

(Continued)

Table 8.1 (Continued) Variation of T_v with U_{av}

	Value of T_v	
	$u_i = u_0 = const$ (Figure 8.6a through c)	
$U_{av}(\%)$	$u_i = u_0 - u_1\left(\dfrac{H-z}{H}\right)$ (Figure 8.6d)	$u_i = u_0 \sin\dfrac{\pi z}{2H}$ (Figure 8.6e)
36	0.102	0.181
37	0.107	0.187
38	0.113	0.194
39	0.119	0.200
40	0.126	0.207
41	0.132	0.214
42	0.138	0.221
43	0.145	0.228
44	0.152	0.235
45	0.159	0.242
46	0.166	0.250
47	0.173	0.257
48	0.181	0.265
49	0.188	0.273
50	0.196	0.281
51	0.204	0.289
52	0.212	0.297
53	0.221	0.306
54	0.230	0.315
55	0.239	0.324
56	0.248	0.333
57	0.257	0.342
58	0.267	0.352
59	0.276	0.361
60	0.286	0.371
61	0.297	0.382
62	0.307	0.392
63	0.318	0.403
64	0.329	0.414
65	0.304	0.425
66	0.352	0.437
67	0.364	0.449
68	0.377	0.462
69	0.390	0.475
70	0.403	0.488
71	0.417	0.502

(Continued)

Table 8.1 (Continued) Variation of T_v with U_{av}

	Value of T_v	
	$u_i = u_0 = const$ (Figure 8.6a through c)	
U_{av}(%)	$u_i = u_0 - u_1\left(\dfrac{H-z}{H}\right)$ (Figure 8.6d)	$u_i = u_0 \sin\dfrac{\pi z}{2H}$ (Figure 8.6e)
72	0.431	0.516
73	0.446	0.531
74	0.461	0.546
75	0.477	0.562
76	0.493	0.578
77	0.511	0.600
78	0.529	0.614
79	0.547	0.632
80	0.567	0.652
81	0.588	0.673
82	0.610	0.695
83	0.633	0.718
84	0.658	0.743
85	0.684	0.769
86	0.712	0.797
87	0.742	0.827
88	0.774	0.859
89	0.809	0.894
90	0.848	0.933
91	0.891	0.976
92	0.938	1.023
93	0.993	1.078
94	1.055	1.140
95	1.129	1.214
96	1.219	1.304
97	1.336	1.420
98	1.500	1.585
99	1.781	1.866
100	∞	∞

The solution for the average degree of consolidation for this type of excess pore water pressure distribution is of the form

$$U_{av} = 1 - \exp\left(\frac{-\pi^2 T_v}{4}\right) \tag{8.49}$$

The variation of U_{av} for various values of T_v is given in Table 8.1.

8.2.4 Other types of pore water pressure variation

Figure 8.6f through k shows several other types of pore water pressure variation. Table 8.2 gives the relationships for the initial excess pore water pressure variation (u_i) and the boundary conditions. These could be solved to provide the variation of U_{av} with T_v and they are shown in Figure 8.8.

Example 8.2

Owing to certain loading conditions, the excess pore water pressure in a clay layer (drained at top and bottom) increased in the manner shown in Figure 8.9a. For a time factor $T_v = 0.3$, calculate the average degree of consolidation.

Solution

The excess pore water pressure diagram shown in Figure 8.9a can be expressed as the difference of two diagrams, as shown in Figure 8.9b and c. The excess pore water pressure diagram in Figure 8.9b shows a case where u_i varies linearly with depth. Figure 8.9c can be approximated as a sinusoidal variation.

Table 8.2 Relationships for u_i and boundary conditions

Figure	u_i	Boundary conditions
8.6f	$u_0 \cos \dfrac{\pi z}{4H}$	Time $t = 0, u = u_i$ $u = 0$ at $z = 2H$ $u = 0$ at $z = 0$
8.6g	For $z \leq H$, $\dfrac{u_0}{H} z$ For $z \geq H$, $2u_0 - \dfrac{u_0}{H} z$	$t = 0, u = u_i$ $u = 0$ at $z = 2H$ $u = 0$ at $z = 0$
8.6h	$u_0 - \dfrac{u_0}{H} z$	$t = 0, u = u_i$ $u = 0$ at $z = H$ $u = u_0$ at $z = 0$
8.6i	$\dfrac{u_0}{H} z$	$t = 0, u = u_i$ $u = u_0$ at $z = H$ $u = 0$ at $z = 0$
8.6j	$\dfrac{u_0}{H} z$	$t = 0, u = u_i$ $u = u_0$ at $z = H$ $u = 0$ at $z = 0$
8.6k	$u_0 - \dfrac{u_0}{H} z$	$t = 0, u = u_i$ $u = 0$ at $z = H$ $u = u_0$ at $z = 0$

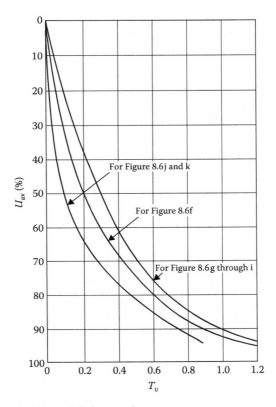

Figure 8.8 Variation of U_{av} with T_v for initial excess pore water pressure diagrams shown in Figure 8.6.

The area of the diagram in Figure 8.9b is

$$A_1 = 6\left(\frac{1}{2}\right)(15+5) = 60 \text{ kN/m}$$

The area of the diagram in Figure 8.9c is

$$A_2 = \sum_{z=0}^{z=6} 2\sin\frac{\pi z}{2H}\,dz = \int_0^6 2\sin\frac{\pi z}{6}\,dz$$

$$= (2)\left(\frac{6}{\pi}\right)\left(-\cos\frac{\pi z}{6}\right)_0^6 = \frac{12}{\pi}(2) = \frac{24}{\pi} = 7.64 \text{ kN/m}$$

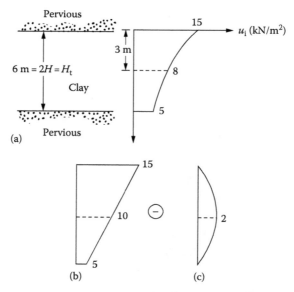

Figure 8.9 Calculation of average degree of consolidation ($T_v = 0.3$): (a) soil profile and excess pore water pressure; (b) excess pore water pressure as a linear distribution; (c) excess pore water pressure as a sinusoidal distribution.

The average degree of consolidation can now be calculated as follows:

| For Figure 8.9b | For Figure 8.9c |

$$U_{av}(T_v = 0.3) = \frac{U_{av}(T_v = 0.3)A_1 - U_{av}(T_v = 0.3)A_2}{A_1 - A_2}$$

For Figure 8.9a   Net area of Figure 8.9a

From Table 8.1 for $T_v = 0.3$, $U_{av} \approx 61\%$ for area A_1; $U_{av} \approx 52.3\%$ for area A_2.
So

$$U_{av} = \frac{61(60) - (7.64)52.3}{60 - 7.64} = \frac{3260.43}{52.36} = 62.3\%$$

Example 8.3

Due to a certain type of loading condition, the initial excess pore water pressure distribution in a 4-m-thick layer is as shown in Figure 8.10. Given $C_v = 0.3$ mm²/s. Determine the degree of consolidation after 100 days of load application.

Solution

From Figure 8.10,
  Area of rectangle:

$$A_1 = 4 \times 100 = 400 \text{ kN/m}$$

Area of half sinusoidal wave:

$$A_2 = \int_{z=0}^{z=4} 80 \sin\left(\frac{\pi z}{2H_t}\right) dz = 203.6 \text{ kN/m}$$

$$T_v = \frac{C_v t}{H^2} = \frac{(0.3)(100 \times 24 \times 60 \times 60)}{\left(\dfrac{4 \times 1000}{2}\right)^2} = 0.648$$

From Table 8.1, for area A_1, $U_{av(1)} = 83.5\%$. From Figure 8.8, for area A_2 (see Figure 8.6f), $U_{av(2)} = 82.5\%$.

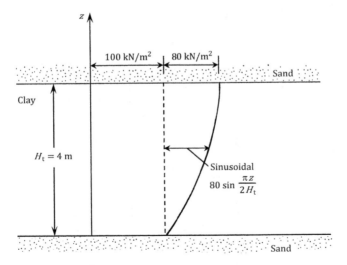

Figure 8.10 Variation of initial excess pore water pressure with depth in a clay layer.

$$U_{av} = \frac{A_1 U_{av(1)} + A_2 U_{av(2)}}{A_1 + A_2}$$

$$= \frac{(400)(83.5) + (203.6)(82.5)}{400 + 203.6} = 83.2\%$$

Example 8.4

A 25-mm total consolidation settlement of the two clay layers shown in Figure 8.11 is expected owing to the application of the uniform surcharge q. Find the duration after the load application at which 12.5 mm of total settlement would take place.

Solution

$$U_{av} = \frac{12.5 \text{ mm}}{25 \text{ mm}} = 50\%$$

$$(3 \text{ m} + 1.5 \text{ m})(0.5) = 3U_{v(1)} + 1.5U_{v(2)} \qquad \text{(a)}$$

$$T_{v(1)} = \frac{C_{v(1)}t}{H_1^2} = \frac{(0.13 \text{ cm}^2/\text{min})t}{\left(\dfrac{300 \text{ cm}}{2}\right)^2} = 5.78 \times 10^{-6} \ t$$

$$T_{v(2)} = \frac{C_{v(2)}t}{H_2^2} = \frac{(0.13 \text{ cm}^2/\text{min})t}{\left(\dfrac{150 \text{ cm}}{2}\right)^2} = 2.31 \times 10^{-5} \ t$$

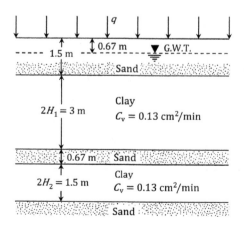

Figure 8.11 Consolidation settlement of two clay layers.

Hence,

$$T_{v(2)} = 4T_{v(1)} \tag{b}$$

Now, trial and error will have to be used. Let $T_{v(1)} = 0.1125$. From Equation (b), $T_{v(2)} = 0.45$. From Table 8.1, $U_{v(1)} \approx 0.385$ and $U_{v(2)} \approx 0.73$.

Equation (a) gives $2.25 = 3U_{v(1)} + 1.5U_{v(2)}$. Substituting the values of $U_{v(1)}$ and $U_{v(2)}$ in Equation (a) we obtain,

$$3U_{v(1)} + 1.5U_{v(2)} = 3(0.385) + 1.5(0.73) = 2.25 \text{ m} --\text{O.K.}$$

$$T_{v(1)} = 5.78 \times 10^{-6} \, t$$

$$t = \frac{T_{v(1)}}{5.78 \times 10^{-6}} = \frac{0.1125}{5.78 \times 10^{-6}} = 19,463 \text{ min} = \mathbf{13.5 \text{ days}}$$

Example 8.5

Starting from Equation 8.47a, solve the average degree of consolidation for linearly varying initial excess pore water pressure distribution for a clay layer with two-way drainage for $T_v = 0.6$ (Figure 8.6d).

Solution

From Equation 8.41:

$$U_{av} = 1 - \sum_{m=0}^{m=\infty} \frac{2}{M^2} \exp\left(-M^2 T_v\right)$$

Now, the following table can be prepared.

T_v	0.6	0.6	0.6	0.6
m	0	1	2	3
$M = (2m+1)(\pi/2)$	$\pi/2$	$3\pi/2$	$5\pi/2$	$7\pi/2$
$2/M^2$	0.811	0.09	0.32	0.017
$e^{-M^2 T_v}$	0.228	0.16×10^{-5}	≈ 0	≈ 0
$(2/M^2)e^{-M^2 T_v}$	0.185	≈ 0	≈ 0	≈ 0

$$\Sigma (2/M^2) e^{-M^2 T_v} = 0.185$$

$$U_{av} = 1 - 0.185 = 0.815 = \mathbf{81.5\%}$$

Example 8.6

A uniform surcharge of $q = 100$ kN/m² is applied on the ground surface as shown in Figure 8.12a.

   a. Determine the initial excess pore water pressure distribution in the clay layer.
   b. Plot the distribution of the excess pore water pressure with depth in the clay layer at a time for which $T_v = 0.5$.

Solution

Part a: The initial excess pore water pressure will be 100 kN/m² and will be the same throughout the clay layer (Figure 8.12a).

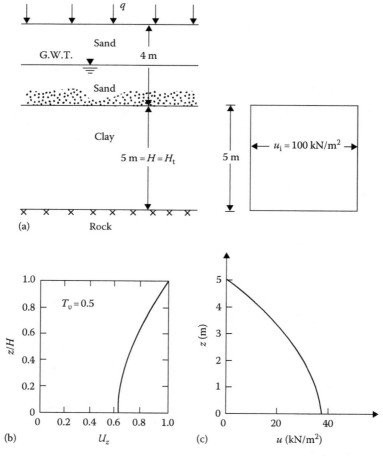

Figure 8.12 Excess pore water pressure distribution: (a) soil profile and plot of initial excess pore water pressure with depth; (b) plot of U_z with z/H at $T_v = 0.5$; (c) plot of u with z at $T_v = 0.5$.

Part b: From Equation 8.38, $U_z = 1 - u/u_i$, or $u = u_i(1 - U_z)$. For $T_v = 0.5$, the values of U_z can be obtained from the top half of Figure 8.7 as shown in Figure 8.12b, and then the following table can be prepared:

z/H	z (m)	U_z	$u = u_i(1 - U_z)$ (kN/m²)
0	0	0.63	37
0.2	1	0.65	35
0.4	2	0.71	29
0.6	3	0.78	22
0.8	4	0.89	11
1.0	5	1	0

Figure 8.12c shows the variation of excess pore water pressure with depth.

Example 8.7

Refer to Figure 8.6e. For the sinusoidal initial excess pore water pressure distribution, given

$$u_i = 50 \sin\left(\frac{\pi z}{2H}\right) \text{kN/m}^2$$

Assume $H_i = 2H = 5$ m. Calculate the excess pore water pressure at the midheight of the clay layer for $T_v = 0.2, 0.4, 0.6,$ and 0.8.

Solution

From Equation 8.35

$$u = \sum_{n=1}^{n=\infty} \underbrace{\left(\frac{1}{H} \int_0^{2H} u_i \sin\frac{n\pi z}{2H} dz \right)}_{\text{term } A} \left(\sin\frac{n\pi z}{2H} \right) \exp\left(\frac{-n^2 \pi T_v}{4} \right)$$

Let us evaluate the term A

$$A = \frac{1}{H} \int_0^{2H} u_i \sin\frac{n\pi z}{2H} dz$$

or

$$A = \frac{1}{H} \int_0^{2H} 50 \sin\frac{\pi z}{2H} \sin\frac{n\pi z}{2H} dz$$

Note that the integral mentioned earlier is zero if $n \neq 1$, and so the only nonzero term is obtained when $n = 1$. Therefore

$$A = \frac{50}{H} \int_0^{2H} \sin^2 \frac{\pi z}{2H} dz = \frac{50}{H} H = 50$$

Since only for $n = 1$ is A not zero

$$u = 50 \sin \frac{\pi z}{2H} \exp\left(\frac{-\pi^2 T_v}{4}\right)$$

At the midheight of the clay layer, $z = H$, and so

$$u = 50 \sin \frac{\pi}{2} \exp\left(\frac{-\pi^2 T_v}{4}\right) = 50 \exp\left(\frac{-\pi^2 T_v}{4}\right)$$

The values of the excess pore water pressure are tabulated as follows:

T_v	$u = 50 \exp\left(\dfrac{-\pi^2 T_v}{4}\right)$ (kN/m²)
0.2	30.52
0.4	18.64
0.6	11.38
0.8	6.95

8.3 DEGREE OF CONSOLIDATION UNDER TIME-DEPENDENT LOADING

Olson (1977) presented a mathematical solution for one-dimensional consolidation due to a single ramp load. Olson's solution can be explained with the help of Figure 8.13, in which a clay layer is drained at the top and at the bottom (H is the drainage distance). A uniformly distributed load q is applied at the ground surface. Note that q is a function of time, as shown in Figure 8.13b.

The expression for the excess pore water pressure for the case where $u_i = u_0$ is given in Equation 8.37 is

$$u = \sum_{m=0}^{m=\infty} \frac{2u_0}{M} \sin \frac{Mz}{H} \exp(-M^2 T_v)$$

where $T_v = C_v t / H^2$.

As stated earlier, the applied load is a function of time:

$$q = f(t_a) \tag{8.50}$$

where t_a is the time of application of any load.

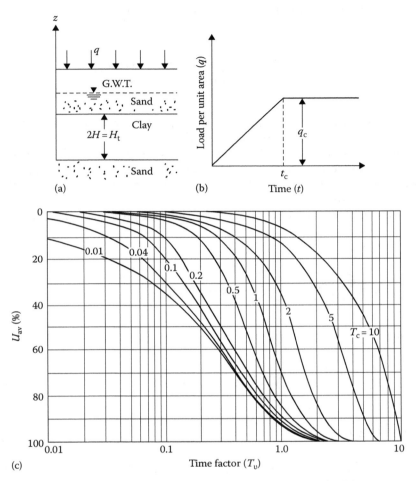

Figure 8.13 One-dimensional consolidation due to single ramp load: (a) soil profile; (b) ramp loading; (c) variation of U_{av} (%) with T_v and T_c. [After Olson, R. E., J. Geotech. Eng. Div., ASCE, 103(GT1), 55, 1977.]

For a differential load dq applied at time t_a, the instantaneous pore pressure increase will be $du_i = dq$. At time t, the remaining excess pore water pressure du at a depth z can be given by the expression

$$du = \sum_{m=0}^{m=\infty} \frac{2du_i}{M} \sin \frac{Mz}{H} \exp\left[\frac{-M^2 C_v(t - t_a)}{H^2} \right]$$

$$= \sum_{m=0}^{m=\infty} \frac{2dq}{M} \sin \frac{Mz}{H} \exp\left[\frac{-M^2 C_v(t - t_a)}{H^2} \right] \tag{8.51}$$

The average degree of consolidation can be defined as

$$U_{av} = \frac{\alpha q_c - (1/H_t)\int_0^{H_t} u\,dz}{q_c} = \frac{\text{Settlement at time } t}{\text{Settlement at time } t = \infty} \tag{8.52}$$

where αq_c is the total load per unit area applied at the time of the analysis. The settlement at time $t = \infty$ is, of course, the ultimate settlement. Note that the term q_c in the denominator of Equation 8.52 is equal to the instantaneous excess pore water pressure ($u_i = q_c$) that might have been generated throughout the clay layer had the stress q_c been applied instantaneously.

Proper integration of Equations 8.51 and 8.52 gives the following:

For $T_v \leq T_c$

$$u = \sum_{m=0}^{m=\infty} \frac{2q_c}{M^3 T_c} \sin \frac{Mz}{H}[1 - \exp(-M^2 T_v)] \tag{8.53}$$

and

$$U_{av} = \frac{T_v}{T_c}\left\{1 - \frac{2}{T_v}\sum_{m=0}^{m=\infty}\frac{1}{M^4}[1 - \exp(-M^2 T_v)]\right\} \tag{8.54}$$

For $T_v \geq T_c$

$$u = \sum_{m=0}^{m=\infty} \frac{2q_c}{M^3 T_c}[\exp(M^2 T_c) - 1]\sin \frac{Mz}{H}\exp(-M^2 T_v) \tag{8.55}$$

and

$$U_{av} = 1 - \frac{2}{T_c}\sum_{m=0}^{m=\infty}\frac{1}{M^4}[\exp(M^2 T_c) - 1]\exp(-M^2 T_c) \tag{8.56}$$

where

$$T_c = \frac{C_v t_c}{H^2} \tag{8.57}$$

Figure 8.13c shows the plot of U_{av} against T_v for various values of T_c.

Example 8.8

Based on one-dimensional consolidation test results on a clay, the coefficient of consolidation for a given pressure range was obtained as 8×10^{-3} mm^2/s. In the field, there is a 2 m-thick layer of the same clay with two-way drainage. Based on the assumption that a uniform surcharge of 70 kN/m^2 was to be applied instantaneously, the total consolidation settlement was estimated to be 150 mm. However, during the construction, the loading was gradual; the resulting surcharge can be approximated as

$$q \,(\text{kN/m}^2) = \frac{70}{60} t \,(\text{days}) \quad \text{for } t \leq 60 \text{ days}$$

and

$$q = 70 \text{ kN/m}^2 \quad \text{for } t \geq 60 \text{ days}$$

Estimate the settlement at t = 30 and 120 days.

Solution

$$T_c = \frac{C_v t_c}{H^2} \tag{8.57}$$

Now, t_c = 60 days = 60 × 24 × 60 × 60 s; also, H_t = 2 m = $2H$ (two-way drainage), and so H = 1 m = 1000 mm. Hence

$$T_c = \frac{(8 \times 10^{-3})(60 \times 24 \times 60 \times 60)}{(1000)^2} = 0.0414$$

At t = 30 days

$$T_v = \frac{C_v t}{H^2} = \frac{(8 \times 10^{-3})(30 \times 24 \times 60 \times 60)}{(1000)^2} = 0.0207$$

From Figure 8.13c, for T_v = 0.0207 and T_c = 0.0414, $U_{av} \approx 5\%$. So

Settlement = (0.05)(150) = 7.5 mm

At t = 120 days

$$T_v = \frac{(8 \times 10^{-3})(120 \times 24 \times 60 \times 60)}{(1000)^2} = 0.083$$

From Figure 8.13c for T_v = 0.083 and T_c = 0.0414, $U_{av} \approx 27\%$. So

Settlement = (0.27)(150) = 40.5 mm

8.3.1 Solution of Hanna et al. (2013) and Sivakugan et al. (2014)

Hanna et al. (2013) and Sivakugan et al. (2014) have shown that, during ramp loading (i.e., $t \le t_c$ in Figure 8.13b), the average degree of consolidation can be expressed as

$$U_{av} = 1 - \frac{1}{T_v} \left[\sum_{m=0}^{\infty} \left(\frac{2}{M^4} \right) (1 - e^{-M^2 T_v}) \right]$$

(8.58)

The variation of U_{av} with T_v is shown in Figure 8.14. Also shown in the figure are the $U_{av} - T_v$ variations for the instantaneous loading (Equation 8.41), which is quite different.

Denoting the degree of consolidation at the end of construction (i.e., at $t \le t_c$) as $U_{av(0)}$, the remaining excess pore water pressure can be considered

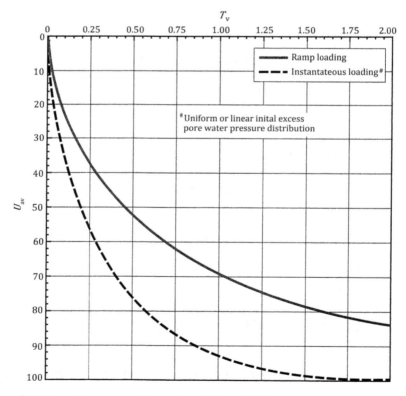

Figure 8.14 Plot of U_{av} versus T_v (Equations 8.58 and 8.41).

as being applied instantaneously at time t_c. The average degree of consolidation at time t ($>t_c$) can then be computed as

$$U_{av(t)} = U_{av(0)} + [1 - U_{av(0)}]U_{av(t-t_c)} \qquad (8.59)$$

where $U_{av(t-t_c)}$ is the average degree of consolidation for duration of $t - t_c$, assuming instantaneous loading at t_c.

Example 8.9

Solve Example 8.8 using the procedure of Hanna et al. (2013) and Sivakugan et al. (2014).

Solution

Given $t_c = 60$ days.
For $t = 30$ days: Equation 8.58 can be used.

$$T_v = \frac{C_v t}{H^2} = \frac{(8 \times 10^{-3}\ mm^2/s)(30 \times 24 \times 60 \times 60)}{\left(\dfrac{2000\ mm}{2}\right)^2} = 0.0207$$

From Figure 8.14, for ramp loading, $U_{av} \approx 5\%$. So,

Settlement $= (0.05)(150) = 7.52$ mm

For $t = 120$ days: At time $t = t_c$, the magnitude of

$$T_v = \frac{C_v t_c}{H^2} = \frac{(8 \times 10^{-3}\ mm^2/s)(60 \times 24 \times 60 \times 60)}{\left(\dfrac{2000\ mm}{2}\right)^2} = 0.0414$$

From Figure 8.14, for ramp loading, $U_{av(0)} \approx 12\%$.

At $t = 120$ days, $t - t_c = 120 - 60 = 60$ days. So

$$T_{v(t-t_c)} = \frac{C_v t_{(t-t_c)}}{H^2} = \frac{(8 \times 10^{-3})(60 \times 24 \times 60 \times 60)}{\left(\dfrac{2000}{2}\right)^2} = 0.0414$$

For instantaneous loading with $T_{v(t-t_c)} = 0.0414$, Figure 8.14 gives $U_{av(t-t_c)} \approx 20\%$.
From Equation 8.59,

$$\begin{aligned} U_{av(t)} &= U_{av(0)} + [1 - U_{av(0)}]U_{av(t-t_c)} \\ &= 0.12 + (1 - 0.12)(0.20) = 0.296 = 29.6\% \end{aligned}$$

Settlement $= (150)(0.296) = 44.4$ mm

8.4 NUMERICAL SOLUTION FOR ONE-DIMENSIONAL CONSOLIDATION

8.4.1 Finite difference solution

The principles of finite difference solutions were introduced in Section 7.8. In this section, we will consider the finite difference solution for one-dimensional consolidation, starting from the basic differential equation of Terzaghi's consolidation theory:

$$\frac{\partial u}{\partial t} = C_v \frac{\partial^2 u}{\partial z^2} \tag{8.60}$$

Let u_R, t_R, and z_R be any arbitrary reference excess pore water pressure, time, and distance, respectively. From these, we can define the following nondimensional terms:

Nondimensional excess pore water pressure: $\bar{u} = \dfrac{u}{u_R}$ $\hspace{2cm}$ (8.61)

Nondimensional time: $\bar{t} = \dfrac{t}{t_R}$ $\hspace{2cm}$ (8.62)

Nondimensional depth: $\bar{z} = \dfrac{z}{z_R}$ $\hspace{2cm}$ (8.63)

From Equations 8.61, 8.63, and the left-hand side of Equation 8.60

$$\frac{\partial u}{\partial t} = \frac{u_R}{t_R} \frac{\partial \bar{u}}{\partial \bar{t}} \tag{8.64}$$

Similarly, from Equations 8.61, 8.63, and the right-hand side of Equation 8.60

$$C_v \frac{\partial^2 u}{\partial z^2} = C_v \frac{u_R}{z_R^2} \frac{\partial^2 \bar{u}}{\partial \bar{z}^2} \tag{8.65}$$

From Equations 8.64 and 8.65

$$\frac{u_R}{t_R} \frac{\partial \bar{u}}{\partial \bar{t}} = C_v \frac{u_R}{z_R^2} \frac{\partial^2 \bar{u}}{\partial \bar{z}^2}$$

or

$$\frac{1}{t_R}\frac{\partial \bar{u}}{\partial \bar{t}} = \frac{C_v}{z_R^2}\frac{\partial^2 \bar{u}}{\partial \bar{z}^2} \tag{8.66}$$

If we adopt the reference time in such a way that $t_R = z_R^2/C_v$, then Equation 8.66 will be of the form

$$\frac{\partial \bar{u}}{\partial \bar{t}} = \frac{\partial^2 \bar{u}}{\partial \bar{z}^2} \tag{8.67}$$

The left-hand side of Equation 8.67 can be written as

$$\frac{\partial \bar{u}}{\partial \bar{t}} = \frac{1}{\Delta \bar{t}}(\bar{u}_{0,\bar{t}+\Delta\bar{t}} - \bar{u}_{0,\bar{t}}) \tag{8.68}$$

where $\bar{u}_{0,\bar{t}}$ and $\bar{u}_{0,\bar{t}+\Delta\bar{t}}$ are the nondimensional pore water pressures at point 0 (Figure 8.15a) at nondimensional times t and $t + \Delta t$. Again, similar to Equation 7.56:

$$\frac{\partial^2 \bar{u}}{\partial \bar{z}^2} = \frac{1}{(\Delta \bar{z})^2}(\bar{u}_{1,\bar{t}} + \bar{u}_{3,\bar{t}} - 2\bar{u}_{0,\bar{t}}) \tag{8.69}$$

Equating the right sides of Equations 8.68 and 8.69 gives

$$\frac{1}{\Delta \bar{t}}(\bar{u}_{0,\bar{t}+\Delta\bar{t}} - \bar{u}_{0,\bar{t}}) = \frac{1}{(\Delta \bar{z})^2}(\bar{u}_{1,\bar{t}} + \bar{u}_{3,\bar{t}} - 2\bar{u}_{0,\bar{t}})$$

or

$$\bar{u}_{0,\bar{t}+\Delta\bar{t}} = \frac{\Delta \bar{t}}{(\Delta \bar{z})^2}(\bar{u}_{1,\bar{t}} + \bar{u}_{3,\bar{t}} - 2\bar{u}_{0,\bar{t}}) + \bar{u}_{0,\bar{t}} \tag{8.70}$$

For Equation 8.70 to converge, $\Delta \bar{t}$ and $\Delta \bar{z}$ must be chosen such that $\Delta \bar{t}/(\Delta \bar{z})^2$ is less than 0.5.

When solving for pore water pressure at the interface of a clay layer and an impervious layer, Equation 8.70 can be used. However, we need to take point 3 as the mirror image of point 1 (Figure 8.15b); thus $\bar{u}_{1,\bar{t}} = \bar{u}_{3,\bar{t}}$. So, Equation 8.70 becomes

$$\bar{u}_{0,\bar{t}+\Delta\bar{t}} = \frac{\Delta \bar{t}}{(\Delta \bar{z})^2}(2\bar{u}_{1,\bar{t}} - 2\bar{u}_{0,\bar{t}}) + \bar{u}_{0,\bar{t}} \tag{8.71}$$

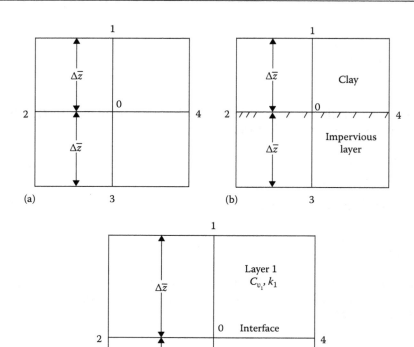

Figure 8.15 Numerical solution for consolidation: (a) derivation of Equation 8.70; (b) derivation of Equation 8.71; (c) derivation of Equation 8.75.

8.4.2 Consolidation in a layered soil

It is not always possible to develop a closed-form solution for consolidation in layered soils. There are several variables involved, such as different coefficients of permeability, the thickness of layers, and different values of coefficient of consolidation. Figure 8.16 shows the nature of the degree of consolidation of a two-layered soil.

In view of the earlier description, numerical solutions provide a better approach. If we are involved with the calculation of excess pore water pressure at the interface of two different types (i.e., different values of C_v)

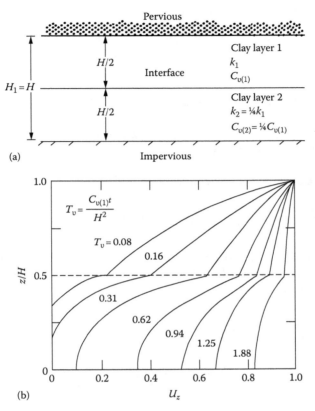

Figure 8.16 Degree of consolidation in two-layered soil: (a) soil profile; (b) variation of U_z with z/H and T_v. [After Luscher, U., *J. Soil Mech. Found. Div.*, ASCE, 91(SM1), 190, 1965.]

of clayey soils, Equation 8.70 will have to be modified to some extent. Referring to Figure 8.15c, this can be achieved as follows (Scott, 1963). From Equation 8.21

$$\frac{k}{C_v}\frac{\partial u}{\partial t} = k \qquad \frac{\partial^2 u}{\partial z^2}$$

↑   ↑

Change   Difference between
in volume   the rate of flow

Based on the derivations of Equation 7.76

$$k\frac{\partial^2 u}{\partial z^2} = \frac{1}{2}\left[\frac{k_1}{(\Delta z)^2} + \frac{k_2}{(\Delta z)^2}\right]\left(\frac{2k_1}{k_1+k_2}u_{1,t} + \frac{2k_2}{k_1+k_2}u_{3,t} - 2u_{0,t}\right) \tag{8.72}$$

where
$\quad$ k_1 and k_2 are the coefficients of permeability in layers 1 and 2, respectively
$\quad$ $u_{0,t}$, $u_{1,t}$, and $u_{3,t}$ are the excess pore water pressures at time t for points 0, 1, and 3, respectively

Also, the average volume change for the element at the boundary is

$$\frac{k}{C_v}\frac{\partial u}{\partial t} = \frac{1}{2}\left(\frac{k_1}{C_{v_1}} + \frac{k_2}{C_{v_2}}\right)\frac{1}{\Delta t}(u_{0,t+\Delta t} - u_{0,t}) \tag{8.73}$$

where $u_{0,t}$ and $u_{0,t+\Delta t}$ are the excess pore water pressures at point 0 at times t and $t + \Delta t$, respectively. Equating the right-hand sides of Equations 8.72 and 8.73, we get

$$\left(\frac{k_1}{C_{v_1}} + \frac{k_2}{C_{v_2}}\right)\frac{1}{\Delta t}(u_{0,t+\Delta t} - u_{0,t})$$

$$= \frac{1}{(\Delta z)^2}(k_1 + k_2)\left(\frac{2k_1}{k_1+k_2}u_{1,t} + \frac{2k_2}{k_1+k_2}u_{3,t} - 2u_{0,t}\right)$$

or

$$u_{0,t+\Delta t} = \frac{\Delta t}{(\Delta z)^2}\frac{k_1+k_2}{k_1/C_{v_1}+k_2/C_{v_2}}\left(\frac{2k_1}{k_1+k_2}u_{1,t} + \frac{2k_2}{k_1+k_2}u_{3,t} - 2u_{0,t}\right) + u_{0,t}$$

or

$$u_{0,t+\Delta t} = \frac{\Delta t C_{v_1}}{(\Delta z)^2}\frac{1+k_2/k_1}{1+(k_2/k_1)(C_{v_1}/C_{v_2})}\left(\frac{2k_1}{k_1+k_2}u_{1,t} + \frac{2k_2}{k_1+k_2}u_{3,t} - 2u_{0,t}\right) + u_{0,t}$$

$$\tag{8.74}$$

Assuming $1/t_R = C_{v_1}/z_R^2$ and combining Equations 8.61 through 8.63 and 8.74, we get

$$\bar{u}_{0,\bar{t}+\Delta\bar{t}} = \frac{1+k_2/k_1}{1+(k_2/k_1)(C_{v_1}/C_{v_2})}\frac{\Delta\bar{t}}{(\Delta\bar{z})^2}$$

$$\times\left(\frac{2k_1}{k_1+k_2}\bar{u}_{1,\bar{t}} + \frac{2k_2}{k_1+k_2}\bar{u}_{3,t} - 2\bar{u}_{0,\bar{t}}\right) + \bar{u}_{0,\bar{t}} \tag{8.75}$$

Example 8.10

A uniform surcharge of q = 150 kN/m² is applied at the ground surface of the soil profile shown in Figure 8.17a. Using the numerical method, determine the distribution of excess pore water pressure for the clay layers after 10 days of load application.

Solution

Since this is a uniform surcharge, the excess pore water pressure immediately after the load application will be 150 kN/m² through-out the clay layers. However, owing to the drainage conditions, the excess pore water pressures at the top of layer 1 and bottom of layer 2 will immediately become zero. Now, let z_R = 8 m and u_R = 1.5 kN/m². So, $\bar{z}$ = (8 m)/(8 m) = 1 and $\bar{u}$ = (150 kN/m²)/(1.5 kN/m²) = 100.

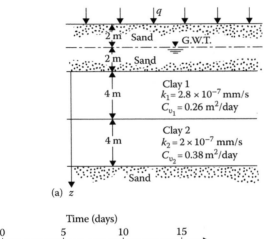

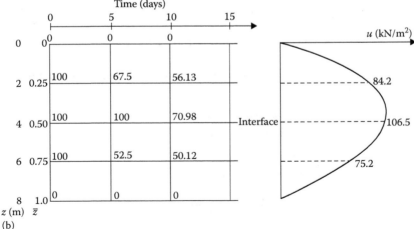

Figure 8.17 Numerical solution for consolidation in layered soil: (a) soil profile; (b) variation of pore water pressure with depth and time.

Figure 8.17b shows the distribution of $\bar{u}$ at time $t = 0$; note that $\Delta\bar{z} = 2/8 = 0.25$. Now

$$t_R = \frac{z_R^2}{C_v} \quad \bar{t} = \frac{t}{t_R} \quad \frac{\Delta t}{\Delta\bar{t}} = \frac{z_R^2}{C_v} \quad \text{or} \quad \Delta\bar{t} = \frac{C_v \Delta t}{z_R^2}$$

Let $\Delta t = 5$ days for both layers. So, for layer 1

$$\Delta\bar{t}_{(1)} = \frac{C_v \Delta t}{z_R^2} = \frac{0.26(5)}{8^2} = 0.0203$$

$$\frac{\Delta\bar{t}_{(1)}}{(\Delta\bar{z})^2} = \frac{0.0203}{0.25^2} = 0.325 \quad (<0.5)$$

For layer 2

$$\Delta\bar{t}_{(2)} = \frac{C_{v_2} \Delta t}{z_R^2} = \frac{0.38(5)}{8^2} = 0.0297$$

$$\frac{\Delta\bar{t}_{(2)}}{(\Delta\bar{z})^2} = \frac{0.0297}{0.25^2} = 0.475 \quad (<0.5)$$

For $t = 5$ days
   At $\bar{z} = 0$

$$\bar{u}_{0,\bar{t}+\Delta\bar{t}} = 0$$

   At $\bar{z} = 0.25$

$$\bar{u}_{0,\bar{t}+\Delta\bar{t}} = \frac{\Delta\bar{t}_{(1)}}{(\Delta\bar{z})^2}(\bar{u}_{1,\bar{t}} + \bar{u}_{3,\bar{t}} - 2\bar{u}_{0,\bar{t}}) + \bar{u}_{0,\bar{t}}$$

$$= 0.325[0 + 100 - 2(100)] + 100 = 67.5 \tag{8.70}$$

At $\bar{z} = 0.5$ (Note: this is the boundary of two layers, so we will use Equation 8.75)

$$\bar{u}_{0,\bar{t}+\Delta\bar{t}} = \frac{1 + k_2/k_1}{1 + (k_2/k_1)(C_{v_1}/C_{v_2})} \frac{\Delta\bar{t}_{(1)}}{(\Delta\bar{z})^2}$$

$$\times \left(\frac{2k_1}{k_1 + k_2}\bar{u}_{1,\bar{t}} + \frac{2k_2}{k_1 + k_2}\bar{u}_{3,\bar{t}} - 2\bar{u}_{0,\bar{t}} \right) + \bar{u}_{0,\bar{t}}$$

$$= \frac{1 + (2/2.8)}{1 + (2 \times 0.26)/(2.8 \times 0.38)}(0.325)$$

$$\times \left[\frac{2 \times 2.8}{2 + 2.8}(100) + \frac{2 \times 2}{2 + 2.8}(100) - 2(100) \right] + 100$$

or

$$\bar{u}_{0,\bar{t}+\Delta\bar{t}} = (1.152)(0.325)(116.67 + 83.33 - 200) + 100 = 100$$

At $\bar{z} = 0.75$

$$\bar{u}_{0,\bar{t}+\Delta\bar{t}} = \frac{\Delta\bar{t}_{(2)}}{(\Delta\bar{z})^2}(\bar{u}_{1,\bar{t}} + \bar{u}_{3,\bar{t}} - 2\bar{u}_{0,\bar{t}}) + \bar{u}_{0,\bar{t}}$$

$$= 0.475[100 + 0 - 2(100)] + 100 = 52.5$$

At $\bar{z} = 1.0$

$$\bar{u}_{0,\bar{t}+\Delta\bar{t}} = 0$$

For $t = 10$ days
At $\bar{z} = 0$

$$\bar{u}_{0,\bar{t}+\Delta\bar{t}} = 0$$

At $\bar{z} = 0.25$

$$\bar{u}_{0,\bar{t}+\Delta\bar{t}} = 0.325[0 + 100 - 2(67.5)] + 67.5 = 56.13$$

At $\bar{z} = 0.5$

$$\bar{u}_{0,\bar{t}+\Delta\bar{t}} = (1.152)(0.325)\left[\frac{2 \times 2.8}{2 + 2.8}(67.5) + \frac{2 \times 2}{2 + 2.8}(52.5) - 2(100)\right] + 100$$

$$= (1.152)(0.325)(78.75 + 43.75 - 200) + 100 = 70.98$$

At $\bar{z} = 0.75$

$$\bar{u}_{0,\bar{t}+\Delta\bar{t}} = 0.475[100 + 0 - 2(52.5)] + 52.5 = 50.12$$

At $\bar{z} = 1.0$

$$\bar{u}_{0,\bar{t}+\Delta\bar{t}} = 0$$

The variation of the nondimensional excess pore water pressure is shown in Figure 8.17b. Knowing $u = (\bar{u})(u_R) = \bar{u}\ (1.5)$ kN/m^2, we can plot the variation of u with depth.

Example 8.11

A uniform surcharge of 100 kN/m^2 is applied to the ground surface of a soil profile, as shown in Figure 8.18. Determine the distribution of the excess pore water pressure in the 3-m-thick clay layer after 1 year of load application. Use the numerical method of calculation given in

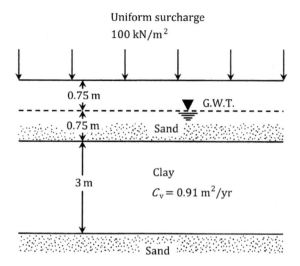

Figure 8.18 Uniform surcharge on the surface of a soil profile.

Section 8.4. Also calculate the average degree of consolidation at that time using the above results.

Solution

Let $z_R = 3$ m; $\Delta z = 0.75$ m; $\Delta \bar{z} = 0.75/3 = 0.25$; $\Delta t = 3$ months

Let $u_R = 1$ kN/m^2

$$\Delta \bar{t} = \frac{C_v \Delta t}{z_R^2} = \left[\frac{0.91 \times (3/12)}{3^3} \right] = 0.025$$

$$\frac{\Delta \bar{t}}{(\Delta \bar{z})^2} = \frac{0.025}{(0.25)^2} = 0.4 > 0.5 --O.K.$$

From Equation 8.70,

$$\bar{u}_{0,\bar{i}+\Delta\bar{t}} = \frac{\Delta \bar{t}}{(\Delta \bar{z})^2} (\bar{u}_{1,\bar{i}} + \bar{u}_{3,\bar{i}} + 2\bar{u}_{0,\bar{i}}) + \bar{u}_{0,\bar{i}}$$

The distribution of $\bar{u}$ with time and $\bar{z}$ is shown in Figure 8.19.

$$u_R$$
$$\downarrow$$
$$U_{av} = \frac{(100 \times 3) - [(29.6 + 41.12 + 29.6)](1)(0.75)}{100 \times 3} = 74.9\%$$

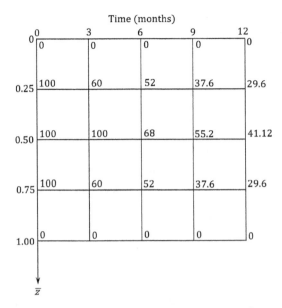

Figure 8.19 Variation of nondimensional pore water pressure with $\bar{z}$ and time.

Example 8.12

For Example 8.10, assume that the surcharge q is applied gradually. The relation between time and q is shown in Figure 8.20a. Using the numerical method, determine the distribution of excess pore water pressure after 15 days from the start of loading.

Solution

As mentioned earlier, in Example 8.10, $z_R = 8$ m, $u_R = 1.5$ kN/m². For $\Delta t = 5$ days

$$\frac{\Delta \bar{t}_{(1)}}{(\Delta \bar{z})^2} = 0.325 \quad \frac{\Delta \bar{t}_{(2)}}{(\Delta \bar{z})^2} = 0.475$$

The continuous loading can be divided into step loads such as 60 kN/m² from 0 to 10 days and an added 90 kN/m² from the tenth day on. This is shown by dashed lines in Figure 8.20a.
At $t = 0$ days

$\bar{z} = 0 \quad \bar{u} = 0$
$\bar{z} = 0.25 \quad \bar{u} = 60/1.5 = 40$
$\bar{z} = 0.5 \quad \bar{u} = 40$
$\bar{z} = 0.75 \quad \bar{u} = 40$
$\bar{z} = 1 \quad \bar{u} = 0$

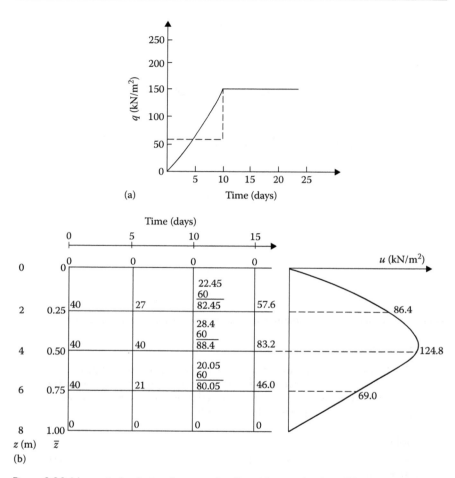

Figure 8.20 Numerical solution for ramp loading: (a) ramp loading; (b) variation of pore water pressure with depth and time.

At $t = 5$ days
  At $\bar{z} = 0$

    $\bar{u} = 0$

  At $\bar{z} = 0.25$, from Equation 8.70

    $\bar{u}_{0,\bar{t}+\Delta\bar{t}} = 0.325[0 + 40 - 2(40)] + 40 = 27$

At $\bar{z} = 0.5$, from Equation 8.75

$$\bar{u}_{0,\bar{t}+\Delta\bar{t}} = (1.532)(0.325)$$

$$\times \left[\frac{2 \times 2.8}{2 + 2.8}(40) + \frac{2 \times 2}{2 + 2.8}(40) - 2(40) \right] + 40 = 40$$

At $\bar{z} = 0.75$, from Equation 8.70

$$\bar{u}_{0,\bar{t}+\Delta\bar{t}} = 0.475[40 + 0 - 2(40)] + 40 = 21$$

At $\bar{z} = 1$

$$\bar{u}_{0,\bar{t}+\Delta\bar{t}} = 0$$

At $t = 10$ days
At $\bar{z} = 0$

$$\bar{u} = 0$$

At $\bar{z} = 0.25$, from Equation 8.70

$$\bar{u}_{0,\bar{t}+\Delta\bar{t}} = 0.325[0 + 40 - 2(27)] + 27 = 22.45$$

At this point, a new load of 90 kN/m^2 is added, so $\bar{u}$ will increase by an amount 90/1.5 = 60. The new $\bar{u}_{0,\bar{t}+\Delta\bar{t}}$ is 60 + 22.45 = 82.45. At $\bar{z} = 0.5$, from Equation 8.75

$$\bar{u}_{0,\bar{t}+\Delta\bar{t}} = (1.152)(0.325)$$

$$\times \left[\frac{2 \times 2.8}{2 + 2.8}(27) + \frac{2 \times 2}{2 + 2.8}(21) - 2(40) \right] + 40 = 28.4$$

New $\bar{u}_{0,\bar{t}+\Delta\bar{t}}$ = 28.4 + 60 = 88.4

At $\bar{z} = 0.75$, from Equation 8.70

$$\bar{u}_{0,\bar{t}+\Delta\bar{t}} = 0.475[40 + 0 - 2(21)] + 21 = 20.05$$

New $\bar{u}_{0,\bar{t}+\Delta\bar{t}}$ = 60 + 20.05 = 80.05

At $\bar{z} = 1$

$$\bar{u} = 0$$

At $t = 15$ days
At $\bar{z} = 0$

$$\bar{u} = 0$$

At $\bar{z} = 0.25$

$$\bar{u}_{0,\bar{t}+\Delta\bar{t}} = 0.325[0 + 88.4 - 2(82.45)] + 82.45 = 57.6$$

At $\bar{z} = 0.5$

$$\bar{u}_{0,\bar{t}+\Delta\bar{t}} = (1.152)(0.325)$$

$$\times\left[\frac{2\times2.8}{2+2.8}(82.45)+\frac{2\times2}{2+2.8}(80.05)-2(88.4)\right]+88.4=83.2$$

At $\bar{z} = 0.75$

$$\bar{u}_{0,\bar{t}+\Delta\bar{t}} = 0.475[88.4 + 0 - 2(80.05)]+ 80.05 = 46.0$$

At $\bar{z} = 1$

$$\bar{u} = 0$$

The distribution of excess pore water pressure is shown in Figure 8.20b.

8.5 STANDARD ONE-DIMENSIONAL CONSOLIDATION TEST AND INTERPRETATION

The standard one-dimensional consolidation test is usually carried out on saturated specimens about 25.4 mm thick and 63.5 mm in diameter (Figure 8.21). The soil specimen is kept inside a metal ring, with a porous stone at the top and another at the bottom. The load P on the specimen is applied through a lever arm, and the compression of the specimen is measured by a micrometer dial gauge. The load is usually doubled every 24 h. The specimen is kept under water throughout the test.

For each load increment, the specimen deformation and the corresponding time t are plotted on semilogarithmic graph paper. Figure 8.22a shows a

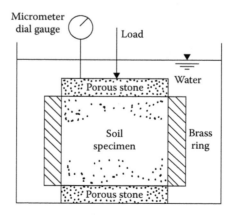

Figure 8.21 Consolidometer.

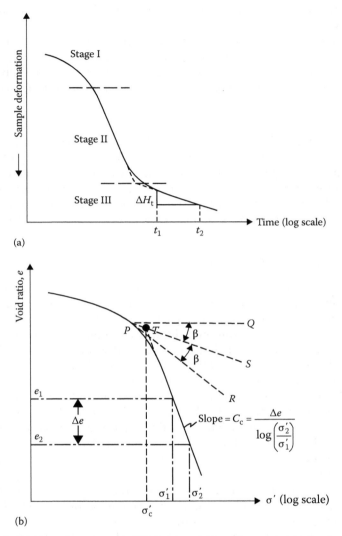

Figure 8.22 (a) Typical specimen deformation versus log-of-time plot for a given load increment and (b) typical e versus log σ' plot showing procedure for determination of σ'_c and C_c.

typical deformation versus log t graph. The graph consists of three distinct parts:

1. Upper curved portion (stage I). This is mainly the result of precompression of the specimen.
2. A straight-line portion (stage II). This is referred to as primary consolidation. At the end of the primary consolidation, the excess pore

water pressure generated by the incremental loading is dissipated to a large extent.

3. A lower straight-line portion (stage III). This is called secondary consolidation. During this stage, the specimen undergoes small deformation with time. In fact, there must be immeasurably small excess pore water pressure in the specimen during secondary consolidation.

Note that at the end of the test, for each incremental loading, the stress on the specimen is the effective stress σ'. Once the specific gravity of the soil solids, the initial specimen dimensions, and the specimen deformation at the end of each load have been determined, the corresponding void ratio can be calculated. A typical void ratio versus effective pressure relation plotted on semilogarithmic graph paper is shown in Figure 8.22b.

8.5.1 Preconsolidation pressure

In the typical e versus log σ' plot shown in Figure 8.22b, it can be seen that the upper part is curved; however, at higher pressures, e and log σ' bear a linear relation. The upper part is curved because when the soil specimen was obtained from the field, it was subjected to a certain maximum effective pressure. During the process of soil exploration, the pressure is released. In the laboratory, when the soil specimen is loaded, it will show relatively small decrease of void ratio with load up to the maximum effective stress to which the soil was subjected in the past. This is represented by the upper curved portion in Figure 8.22b. If the effective stress on the soil specimen is increased further, the decrease of void ratio with stress level will be larger. This is represented by the straight-line portion in the e versus log σ' plot. The effect can also be demonstrated in the laboratory by unloading and reloading a soil specimen, as shown in Figure 8.23. In this figure, cd is the void ratio–effective stress relation as the specimen is unloaded, and $dfgh$ is the reloading branch. At d, the specimen is being subjected to a lower effective stress than the maximum stress σ'_1 to which the soil was ever subjected. So, df will show a flatter curved portion. Beyond point f, the void ratio will decrease at a larger rate with effective stress, and gh will have the same slope as bc.

Based on the previous explanation, we can now define the two conditions of a soil:

1. *Normally consolidated.* A soil is called normally consolidated if the present effective overburden pressure is the maximum to which the soil has ever been subjected, that is, $\sigma'_{present} \geq \sigma'_{past\ maximum}$.

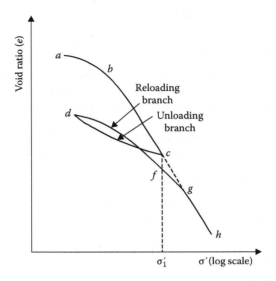

Figure 8.23 Plot of void ratio versus effective pressure showing unloading and reloading branches.

2. *Overconsolidated.* A soil is called overconsolidated if the present effective overburden pressure is less than the maximum to which the soil was ever subjected in the past, that is, $\sigma'_{present} < \sigma'_{past\ maximum}$.

In Figure 8.23, the branches *ab*, *cd*, and *df* are the overconsolidated state of a soil, and the branches *bc* and *fh* are the normally consolidated state of a soil.

In the natural condition in the field, a soil may be either normally consolidated or overconsolidated. A soil in the field may become overconsolidated through several mechanisms, some of which are listed in the following (Brummund et al., 1976):

- Removal of overburden pressure
- Past structures
- Glaciation
- Deep pumping
- Desiccation due to drying
- Desiccation due to plant lift
- Change in soil structure due to secondary compression
- Change in pH
- Change in temperature
- Salt concentration
- Weathering
- Ion exchange
- Precipitation of cementing agents

8.5.1.1 Preconsolidation pressure determination from laboratory test results

Casagrande's procedure (1936)

The preconsolidation pressure from an e versus log σ' plot is generally determined by a graphical procedure suggested by Casagrande (1936), as shown in Figure 8.22b. The steps are as follows:

1. Visually determine the point P (on the upper curved portion of the e versus log σ' plot) that has the maximum curvature.
2. Draw a horizontal line PQ.
3. Draw a tangent PR at P.
4. Draw the line PS bisecting the angle QPR.
5. Produce the straight-line portion of the e versus log σ' plot backward to intersect PS at T.
6. The effective pressure corresponding to point T is the preconsolidation pressure σ'_c.

Log-log method (1989)

This method was proposed by Jose et al. (1989). In this method, a plot is prepared for log e vs. log σ' (Figure 8.24). Straight lines are fitted, giving emphasis for the initial and final portions of the log e – log σ' plot. The intersection of the two straight lines gives σ'_c.

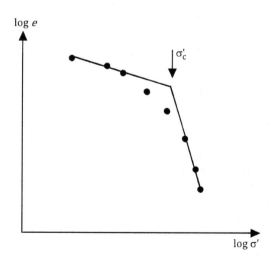

Figure 8.24 Log-log method of determine σ'_c.

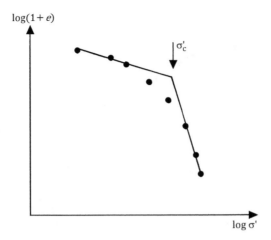

Figure 8.25 Oikawa's method to determine σ'_c.

Oikawa's method (1987)

Oikawa (1987) suggested that a better result of σ'_c may be obtained by plotting $\log(1 + e)$ vs. $\log \sigma'$ (Figure 8.25) and fitting straight lines for the initial and final portions of the $\log(1 + e) - \log \sigma'$ plot. The intersection of the two lines gives σ'_c. Umar and Sadrekarimi (2017) compared several methods proposed to determine σ'_c and concluded that Oikawa's method provides more accurate results.

In the field, the overconsolidation ratio (OCR) can be defined as

$$OCR = \frac{\sigma'_c}{\sigma'_o} \tag{8.76}$$

where σ'_o = present effective overburden pressure.

Example 8.13

Following are the results of a laboratory consolidation test.

Pressure, σ' (kN/m²)	Void ratio, e
50	0.840
100	0.826
200	0.774
400	0.696
800	0.612
1000	0.528

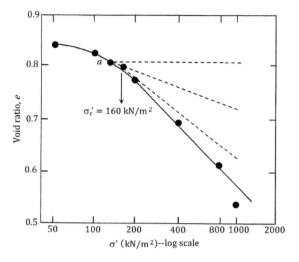

Figure 8.26 Casagrande's procedure to estimate σ_c'.

Using Casagrande's procedure, determine the preconsolidation pressure σ_c'.

Solution

Figure 8.26 shows the e–log σ' plot. In this plot, a is the point where the radius of curvature is minimum. The preconsolidation pressure is determined using the procedure shown in Figure 8.22. From the plot, $\sigma_c' = 160 \text{ kN/m}^2$.

Example 8.14

Redo Example 8.13 using Oikawa's method.

Solution

The following table can be prepared:

Pressure, σ' (kN/m²)	e	log(1 + e)
50	0.840	0.265
100	0.826	0.262
200	0.774	0.249
400	0.696	0.229
800	0.612	0.207
1000	0.528	0.184

Figure 8.27 shows a plot of log σ' vs. $\log(1 + e)$ from which $\sigma_c' = 170 \text{ kN/m}^2$.

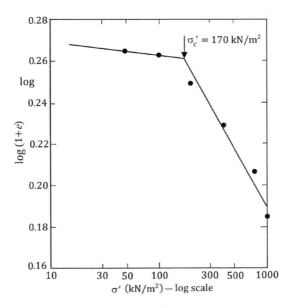

Figure 8.27 Oikawa's procedure to estimate σ'_c.

8.5.1.2 Empirical correlations for preconsolidation pressure

There are some empirical correlations presently available in the literature to estimate the preconsolidation pressure in the field. Following are a few of these relationships. However, they should be used cautiously.

Stas and Kulhawy (1984)

$$\frac{\sigma'_c}{p_a} = 10^{(1.11-1.62\mathrm{LI})} \text{ (for clays with sensitivity between 1 and 10)} \quad (8.77)$$

where
   p_a is the atmospheric pressure ($\approx$100 kN/m²)
   LI is the liquidity index

Hansbo (1957)

$$\sigma'_c = \alpha_{(VST)}S_{u(VST)} \quad (8.78)$$

where
   $S_{u(VST)}$ = undrained shear strength based on the vane shear test
   $\alpha_{(VST)}$ = an empirical coefficient = $\dfrac{222}{\mathrm{LL}(\%)}$

where LL is the liquid limit.

Mayne and Mitchell (1988) gave a correlation for $\alpha_{(VST)}$ as

$$\alpha_{(VST)} = 22PI^{-0.48} \tag{8.79}$$

where PI is the plasticity index (%).
 Nagaraj and Murty (1985)

$$\log \sigma_c' = \frac{1.322\,(e_o/e_L) - 0.0463\log \sigma_o'}{0.188} \tag{8.80}$$

where
 e_o is the void ratio at the present effective overburden pressure, σ'
 e_L is the void ratio of the soil at liquid limit
 σ_c' and σ_o' are in kN/m²

$$e_L = \left[\frac{LL(\%)}{100}\right]G_s$$

G_s is the specific gravity of soil solids

8.5.1.3 Empirical correlations for overconsolidation ratio

Similar to the preceding correlations for preconsolidation pressure, the overconsolidation ratio (OCR) in the field has been empirically correlated by various investigators. Some of those correlations are summarized in the following.

The overconsolidation has been correlated to field vane shear strength $[S_{u(VST)}]$ as

$$OCR = \beta\frac{S_{u(VST)}}{\sigma_o'} \tag{8.81}$$

where σ_o' is the effective overburden pressure.
 The magnitudes of β developed by various investigators are given in the following:

• Mayne and Mitchell (1988)

$$\beta = 22[PI(\%)]^{-0.48} \tag{8.82}$$

where PI is the plasticity index.

- Hansbo (1957)

$$\beta = \frac{222}{w(\%)} \qquad (8.83)$$

where w is the moisture content.

- Larsson (1980)

$$\beta = \frac{1}{0.08 + 0.0055(\text{PI})} \qquad (8.84)$$

Kulhawy and Mayne (1990) have also presented the following three correlations:

$$\text{OCR} = \left(\frac{p_a}{\sigma'_o}\right) 10^{(1.11-1.62\text{LI})} \qquad (8.85)$$

$$\text{OCR} = 10^{[1-2.5\text{LI}-1.25\log(\sigma'_o/p_a)]} \qquad (8.86)$$

where LI is the liquidity index.

$$\text{OCR} = 0.58N\left(\frac{p_a}{\sigma'_o}\right) \qquad (8.87)$$

where N is the field standard penetration resistance.

Mayne and Kemper (1988) provided a correlation between OCR and the cone penetration resistance q_c in the form

$$\text{OCR} = 0.37\left(\frac{q_c - \sigma_o}{\sigma'_o}\right)^{1.01} \qquad (8.88)$$

where σ_o and σ'_o are total and effective vertical stress, respectively.

8.5.2 Compression index

The slope of the e versus $\log \sigma'$ plot for normally consolidated soil is referred to as the compression index C_c. From Figure 8.22b

$$C_c = \frac{e_1 - e_2}{\log \sigma'_2 - \log \sigma'_1} = \frac{\Delta e}{\log(\sigma'_2/\sigma'_1)} \qquad (8.89)$$

For undisturbed normally consolidated clays, Terzaghi and Peck (1967) gave a correlation for the compression index as

$$C_c = 0.009(\text{LL} - 10)$$

Based on the laboratory test results, several empirical relations for C_c have been proposed, some of which are given in Table 8.3.

Based on the modified Cam clay model, Wroth and Wood (1978) have shown that

$$C_c \approx 0.5 G_s \frac{[PI(\%)]}{100} \tag{8.90}$$

where PI is the plasticity index.

If an average value of G_s is taken to be about 2.7 (Kulhawy and Mayne, 1990)

$$C_c \approx \frac{PI}{74} \tag{8.91}$$

Table 8.3 Empirical relations for C_c

Reference	Relation	Comments
Terzaghi and Peck (1967)	$C_c = 0.009(LL - 10)$	Undisturbed clay
	$C_c = 0.007(LL - 10)$	Remolded clay
	LL = liquid limit (%)	
Azzouz et al. (1976)	$C_c = 0.01 w_N$	Chicago clay
	w_N = natural moisture content (%)	
	$C_c = 0.0046(LL - 9)$	Brazilian clay
	LL = liquid limit (%)	
	$C_c = 1.21 + 1.005(e_0 - 1.87)$	Motley clays from Sao Paulo city
	e_0 = in situ void ratio	
	$C_c = 0.208 e_0 + 0.0083$	Chicago city
	e_0 = in situ void ratio	
	$C_c = 0.0115 w_N$	Organic soil, peat
	w_N = natural moisture content (%)	
Nacci et al. (1975)	$C_c = 0.02 + 0.014(PI)$	North Atlantic clay
	PI = plasticity index (%)	
Rendon-Herrero (1983)	$C_c = 0.141 \, G_s^{1.2} \left(\dfrac{1 + e_0}{G_s} \right)^{2.38}$	
	G_s = specific gravity of soil solids	
	e_0 = in situ void ratio	
Nagaraj and Murty (1985)	$C_c = 0.2343 \left(\dfrac{LL}{100} \right) G_s$	
	G_s = specific gravity of soil solids	
	LL = liquid limit (%)	
Park and Koumoto (2004)	$C_c = \dfrac{n_o}{371.747 - 4.275 n_o}$	
	n_o = in situ porosity of soil (%)	

Burland (1990) showed that there exists a good relationship between e_L and C_c^* in the form

$$C_c^* = 0.256e_L - 0.04 \qquad (8.92)$$

where e_L is the void ratio at liquid limit (LL).

$$C_c^* = \frac{e_{100} - e_{1000}}{\log\left(\dfrac{1000 \text{ kN/m}^2}{100 \text{ kN/m}^2}\right)} = e_{100} - e_{1000} \qquad (8.93)$$

where e_{100} and e_{1000} are void ratios at vertical effective pressures of 100 and 1000 kN/m², respectively.

Example 8.15

For a clay soil, given LL = 54% and specific gravity of soil solids G_s = 2.71. Determine the value of C_c^* based on Equation 8.92.

Solution

From Equation 1.44

$$e = wG_s$$

For $e = e_L$, w = LL. Hence

$$e_L = \left(\frac{LL(\%)}{100}\right)(G_s) = \left(\frac{54}{100}\right)(2.71) = 1.463$$

From Equation 8.83

$$C_c^* = 0.256e_L - 0.04 = (0.256)(1.463) - 0.04 = 0.335$$

8.6 EFFECT OF SAMPLE DISTURBANCE ON THE e VERSUS LOG σ' CURVE

Soil samples obtained from the field are somewhat disturbed. When consolidation tests are conducted on these specimens, we obtain e versus log σ' plots that are slightly different from those in the field. This is demonstrated in Figure 8.28.

Curve I in Figure 8.28a shows the nature of the e versus log σ' variation that an undisturbed normally consolidated clay (present effective overburden pressure σ_0'; void ratio e_0) in the field would exhibit. This is called the *virgin compression curve*. A laboratory consolidation test on a carefully

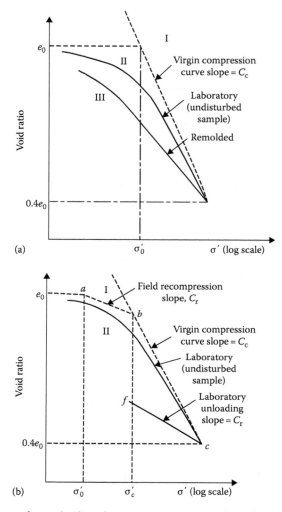

Figure 8.28 Effect of sample disturbance on the e versus log σ' curve: (a) normally consolidated soil; (b) overconsolidated soil.

recovered specimen would result in an e versus log σ' plot such as curve II. If the same soil is completely remolded and then tested in a consolidometer, the resulting void ratio–pressure plot will be like curve III. The virgin compression curve (curve I) and the laboratory e versus log σ' curve obtained from a carefully recovered specimen (curve II) intersect at a void ratio of about $0.4e_0$ (Terzaghi and Peck, 1967).

Curve I in Figure 8.28b shows the nature of the field consolidation curve of an overconsolidated clay. Note that the present effective

Table 8.4 Typical values of C_c and C_r of some natural clays

Soil	C_c	C_r	C_c/C_r
Boston blue clay	0.35	0.07	5
Chicago clay	0.4	0.07	5.7
New Orleans clay	0.3	0.05	6
Montana clay	0.21	0.05	4.2

overburden pressure is σ_0', the corresponding void ratio e_0, σ_c' the pre-consolidation pressure, and bc a part of the virgin compression curve. Curve II is the corresponding laboratory consolidation curve. After careful testing, Schmertmann (1953) concluded that the field recompression branch (ab in Figure 8.28b) has approximately the same slope as the laboratory unloading branch, cf. The slope of the laboratory unloading branch is referred to as C_r (recompression index). The range of C_r is approximately from one-fifth to one-tenth of C_c. Table 8.4 gives typical values of C_c and C_r of some natural clays.

Based on the modified Cam clay model, Kulhawy and Mayne (1990) have shown that

$$C_r \approx \frac{PI}{370} \tag{8.94}$$

Table 8.5 provides some correlations presently available in the literature for C_r.

Table 8.5 Correlations for C_r

Correlation	Source
$C_r = 0.046e_L$ (remolded clay)	Nagaraj and Murty (1985)
$C_r = 0.0025(w_n - 15)$	Kempfort and Soumaya (2004)
$C_r = 0.0019$ (PI − 4.6) (reconstituted marine soil)	Nakase et al. (1988)
$C_r = 0.054$ ($e_0 - 0.3$)	Kootahi (2017)
$C_r = 0.0014$ ($w_n - 10$)	
$C_r = 0.0017$ (LL − 21)	
$C_r = 0.0025$ (PI − 3)	
$C_r = 0.021$ ($e_0 + 0.06$LL − 1.6)	
$C_r = 0.033$ ($e_0 + 0.04$PI − 0.55) (undisturbed samples)	

Note: e_0 = initial void ratio; e_L = void ratio at liquid limit; w_n = natural water content; LL = liquid limit; PI = plasticity index.

8.7 SECONDARY CONSOLIDATION

It has been pointed out previously that clays continue to settle under sustained loading at the end of primary consolidation, and this is due to the continued readjustment of clay particles. Several investigations have been carried out for qualitative and quantitative evaluation of secondary consolidation. The magnitude of secondary consolidation is often defined by (Figure 8.22a)

$$C_\alpha = \frac{\Delta H_t / H_t}{\log t_2 - \log t_1} \tag{8.95}$$

where C_α is the coefficient of secondary consolidation.

Mesri (1973) published an extensive list of the works of various investigators in this area. Figure 8.29 details the general range of the coefficient of secondary consolidation observed in a number of clayey soils. Secondary compression is greater in plastic clays and organic soils. Based on the coefficient of secondary consolidation, Mesri (1973) classified the secondary compressibility, and this is summarized as follows:

C_α	Secondary compressibility
<0.002	Very low
0.002–0.004	Low
0.004–0.008	Medium
0.008–0.016	High
0.016–0.032	Very high

In order to study the effect of remolding and preloading on secondary compression, Mesri (1973) conducted a series of one-dimensional consolidation tests on an organic Paulding clay. Figure 8.30 shows the results in the form of a plot of $\Delta e/(\Delta \log t)$ versus consolidation pressure. For these tests, each specimen was loaded to a final pressure with load increment ratios of 1 and with only sufficient time allowed for excess pore water pressure dissipation. Under the final pressure, secondary compression was observed for a period of 6 months. The following conclusions can be drawn from the results of these tests:

1. For sedimented (undisturbed) soils, $\Delta e/(\Delta \log t)$ decreases with the increase of the final consolidation pressure.
2. Remolding of clays creates a more dispersed fabric. This results in a decrease of the coefficient of secondary consolidation at lower consolidation pressures as compared to that for undisturbed samples. However, it increases with consolidation pressure to a maximum value and then decreases, finally merging with the values for normally consolidated undisturbed samples.

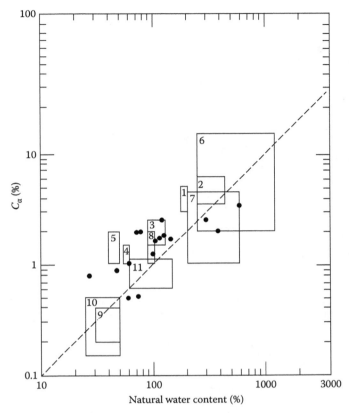

Figure 8.29 Coefficient of secondary consolidation for natural soil deposits: 1, Whangamarino clay; 2, Mexico City clay; 3, calcareous organic silt; 4, Leda clay; 5, Norwegian plastic clay; 6, amorphous and fibrous peat; 7, Canadian muskeg; 8, organic marine deposits; 9, boston blue clay; 10, Chicago blue clay; 11, organic silty clay; •, organic silt, etc. [After Mesri, G., *J. Soil Mech. Found. Div.,* ASCE, 99(SMI), 123, 1973.]

3. Precompressed clays show a smaller value of coefficient of secondary consolidation. The degree of reduction appears to be a function of the degree of precompression.

Mesri and Godlewski (1977) compiled the values of C_α/C_c for a number of naturally occurring soils. A summary of this is given in Table 8.6. From this study, it appears that, in general,

- $C_\alpha/C_c \approx 0.04 \pm 0.01$ (for inorganic clays and silts)
- $C_\alpha/C_c \approx 0.05 \pm 0.01$ (for organic clays and silts)
- $C_\alpha/C_c \approx 0.075 \pm 0.01$ (for peats)

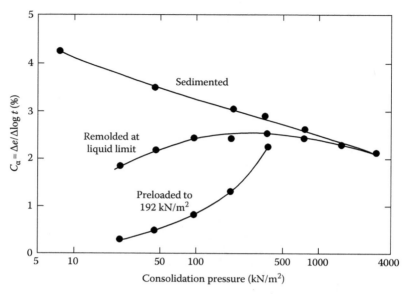

Figure 8.30 Coefficient of secondary compression for organic Paulding clay. [After Mesri, G., *J. Soil Mech. Found. Div.*, ASCE, 99(SMI), 123, 1973.]

Table 8.6 Values of C_α/C_c for natural oils

Soil	C_α/C_c
Whangamarino clay	0.03–0.04
Calcareous organic silt	0.035–0.06
Amorphous and fibrous peat	0.035–0.083
Canadian muskeg	0.09–0.10
Leda clay	0.03–0.055
Peat	0.075–0.085
Post-glacial organic clay	0.05–0.07
Soft blue clay	0.026
Organic clays and silts	0.04–0.06
Sensitive clay, Portland	0.025–0.055
Peat	0.05–0.08
San Francisco Bay mud	0.04–0.06
New Liskeard varved clay	0.03–0.06
Nearshore clays and silts	0.055–0.075
Fibrous peat	0.06–0.085
Mexico City clay	0.03–0.035
Hudson River silt	0.03–0.06
Leda clay	0.025–0.04
New Haven organic clay silt	0.04–0.075

Source: Compiled from Mesri, G. and Godlewski, P. M., *J. Geotech. Eng.*, ASCE, 103(5), 417, 1977.

8.8 GENERAL COMMENTS ON CONSOLIDATION TESTS

Standard one-dimensional consolidation tests as described in Section 8.5 are conducted with a soil specimen having a thickness of 25.4 mm in which the load on the specimen is doubled every 24 h. This means that $\Delta\sigma/\sigma'$ is kept at 1 ($\Delta\sigma$ is the step load increment, and σ' the effective stress on the specimen before the application of the incremental step load). Following are some general observations as to the effect of any deviation from the standard test procedure.

Effect of load-increment ratio $\Delta\sigma/\sigma'$. Striking changes in the shape of the compression–time curves for one-dimensional consolidation tests are generally noticed if the magnitude of $\Delta\sigma/\sigma'$ is reduced to less than about 0.25. Leonards and Altschaeffl (1964) conducted several tests on Mexico City clay in which they varied the value of $\Delta\sigma/\sigma'$ and then measured the excess pore water pressure with time. The general nature of specimen deformation with time is shown in Figure 8.31a. From this figure it may be seen that, for $\Delta\sigma/\sigma' < 0.25$, the position of the end of primary consolidation (i.e., zero excess pore water pressure due to incremental load) is somewhat difficult to resolve. Furthermore, the load-increment ratio has a high influence on consolidation of clay. Figure 8.31b shows the nature of the e versus log σ' curve for various values of $\Delta\sigma/\sigma'$. If $\Delta\sigma/\sigma'$ is small, the ability of individual clay particles to readjust to their positions of equilibrium is small, which results in a smaller compression compared to that for larger values of $\Delta\sigma/\sigma'$.

Effect of load duration. In conventional testing, in which the soil specimen is left under a given load for about a day, a certain amount of secondary consolidation takes place before the next load increment is added.

If the specimen is left under a given load for more than a day, additional secondary consolidation settlement will occur. This additional amount of secondary consolidation will have an effect on the e versus log σ' plot, as shown in Figure 8.32. Curve a is based on the results at the end of primary consolidation. Curve b is based on the standard 24 h load-increment duration. Curve c refers to the condition for which a given load is kept for more than 24 h before the next load increment is applied. The strain for a given value of σ' is calculated from the total deformation that the specimen has undergone before the next load increment is applied. In this regard, Crawford (1964) provided experimental results on Leda clay. For his study, the preconsolidation pressure obtained from the end of primary e versus log σ' plot was about twice that obtained from the e versus log σ' plot where each load increment was kept for a week.

Effect of specimen thickness. Other conditions remaining the same, the proportion of secondary to primary compression increases with the decrease of specimen thickness for similar values of $\Delta\sigma/\sigma'$.

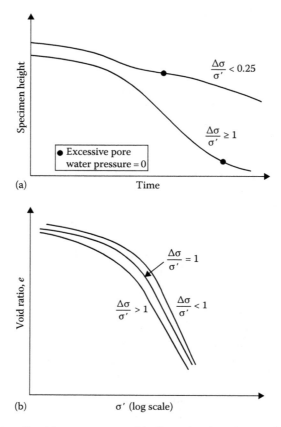

Figure 8.31 Effect of load-increment ratio: (a) effect of $\Delta\sigma/\sigma'$ on the consolidation curve; (b) effect of $\Delta\sigma/\sigma'$ on e-log σ' plot.

Effect of secondary consolidation. The continued secondary consolidation of a natural clay deposit has some influence on the preconsolidation pressure σ'_c. This fact can be further explained by the schematic diagram shown in Figure 8.33.

A clay that has recently been deposited and comes to equilibrium by its own weight can be called a "young, normally consolidated clay." If such a clay, with an effective overburden pressure of σ'_0 at an equilibrium void ratio of e_0, is now removed from the ground and tested in a consolidometer, it will show an e versus log σ' curve like that marked curve a in Figure 8.33. Note that the preconsolidation pressure for curve a is σ'_0. On the contrary, if the same clay is allowed to remain undisturbed for 10,000 years, for example, under the same effective overburden pressure σ'_0, there will be creep or secondary consolidation. This will reduce the void ratio to e_1. The clay may now be called an "aged, normally consolidated clay." If this clay, at a void

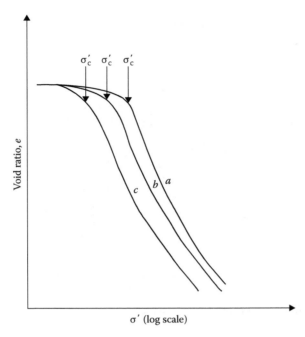

Figure 8.32 Effect of load duration on the e versus log σ′ plot: (a) end of primary consolidation, (b) 24 h load increment duration, and (c) more than 24 h load duration.

ratio of e_1 and effective overburden pressure of σ'_0, is removed and tested in a consolidometer, the e versus $\log \sigma'$ curve will be like curve b. The preconsolidation pressure, when determined by standard procedure, will be σ'_1. Now, $\sigma'_c = \sigma'_1 > \sigma'_0$. This is sometimes referred to as a quasi-preconsolidation effect. The effect of preconsolidation is pronounced in most plastic clays. Thus, it may be reasoned that, under similar conditions, the ratio of the quasi-preconsolidation pressure to the effective overburden pressure σ'_c/σ'_0 will increase with the plasticity index of the soil. Bjerrum (1972) gave an estimate of the relation between the plasticity index and the ratio of quasi-preconsolidation pressure to effective overburden pressure (σ'_c/σ'_0) for late glacial and postglacial clays. This relation is shown as follows:

Plasticity index	$\approx \sigma'_c/\sigma'_0$
20	1.4
40	1.65
60	1.75
80	1.85
100	1.90

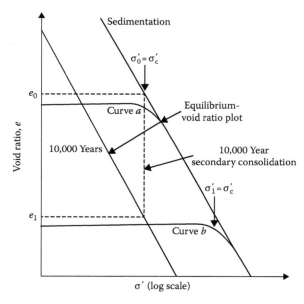

Figure 8.33 Effect of secondary consolidation.

8.9 CALCULATION OF ONE-DIMENSIONAL CONSOLIDATION SETTLEMENT

The basic principle of one-dimensional consolidation settlement calculation is demonstrated in Figure 8.34. If a clay layer of total thickness H_t is subjected to an increase of average effective overburden pressure from σ_0' to σ_1', it will undergo a consolidation settlement of ΔH_t. Hence, the strain can be given by

$$\epsilon = \frac{\Delta H_t}{H_t} \tag{8.96}$$

where ϵ is strain. Again, if an undisturbed laboratory specimen is subjected to the same effective stress increase, the void ratio will decrease by Δe. Thus, the strain is equal to

$$\epsilon = \frac{\Delta e}{1 + e_0} \tag{8.97}$$

where e_0 is the void ratio at an effective stress of σ_0'.

Thus, from Equations 8.96 and 8.97

$$\Delta H_t = \frac{\Delta e H_t}{1 + e_0} \tag{8.98}$$

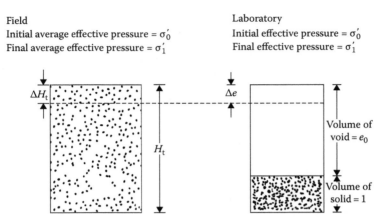

Field
Initial average effective pressure = σ'_0
Final average effective pressure = σ'_1

Laboratory
Initial effective pressure = σ'_0
Final effective pressure = σ'_1

Figure 8.34 Calculation of one-dimensional consolidation settlement.

For a normally consolidated clay in the field (Figure 8.35a)

$$\Delta e = C_c \log \frac{\sigma'_1}{\sigma'_0} = C_c \log \frac{\sigma'_0 + \Delta \sigma}{\sigma'_0} \qquad (8.99)$$

For an overconsolidated clay, (1) if $\sigma'_1 < \sigma'_c$ (i.e., overconsolidation pressure) (Figure 8.35b)

$$\Delta e = C_r \log \frac{\sigma'_1}{\sigma'_0} = C_r \log \frac{\sigma' + \Delta \sigma}{\sigma'_0} \qquad (8.100)$$

and (2) if $\sigma'_0 < \sigma'_c < \sigma'_1$ (Figure 8.35c)

$$\Delta e = \Delta e_1 + \Delta e_2 = C_r \log \frac{\sigma'_c}{\sigma'_0} + C_c \log \frac{\sigma'_0 + \Delta \sigma}{\sigma'_c} \qquad (8.101)$$

The procedure for calculation of one-dimensional consolidation settlement is described in more detail in Chapter 11.

8.10 COEFFICIENT OF CONSOLIDATION

For a given load increment, the coefficient of consolidation C_v can be determined from the laboratory observations of time versus dial reading. There are several procedures presently available to estimate the coefficient of consolidation, some of which are described later.

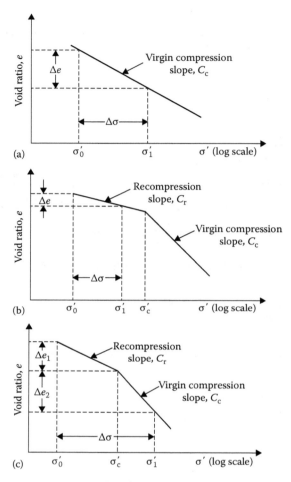

Figure 8.35 Calculation of Δe: (a) Equation 8.99; (b) Equation 8.100; (c) Equation 8.101.

8.10.1 Logarithm-of-time method

The logarithm-of-time method was originally proposed by Casagrande and Fadum (1940) and can be explained by referring to Figure 8.36.

1. Plot the dial readings for specimen deformation for a given load increment against time on semilog graph paper as shown in Figure 8.36.
2. Plot two points, P and Q, on the upper portion of the consolidation curve, which correspond to time t_1 and t_2, respectively. Note that $t_2 = 4t_1$.

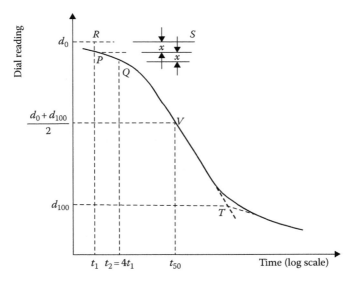

Figure 8.36 Logarithm-of-time method for determination of C_v.

3. The difference of dial readings between P and Q is equal to x. Locate point R, which is at a distance x above point P.
4. Draw the horizontal line RS. The dial reading corresponding to this line is d_0, which corresponds to 0% consolidation.
5. Project the straight-line portions of the primary consolidation and the secondary consolidation to intersect at T. The dial reading corresponding to T is d_{100}, that is, 100% primary consolidation.
6. Determine the point V on the consolidation curve that corresponds to a dial reading of $(d_0 + d_{100})/2 = d_{50}$. The time corresponding to point V is t_{50}, that is, time for 50% consolidation.
7. Determine C_v from the equation $T_v = C_v t/H^2$. The value of T_v for $U_{av} = 50\%$ is 0.197 (Table 8.1). So

$$C_v = \frac{0.197 H^2}{t_{50}} \tag{8.102}$$

8.10.2 Square-root-of-time method

The steps for the square-root-of-time method (Taylor, 1942) are as follows:

1. Plot the dial reading and the corresponding square-root-of-time $\sqrt{t}$ as shown in Figure 8.37.
2. Draw the tangent PQ to the early portion of the plot.
3. Draw a line PR such that $OR = (1.15)(OQ)$.

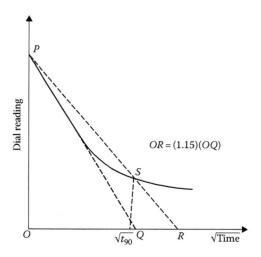

Figure 8.37 Square-root-of-time method for determination of C_v.

4. The abscissa of the point S (i.e., the intersection of PR and the consolidation curve) will give $\sqrt{t_{90}}$ (i.e., the square root of time for 90% consolidation).
5. The value of T_v for $U_{av} = 90\%$ is 0.848. So

$$C_v = \frac{0.848H^2}{t_{90}}. \tag{8.103}$$

8.10.3 Su's maximum-slope method

1. Plot the dial reading against time on semilog graph paper as shown in Figure 8.38.
2. Determine d_0 in the same manner as in the case of the logarithm-of-time method (steps 2–4).
3. Draw a tangent PQ to the steepest part of the consolidation curve.
4. Find h, which is the slope of the tangent PQ.
5. Find d_u as

$$d_u = d_0 + \frac{h}{0.688}U_{av} \tag{8.104}$$

where d_u is the dial reading corresponding to any given average degree of consolidation, U_{av}.

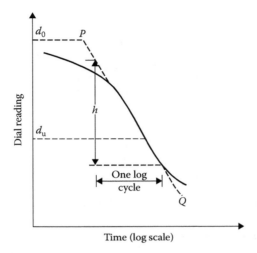

Figure 8.38 Maximum-slope method for determination of C_v.

6. The time corresponding to the dial reading d_u can now be determined, and

$$C_v = \frac{T_v H^2}{t} \qquad (8.105)$$

Su's method (1958) is more applicable for consolidation curves that do not exhibit the typical S-shape.

8.10.4 Computational method

The computational method of Sivaram and Swamee (1977) is explained in the following steps:

1. Note two dial readings, d_1 and d_2, and their corresponding times, t_1 and t_2, from the early phase of consolidation. ("Early phase" means that the degree of consolidation should be less than 53%.)
2. Note a dial reading, d_3, at time t_3 after considerable settlement has taken place.
3. Determine d_0 as

$$d_0 = \frac{d_1 - d_2 \sqrt{\dfrac{t_1}{t_2}}}{1 - \sqrt{\dfrac{t_1}{t_2}}} \qquad (8.106)$$

4. Determine d_{100} as

$$d_{100} = d_0 - \cfrac{d_0 - d_3}{\left\{ 1 - \left[\cfrac{(d_0 - d_3)(\sqrt{t_2} - \sqrt{t_1})}{(d_1 - d_2)\sqrt{t_3}} \right]^{5.6} \right\}^{0.179}} \qquad (8.107)$$

5. Determine C_v as

$$C_v = \frac{\pi}{4} \left(\frac{d_1 - d_2}{d_0 - d_{100}} \frac{H}{\sqrt{t_2} - \sqrt{t_1}} \right)^2 \qquad (8.108)$$

where H is the length of the maximum drainage path.

8.10.5 Empirical correlation

Based on laboratory tests, Raju et al. (1995) proposed the following empirical relation to predict the coefficient of consolidation of normally consolidated uncemented clayey soils:

$$C_v = \left[\frac{1 + e_L(1.23 - 0.276 \log \sigma_0')}{e_L} \right] \left[\frac{10^{-3}}{(\sigma_0')^{0.353}} \right] \qquad (8.109)$$

where
  C_v is the coefficient of consolidation (cm^2/s)
  σ_0' is the effective overburden pressure (kN/m^2)
  e_L is the void ratio at liquid limit

Note that

$$e_L = \left[\frac{LL(\%)}{100} \right] G_s \qquad (8.110)$$

where
  LL is the liquid limit
  G_s is the specific gravity of soil solids

8.10.6 Rectangular hyperbola method

The rectangular hyperbola method (Sridharan and Prakash, 1985) can be illustrated as follows. Based on Equations 8.39 and 8.41, it can be shown that the plot of T_v/U_{av} versus T_v will be of the type shown in Figure 8.39a. In the range of $60\% \leq U_{av} \leq 90\%$, the relation is linear and can be expressed as

$$\frac{T_v}{U_{av}} = 8.208 \times 10^{-3} T_v + 2.44 \times 10^{-3} \tag{8.111}$$

Using the same analogy, the consolidation test results can be plotted in graphical form as $t/\Delta H_t$ versus t (where t is time and ΔH_t is specimen deformation), which will be of the type shown in Figure 8.39b. Now the following procedure can be used to estimate C_v.

1. Identify the straight-line portion, bc, and project it back to d. Determine the intercept, D.
2. Determine the slope m of the line bc.
3. Calculate C_v as

$$C_v = 0.3 \left(\frac{mH^2}{D} \right)$$

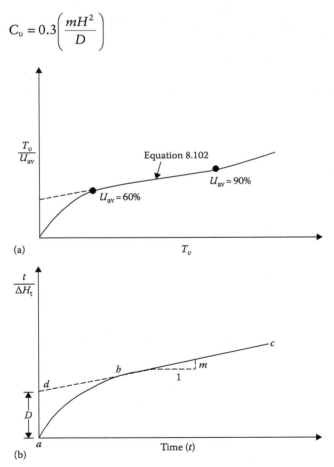

Figure 8.39 Rectangular hyperbola method for determination of C_v: (a) plot of T_v/U_{av} vs. T_v; (b) plot of $t/\Delta H_t$ vs. t.

where H is the length of maximum drainage path. Note that the unit of m is L^{-1} and the unit of D is TL^{-1}. Hence, the unit of C_v is

$$\frac{(L^{-1})(L^2)}{TL^{-1}} = L^2T^{-1}$$

8.10.7 $\Delta H_t - t/\Delta H_t$ method

According to the $\Delta H_t - t/\Delta H_t$ method (Sridharan and Prakash, 1993),

1. Plot the variation of ΔH_t versus $t/\Delta H_t$ as shown in Figure 8.40. (Note: t is time and ΔH_t compression of specimen at time t.)
2. Draw the tangent PQ to the early portion of the plot.
3. Draw a line PR such that

$$OR = (1.33)(OQ)$$

4. Determine the abscissa of point S, which gives $t_{90}/\Delta H_t$ from which t_{90} can be calculated.
5. Calculate C_v as

$$C_v = \frac{0.848H^2}{t_{90}} \tag{8.112}$$

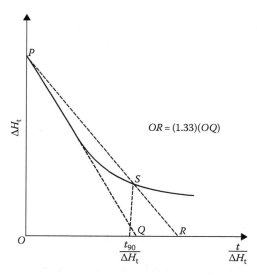

Figure 8.40 $\Delta H_t - t/\Delta H_t$ method for determination of C_v.

8.10.8 Early-stage log t method

The early-stage log t method (Robinson and Allam, 1996), an extension of the logarithm-of-time method, is based on specimen deformation against log-of-time plot as shown in Figure 8.41. According to this method, follow the logarithm-of-time method to determine d_0. Draw a horizontal line DE through d_0. Then, draw a tangent through the point of inflection F. The tangent intersects line DE at point G. Determine the time t corresponding to G, which is the time at $U_{av} = 22.14\%$. So

$$C_v = \frac{0.0385H^2}{t_{22.14}}$$

In most cases, for a given soil and pressure range, the magnitude of C_v determined using the *logarithm-of-time method* provides the *lowest value*. The *highest value* is obtained from the *early stage log t method*. The primary reason is that the early-stage log t method uses the earlier part of the consolidation curve, whereas the logarithm-of-time method uses the lower portion of the consolidation curve. When the lower portion of the consolidation curve is taken into account, the effect of secondary consolidation plays a role in the magnitude of C_v. This fact is demonstrated for several soils in Table 8.7.

Several investigators have also reported that the C_v value obtained from the field is substantially higher than that obtained from laboratory tests conducted using conventional testing methods (i.e., logarithm-of-time and square-root-of-time methods). Table 8.8 provides some examples of this as summarized by Leroueil (1988). Hence, the early-stage log t method may provide a more realistic value of fieldwork.

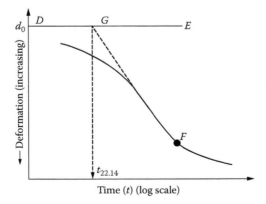

Figure 8.41 Early stage log t method.

Table 8.7 Comparison of C_v obtained from various methods (based on the results of Robinson and Allam, 1996) for the pressure range σ' between 400 and 800 kN/m^2

Soil	$C_{v(esm)}$ (cm^2/s)	$\dfrac{C_{v(esm)}}{C_{v(ltm)}}$	$\dfrac{C_{v(esm)}}{C_{v(stm)}}$
Red earth	12.80×10^{-4}	1.58	1.07
Brown soil	1.36×10^{-4}	1.05	0.94
Black cotton soil	0.79×10^{-4}	1.41	1.23
Illite	6.45×10^{-4}	1.55	1.1
Bentonite	0.022×10^{-4}	1.47	1.29
Chicago clay	7.41×10^{-4}	1.22	1.15

Note: esm: early-stage log t method; ltm: logarithm-of-time method; stm: square-root-of-time method.

Table 8.8 Comparison between the coefficients of consolidation determined in the laboratory and those deduced from embankment settlement analysis as observed by Leroueil (1988)

Site	$C_{v(lab)}$ (m^2/s)	$C_{v(in\ situ)}$ (m^2/sec)	$C_{v(lab)}/C_{v(in\ situ)}$
Ska-Edeby IV	5.0×10^{-9}	1.0×10^{-7}	20
Oxford (1)			4–57
Donnington			4–7
Oxford (2)			3–36
Avonmouth			6–47
Tickton			7–47
Over causeway			3–12
Melbourne			200
Penang	1.6×10^{-8}	1.1×10^{-6}	70
Cubzac B	2.0×10^{-8}	2.0×10^{-7}	10
Cubzac C	1.4×10^{-8}	4.3×10^{-7}	31
A-64	7.5×10^{-8}	2.0×10^{-6}	27
Saint-Alban	1.0×10^{-8}	8.0×10^{-8}	8
R-7	6.0×10^{-9}	2.8×10^{-7}	47
Matagami	8.0×10^{-9}	8.5×10^{-8}	10
Berthierville		4.0×10^{-8}	3–10

Example 8.16

The results of an oedometer test on a normally consolidated clay are given as follows (two-way drainage):

σ' (kN/m^2)	e
50	1.01
100	0.90

The time for 50% consolidation for the load increment from 50 to 100 kN/m² was 12 min, and the average thickness of the sample was 24 mm. Determine the coefficient of permeability and the compression index.

Solution

$$T_\upsilon = \frac{C_\upsilon t}{H^2}$$

For $U_{av} = 50\%$, $T_\upsilon = 0.197$. Hence

$$0.197 = \frac{C_\upsilon(12)}{(2.4/2)^2} \quad C_\upsilon = 0.0236 \text{ cm}^2/\text{min} = 0.0236 \times 10^{-4} \text{ m}^2/\text{min}$$

$$C_\upsilon = \frac{k}{m_\upsilon \gamma_w} = \frac{k}{[\Delta e/\Delta\sigma(1+e_{av})]\gamma_w}$$

For the given data, $\Delta e = 1.01 - 0.90 = 0.11$; $\Delta\sigma = 100 - 50 = 50$ kN/m²

$$\gamma_w = 9.81 \; kN/m^3; \quad \text{and} \quad e_{av} = (1.01 + 0.9)/2 = 0.955. \quad \text{So}$$

$$k = C_\upsilon \frac{\Delta e}{\Delta\sigma(1+e_{av})} \gamma_w = (0.0236 \times 10^{-4}) \left[\frac{0.11}{50(1+0.955)} \right](9.81)$$

$$= 0.2605 \times 10^{-7} \text{ m/min}$$

$$\text{Compression index} = C_c = \frac{\Delta e}{\log(\sigma_2'/\sigma_1')} = \frac{1.01 - 0.9}{\log(100/50)} = 0.365$$

Example 8.17

During a laboratory consolidation test, the time and dial gauge readings obtained from an increase in pressure on the specimen from 50 kN/m² to 100 kN/m² are as follow:

Time (min)	Dial gauge reading (cm)	Time (min)	Dial gauge reading (cm)
0	0.3975	16.0	0.4572
0.1	0.4082	30.0	0.4737
0.25	0.4102	60.0	0.4923
0.5	0.4128	120.0	0.5080
1.0	0.4166	240.0	0.5207
2.0	0.4224	180.0	0.5283
1.0	0.4298	960.0	0.5334
8.0	0.4420	1440.0	0.5364

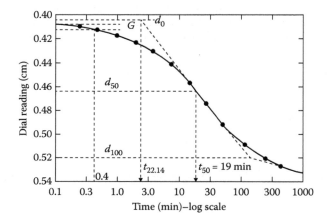

Figure 8.42 Plot of dial reading versus time (log scale).

Using the logarithm-of-time method, determine C_v. The average height of the specimen during consolidation was 2.24 cm, and it was drained at the top and bottom.

Solution

The semilogarithmic plot of dial reading versus time is shown in Figure 8.42. For this, $t_1 = 0.1$ min and $t_2 = 0.4$ min have been used to determine d_0. Following the procedure outlined in Figure 8.36, $t_{50} \approx$ 19 min. From Equation 8.102,

$$C_v = \frac{0.197H^2}{t_{50}} = \frac{0.197\left(\dfrac{2.24}{2}\right)^2}{19} = 0.013 \text{ cm}^2/\text{min} = 2.17 \times 10^{-4} \text{ cm}^2/\text{s}$$

Example 8.18

Refer to the laboratory test results of a consolidation test given in Example 8.17. Using the rectangular hyperbola methods, determine C_v.

Solution

The following table can now be prepared.

Time, t (min)	Dial reading (cm)	ΔH (cm)	$\dfrac{t}{\Delta H}$ (min/cm)
0	0.3975	0	0
0.10	0.4082	0.0107	9.346
0.25	0.4102	0.0127	19.89
0.50	0.4128	0.0153	32.68

Time, t (min)	Dial reading (cm)	ΔH (cm)	$\dfrac{t}{\Delta H}$ (min/cm)
1.00	0.4166	0.0191	52.36
2.00	0.4224	0.0249	80.32
4.00	0.4298	0.0323	123.84
8.00	0.4420	0.0445	179.78
16.00	0.4572	0.0597	268.00
30.00	0.4737	0.0762	393.70
60.00	0.4923	0.0948	623.91
120.00	0.5080	0.1105	1085.97

The plot of $t/\Delta H$ versus time (t) is shown in Figure 8.43. From this plot,

$$D \approx 180$$

$$m = \frac{1085.97 - 623.91}{60} \approx 7.7$$

$$C_v = \frac{0.3mH^2}{D} = \frac{(0.3)(7.7)\left(\dfrac{2.24}{2}\right)^2}{180} = 0.0161 \text{ cm}^2/\text{min} = 2.68 \times 10^{-4} \text{ cm}^2/\text{s}$$

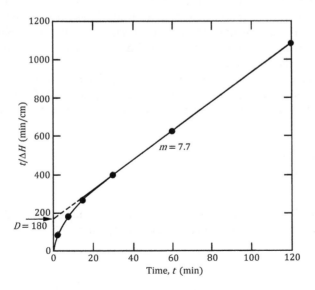

Figure 8.43 Plot of $t/\Delta H$ versus time (t).

Example 8.19

Refer to the laboratory test results of a consolidation test given in Example 8.17. Using the early stage log-t method, determine C_v.

Solution

Refer to Figure 8.42. A tangent is drawn through the point of inflection. It intersects the d_0 line at G. The time corresponding to point G is 2.57 min. So,

$$C_v = \frac{0.0385(H)^2}{t_{22.14}} = \frac{(0.0385)\left(\dfrac{2.24}{2}\right)^2}{2.57} = 0.01879 \ \text{cm}^2/\text{min}$$

$$= 3.13 \times 10^{-4} \ \text{cm}^2/\text{s}$$

8.11 ONE-DIMENSIONAL CONSOLIDATION WITH VISCOELASTIC MODELS

The theory of consolidation we have studied thus far is based on the assumption that the effective stress and the volumetric strain can be described by linear elasticity. Since Terzaghi's founding work on the theory of consolidation, several investigators (Taylor and Merchant, 1940; Taylor, 1942; Tan, 1957; Gibson and Lo, 1961; Schiffman et al., 1964; Barden, 1965, 1968) have used viscoelastic models to study one-dimensional consolidation. This gives an insight into the secondary consolidation phenomenon that the Terzaghi's theory does not explain. In this section, the work of Barden is briefly outlined.

The rheological model for soil chosen by Barden consists of a linear spring and nonlinear dashpot as shown in Figure 8.44. The equation of continuity for one-dimensional consolidation is given in Equation 8.9 as

$$\frac{k(1+e)}{\gamma_w} \frac{\partial^2 u}{\partial z^2} = \frac{\partial e}{\partial t}$$

Figure 8.45 shows the typical nature of the variation of void ratio with effective stress. From this figure, we can write that

$$\frac{e_1 - e_2}{a_v} = \frac{e_1 - e}{a_v} + u + \tau \tag{8.113}$$

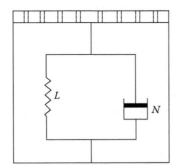

Figure 8.44 Rheological model for soil. *L*: Linear spring; *N*: Nonlinear dashpot.

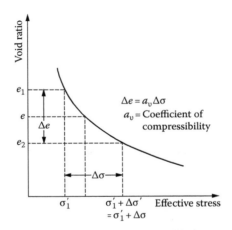

Figure 8.45 Nature of variation of void ratio with effective stress.

where

$\dfrac{e_1 - e_2}{a_v} = \Delta\sigma' =$ total effective stress increase the soil will be subjected to
at the end of consolidation

$\dfrac{e_1 - e}{a_v} = \textit{effective}$ stress increase in the soil at some stage of consolidation
(i.e., the stress carried by the soil grain bond, represented by the spring in Figure 8.44)

u is the excess pore water pressure

τ is the strain carried by film bond (represented by the dashpot in Figure 8.44)

The strain τ can be given by a power-law relation:

$$\tau = b\left(\frac{\partial e}{\partial t}\right)^{1/n}$$

where $n > 1$, and b is assumed to be a constant over the pressure range $\Delta\sigma$. Substitution of the preceding power-law relation for τ in Equation 8.113 and simplification gives

$$e - e_2 = a_v\left[u + b\left(\frac{\partial e}{\partial t}\right)^{1/n}\right] \tag{8.114}$$

Now, let $e - e_2 = e'$. So

$$\frac{\partial e'}{\partial t} = \frac{\partial e}{\partial t} \tag{8.115}$$

$$\bar{z} = \frac{z}{H} \tag{8.116}$$

where H is the length of maximum drainage path, and

$$\bar{u} = \frac{u}{\Delta\sigma'} \tag{8.117}$$

The degree of consolidation is

$$U_z = \frac{e_1 - e}{e_1 - e_2} \tag{8.118}$$

and

$$\lambda = 1 - U_z = \frac{e - e_2}{e_1 - e_2} = \frac{e'}{a_v\Delta\sigma'} \tag{8.119}$$

Elimination of u from Equations 8.16 and 8.114 yields

$$\frac{k(1+e)}{\gamma_w}\frac{\partial^2}{\partial z^2}\left[\frac{e'}{a_v} - b\left(\frac{\partial e'}{\partial t}\right)^{1/n}\right] = \frac{\partial e'}{\partial t} \tag{8.120}$$

Combining Equations 8.116, 8.117, and 8.120, we obtain

$$\frac{\partial^2}{\partial z^2}\left\{\lambda - \left[a_v b^n (\Delta\sigma')^{1-n}\frac{\partial\lambda}{\partial t}\right]^{1/n}\right\} = \frac{a_v H^2 \gamma_w}{k(1+e)}\frac{\partial\lambda}{\partial t} = \frac{m_v H^2 \gamma_w}{k(1+e)}\frac{\partial\lambda}{\partial t} = \frac{H^2}{C_v}\frac{\partial\lambda}{\partial t}$$

$$(8.121)$$

where

m_v is the volume coefficient of compressibility
C_v the coefficient of consolidation

The right-hand side of Equation 8.121 can be written in the form

$$\frac{\partial\lambda}{\partial T_v} = \frac{H^2}{C_v}\frac{\partial\lambda}{\partial t}$$

$$(8.122)$$

where T_v is the nondimensional time factor and is equal to $C_v t / H^2$.
Similarly defining

$$T_s = \frac{t(\Delta\sigma')^{n-1}}{a_v b^n}$$

$$(8.123)$$

we can write

$$\left[a_v b^n (\Delta\sigma')^{1-n}\frac{\partial\lambda}{\partial t}\right]^{1/n} = \left(\frac{\partial\lambda}{\partial T_s}\right)^{1/n}$$

$$(8.124)$$

T_s in Equations 8.123 and 8.124 is defined as structural viscosity.
It is useful now to define a nondimensional ratio R as

$$R = \frac{T_v}{T_s} = \frac{C_v a_v}{H^2}\frac{b^n}{(\Delta\sigma')^{n-1}}$$

$$(8.125)$$

From Equations 8.121, 8.122, and 8.124

$$\frac{\partial^2}{\partial z^2}\left[\lambda - \left(\frac{\partial\lambda}{\partial T_s}\right)^{1/n}\right] = \frac{\partial\lambda}{\partial T_v}$$

$$(8.126)$$

Note that Equation 8.126 is nonlinear. For that reason, Barden suggested solving the two simultaneous equations obtained from the basic Equation 8.16.

$$\frac{\partial^2 \bar{u}}{\partial z^2} = \frac{\partial\lambda}{\partial T_v}$$

$$(8.127)$$

and

$$-\frac{1}{R}(\lambda - \bar{u})^n = \frac{\partial \lambda}{\partial T_v} \tag{8.128}$$

Finite-difference approximation is employed for solving the previous two equations. Figure 8.46 shows the variation of λ and $\bar{u}$ with depth for a clay layer of height $H_t = 2H$ and drained both at the top and bottom (for $n = 5$, $R = 10^{-4}$). Note that for a given value of T_v (i.e., time t), the nondimensional excess pore water pressure decreases more than λ (i.e., void ratio).

For a given value of T_v, R, and n, the average degree of consolidation can be determined as (Figure 8.46)

$$U_{av} = 1 - \int_0^1 \lambda \, d\bar{z} \tag{8.129}$$

Figure 8.47 shows the variation of U_{av} with T_v (for $n = 5$). Similar results can be obtained for other values of n. Note that in this figure the beginning of secondary consolidation is assumed to start after the *midplane excess pore water pressure* falls below an arbitrary value of $u = 0.01 \, \Delta\sigma$. Several other observations can be made concerning this plot:

1. Primary and secondary consolidation are continuous processes and depend on the structural viscosity (i.e., R or T_s).
2. The proportion of the total settlement associated with the secondary consolidation increases with the increase of R.
3. In the conventional consolidation theory of Terzaghi, $R = 0$. Thus, the average degree of consolidation becomes equal to 100% at the end of primary consolidation.
4. As defined in Equation 8.125

$$R = \frac{C_v a_v}{H^2} \frac{b^n}{(\Delta\sigma')^{n-1}}$$

The term b is a complex quantity and depends on the electrochemical environment and structure of clay. The value of b increases with the increase of effective pressure σ' on the soil. When the ratio $\Delta\sigma'/\sigma'$ is small, it will result in an increase of R, and thus in the proportion of secondary to primary consolidation. Other factors remaining constant, R will also increase with decrease of H, which is the length of the maximum drainage path, and thus so will the ratio of secondary to primary consolidation.

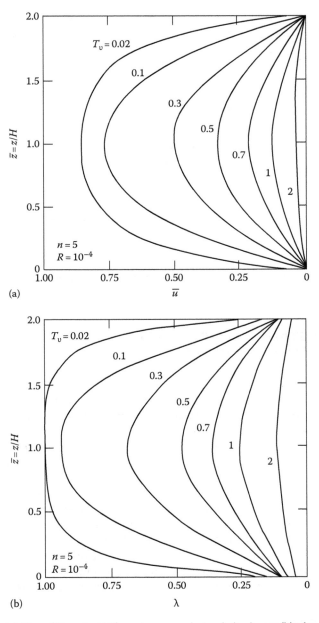

Figure 8.46 (a) Plot of $\bar{z}$ against $\bar{u}$ for a two-way drained clay layer; (b) plot of $\bar{z}$ against λ for a two-way drained clay layer. [After Barden, L., *Geotechnique*, 15(4), 345, 1965.]

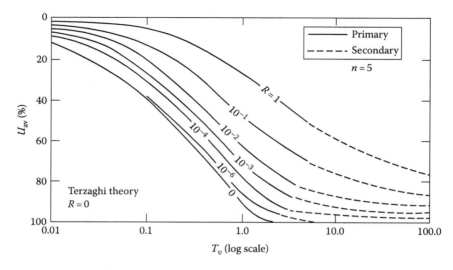

Figure 8.47 Plot of degree of consolidation versus T_v for various values of R (n = 5). [After Barden, L., *Geotechnique*, 15(4), 345, 1965.]

8.12 CONSTANT RATE-OF-STRAIN CONSOLIDATION TESTS

The standard one-dimensional consolidation test procedure discussed in Section 8.5 is time-consuming. At least two other one-dimensional consolidation test procedures have been developed in the past that are much faster yet give reasonably good results. The methods are (1) the constant rate-of-strain consolidation test and (2) the constant-gradient consolidation test. The fundamentals of these test procedures are described in this and the next sections.

The constant rate-of-strain method was developed by Smith and Wahls (1969). A soil specimen is taken in a fixed-ring consolidometer and saturated. For conducting the test, drainage is permitted at the top of the specimen, but not at the bottom. A continuously increasing load is applied to the top of the specimen so as to produce a constant rate of compressive strain, and the excess pore water pressure u_b (generated by the continuously increasing stress σ at the top) at the bottom of the specimen is measured.

8.12.1 Theory

The mathematical derivations developed by Smith and Wahls for obtaining the void ratio–effective pressure relation and the corresponding coefficient of consolidation are given later.

The basic equation for continuity of flow through a soil element is given in Equation 8.16 as

$$\frac{k}{\gamma_w} \frac{\partial^2 u}{\partial z^2} = \frac{1}{1+e} \frac{\partial e}{\partial t}$$

The coefficient of permeability at a given time is a function of the average void ratio $\bar{e}$ in the specimen. The average void ratio is, however, continuously changing owing to the constant rate of strain. Thus

$$k = k(\bar{e}) = f(t) \tag{8.130}$$

The average void ratio is given by

$$\bar{e} = \frac{1}{H} \int_0^H e \, dz$$

where H (= H_t) is the sample thickness. (Note: $z = 0$ is the top of the specimen and $z = H$ is the bottom of the specimen.)

In the constant rate-of-strain type of test, the rate of change of volume is constant, or

$$\frac{dV}{dt} = -RA \tag{8.131}$$

where
  V is the volume of the specimen
  A is the area of cross-section of the specimen
  R is the constant rate of deformation of upper surface

The rate of change of average void ratio $\bar{e}$ can be given by

$$\frac{d\bar{e}}{dt} = \frac{1}{V_s} \frac{dV}{dt} = -\frac{1}{V_s} RA = -r \tag{8.132}$$

where r is a constant.

Based on the definition of $\bar{e}$ and Equation 8.130, we can write

$$e_{(z,t)} = g(z)t + e_0 \tag{8.133}$$

where
  $e_{(z,t)}$ is the void ratio at depth z and time t
  e_0 is the initial void ratio at the beginning of the test
  $g(z)$ is a function of depth only

The function $g(z)$ is difficult to determine. We will assume it to be a linear function of the form

$$-r\left[1 - \frac{b}{r}\left(\frac{z - 0.5H}{H}\right)\right]$$

where b is a constant. Substitution of this into Equation 8.133 gives

$$e_{(z,t)} = e_0 - rt\left[1 - \frac{b}{r}\left(\frac{z - 0.5H}{H}\right)\right] \qquad (8.134)$$

Let us consider the possible range of variation of b/r as given in Equation 8.134:

1. If $b/r = 0$,

$$e_{(z,t)} = e_0 - rt \qquad (8.135)$$

   This indicates that the void is constant with depth and changes with time only. In reality, this is not the case.
2. If $b/r = 2$, the void ratio at the base of the specimen, that is, at $z = H$, becomes

$$e_{(H,t)} = e_0 \qquad (8.136)$$

This means that the void ratio at the base does not change with time at all, which is not realistic.

So the value of b/r is somewhere between 0 and 2 and may be taken as about 1.

Assuming $b/r \neq 0$ and using the definition of the void ratio as given by Equation 8.134, we can integrate Equation 8.16 to obtain an equation for the excess pore water pressure. The boundary conditions are as follows: at $z = 0$, $u = 0$ (at any time); and at $z = H$, $\partial u/\partial z = 0$ (at any time). Thus

$$u = \frac{\gamma_w r}{k}\left\{zH\left[\frac{1 + e_0 - bt}{rt(bt)}\right] + \frac{z^2}{2rt} - \left[\frac{H(1 + e_0)}{rt(bt)}\right]\right.$$

$$\left. \times \left[\frac{H(1 + e)}{bt}\ln(1 + e) - z\ln(1 + e_B) - \frac{H(1 + e_T)}{bt}\ln(1 + e_T)\right]\right\} \qquad (8.137)$$

where

$$e_B = e_0 - rt\left(1 - \frac{1}{2}\frac{b}{r}\right) \tag{8.138}$$

$$e_T = e_0 - rt\left(1 + \frac{1}{2}\frac{b}{r}\right) \tag{8.139}$$

Equation 8.137 is very complicated. Without losing a great deal of accuracy, it is possible to obtain a simpler form of expression for u by assuming that the term $1 + e$ in Equation 8.16 is approximately equal to $1 + \bar{e}$ (note that this is not a function of z). So, from Equations 8.16 and 8.134

$$\frac{\partial^2 u}{\partial z^2} = \left[\frac{\gamma_w}{k(1 + \bar{e})}\right]\frac{\partial}{\partial t}\left\{e_0 - rt\left[1 - \frac{b}{r}\left(\frac{z - 0.5H}{H}\right)\right]\right\} \tag{8.140}$$

Using the boundary condition $u = 0$ at $z = 0$ and $\partial u/\partial t = 0$ at $z = H$, Equation 8.140 can be integrated to yield

$$u = \left[\frac{\gamma_w r}{k(1 + \bar{e})}\right]\left[\left(Hz - \frac{z^2}{2}\right) - \frac{b}{r}\left(\frac{z^2}{4} - \frac{z^3}{6H}\right)\right] \tag{8.141}$$

The pore pressure at the base of the specimen can be obtained by substituting $z = H$ in Equation 8.141:

$$u_{z=H} = \frac{\gamma_w r H^2}{k(1 + \bar{e})}\left(\frac{1}{2} - \frac{1}{12}\frac{b}{r}\right) \tag{8.142}$$

The average effective stress corresponding to a given value of $u_{z=H}$ can be obtained by writing

$$\sigma'_{av} = \sigma - \frac{u_{av}}{u_{z=H}}u_{z=H} \tag{8.143}$$

where
  σ'_{av} is the average effective stress on the specimen at any time
  σ is the total stress on the specimen
  u_{av} is the corresponding average pore water pressure

$$\frac{u_{av}}{u_{z=H}} = \frac{\dfrac{1}{H}\displaystyle\int_0^H u\,dz}{u_{z=H}} \tag{8.144}$$

Substitution of Equations 8.141 and 8.142 into 8.144 and further simplification gives

$$\frac{u_{av}}{u_{z=H}} = \frac{\dfrac{1}{3} - \dfrac{1}{24}(b/r)}{\dfrac{1}{2} - \dfrac{1}{12}(b/r)} \tag{8.145}$$

Note that for $b/r = 0$, $u_{av}/u_{z=H} = 0.667$; and for $b/r = 1$, $u_{av}/u_{z=H} = 0.700$. Hence, for $0 \le b/r \le 1$, the values of $u_{av}/u_{z=H}$ do not change significantly. So, from Equations 8.143 and 8.145

$$\sigma'_{av} = \sigma - \left[\frac{\dfrac{1}{3} - \dfrac{1}{24}(b/r)}{\dfrac{1}{2} - \dfrac{1}{12}(b/r)} \right] u_{z=H} \tag{8.146}$$

8.12.2 Coefficient of consolidation

The coefficient of consolidation was defined previously as

$$C_v = \frac{k(1+e)}{a_v \gamma_w}$$

We can assume $1 + e \approx 1 + \bar{e}$, and from Equation 8.142

$$k = \frac{\gamma_w r H^2}{(1+\bar{e})u_{z=H}}\left(\frac{1}{2} - \frac{1}{12}\frac{b}{r} \right) \tag{8.147}$$

Substitution of these into the expression for C_v gives

$$C_v = \frac{r H^2}{a_v u_{z=H}}\left(\frac{1}{2} - \frac{1}{12}\frac{b}{r} \right) \tag{8.148}$$

8.12.3 Interpretation of experimental results

The following information can be obtained from a constant rate-of-strain consolidation test:

1. Initial height of specimen, H_i
2. Value of A
3. Value of V_s
4. Strain rate R
5. A continuous record of $u_{z=H}$
6. A corresponding record of σ (total stress applied at the top of the specimen)

The plot of e versus σ'_{av} can be obtained in the following manner:

1. Calculate $r = RA/V_s$.
2. Assume $b/r \approx 1$.
3. For a given value of $u_{z=H}$, the value of σ is known (at time t from the start of the test), and so σ'_{av} can be calculated from Equation 8.146.
4. Calculate $\Delta H = Rt$ and then the change in void ratio that has taken place during time t

$$\Delta e = \frac{\Delta H}{H_i}(1 + e_0)$$

   where H_i is the initial height of the specimen.
5. The corresponding void ratio (at time t) is $e = e_0 - \Delta e$.
6. After obtaining a number of points of σ'_{av} and the corresponding e, plot the graph of e versus $\log \sigma'_{av}$.
7. For a given value of σ'_{av} and e, the coefficient of consolidation C_v can be calculated by using Equation 8.148. (Note that H in Equation 8.148 is equal to $H_i - \Delta H$.)

Smith and Wahls (1969) provided the results of constant rate-of-strain consolidation tests on two clays—Massena clay and calcium montmorillonite. The tests were conducted at various rates of strain (0.0024%/min–0.06%/min) and the e versus $\log \sigma'$ curves obtained were compared with those obtained from the conventional tests.

Figures 8.48 and 8.49 show the results obtained from tests conducted with Massena clay.

This comparison showed that, for higher rates of strain, the e versus $\log \sigma'$ curves obtained from these types of tests may deviate considerably from those obtained from conventional tests. For that reason, it is recommended that the strain rate for a given test should be chosen such that

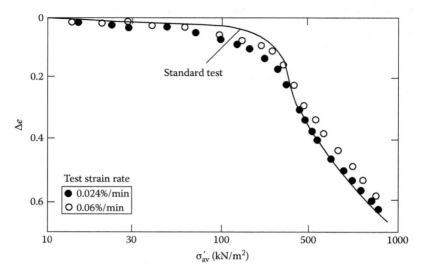

Figure 8.48 CRS tests on Massena clay—plot of Δe versus σ'_{av}. [After Smith, R. E. and Wahls, H. E., *J. Soil Mech. Found. Div.*, ASCE, 95(SM2), 519, 1969.]

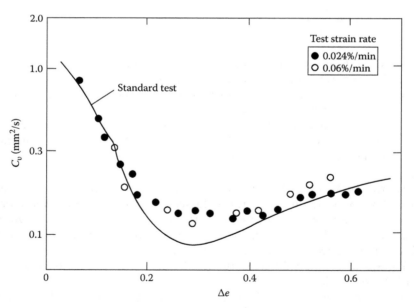

Figure 8.49 CRS tests on Massena clay—plot of C_v versus Δe. [After Smith, R. E. and Wahls, H. E., *J. Soil Mech. Found. Div.*, ASCE, 95(SM2), 519, 1969.]

the value of $u_{z=H}/\sigma$ at the end of the test does not exceed 0.5. However, the value should be high enough that it can be measured with reasonable accuracy.

8.13 CONSTANT-GRADIENT CONSOLIDATION TEST

The constant-gradient consolidation test was developed by Lowe et al. (1969). In this procedure, a saturated soil specimen is taken in a consolidation ring. As in the case of the constant rate-of-strain type of test, drainage is allowed at the top of the specimen and pore water pressure is measured at the bottom. A load P is applied on the specimen, which increases the excess pore water pressure in the specimen by an amount Δu (Figure 8.50a). After a small lapse of time t_1, the excess pore water pressure at the top of the specimen will be equal to zero (since drainage is permitted). However, at the bottom of the specimen, the excess pore water pressure will still be approximately Δu (Figure 8.50b). From this point on, the load P is increased slowly in such a way that the difference between the pore water pressures at the top and bottom of the specimen remains constant, that is, the difference is maintained at a constant Δu (Figure 8.50c and d). When the desired value of P is reached, say at time t_3, the loading is stopped and the excess pore water pressure is allowed to dissipate. The elapsed time t_4 at which the pore water pressure at the bottom of the specimen reaches a value of $0.1\Delta u$ is recorded. During the entire

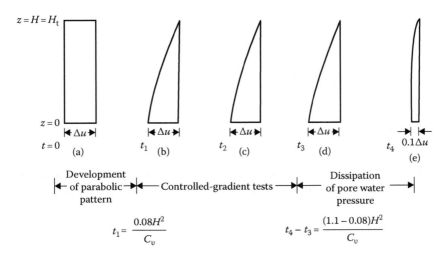

Figure 8.50 Stages in controlled-gradient test.

test, the compression ΔH_t that the specimen undergoes is recorded. For complete details of the laboratory test arrangement, the reader is referred to the original paper of Lowe et al. (1969).

8.13.1 Theory

From the basic Equations 8.16 and 8.17, we have

$$\frac{k}{\gamma_w}\frac{\partial^2 u}{\partial z^2} = -\frac{a_v}{1+e}\frac{\partial \sigma'}{\partial t} \tag{8.149}$$

or

$$\frac{\partial \sigma'}{\partial t} = -\frac{k}{\gamma_w m_v}\frac{\partial^2 u}{\partial z^2} = -C_v\frac{\partial^2 u}{\partial z^2} \tag{8.150}$$

Since $\sigma' = \sigma - u$

$$\frac{\partial \sigma'}{\partial t} = \frac{\partial \sigma}{\partial t} - \frac{\partial u}{\partial t} \tag{8.151}$$

For the controlled-gradient tests (i.e., during the time t_1 to t_3 in Figure 8.50), $\partial u/\partial t = 0$. So

$$\frac{\partial \sigma'}{\partial t} = \frac{\partial \sigma}{\partial t} \tag{8.152}$$

Combining Equations 8.150 and 8.152

$$\frac{\partial \sigma}{\partial t} = -C_v\frac{\partial^2 u}{\partial z^2} \tag{8.153}$$

Note that the left-hand side of Equation 8.153 is independent of the variable z and the right-hand side is independent of the variable t. So both sides should be equal to a constant, say A_1. Thus

$$\frac{\partial \sigma}{\partial t} = A_1 \tag{8.154}$$

and

$$\frac{\partial^2 u}{\partial z^2} = -\frac{A_1}{C_v} \tag{8.155}$$

Integration of Equation 8.155 yields

$$\frac{\partial u}{\partial z} = -\frac{A_1}{C_v} z + A_2 \tag{8.156}$$

and

$$u = -\frac{A_1}{C_v} \frac{z^2}{2} + A_2 z + A_3 \tag{8.157}$$

The boundary conditions are as follows (note that $z = 0$ is at the bottom of the specimen):

   1. At $z = 0$, $\partial u/\partial z = 0$
   2. At $z = H$, $u = 0$ (note that $H = H_t$; one-way drainage)
   3. At $z = 0$, $u = \Delta u$

From the first boundary condition and Equation 8.156, we find that $A_2 = 0$. So

$$u = -\frac{A_1}{C_v} \frac{z^2}{2} + A_3 \tag{8.158}$$

From the second boundary condition and Equation 8.158

$$A_3 = \frac{A_1 H^2}{2C_v} \tag{8.159}$$

$$\text{or} \quad u = -\frac{A_1}{C_v} \frac{z^2}{2} + \frac{A_1}{C_v} \frac{H^2}{2} \tag{8.160}$$

From the third boundary condition and Equation 8.160

$$\Delta u = \frac{A_1}{C_v} \frac{H^2}{2}$$

or

$$A_1 = \frac{2C_v \Delta u}{H^2} \tag{8.161}$$

Substitution of this value of A_1 into Equation 8.160 yields

$$u = \Delta u \left(1 - \frac{z^2}{H^2} \right) \tag{8.162}$$

Equation 8.162 shows a parabolic pattern of excess pore water pressure distribution, which remains constant during the controlled-gradient test (time $t_1 - t_3$ in Figure 8.50). This closely corresponds to Terzaghi isochrone (Figure 8.4) for $T_v = 0.08$.

Combining Equations 8.154 and 8.161, we obtain

$$\frac{\partial \sigma}{\partial t} = A_1 = \frac{2C_v \Delta u}{H^2}$$

$$\text{or } C_v = \frac{\partial \sigma}{\partial t} \frac{H^2}{2 \Delta u} \tag{8.163}$$

8.13.2 Interpretation of experimental results

The following information will be available from the constant-gradient test:

1. Initial height of the specimen H_i and height H_t at any time during the test
2. Rate of application of the load P and thus the rate of application of stress $\partial \sigma / \partial t$ on the specimen
3. Differential pore pressure Δu
4. Time t_1
5. Time t_3
6. Time t_4

The plot of e versus σ'_{av} can be obtained in the following manner:

1. Calculate the initial void ratio e_0.
2. Calculate the change in void ratio at any other time t during the test as

$$\Delta e = \frac{\Delta H}{H_i}(1 + e_0) = \frac{\Delta H_t}{H_i}(1 + e_0)$$

where $\Delta H = \Delta H_t$ is the total change in height from the beginning of the test. So, the average void ratio at time t is $e = e_0 - \Delta e$.

3. Calculate the average effective stress at time t using the known total stress σ applied on the specimen at that time:

$$\sigma'_{av} = \sigma - u_{av}$$

where u_{av} is the average excess pore water pressure in the specimen, which can be calculated from Equation 8.162.

Calculation of the coefficient of consolidation is as follows:

1. At time t_1

$$C_v = \frac{0.08H^2}{t_1}$$

2. At time $t_1 < t < t_3$

$$C_v = \frac{\Delta\sigma}{\Delta t} \frac{H^2}{2\Delta u} \qquad (8.163)$$

Note that $\Delta\sigma/\Delta t$, H, and Δu are all known from the tests.

3. Between time t_3 and t_4

$$C_v = \frac{(1.1-0.08)H^2}{t_3 - t_4} = \frac{1.02H^2}{t_3 - t_4}$$

8.14 SAND DRAINS

In order to accelerate the process of consolidation settlement for the construction of some structures, the useful technique of building sand drains can be used. Sand drains are constructed by driving down casings or hollow mandrels into the soil. The holes are then filled with sand, after which the casings are pulled out. When a surcharge is applied at ground surface, the pore water pressure in the clay will increase, and there will be drainage in the vertical and horizontal directions (Figure 8.51a). The horizontal drainage is induced by the sand drains. Hence, the process of dissipation of excess pore water pressure created by the loading (and hence the settlement) is accelerated.

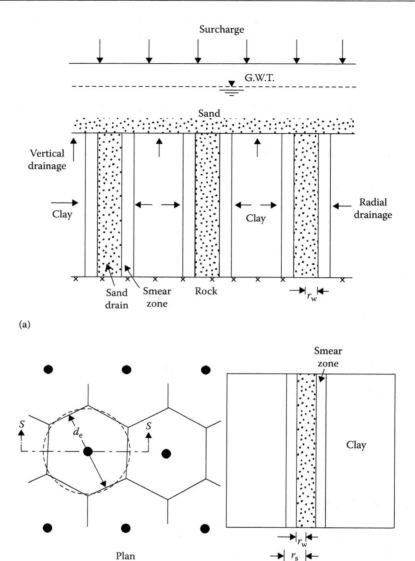

(a)

(b)   Cross section at S–S

Figure 8.51 (a) Sand drains and (b) layout of sand drains.

The basic theory of sand drains was presented by Rendulic (1935) and Barron (1948) and later summarized by Richart (1959). In the study of sand drains, we have two fundamental cases:

1. *Free-strain case.* When the surcharge applied at the ground surface is of a flexible nature, there will be equal distribution of surface load. This will result in an uneven settlement at the surface.
2. *Equal-strain case.* When the surcharge applied at the ground surface is rigid, the surface settlement will be the same all over. However, this will result in an unequal distribution of stress.

Another factor that must be taken into consideration is the effect of "smear." A smear zone in a sand drain is created by the remolding of clay during the drilling operation for building it (see Figure 8.51a). This remolding of the clay results in a decrease of the coefficient of permeability in the horizontal direction.

The theories for free-strain and equal-strain consolidation are given later. In the development of these theories, it is assumed that drainage takes place only in the radial direction, that is, *no dissipation of excess pore water pressure in the vertical direction.*

8.14.1 Free-strain consolidation with no smear

Figure 8.51b shows the general pattern of the layout of sand drains. For triangular spacing of the sand drains, the zone of influence of each drain is hexagonal in plan. This hexagon can be approximated as an equivalent circle of diameter d_e. Other notations used in this section are as follows:

1. r_e = radius of the equivalent circle = $d_e/2$
2. r_w = radius of the sand drain well
3. r_s = radial distance from the centerline of the drain well to the farthest point of the smear zone. Note that, in the no-smear case, $r_w = r_s$

The basic differential equation of Terzaghi's consolidation theory for flow in the vertical direction is given in Equation 8.21. For radial drainage, this equation can be written as

$$\frac{\partial u}{\partial t} = C_{vr}\left(\frac{\partial^2 u}{\partial r^2} + \frac{1}{r}\frac{\partial u}{\partial r}\right) \tag{8.164}$$

where
  u is the excess pore water pressure
  r is the radial distance measured from the center of the drain well
  C_{vr} is the coefficient of consolidation in radial direction

For solution of Equation 8.164, the following boundary conditions are used:

1. At time $t = 0$, $u = u_i$
2. At time $t > 0$, $u = 0$ at $r = r_w$
3. At $r = r_e$, $\partial u/\partial r = 0$

With the aforementioned boundary conditions, Equation 8.164 yields the solution for excess pore water pressure at any time t and radial distance r:

$$u = \sum_{\alpha_1,\alpha_2,\ldots}^{\alpha=\infty} \frac{-2U_1(\alpha)U_0(\alpha r/r_w)}{\alpha\left[n^2 U_0^2(\alpha n) - U_1^2(\alpha)\right]} \exp(-4\alpha^2 n^2 T_r) \tag{8.165}$$

In Equation 8.165

$$n\frac{r_e}{r_w} \tag{8.166}$$

$$U_1(\alpha) = J_1(\alpha)Y_0(\alpha) - Y_1(\alpha)J_0(\alpha) \tag{8.167}$$

$$U_0(\alpha n) = J_0(\alpha n)Y_0(\alpha) - Y_0(\alpha n)J_0(\alpha) \tag{8.168}$$

$$U_0\left(\frac{\alpha r}{r_w}\right) = J_0\left(\frac{\alpha r}{r_w}\right)Y_0(\alpha) - Y_0\left(\frac{\alpha r}{r_w}\right)J_0(\alpha) \tag{8.169}$$

where
  J_0 is the Bessel function of first kind of zero order
  J_1 is the Bessel function of first kind of first order
  Y_0 is the Bessel function of second kind of zero order
  Y_1 is the Bessel function of second kind of first order
  α_1, α_2,... are roots of Bessel function that satisfy $J_1(\alpha n)Y_0(\alpha) - Y_1(\alpha n)$
    $J_0(\alpha) = 0$

$$T_r = \text{Time factor for radial flow} = \frac{C_{vr}t}{d_e^2} \tag{8.170}$$

In Equation 8.170

$$C_{vr} = \frac{k_h}{m_v\gamma_w} \tag{8.171}$$

where k_h is the coefficient of permeability in the horizontal direction.

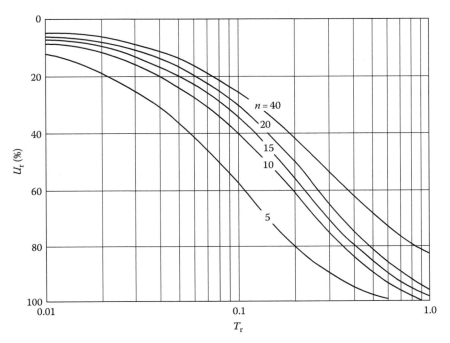

Figure 8.52 Free strain—variation of degree of consolidation U_r with time factor T_r.

The average pore water pressure u_{av} throughout the soil mass may now be obtained from Equation 8.165 as

$$u_{av} = u_i \sum_{\alpha_1, \alpha_2, \ldots}^{\alpha = \infty} \frac{4U_1^2(\alpha)}{\alpha^2(n^2 - 1)[n^2 U_0^2(\alpha n) - U_1^2(\alpha)]} \exp(-4\alpha^2 n^2 T_r) \qquad (8.172)$$

The average degree of consolidation U_r can be determined as

$$U_r = 1 - \frac{u_{av}}{u_i} \qquad (8.173)$$

Figure 8.52 shows the variation of U_r with the time factor T_r.

8.14.2 Equal-strain consolidation with no smear

The problem of equal-strain consolidation with no smear ($r_w = r_s$) was solved by Barron (1948). The results of the solution are described later (refer to Figure 8.51).

The excess pore water pressure at any time t and radial distance r is given by

$$u = \frac{4u_{av}}{d_e^2 F(n)} \left[r_e^2 \ln\left(\frac{r}{r_w}\right) - \frac{r^2 - r_w^2}{2} \right] \tag{8.174}$$

where

$$F(n) = \frac{n^2}{n^2 - 1} \ln(n) - \frac{3n^2 - 1}{4n^2} \tag{8.175}$$

u_{av} = average value of pore water pressure throughout the clay layer

$$= u_i e^{\lambda} \tag{8.176}$$

$$\lambda = \frac{-8T_r}{F(n)} \tag{8.177}$$

The average degree of consolidation due to radial drainage is

$$U_r = 1 - \exp\left[\frac{-8T_r}{F(n)}\right] \tag{8.178}$$

Table 8.9 gives the values of the time factor T_r for various values of U_r. For $r_e/r_w > 5$, the free-strain and equal-strain solutions give approximately the same results for the average degree of consolidation.

Olson (1977) gave a solution for the average degree of consolidation U_r for time-dependent loading (ramp load) similar to that for vertical drainage, as described in Section 8.3.

Referring to Figure 8.13b, the surcharge increases from zero at time $t = 0$ to $q = q_c$ at time $t = t_c$. For $t \geq t_c$, the surcharge is equal to q_c. For this case

$$T_r' = \frac{C_{vr}t}{r_e^2} = 4T_r \tag{8.179}$$

and

$$T_{rc}' = \frac{C_{vr}t_c}{r_e^2} \tag{8.180}$$

For $T_r' \leq T_{rc}'$

$$U_r = \frac{T_r' - \frac{1}{A}[1 - \exp(AT_r')]}{T_{rc}'} \tag{8.181}$$

Table 8.9 Solution for radial-flow equation (equal vertical strain)

Degree of consolidation U_r(%)	Time factor T_r for value of $n(= r_e/r_w)$				
	5	10	15	20	25
0	0	0	0	0	0
1	0.0012	0.0020	0.0025	0.0028	0.0031
2	0.0024	0.0040	0.0050	0.0057	0.0063
3	0.0036	0.0060	0.0075	0.0086	0.0094
4	0.0048	0.0081	0.0101	0.0115	0.0126
5	0.0060	0.0101	0.0126	0.0145	0.0159
6	0.0072	0.1222	0.0153	0.0174	0.0191
7	0.0085	0.0143	0.0179	0.0205	0.0225
8	0.0098	0.0165	0.0206	0.0235	0.0258
9	0.0110	0.0186	0.0232	0.0266	0.0292
10	0.0123	0.0208	0.0260	0.0297	0.0326
11	0.0136	0.0230	0.0287	0.0328	0.0360
12	0.0150	0.0252	0.0315	0.0360	0.0395
13	0.0163	0.0275	0.0343	0.0392	0.0431
14	0.0177	0.0298	0.0372	0.0425	0.0467
15	0.0190	0.0321	0.0401	0.0458	0.0503
16	0.0204	0.0344	0.0430	0.0491	0.0539
17	0.0218	0.0368	0.0459	0.0525	0.0576
18	0.0232	0.0392	0.0489	0.0559	0.0614
19	0.0247	0.0416	0.0519	0.0594	0.0652
20	0.0261	0.0440	0.0550	0.0629	0.0690
21	0.0276	0.0465	0.0581	0.0664	0.0729
22	0.0291	0.0490	0.0612	0.0700	0.0769
23	0.0306	0.0516	0.0644	0.0736	0.0808
24	0.0321	0.0541	0.0676	0.0773	0.0849
25	0.0337	0.0568	0.0709	0.0811	0.0890
26	0.0353	0.0594	0.0742	0.0848	0.0931
27	0.0368	0.0621	0.0776	0.0887	0.0973
28	0.0385	0.0648	0.810	0.0926	0.1016
29	0.0401	0.0676	0.0844	0.0965	0.1059
30	0.0418	0.0704	0.0879	0.1005	0.1103
31	0.0434	0.0732	0.0914	0.1045	0.1148
32	0.0452	0.0761	0.0950	0.1087	0.1193
33	0.0469	0.0790	0.0987	0.1128	0.1239
34	0.0486	0.0820	0.1024	0.1171	0.1285
35	0.0504	0.0850	0.1062	0.1214	0.1332
36	0.0522	0.0881	0.1100	0.1257	0.1380
37	0.0541	0.0912	0.1139	0.1302	0.1429

(Continued)

Table 8.9 (Continued) Solution for radial-flow equation (equal vertical strain)

Degree of consolidation U_r(%)	Time factor T_r for value of $n(= r_e/r_w)$				
	5	10	15	20	25
38	0.0560	0.0943	0.1178	0.1347	0.1479
39	0.579	0.0975	0.1218	0.1393	0.1529
40	0.0598	0.1008	0.1259	0.1439	0.1580
41	0.0618	0.1041	0.1300	0.1487	0.1632
42	0.0638	0.1075	0.1342	0.1535	0.1685
43	0.0658	0.1109	0.1385	0.1584	0.1739
44	0.0679	0.1144	0.1429	0.1634	0.1793
45	0.0700	0.1180	0.1473	0.1684	0.1849
46	0.0721	0.1216	0.1518	0.1736	0.1906
47	0.0743	0.1253	0.1564	0.1789	0.1964
48	0.0766	0.1290	0.1611	0.1842	0.2023
49	0.0788	0.1329	0.1659	0.1897	0.2083
50	0.0811	0.1368	0.1708	0.1953	0.2144
51	0.0835	0.1407	0.1758	0.2020	0.2206
52	0.0859	0.1448	0.1809	0.2068	0.2270
53	0.0884	0.1490	0.1860	0.2127	0.2335
54	0.0909	0.1532	0.1913	0.2188	0.2402
55	0.0935	0.1575	0.1968	0.2250	0.2470
56	0.0961	0.1620	0.2023	0.2313	0.2539
57	0.0988	0.1665	0.2080	0.2378	0.2610
58	0.1016	0.1712	0.2138	0.2444	0.2683
59	0.1044	0.1759	0.2197	0.2512	0.2758
60	0.1073	0.1808	0.2258	0.2582	0.2834
61	0.1102	0.1858	0.2320	0.2653	0.2912
62	0.1133	0.1909	0.2384	0.2726	0.2993
63	0.1164	0.1962	0.2450	0.2801	0.3075
64	0.1196	0.2016	0.2517	0.2878	0.3160
65	0.1229	0.2071	0.2587	0.2958	0.3247
66	0.1263	0.2128	0.2658	0.3039	0.3337
67	0.1298	0.2187	0.2732	0.3124	0.3429
68	0.1334	0.2248	0.2808	0.3210	0.3524
69	0.1371	0.2311	0.2886	0.3300	0.3623
70	0.1409	0.2375	0.2967	0.3392	0.3724
71	0.1449	0.2442	0.3050	0.3488	0.3829
72	0.1490	0.2512	0.3134	0.3586	0.3937
73	0.1533	0.2583	0.3226	0.3689	0.4050
74	0.1577	0.2658	0.3319	0.3795	0.4167

(Continued)

Table 8.9 (Continued) Solution for radial-flow equation (equal vertical strain)

Degree of consolidation U_r (%)	Time factor T_r for value of $n(= r_e/r_w)$				
	5	10	15	20	25
75	0.1623	0.2735	0.3416	0.3906	0.4288
76	0.1671	0.2816	0.3517	0.4021	0.4414
77	0.1720	0.2900	0.3621	0.4141	0.4546
78	0.1773	0.2988	0.3731	0.4266	0.4683
79	0.1827	0.3079	0.3846	0.4397	0.4827
80	0.1884	0.3175	0.3966	0.4534	0.4978
81	0.1944	0.3277	0.4090	0.4679	0.5137
82	0.2007	0.3383	0.4225	0.4831	0.5304
83	0.2074	0.3496	0.4366	0.4992	0.5481
84	0.2146	0.3616	0.4516	0.5163	0.5668
85	0.2221	0.3743	0.4675	0.5345	0.5868
86	0.2302	0.3879	0.4845	0.5539	0.6081
87	0.2388	0.4025	0.5027	0.5748	0.6311
88	0.2482	0.4183	0.5225	0.5974	0.6558
89	0.2584	0.4355	0.5439	0.6219	0.6827
90	0.2696	0.4543	0.5674	0.6487	0.7122
91	0.2819	0.4751	0.5933	0.6784	0.7448
92	0.2957	0.4983	0.6224	0.7116	0.7812
93	0.3113	0.5247	0.6553	0.7492	0.8225
94	0.3293	0.5551	0.6932	0.7927	0.8702
95	0.3507	0.5910	0.7382	0.8440	0.9266
96	0.3768	0.6351	0.7932	0.9069	0.9956
97	0.4105	0.6918	0.8640	0.9879	1.0846
98	0.4580	0.7718	0.9640	1.1022	1.2100
99	0.5391	0.9086	1.1347	1.2974	1.4244

For $T_r' \geq T_{rc}'$

$$U_r = 1 - \frac{1}{AT_{rc}'}\left[\exp(AT_{rc}') - 1\right]\exp(-AT_r') \tag{8.182}$$

where

$$A = \frac{2}{F(n)} \tag{8.183}$$

Figure 8.53 shows the variation of U_r with T_r' and T_{rc}' for $n = 5$ and 10.

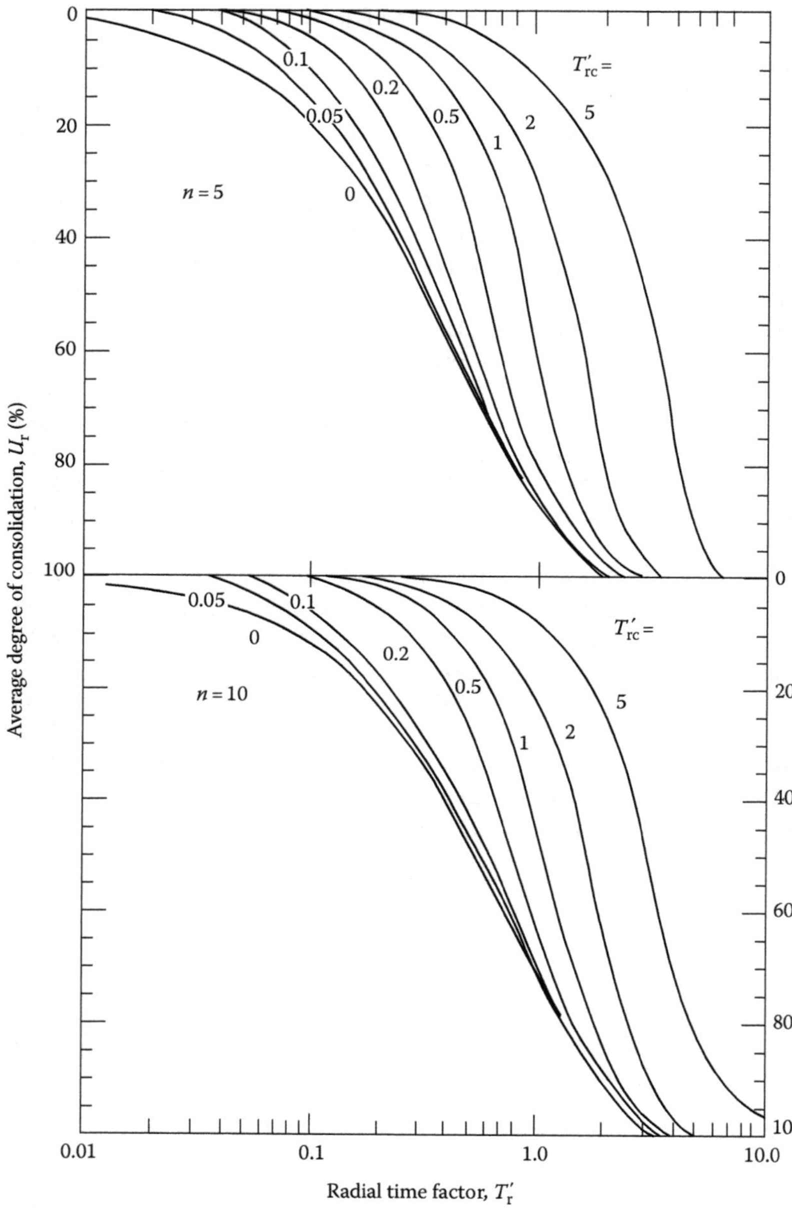

Figure 8.53 Olson's solution for radial flow under single ramp loading for n = 5 and 10 (Equations 8.181 and 8.182).

8.14.3 Effect of smear zone on radial consolidation

Barron (1948) also extended the analysis of equal-strain consolidation by sand drains to account for the smear zone. The analysis is based on the assumption that the clay in the smear zone will have one boundary with zero excess pore water pressure and the other boundary with an excess pore water pressure that will be time-dependent. Based on this assumption

$$u = \frac{1}{m'} u_{av} \left[\ln\left(\frac{r}{r_e} \right) - \frac{r^2 - r_s^2}{2r_e^2} + \frac{k_h}{k_s} \left(\frac{n^2 - S^2}{n^2} \right) \ln S \right] \qquad (8.184)$$

where k_s is the coefficient of permeability of the smeared zone.

$$S = \frac{r_s}{r_w} \qquad (8.185)$$

where r_s is the radius of the smeared zone.

$$m' = \frac{n^2}{n^2 - S^2} \ln\left(\frac{n}{S} \right) - \frac{3}{4} + \frac{S^2}{4n^2} + \frac{k_h}{k_s} \left(\frac{n^2 - S^2}{n^2} \right) \ln S \qquad (8.186)$$

$$u_{av} = u_i \exp\left(\frac{-8T_r}{m'} \right) \qquad (8.187)$$

The average degree of consolidation is given by the relation

$$U_r = 1 - \frac{u_{av}}{u_i} = 1 - \exp\left(\frac{-8T_r}{m'} \right) \qquad (8.188)$$

8.15 NUMERICAL SOLUTION FOR RADIAL DRAINAGE (SAND DRAIN)

As shown previously for vertical drainage (Section 8.4), we can adopt the finite-difference technique for solving consolidation problems in the case of radial drainage. From Equation 8.164

$$\frac{\partial u}{\partial t} = C_{vr} \left(\frac{\partial^2 u}{\partial r^2} + \frac{1}{r} \frac{\partial u}{\partial r} \right)$$

Let u_R, t_R, and r_R be any reference excess pore water pressure, time, and radial distance, respectively. So

$$\text{Nondimensional excess pore water pressure} = \bar{u} = \frac{u}{u_R} \tag{8.189}$$

$$\text{Nondimensional time} = \bar{t} = \frac{t}{t_R} \tag{8.190}$$

$$\text{Nondimensional radial distance} = \bar{r} = \frac{r}{r_R} \tag{8.191}$$

Substituting Equations 8.189 through 8.191 into 8.164, we get

$$\frac{1}{t_R}\frac{\partial \bar{u}}{\partial \bar{t}} = \frac{C_{vr}}{r_R^2}\left(\frac{\partial^2 \bar{u}}{\partial \bar{r}^2} + \frac{1}{\bar{r}}\frac{\partial \bar{u}}{\partial \bar{r}}\right) \tag{8.192}$$

Referring to Figure 8.54

$$\frac{\partial \bar{u}}{\partial \bar{t}} = \frac{1}{\Delta \bar{t}}(\bar{u}_{0,\bar{t}+\Delta \bar{t}} - \bar{u}_{0,\bar{t}}) \tag{8.193}$$

$$\frac{\partial^2 \bar{u}}{\partial \bar{r}^2} = \frac{1}{(\Delta \bar{r})^2}(\bar{u}_{1,\bar{t}} + \bar{u}_{3,\bar{t}} - 2\bar{u}_{0,\bar{t}}) \tag{8.194}$$

and

$$\frac{1}{\bar{r}}\frac{\partial \bar{u}}{\partial \bar{r}} = \frac{1}{\bar{r}}\left(\frac{u_{3,\bar{t}} - \bar{u}_{1,\bar{t}}}{2\Delta \bar{r}}\right) \tag{8.195}$$

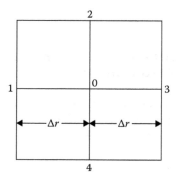

Figure 8.54 Numerical solution for radial drainage.

If we adopt t_R in such a way that $1/t_R = C_{vr}/r_R^2$ and then substitute Equations 8.193 through 8.195 into 8.192, then

$$\bar{u}_{0,\bar{t}+\Delta\bar{t}} = \frac{\Delta\bar{t}}{(\Delta\bar{r})^2}\left[\bar{u}_{1,\bar{t}} + \bar{u}_{3,\bar{t}} + \frac{\bar{u}_{3,\bar{t}} - \bar{u}_{1,\bar{t}}}{2(\bar{r}/\Delta\bar{r})} - 2\bar{u}_{0,\bar{t}}\right] + \bar{u}_{0,\bar{t}} \qquad (8.196)$$

Equation 8.196 is the basic finite-difference equation for solution of the excess pore water pressure (for radial drainage only).

Example 8.20

For a sand drain, the following data are given: $r_w = 0.38$ m, $r_e = 1.52$ m, $r_w = r_s$, and $C_{vr} = 46.2 \times 10^{-4}$ m²/day. A uniformly distributed load of 50 kN/m² is applied at the ground surface. Determine the distribution of excess pore water pressure after 10 days of load application assuming radial drainage only.

Solution

Let $r_R = 0.38$ m, $\Delta r = 0.38$ m, and $\Delta t = 5$ days. So $\bar{r}_e = r_e/r_R = 1.52/0.38 = 4$; $\Delta\bar{r} = \Delta r/r_R = 0.38/0.38 = 1$

$$\Delta\bar{t} = \frac{C_{vr}\Delta t}{r_R^2} = \frac{(46.2 \times 10^{-4})(5)}{(0.38)^2} = 0.16$$

$$\frac{\Delta\bar{t}}{(\Delta\bar{r})^2} = \frac{0.16}{(1)^2} = 0.16$$

Let $u_R = 0.5$ kN/m². So, immediately after load application, $\bar{u} = 50/0.5 = 100$.

Figure 8.55 shows the initial nondimensional pore water pressure distribution at time $t = 0$. (Note that at $\bar{r} = 1$, $\bar{u} = 0$ owing to the drainage face.)

At 5 days: $\bar{u} = 0$, $\bar{r} = 1$. From Equation 8.196

$$\bar{u}_{0,\bar{t}+\Delta\bar{t}} = \frac{\Delta\bar{t}}{(\Delta\bar{r})^2}\left[\bar{u}_{1,\bar{t}} + \bar{u}_{3,\bar{t}} + \frac{\bar{u}_{3,\bar{t}} - \bar{u}_{1,\bar{t}}}{2(\bar{r}/\Delta\bar{r})} - 2\bar{u}_{0,\bar{t}}\right] + \bar{u}_{0,\bar{t}}$$

At $\bar{r} = 2$

$$\bar{u}_{0,\bar{t}+\Delta\bar{t}} = 0.16\left[0 + 100 + \frac{100-0}{2(2/1)} - 2(100)\right] + 100 = 88$$

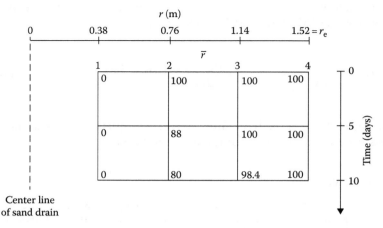

Figure 8.55 Excess pore water pressure variation with time for radial drainage.

At $\bar{r} = 3$

$$\bar{u}_{0,\bar{t}+\Delta\bar{t}} = 0.16\left[100+100+\frac{100-100}{2(3/1)}-2(100)\right]+100 = 100$$

Similarly at $\bar{r} = 4$

$$\bar{u}_{0,\bar{t}+\Delta\bar{t}} = 100$$

(note that, here, $\bar{u}_{3,\bar{r}} = \bar{u}_{1,\bar{r}}$)
   At 10 days, at $\bar{r} = 1$, $\bar{u} = 0$.
   At $\bar{r} = 2$

$$\bar{u}_{0,\bar{t}+\Delta\bar{t}} = 0.16\left[0+100+\frac{100-0}{2(2/1)}-2(88)\right]+88 = 79.84 \cong 80$$

At $\bar{r} = 3$

$$\bar{u}_{0,\bar{t}+\Delta\bar{t}} = 0.16\left[88+100+\frac{100-88}{2(3/1)}-2(100)\right]+100 = 98.4$$

At $\bar{r} = 4$

$$\bar{u} = 100$$

$$u = \bar{u} \times u_R = 0.5\,\bar{u}\ \text{kN/m}^2$$

The distribution of nondimensional excess pore water pressure is shown in Figure 8.55.

8.16 GENERAL COMMENTS ON SAND DRAIN PROBLEMS

Figure 8.51b shows a triangular pattern of the layout of sand drains. In some instances, the sand drains may also be laid out in a square pattern. For all practical purposes, the magnitude of the radius of equivalent circles can be given as follows:
Triangular pattern

$$r_e = (0.525)(\text{drain spacing}) \tag{8.197}$$

Square pattern

$$r_e = (0.565)(\text{drain spacing}) \tag{8.198}$$

Prefabricated vertical drains (PVDs), also referred to as *wick* or *strip* drains, were originally developed as a substitute for the commonly used sand drain. With the advent of materials science, these drains began to be manufactured from synthetic polymers such as polypropylene and high-density polyethylene. PVDs are normally manufactured with a corrugated or channeled synthetic core enclosed by a geotextile filler as shown schematically in Figure 8.56. Installation rates reported in the literature are on the order of 0.1–0.3 m/s, excluding equipment mobilization and setup time. PVDs have been used extensively in the past for expedient consolidation of low-permeability soils under surface surcharge. The main advantage of

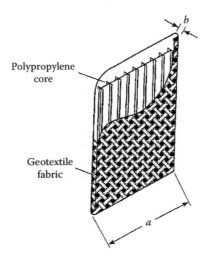

Figure 8.56 Prefabricated vertical drain (PVD).

PVDs over sand drains is that they do not require drilling; thus, installation is much faster. For rectangular flexible drains, the radius of the equivalent circles can be given as

$$r_{\text{w}} = \frac{(b+a)}{\pi} \tag{8.199}$$

where
  a is width of the drain
  b is the thickness of the drain

The relation for average degree of consolidation for *vertical drainage only* was presented in Section 8.2. Also the relations for the degree of consolidation due to *radial drainage only* were given in Sections 8.14 and 8.15. In reality, the drainage for the dissipation of excess pore water pressure takes place in both directions simultaneously. For such a case, Carrillo (1942) has shown that

$$U_{\text{v,r}} = 1 - (1 - U_{\text{v}})(1 - U_{\text{r}}) \tag{8.200}$$

where
  $U_{\text{v,r}}$ is the average degree of consolidation for simultaneous vertical and radial drainage
  U_{v} is the average degree of consolidation calculated on the assumption that only vertical drainage exists (note the notation U_{av} was used before in this chapter)
  U_{r} is the average degree of consolidation calculated on the assumption that only radial drainage exists

Example 8.21

A 6-m-thick clay layer is drained at the top and bottom and has some sand drains. The given data are C_{v} (for vertical drainage) = 49.51 × 10^{-4} m/day; $k_{\text{v}} = k_{\text{h}}$; $d_{\text{w}} = 0.45$ m; $d_{\text{e}} = 3$ m; $r_{\text{w}} = r_{\text{s}}$ (i.e., no smear at the periphery of drain wells).

It has been estimated that a given uniform surcharge would cause a total consolidation settlement of 250 mm without the sand drains. Calculate the consolidation settlement of the clay layer with the same surcharge and sand drains at time $t = 0, 0.2, 0.4, 0.6, 0.8,$ and 1 year.

Solution

Vertical drainage: $C_{\text{v}} = 49.51 \times 10^{-4}$ m/day = 1.807 m/year.

$$T_{\text{v}} = \frac{C_{\text{v}}t}{H^2} = \frac{1.807 \times t}{(6/2)^2} = 0.2008t \cong 0.2t \tag{E8.1}$$

Table 8.10 Steps in calculation of consolidation settlement

t (Year)	T_v (Equation E8.1)	U_v (Table 8.1)	$1-U_v$	T_r (Equation E8.2)	$1 - \exp[-8T_r/F(n)] = U_r$	$1-U_r$	$U_{v,r} = 1 - (1 - U_v)(1 - U_r)$	$S_c = 250 \times U_{v,r}$ (mm)
0	0	0	1	0	0	1	0	0
0.2	0.04	0.22	0.78	0.04	0.235	0.765	0.404	101
0.4	0.08	0.32	0.68	0.08	0.414	0.586	0.601	150.25
0.6	0.12	0.39	0.61	0.12	0.552	0.448	0.727	181.75
0.8	0.16	0.45	0.55	0.16	0.657	0.343	0.812	203
1	0.2	0.505	0.495	0.2	0.738	0.262	0.870	217.5

Radial drainage:

$$\frac{r_e}{r_w} = \frac{1.5 \text{ m}}{0.225 \text{ m}} = 6.67 = n$$

$$F_n = \frac{n^2}{n^2-1}\ln(n) - \frac{3n^2-1}{4n^2} \quad \text{(equal strain case)}$$

$$= \left[\frac{(6.67)^2}{(6.67)^2-1}\ln(6.67) - \frac{3(6.67)^2-1}{4(6.67)^2}\right]$$

$$= 1.94 - 0.744 = 1.196$$

Since $k_v = k_h$, $C_v = C_{vr}$. So

$$T_r = \frac{C_{vr}t}{d_e^2} = \frac{1.807 \times t}{3^2} = 0.2t \tag{E8.2}$$

The steps in the calculation of the consolidation settlement are shown in Table 8.10. From Table 8.10, the consolidation settlement at $t = 1$ year is 217.5 mm. Without the sand drains, the consolidation settlement at the end of 1 year would have been only 126.25 mm.

8.17 DESIGN CURVES FOR PREFABRICATED VERTICAL DRAINS

The relationships for the average degree of consolidation due to radial drainage into sand drains were given in Equations 8.170, 8.171, 8.186 and 8.188, and they are for equal strain cases. Yeung (1997) used these relationships to develop curves for PVDs. The theoretical developments used by Yeung are given below.

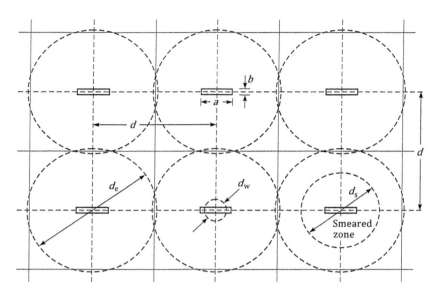

Figure 8.57 Prefabricated vertical drains—square-grid layout.

Figure 8.57 shows the layout of a square-grid pattern of prefabricated vertical drains (also see Figure 8.56 for the definition of a and b). From Equation 8.199, the equivalent diameter of a PVD can be given as,

$$d_w = \frac{2(a+b)}{\pi} \tag{8.201}$$

Now, Equation 8.188 can be rewritten as

$$U_r = 1 - \exp\left(-\frac{8C_{vr}t}{d_w^2}\frac{d_w^2}{d_e^2 m'}\right) = 1 - \exp\left(-\frac{8T_r'}{\alpha'}\right) \tag{8.202}$$

where d_e = diameter of the effective zone of drainage = $2r_e$

$$T_r' = \frac{C_{vr}t}{d_w^2} \tag{8.203}$$

$$\alpha' = n^2 m' = \frac{n^4}{n^2 - S^2}\ln\left(\frac{n}{S}\right) - \left(\frac{3n^2 - S^2}{4}\right) + \frac{k_h}{k_s}(n^2 - S^2)\ln S \tag{8.204}$$

$$n = \frac{d_e}{d_w} \tag{8.205}$$

(Also see Equation 8.166.)

From Equation 8.202,

$$T'_r = -\frac{\alpha'}{8}\ln(1 - U_r)$$

or

$$(T'_r)_1 = \frac{T'_r}{\alpha'} = -\frac{\ln(1 - U_r)}{8} \tag{8.206}$$

Table 8.11 gives the variation of $(T'_r)_1$ with U_r. Figure 8.58 shows the plot of n versus α' from Equation 8.204.

The following is a step-by-step procedure for the design of prefabricated vertical drains:

1. Assume that the magnitude of required $U_{v,r}$ and the corresponding time t is known.
2. With known time t, determine U_v.
3. From Equation 8.200, determine

$$U_r = \frac{1 - U_{v,r}}{1 - U_v} \tag{8.207}$$

4. With known U_r, use Equation 8.206 to obtain $(T'_r)_1$.
5. Determine T'_r from Equation 8.203.
6. Determine $\alpha' = T'_r/(T'_r)_1$.

Table 8.11 Variation of $(T'_r)_1$ with U_r (Equation 8.206)

U_r	$(T'_r)_1$
0.05	0.0064
0.10	0.0132
0.20	0.0279
0.30	0.0446
0.40	0.0639
0.50	0.0866
0.60	0.1145
0.70	0.1505
0.80	0.2012
0.90	0.2878
0.95	0.3745

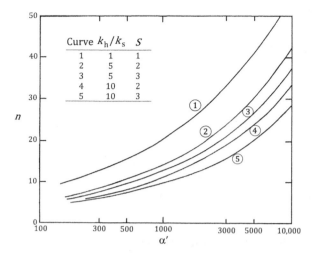

Figure 8.58 Plot of n versus α'. (Based on Yeung, A. T., *J. Geotech. Geoenv. Eng.*, ASCE, 123(8), 755–759, 1997. —Equation 8.204.)

7. Using Figure 8.58 (or a similar figure with varying k_h/k_s and S), determine n.
8. From Equation 8.205

$$d_e = n \underbrace{d_w}_{\substack{\text{Equation} \\ 8.201}}$$

9. Choose the drain spacing (see Equations 8.197 and 8.198).

$$d = \frac{d_e}{1.05} \text{ (for triangular pattern)}$$

$$d = \frac{d_e}{1.128} \text{ (for square pattern)}$$

REFERENCES

Azzouz, A. S., R. J. Krizek, and R. B. Corotis, Regression analysis of soil compressibility, *Soils Found. Tokyo*, 16(2), 19–29, 1976.

Barden, L., Consolidation of clay with non-linear viscosity, *Geotechnique*, 15(4), 345–362, 1965.

Barden, L., Primary and secondary consolidation of clay and peat, *Geotechnique*, 18, 1–14, 1968.

Barron, R. A., Consolidation of fine-grained soils by drain wells, *Trans*, ASCE, 113, 1719, 1948.

Bjerrum, L., Embankments on soft ground, in *Proceedings, Specialty Conference Performance of Earth and Earth-Supported Structures*, vol. 2, 1–54, ASCE, New York, 1972.

Brummund, W. F., E. Jonas, and C. C. Ladd, Estimating in situ maximum past (pre-consolidation) pressure of saturated clays from results of laboratory consolidometer tests, Transportation Research Board, Special Report 163, 4–12, 1976.

Burland, J. B., On the compressibility and shear strength of natural clays, *Geotechnique*, 40(3), 329–378, 1990.

Carrillo, N., Simple two- and three-dimensional cases in theory of consolidation of soils, *J. Math. Phys.*, 21(1), 1–5, 1942.

Casagrande, A., The determination of the preconsolidation load and its practical significance, *Proceedings of the First International Conference on Soil Mechanics and Foundation Engineering*, Cambridge, USA, p. 60, 1936.

Casagrande, A. and R. E. Fadum, *Notes on Soil Testing for Engineering Purposes*, Publication No. 8, Harvard University, Graduate School of Engineering, Cambridge, MA, 1940.

Crawford, C. B., Interpretation of the consolidation tests, *J. Soil Mech. Found. Div.*, ASCE, 90(SM5), 93–108, 1964.

Gibson, R. E. and K. Y. Lo, *A Theory of Consolidation for Soils Exhibiting Secondary Compression*, Publication No. 41, Norwegian Geotechnical Institute, Norway, 1961.

Hanna, D., N. Sivakugan and L. Lovisa, Simple approach to consolidation due to constant rate loading in clays, *Int. J. Geomechanics*, ASCE, New York, 13(2), 193–196, 2013.

Hansbo, S., A new approach to the determination of shear strength of clay by the Fall Cone Test, Report 14, Royal Swedish Geotechnical Institute, Stockholm, Sweden, 1957.

Jose, B. T., A. Sridharan and B. M. Abraham, Log-log method for determination of preconsolidation pressure, *Geotech. Testing J.*, ASTM, 12(3), 230–237, 1989.

Kempfert, H. G. and B. Soumaya, Settlement back-analysis of buildings on soft soil in Southern Germany, *Proc., 5th Int. Conf. Case Histories in Geotech. Eng.*, University of Missouri, Rolla, USA, 2004.

Kootahi, K., Simple index tests for assessing the recompression index of fine-grained soils, *J. Geotech. Geoenv. Eng.*, ASCE, 143(4), 06016027-1, 2017.

Kulhawy, F. F. and P. W. Mayne, *Manual on Estimating Soil Properties for Foundation Design*, EPRI, Palo Alto, CA, 1990.

Larsson, R., Undrained shear strength in stability calculation of embankments and foundations on clay, *Can. Geotech. J.*, 17, 591–602, 1980.

Leonards, G. A. and A. G. Altschaeffl, Compressibility of clay, *J. Soil Mech. Found. Div.*, ASCE, 90(SM5), 133–156, 1964.

Leroueil, S., Tenth Canadian geotechnical colloquium: Recent development in consolidation of natural clays, *Can. Geotech. J.*, 25(1), 85–107, 1988.

Lowe, J., III, E. Jonas, and V. Obrician, Controlled gradient consolidation test, *J. Soil Mech. Found. Div.*, ASCE, 95(SM1), 77–98, 1969.

Luscher, U., Discussion, *J. Soil Mech. Found. Div.*, ASCE, 91(SM1), 190–195, 1965.

Mayne, P. W. and J. B. Kemper, Profiling OCR in stiff clays by CPT and STP, *Geotech. Test. J.*, ASTM, 11(2), 139–147, 1988.

Mayne, P. W. and J. K. Mitchell, Profiling of overconsolidation ratio in clays by field vane, *Can. Geotech. J.*, 25(1), 150–157, 1988.

Mesri, G., Coefficient of secondary compression, *J. Soil Mech. Found. Div.*, ASCE, 99(SMI), 123–137, 1973.

Mesri, G. and P. M. Godlewski, Time and stress-compressibility interrelationship, *J. Geotech. Eng.*, ASCE, 103(5), 417–430, 1977.

Nacci, V. A., M. C. Wang, and K. R. Demars, Engineering behavior of calcareous soils, *Proceedings of the Civil Engineering in the Oceans III*, 1, 380–400, ASCE, New York, 1975.

Nagaraj, T. and B. S. R. Murty, Prediction of the preconsolidation pressure and recompression index of soils, *Geotech. Test. J.*, ASTM, 8(4), 199–202, 1985.

Nakase, A., T. Kamei and O. Kusakabe, Constitutive parameters estimated by plasticity index, *J. Geotech. Eng.*, ASCE, 114(7), 844–858, 1988.

Oikawa, H., Compression curve of soft soils, *Soils and Foundations*, 27(3), 99–104, 1987.

Olson, R. E., Consolidation under time-dependent loading, *J. Geotech. Eng. Div.*, ASCE, 103(GT1), 55–60, 1977.

Park, J. H. and T. Koumoto, New compression index equation, *J. Geotech. Geoenv. Eng.*, ASCE, 130(2), 223–226, 2004.

Raju, P. S. R., N. S. Pandian, and T. S. Nagaraj, Analysis and estimation of coefficient of consolidation, *Geotech. Test. J.*, ASTM, 18(2), 252–258, 1995.

Rendon-Herrero, O., Universal compression index, *J. Geotech. Eng.*, ASCE, 109(10), 1349, 1983.

Rendulic, L., Der Hydrodynamische Spannungsaugleich in Zentral Entwässerten Tonzylindern, *Wasser-wirtsch. Tech.*, 2, 250–253, 269–273, 1935.

Richart, F. E., Review of the theories for sand drains, *Trans.* ASCE, 124, 709–736, 1959.

Robinson, R. G. and M. M. Allam, Determination of coefficient of consolidation from early stage of log t plot, *Geotech. Test. J.*, ASTM, 19(3), 316–320, 1996.

Schiffman, R. L., C. C. Ladd, and A. T. Chen, *The Secondary Consolidation of Clay*, *I.U.T.A.M. Symposium on Rheological Soil Mechanics*, 273 Grenoble, France, 273, 1964.

Schmertmann, J. H., Undisturbed laboratory behavior of clay, *Trans.* ASCE, 120, 1201, 1953.

Scott, R. F., *Principles of Soil Mechanics*, Addison-Wesley, Reading, MA, 1963.

Sivakugan, N., J. Lovisa, J. Ameratunga and B. M. Das, Consolidation settlement due to ramp loading, *Int. J. Geotech. Eng.*, 8(2), 19–196, 2014.

Sivaram, B. and P. Swamee, A computational method for consolidation coefficient, *Soils Found. Tokyo*, 17(2), 48–52, 1977.

Smith, R. E. and H. E. Wahls, Consolidation under constant rate of strain, *J. Soil Mech. Found. Div.*, ASCE, 95(SM2), 519–538, 1969.

Sridharan, A. and K. Prakash, Improved rectangular hyperbola method for the determination of coefficient of consolidation, *Geotech. Test. J.*, ASTM, 8(1), 37–40, 1985.

Sridharan, A. and K. Prakash, $\delta - t/\delta$ Method for the determination of coefficient of consolidation, *Geotech. Test. J.*, ASTM, 16(1), 131–134, 1993.

Stas, C. V. and F. H. Kulhawy, Critical evaluation of design methods for foundations under axial uplift and compression loading, Report EL-3771, EPRI, Palo Alto, CA, 1984.

Su, H. L., Procedure for rapid consolidation test, *J. Soil Mech. Found. Div.*, ASCE, 95, Proc. Pap. 1729, 1958.

Tan, T. K., Discussion, *Proc. 4th Int. Conf. Soil Mech. Found. Eng.*, vol. 3, p. 278, 1957.

Taylor, D. W., *Research on Consolidation of Clays*, Publication No. 82, Massachusetts Institute of Technology, Cambridge, MA, 1942.

Taylor, D. W. and W. Merchant, A theory of clay consolidation accounting for secondary compression, *J. Math. Phys.*, 19, 167, 1940.

Terzaghi, K., *Erdbaumechanik auf Boden-physicalischen Grundlagen*, Deuticke, Vienna, Austria, 1925.

Terzaghi, K., *Theoretical Soil Mechanics*, Wiley, New York, 1943.

Terzaghi, K. and R. B. Peck, *Soil Mechanics in Engineering Practice*, 2nd edn., Wiley, New York, 1967.

Umar, M. and A. Sadrekarimi, Accuracy of determining pre-consolidation pressure from laboratory tests, *Can. Geotech. J.*, 54(3), 441–450, 2017.

Wroth, C. P. and D. M. Wood, The correlation of index properties with some basic engineering properties of soils, *Can. Geotech. J.*, 15(2), 137–145, 1978.

Yeung, A.T., Design curves for prefabricated vertical drains, *J. Geotech. Geoenv. Eng.*, ASCE, 123(8), 755–759, 1997.

Chapter 9

Shear strength of soils

9.1 INTRODUCTION

The shear strength of soils is an important aspect in many foundation engineering problems such as the bearing capacity of shallow foundations and piles, the stability of the slopes of dams and embankments, and lateral earth pressure on retaining walls. In this chapter, we will discuss the shear strength characteristics of granular and cohesive soils and the factors that control them.

9.2 MOHR–COULOMB FAILURE CRITERION

In 1900, Mohr presented a theory for rupture in materials. According to this theory, failure along a plane in a material occurs by a critical combination of normal and shear stresses, and not by normal or shear stress alone. The functional relation between normal and shear stress on the failure plane can be given by

$$s = f(\sigma) \tag{9.1}$$

where
  s is the shear stress at failure
  σ is the normal stress on the failure plane

The failure envelope defined by Equation 9.1 is a curved line, as shown in Figure 9.1.
  In 1776, Coulomb defined the function $f(\sigma)$ as

$$s = c + \sigma \tan \phi \tag{9.2}$$

where
  c is cohesion
  ϕ is the angle of friction of the soil

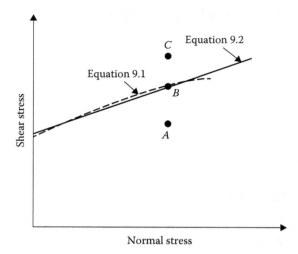

Figure 9.1 Mohr–Coulomb failure criterion.

Equation 9.2 is generally referred to as the Mohr–Coulomb failure criterion. The significance of the failure envelope can be explained using Figure 9.1. If the normal and shear stresses on a plane in a soil mass are such that they plot as point A, shear failure will not occur along that plane. Shear failure along a plane will occur if the stresses plot as point B, which falls on the failure envelope. A state of stress plotting as point C cannot exist, since this falls above the failure envelope; shear failure would have occurred before this condition was reached.

In saturated soils, the stress carried by the soil solids is the effective stress, and so Equation 9.2 must be modified:

$$s = c + (\sigma - u)\tan\phi = c + \sigma'\tan\phi \tag{9.3}$$

where
   u is the pore water pressure
   σ' is the effective stress on the plane

The term ϕ is also referred to as the drained friction angle. For sand, inorganic silts, and normally consolidated clays, $c \approx 0$. The value of c is greater than zero for overconsolidated clays.

The shear strength parameters of granular and cohesive soils will be treated separately in this chapter.

9.3 SHEARING STRENGTH OF GRANULAR SOILS

According to Equation 9.3, the shear strength of a soil can be defined as $s = c + \sigma' \tan \phi$. For granular soils with $c = 0$,

$$s = \sigma' \tan \phi \tag{9.4}$$

The determination of the friction angle ϕ is commonly accomplished by one of two methods: the direct shear test or the triaxial test. The test procedures are given later.

9.3.1 Direct shear test

A schematic diagram of the direct shear test equipment is shown in Figure 9.2. Basically, the test equipment consists of a metal shear box into which the soil specimen is placed. The specimen can be square or circular in plan, about 19–25 cm² in area, and about 25 mm in height. The box is split horizontally into two halves. Normal force on the specimen is applied from the top of the shear box by dead weights. The normal stress on the specimens obtained by the application of dead weights can be as high as 1000 kN/m². Shear force is applied to the side of the top half of the box to cause failure in the soil specimen. (The two porous stones shown in Figure 9.2 are not required for tests on dry soil.) During the test, the shear displacement of the top half of the box and the change in specimen thickness are recorded by the use of horizontal and vertical dial gauges.

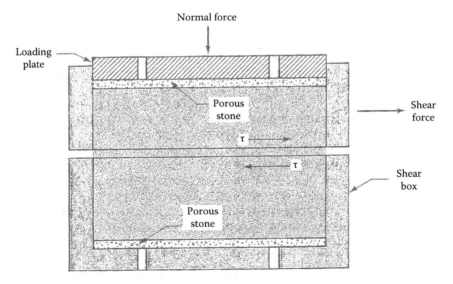

Figure 9.2 Direct shear test arrangement.

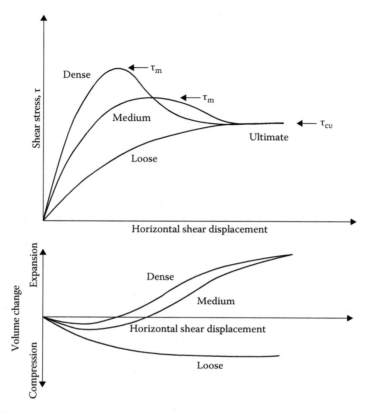

Figure 9.3 Direct shear test results in loose, medium, and dense sands.

Figure 9.3 shows the nature of the results of typical direct shear tests in loose, medium, and dense sands. Based on Figure 9.3, the following observations can be made:

1. In dense and medium sands, shear stress increases with shear displacement to a maximum or peak value τ_m and then decreases to an approximately constant value τ_{cv} at large shear displacements. This constant stress τ_{cv} is the ultimate shear stress.
2. For loose sands, the shear stress increases with shear displacement to a maximum value and then remains constant.
3. For dense and medium sands, the volume of the specimen initially decreases and then increases with shear displacement. At large values of shear displacement, the volume of the specimen remains approximately constant.
4. For loose sands, the volume of the specimen gradually decreases to a certain value and remains approximately constant thereafter.

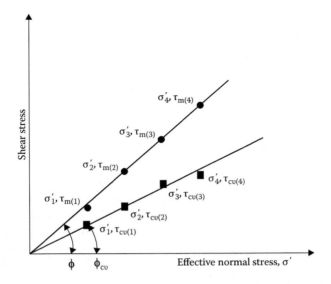

Figure 9.4 Determination of peak and ultimate friction angles from direct shear tests.

If dry sand is used for the test, the pore water pressure u is equal to zero, and so the total normal stress σ is equal to the effective stress σ'. The test may be repeated for several normal stresses. The angle of friction ϕ for the sand can be determined by plotting a graph of the maximum or peak shear stresses versus the corresponding normal stresses, as shown in Figure 9.4. The Mohr–Coulomb failure envelope can be determined by drawing a straight line through the origin and the points representing the experimental results. The slope of this line will give the peak friction angle ϕ of the soil. Similarly, the ultimate friction angle ϕ_{cv} can be determined by plotting the ultimate shear stresses τ_{cv} versus the corresponding normal stresses, as shown in Figure 9.4. The ultimate friction angle ϕ_{cv} represents a condition of shearing at constant volume of the specimen. For loose sands, the peak friction angle is approximately equal to the ultimate friction angle.

If the direct shear test is being conducted on a saturated granular soil, time between the application of the normal load and the shearing force should be allowed for drainage from the soil through the porous stones. Also, the shearing force should be applied at a slow rate to allow complete drainage. Since granular soils are highly permeable, this will not pose a problem. If complete drainage is allowed, the excess pore water pressure is zero, and so $\sigma = \sigma'$.

Some typical values of ϕ and ϕ_{cv} for granular soils are given in Table 9.1. Typical values of peak friction angle ϕ for granular soils suggested by U.S. Navy (1986) are also shown in Figure 9.5.

The strains in the direct shear test take place in two directions, that is, in the vertical direction and in the direction parallel to the applied horizontal

Table 9.1 Typical values of φ and φ$_{cv}$ for granular soils

Type of soil	φ (deg)	φ$_{cv}$ (deg)
Sand: Round grains		
Loose	28–30	
Medium	30–35	26–30
Dense	35–38	
Sand: Angular grains		
Loose	30–35	
Medium	35–40	30–35
Dense	40–45	
Sandy gravel	34–48	33–36

shear force. This is similar to the *plane strain condition*. There are some inherent shortcomings of the direct shear test. The soil is forced to shear in a predetermined plane—that is, the horizontal plane—which is not necessarily the weakest plane. Second, there is an unequal distribution of stress over the shear surface. The stress is greater at the edges than at the center. This type of stress distribution results in progressive failure (Figure 9.6).

In the past, several attempts were made to improve the direct shear test. To that end, the Norwegian Geotechnical Institute developed a *simple shear test device*, which involves enclosing a cylindrical specimen in a rubber membrane reinforced with wire rings. As in the direct shear test, as the end plates move, the specimen distorts, as shown in Figure 9.7a. Although it is an improvement over the direct shear test, the shearing stresses are not uniformly distributed on the specimen. Pure shear as shown in Figure 9.7b only exists at the center of the specimen.

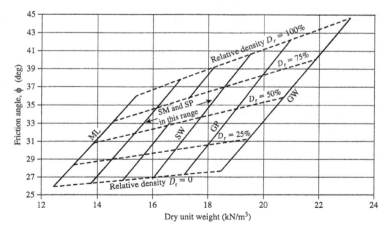

Figure 9.5 Friction angles of granular soils. (Based on U.S. Navy, *Soil Mechanic-Design Manual 7.1*, Department of the Navy, Naval facilities Engineering Command, U.S. Government Printing Press, Washington, DC, 1986.)

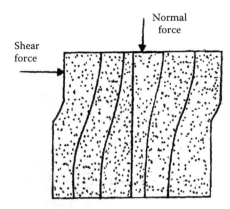

Figure 9.6 Unequal stress distribution in direct shear equipment.

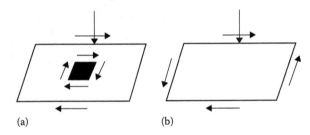

(a)   (b)

Figure 9.7 (a) Simple shear and (b) pure shear.

9.3.2 Triaxial test

A schematic diagram of triaxial test equipment is shown in Figure 9.8. In this type of test, a soil specimen about 38 mm in diameter and 76 mm in length is generally used. The specimen is enclosed inside a thin rubber membrane and placed inside a cylindrical plastic chamber. For conducting the test, the chamber is usually filled with water or glycerin. The specimen is subjected to a confining pressure σ_3 by application of pressure to the fluid in the chamber. (Air can sometimes be used as a medium for applying the confining pressure.) Connections to measure drainage into or out of the specimen or pressure in the pore water are provided. To cause shear failure in the soil, an axial stress $\Delta\sigma$ is applied through a vertical loading ram. This is also referred to as deviator stress. The axial strain is measured during the application of the deviator stress. For determination of ϕ, dry or fully saturated soil can be used. If saturated soil is used, the drainage connection is kept open during the application of the confining pressure and the deviator stress. Thus, during the test, the excess pore water pressure in

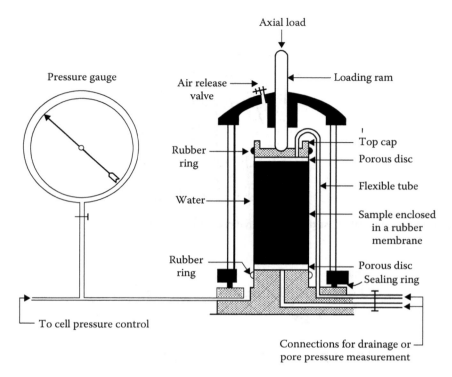

Figure 9.8 Triaxial test equipment. (After Bishop, A. W. and Bjerrum, L., The relevance of the triaxial test to the solution of stability problems, in *Proc. Res. Conf. Shear Strength of Cohesive Soils*, ASCE, pp. 437–501, 1960.)

the specimen is equal to zero. The volume of the water drained from the specimen during the test provides a measure of the volume change of the specimen.

For *drained tests*, the total stress is equal to the effective stress. Thus, the major effective principal stress is $\sigma_1' = \sigma_1 = \sigma_3 + \Delta\sigma$; the minor effective principal stress is $\sigma_3' = \sigma_3$; and the intermediate effective principal stress is $\sigma_2' = \sigma_3'$.

At failure, the major effective principal stress is equal to $\sigma_3 + \Delta\sigma_f$, where $\Delta\sigma_f$ is the deviator stress at failure, and the minor effective principal stress is σ_3. Figure 9.9b shows the nature of the variation of $\Delta\sigma$ with axial strain for loose and dense granular soils. Several tests with similar specimens can be conducted by using different confining pressures σ_3. The value of the soil peak friction angle ϕ can be determined by plotting effective-stress Mohr's circles for various tests and drawing a common tangent to these Mohr's circles passing through the origin. This is shown in

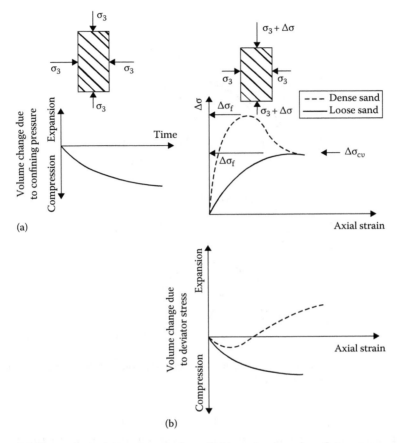

Figure 9.9 Drained triaxial test in granular soil: (a) application of confining pressure and (b) application of deviator stress.

Figure 9.10a. The angle that this envelope makes with the normal stress axis is equal to ϕ. It can be seen from Figure 9.10b that

$$\sin\phi = \frac{\overline{ab}}{\overline{oa}} = \frac{(\sigma_1' - \sigma_3')/2}{(\sigma_1' + \sigma_3')/2}$$

or

$$\phi = \sin^{-1}\left(\frac{\sigma_1' - \sigma_3'}{\sigma_1' + \sigma_3'}\right)_{\text{failure}} \tag{9.5}$$

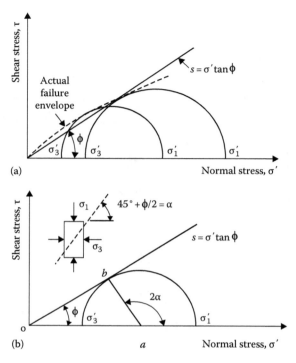

Figure 9.10 Drained triaxial test results: (a) determination of soil friction angle ϕ; (b) derivation of Equation 9.5.

However, it must be pointed out that in Figure 9.10a the failure envelope defined by the equation $s = \sigma'\tan\phi$ is an approximation to the actual curved failure envelope. The ultimate friction angle ϕ_{cv} for a given test can also be determined from the equation

$$\phi_{cv} = \sin^{-1}\left[\frac{\sigma'_{1(cv)} - \sigma'_3}{\sigma'_{1(cv)} + \sigma'_3}\right] \tag{9.6}$$

where $\sigma'_{1(cv)} = \sigma'_3 + \Delta\sigma_{(cv)}$. For similar soils, the friction angle ϕ determined by triaxial tests is slightly lower (0°–3°) than that obtained from direct shear tests.

The axial compression triaxial test described earlier is of the conventional type. However, the loading process on the specimen in a triaxial chamber can be varied in several ways. In general, the tests can be divided into two major groups: axial compression tests and axial extension tests. The following is a brief outline of each type of test (refer to Figure 9.11).

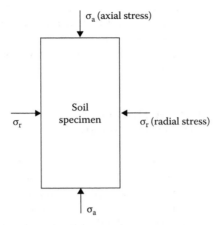

Figure 9.11 Soil specimen subjected to axial and radial stresses.

9.3.3 Axial compression tests

1. Radial confining stress σ_r constant and axial stress σ_a increased. This is the test procedure described earlier.
2. Axial stress σ_a constant and radial confining stress σ_r decreased.
3. Mean principal stress constant and radial stress decreased.

For drained compression tests, σ_a is equal to the major effective principal stress σ'_1, and σ_r is equal to the minor effective principal stress σ'_3, which is equal to the intermediate effective principal stress σ'_2. For the test listed under item 3, the mean principal stress, that is, $(\sigma'_1 + \sigma'_2 + \sigma'_3)/3$, is kept constant. Or, in other words, $\sigma'_1 + \sigma'_2 + \sigma'_3 = J = \sigma_a + 2\sigma_r$ is kept constant by increasing σ_a and decreasing σ_r.

9.3.4 Axial extension tests

1. Radial stress σ_r kept constant and axial stress σ_a decreased.
2. Axial stress σ_a constant and radial stress σ_r increased.
3. Mean principal stress constant and radial stress increased.

For all *drained* extension tests at failure, σ_a is equal to the minor effective principal stress σ'_3, and σ_r is equal to the major effective principal stress σ'_1, which is equal to the intermediate effective principal stress σ'_2.

The detailed procedures for conducting these tests are beyond the scope of this text, and readers are referred to Bishop and Henkel (1969). Several investigations have been carried out to compare the peak friction angles determined by the axial compression tests to those obtained by the axial extension tests. A summary of these investigations is given by Roscoe et al. (1963). Some investigators found no difference in the value of φ from compression and extension tests; however, others reported values of φ determined from the extension tests that were several degrees greater than those obtained by the compression tests.

9.4 RELEVANCE OF LABORATORY TESTS TO FIELD CONDITIONS

Various types of laboratory tests described in Section 9.3 will provide soil friction angles for a given soil which may be slightly different from each other. In the field, under a given structure, different soil elements will be subjected to different boundary conditions. These soil elements may be identified with the various types of laboratory tests and, hence, friction angles thus obtained (also see Kulhawy and Mayne, 1990).

In order to illustrate the above factor, refer to Figure 9.12 which shows the failure surface in soil supporting a circular shallow foundation. The friction angle for the soil element at A can be more appropriately represented by that obtained from *triaxial compression tests*. Similarly, the soil elements at B and C can be represented by the friction angles obtained respectively from *direct simple shear tests* (or direct shear tests) and *triaxial*

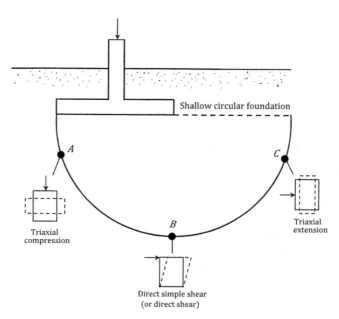

Figure 9.12 Soil elements along the failure surface under a shallow circular foundation.

extension tests. For practical design considerations, however, an average value of the friction angle is considered.

9.5 CRITICAL VOID RATIO

We have seen that for shear tests in dense sands, there is a tendency of the specimen to dilate as the test progresses. Similarly, in loose sand, the volume gradually decreases (Figures 9.3 and 9.9). An increase or decrease of volume means a change in the void ratio of soil. The nature of the change of the void ratio with strain for loose and dense sands is shown in Figure 9.13a. The void ratio for which the change of volume remains constant during shearing is called the *critical void ratio.* Figure 9.13b shows the results of some drained triaxial tests on washed Fort Peck sand. The void ratio after the application of σ_3 is plotted in the ordinate, and the change of volume, ΔV, at the peak point of the stress–strain plot is plotted along the abscissa. For a given σ_3, the void ratio corresponding to $\Delta V = 0$ is the critical void ratio. Note that the critical void ratio is a function of the confining pressure σ_3. It is, however, necessary to recognize that whether the volume of the soil specimen is increasing or decreasing, the critical void ratio is reached only in the shearing zone, even if it is generally calculated on the basis of the total volume change of the specimen.

The concept of critical void ratio was first introduced in 1938 by A. Casagrande to study liquefaction of granular soils. When a natural deposit of saturated sand that has a void ratio greater than the critical void ratio is subjected to a sudden shearing stress (due to an earthquake or to blasting, for example), the sand will undergo a decrease in volume. This will result in an increase in pore water pressure u. At a given depth, the effective stress is given by the relation $\sigma' = \sigma - u$. If σ (i.e., the total stress) remains constant and u increases, the result will be a decrease in σ'. This, in turn, will reduce the shear strength of the soil. If the shear strength is reduced to a value which is less than the applied shear stress, the soil will fail. This is called soil liquefaction. An advanced study of soil liquefaction can be obtained from the work of Seed and Lee (1966).

9.6 CURVATURE OF THE FAILURE ENVELOPE

It was shown in Figure 9.1 that Mohr's failure envelope (Equation 9.1) is actually curved, and the shear strength equation ($s = c + \sigma \tan \phi$) is only a straight-line approximation for the sake of simplicity. For a drained direct shear test on sand, $\phi = \tan^{-1}(\tau_{max}/\sigma')$. Since Mohr's envelope is actually curved, a higher effective normal stress will yield lower values of ϕ. This fact is demonstrated in Figure 9.14, which is a plot of the results of

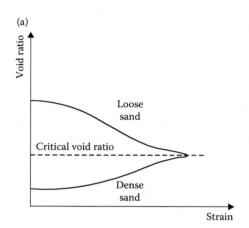

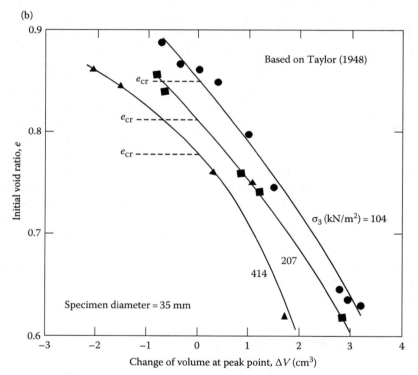

Figure 9.13 (a) definition of critical void ratio; (b) critical void ratio from triaxial test on Fort Peck sand.

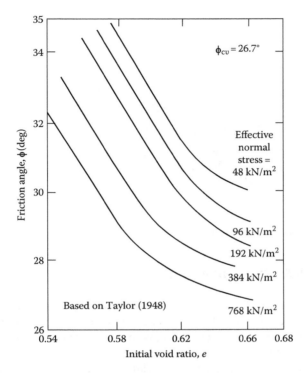

Figure 9.14 Variation of peak friction angle, ϕ, with effective normal stress on standard Ottawa sand.

direct shear tests on standard Ottawa Sand. For loose sand, the value of ϕ decreases from about 30° to less than 27° when the normal stress is increased from 48 to 768 kN/m². Similarly, for dense sand (initial void ratio approximately 0.56), ϕ decreases from about 36° to about 30.5° due to a 16-fold increase of σ′.

For high values of confining pressure (greater than about 400 kN/m²), Mohr's failure envelope sharply deviates from the assumption given by Equation 9.3. This is shown in Figure 9.15. Skempton (1960, 1961) introduced the concept of *angle of intrinsic friction* for a formal relation between shear strength and effective normal stress. Based on Figure 9.15, the shear strength can be defined as

$$s = k + \sigma' \tan \psi \tag{9.7}$$

where ψ is the angle of intrinsic friction. For quartz, Skempton (1961) gave the values of $k \approx 950$ kN/m² and $\psi \approx 13°$.

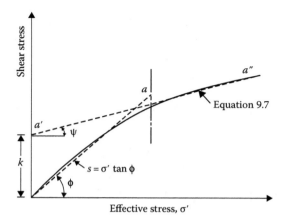

Figure 9.15 Failure envelope at high confining pressure.

9.7 GENERAL COMMENTS ON ϕ_{cv}

As shown in Figure 9.15, the failure envelope in a plot of shear stress at failure and effective normal stress is curved. So, the friction angle ϕ derived from laboratory tests on a dense sand is the secant friction angle $(\phi = \phi_{secant})$ as shown in Figure 9.16(a). At the same time, for a loose sand, $(\phi = \phi_{cv})$. Figure 9.16(b) shows the nature of variation of ϕ_{secant} and ϕ_{cv} with normal effective stress σ'. The difference between ϕ_{secant} and ϕ_{cv} is due to dilation.

Bolten (1986) has analyzed the strength test results of 17 clean sand. Based on this work, he proposed the following relationship to quantify dilatency, or

$$\phi_{secant} - \phi_{cv} = 5I_R \text{ (for plane strain compression)} \tag{9.8}$$

and

$$\phi_{secant} - \phi_{cv} = 3I_R \text{ (for triaxial compression)} \tag{9.9}$$

where

$$I_R = \text{relative dilatency index} = D_r \left[Q - \ln\left(\frac{\sigma'_f}{p_a}\right) \right] - R \tag{9.10}$$

where
   D_r = relative density (fraction)
   p_a = atmospheric pressure (≈ 100 kN/m²)

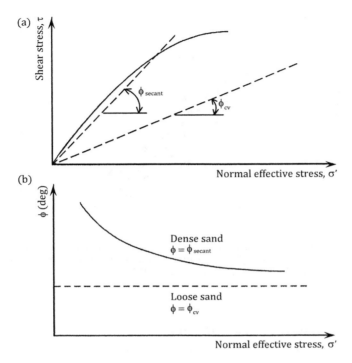

Figure 9.16 (a) Failure envelope of loose and dense sand; (b) Nature of variation of ϕ_{secant} and ϕ_{cv}. (Based on Kulhawy, F. H. and P. W. Mayne, *Manual on Estimating Soil Properties in Foundation Design*, Electric Power Research Institute, Palo Alto, CA, 1990.)

σ'_f = mean effective principal stresses = $\dfrac{\sigma'_1 + \sigma'_2 + \sigma'_3}{3}$

σ'_1, σ'_2, σ'_3 = major, intermediate, and minor principal effective stresses
Q = compressibility coefficient (≈ 10 for quartz and feldspar)
R = filling coefficient (≈ 1 for quartz and feldspar sands)

Hence, for quartz or feldspar sand under triaxial compression test,

$$\phi_{secant} - \phi_{cv} = 3\left\{ D_r\left[10 - \ln\left(100\frac{\sigma'_f}{p_a} \right)\right] - 1 \right\} \tag{9.11}$$

Figure 9.17 compares Equation 9.11 for triaxial compression test results on quartz and feldspar sand ($D_r = 0.8$ and 0.5).
In addition, based on several laboratory test results, Koerner (1970) proposed a relationship for ϕ_{cv} on single mineral soil in the form

$$\phi_{cv} = 36° + \Delta\phi_1 + \Delta\phi_2 + \Delta\phi_3 + \Delta\phi_4 + \Delta\phi_5 \tag{9.12}$$

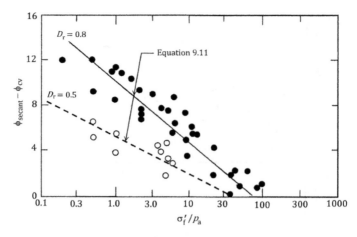

Figure 9.17 Comparison of Equation 9.11 with triaxial compression test results on quartz and feldspar sand. [Redrawn after Bolten, M. D., The strength and dilatancy of sands, *Geotechnique*, 36(1), 65–75, 1986.]

where

$$\Delta\phi_1 = \text{correction for particle shape}$$
$$= -6° \text{ for high sphericity and subrounded shape}$$
$$= +2° \text{ for low sphericity and angular shape}$$

$$\Delta\phi_2 = \text{correction for particle size (Note: } D_{10} = \text{effective size)}$$
$$= -11° \text{ for } D_{10} > 2 \text{ mm (gravel)}$$
$$= -9° \quad \text{for } 2.0 > D_{10} > 0.6 \text{ (coarse sand)}$$
$$= -4° \quad \text{for } 0.6 > D_{10} > 0.2 \text{ (medium sand)}$$
$$= 0 \quad \text{for } 0.2 > D_{10} > 0.06 \text{ (fine sand)}$$

$$\Delta\phi_3 = \text{correction for gradation (Note: uniformity coefficient } C_u)$$
$$= -2° \text{ for } C_u > 2 \text{ (well graded)}$$
$$= -1° \text{ for } C_u = 2 \text{ (medium graded)}$$
$$= 0 \text{ for } C_u < 2 \text{ (poorly graded)}$$

$$\Delta\phi_4 = \text{correction for relative density, } D_r$$
$$= -1° \text{ for } 0 < D_r < 0.5 \text{ (loose)}$$
$$= 0 \quad \text{for } 0.5 < D_r < 0.75 \text{ (intermediate)}$$
$$= 4° \quad \text{for } 0.75 < D_r < 1.0 \text{ (dense)}$$

$\Delta\phi_5$ = correction for mineral type

 = 0 for quartz

 = +4° for feldspar, calcite, chlorite

 = +6° for muscovite mica

9.8 SHEAR STRENGTH OF GRANULAR SOILS UNDER PLANE STRAIN CONDITIONS

The results obtained from triaxial tests are widely used for the design of structures. However, under structures such as continuous wall footings, the soils are actually subjected to a plane strain type of loading, that is, the strain in the direction of the intermediate principal stress is equal to zero. Several investigators have attempted to evaluate the effect of plane strain type of loading (Figure 9.18) on the angle of friction of granular soils. A summary of the results obtained was compiled by Lee (1970). To differentiate the plane strain drained friction angle from the triaxial drained friction angle, the following notations have been used in the discussion in this section:

ϕ_p = drained friction angle obtained from plane strain tests

ϕ_t = drained friction angle obtained from triaxial tests

Lee (1970) also conducted some drained shear tests on a uniform sand collected from the Sacramento River near Antioch, California. Drained triaxial tests were conducted with specimens of diameter 35.56 mm and height

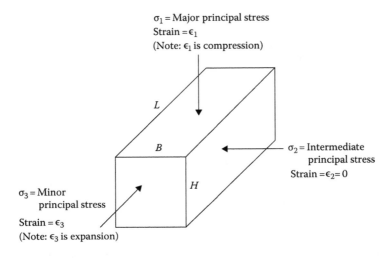

σ_1 = Major principal stress
Strain = ϵ_1
(Note: ϵ_1 is compression)

L

B

σ_2 = Intermediate principal stress
Strain = ϵ_2 = 0

H

σ_3 = Minor principal stress

Strain = ϵ_3

(Note: ϵ_3 is expansion)

Figure 9.18 Plane strain condition.

86.96 mm. Plane strain tests were carried out with rectangular specimens 60.96 mm high and 27.94 × 71.12 mm in cross-sectional area. The plane strain condition was obtained by the use of two lubricated rigid side plates. Loading of the plane strain specimens was achieved by placing them inside a triaxial chamber. All specimens, triaxial and plane strain, were aniso-tropically consolidated with a ratio of major to minor principal stress of 2:

$$k_c = \frac{\sigma_1'(\text{consolidation})}{\sigma_3'(\text{consolidation})} = 2 \qquad (9.13)$$

The results of this study are instructive and are summarized in the following:

1. For loose sand having a relative density of 38%, at low confining pressure, ϕ_p and ϕ_t were determined to be 45° and 38°, respectively. Similarly, for medium-dense sand having a relative density of 78%, ϕ_p and ϕ_t were 48° and 40°, respectively.
2. At higher confining pressure, the failure envelopes (plane strain and triaxial) flatten, and the slopes of the two envelopes become the same.
3. Figure 9.19 shows the results of the initial tangent modulus, E, for various confining pressures. For given values of σ_3', the initial tangent modulus for plane strain loading shows a higher value than that for triaxial loading, although in both cases, E increases exponentially with the confining pressure.
4. The variation of Poisson's ratio v with the confining pressure for plane strain and triaxial loading conditions is shown in Figure 9.20. The values of v were calculated by measuring the change of the volume of specimens and the corresponding axial strains during loading. The

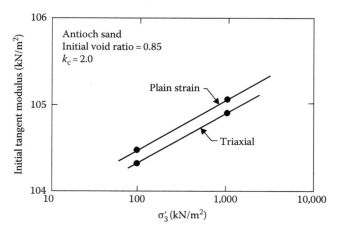

Figure 9.19 Initial tangent modulus from drained tests on Antioch sand. [After Lee, K. L., *J. Soil Mech. Found. Div.*, ASCE, 96(SM3), 901, 1970.]

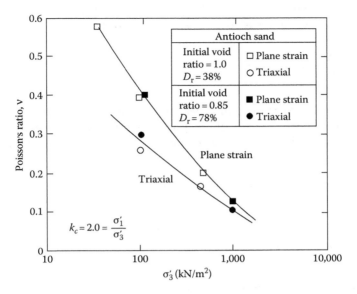

The table shown in the figure:

Antioch sand	
Initial void ratio = 1.0 $D_r = 38\%$	□ Plane strain O Triaxial
Initial void ratio = 0.85 $D_r = 78\%$	■ Plane strain ● Triaxial

$$k_c = 2.0 = \frac{\sigma_1'}{\sigma_3'}$$

Figure 9.20 Poisson's ratio from drained tests on Antioch sand. [After Lee, K. L., J. Soil Mech. Found. Div., ASCE, 96(SM3), 901, 1970.]

derivation of the equations used for finding v can be explained with the aid of Figure 9.18. Assuming compressive strain to be positive, for the stresses shown in Figure 9.18

$$\Delta H = H \, \epsilon_1 \qquad (9.14)$$

$$\Delta B = B \, \epsilon_2 \qquad (9.15)$$

$$\Delta L = L \, \epsilon_3 \qquad (9.16)$$

where
H, L, B are the height, length, and width of the specimen
$\Delta H, \Delta B, \Delta L$ are the changes in height, length, and width of specimen due to application of stresses
$\epsilon_1, \epsilon_2, \epsilon_3$ are the strains in the direction of major, intermediate, and minor principal stresses

The volume of the specimen before load application is equal to $V = LBH$, and the volume of the specimen after the load application is equal to $V - \Delta V$. Thus

$$\Delta V = V - (V - \Delta V) = LBH - (L - \Delta L)(B - \Delta B)(H - \Delta H)$$

$$= LBH - LBH(1 - \epsilon_1)(1 - \epsilon_2)(1 - \epsilon_3) \qquad (9.17)$$

where ΔV is change in volume. Neglecting the higher-order terms such as $\epsilon_1\epsilon_2$, $\epsilon_2\epsilon_3$, $\epsilon_3\epsilon_1$, and $\epsilon_1\epsilon_2\epsilon_3$, Equation 9.17 gives

$$\upsilon = \frac{\Delta V}{V} = \epsilon_1 + \epsilon_2 + \epsilon_3 \tag{9.18}$$

where υ is the change in volume per unit volume of the specimen.

For triaxial tests, $\epsilon_2 = \epsilon_3$, and they are expansions (negative sign). So, $\epsilon_2 = \epsilon_3 = -v\,\epsilon_1$. Substituting this into Equation 9.18, we get $\upsilon = \epsilon_1\,(1 - 2v)$, or

$$v = \frac{1}{2}\left(1 - \frac{\upsilon}{\epsilon_1}\right) \text{ (for triaxial test conditions)} \tag{9.19}$$

With plane strain loading conditions, $\epsilon_2 = 0$ and $\epsilon_3 = -v\,\epsilon_1$. Hence, from Equation 9.18, $\upsilon = \epsilon_1\,(1 - v)$, or

$$v = 1 - \frac{\upsilon}{\epsilon_1} \text{ (for plane strain conditions)} \tag{9.20}$$

Figure 9.20 shows that for a given value of σ_3', Poisson's ratio obtained from plane strain loading is higher than that obtained from triaxial loading.

On the basis of the available information at this time, it can be concluded that ϕ_p exceeds the value of ϕ_t by 0°–8°. The greatest difference is associated with dense sands at low confining pressures. The smaller differences are associated with loose sands at all confining pressures, or dense sand at high confining pressures. Although still disputed, several suggestions have been made to use a value of $\phi \approx \phi_P = 1.1\phi_t$, for calculation of the bearing capacity of strip foundations. For rectangular foundations, the stress conditions on the soil cannot be approximated by either triaxial or plane strain loadings. Meyerhof (1963) suggested for this case that the friction angle to be used for calculation of the ultimate bearing capacity should be approximated as

$$\phi = \left(1.1 - 0.1\frac{B_f}{L_f}\right)\phi_t \tag{9.21}$$

where
L_f is the length of foundation
B_f the width of foundation

After considering several experiment results, Lade and Lee (1976) gave the following approximate relations:

$$\phi_p = 1.5\phi_t - 17 \quad \phi_t > 34° \tag{9.22}$$

$$\phi_p = \phi_t \quad\quad \phi_i \leq 34° \tag{9.23}$$

9.9 SHEAR STRENGTH OF COHESIVE SOILS

The shear strength of cohesive soils can generally be determined in the laboratory by either direct shear test equipment or triaxial shear test equipment; however, the triaxial test is more commonly used. Only the shear strength of saturated cohesive soils will be treated here. The shear strength based on the effective stress can be given by (Equation 9.3) $s = c + \sigma' \tan \phi$. For normally consolidated clays, $c \approx 0$, and for overconsolidated clays, $c > 0$.

The basic features of the triaxial test equipment are shown in Figure 9.8. Three conventional types of tests are conducted with clay soils in the laboratory:

1. Consolidated drained test or drained test (CD test or D test)
2. Consolidated undrained test (CU test)
3. Unconsolidated undrained test (UU test)

Each of these tests will be separately considered in the following sections.

9.9.1 Consolidated drained test

For the consolidated drained test, the saturated soil specimen is first subjected to a confining pressure σ_3 through the chamber fluid; as a result, the pore water pressure of the specimen will increase by u_c. The connection to the drainage is kept open for complete drainage, so that u_c becomes equal to zero. Then, the deviator stress (piston stress) $\Delta\sigma$ is increased at a very slow rate, keeping the drainage valve open to allow complete dissipation of the resulting pore water pressure u_d. Figure 9.21 shows the nature of the variation of the deviator stress with axial strain. From Figure 9.21, it must also be pointed out that, during the application of the deviator stress, the volume of the specimen gradually reduces for normally consolidated clays. However, overconsolidated clays go through some reduction of volume initially but then expand. In a consolidated drained test, the total stress is equal to the effective stress, since the excess pore water pressure is zero. At failure, the maximum *effective* principal stress is $\sigma_1' = \sigma_1 = \sigma_3 + \Delta\sigma_f$, where $\Delta\sigma_f$ is the deviator stress at failure. The minimum effective principal stress is $\sigma_3' = \sigma_3$.

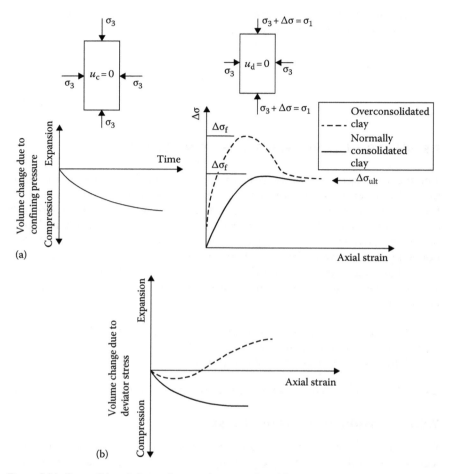

Figure 9.21 Consolidated drained triaxial test in clay: (a) application of confining pressure and (b) application of deviator stress.

From the results of a number of tests conducted using several specimens, Mohr's circles at failure can be plotted as shown in Figure 9.22. The values of c and ϕ are obtained by drawing a common tangent to Mohr's circles, which is the Mohr–Coulomb envelope. For normally consolidated clays (Figure 9.22a), we can see that $c = 0$. Thus, the equation of the Mohr–Coulomb envelope can be given by $s = \sigma' \tan \phi$. The slope of the failure envelope will give us the angle of friction of the soil. As shown by Equation 9.5, for these soils

$$\sin \phi = \left(\frac{\sigma_1' - \sigma_3'}{\sigma_1' + \sigma_3'} \right)_{failure} \quad \text{or} \quad \sigma_1' = \sigma_3' \tan^2 \left(45° + \frac{\phi}{2} \right)$$

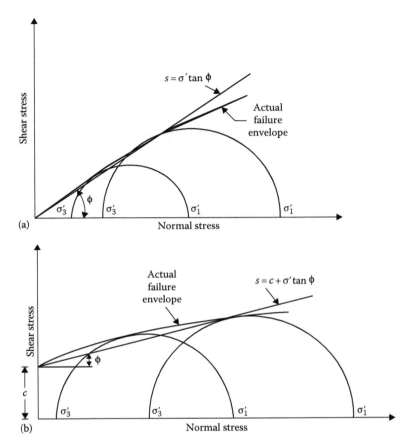

Figure 9.22 Failure envelope for (a) normally consolidated and (b) overconsolidated clays from consolidated drained triaxial tests.

Figure 9.23 shows a modified form of Mohr's failure envelope of pure clay minerals. Note that it is a plot of $(\sigma_1' - \sigma_3')_{\text{failure}}/2$ versus $(\sigma_1' + \sigma_3')_{\text{failure}}/2$.

For overconsolidated clays (Figure 9.22b), $c \neq 0$. So, the shear strength follows the equation $s = c + \sigma' \tan \phi$. The values of c and ϕ can be determined by measuring the intercept of the failure envelope on the shear stress axis and the slope of the failure envelope, respectively. To obtain a general relation between σ_1', σ_3', c, and ϕ, we refer to Figure 9.24 from which

$$\sin\phi = \frac{\overline{ac}}{\overline{bO}+\overline{Oa}} = \frac{(\sigma_1' - \sigma_3')/2}{c \cot\phi + (\sigma_1' + \sigma_3')/2}$$

$$\sigma_1'(1-\sin\phi) = 2c \cos\phi + \sigma_3'(1+\sin\phi) \qquad (9.24)$$

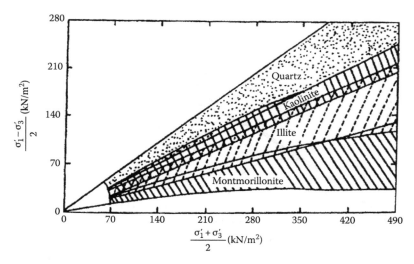

Figure 9.23 Modified Mohr's failure envelope for quartz and clay minerals. [After Olson, R. E., *J. Geotech. Eng. Div.*, ASCE, 100(GTII), 1215, 1974.]

or

$$\sigma'_1 = \sigma'_3 \frac{1+\sin\phi}{1-\sin\phi} + \frac{2c\cos\phi}{1-\sin\phi}$$

$$\sigma'_1 = \sigma'_3 \tan^2\left(45° + \frac{\phi}{2}\right) + 2c\,\tan\left(45° + \frac{\phi}{2}\right) \tag{9.25}$$

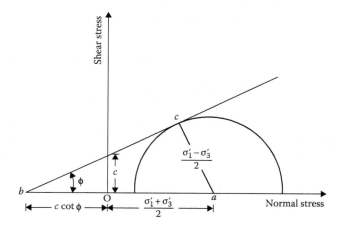

Figure 9.24 Derivation of Equation 9.25.

Note that the plane of failure makes an angle of 45° + φ/2 with the major principal plane.

If a clay is initially consolidated by an encompassing chamber pressure of $\sigma_c = \sigma_c'$ and allowed to swell under a reduced chamber pressure of $\sigma_3 = \sigma_3'$, the specimen will be overconsolidated. The failure envelope obtained from consolidated drained triaxial tests of these types of specimens has two distinct branches, as shown in Figure 9.25. Portion *ab* of the failure envelope has a flatter slope with a cohesion intercept, and portion *bc* represents a normally consolidated stage following the equation $s = \sigma' \tan \phi_{bc}$.

It may also be seen from Figure 9.21 that at very large strains the deviator stress reaches a constant value. The shear strength of clays at very large strains is referred to as *residual shear strength* (i.e., the ultimate shear strength). It has been proved that the residual strength of a given soil is independent of past stress history, and it can be given by the equation (see Figure 9.26)

$$s_{residual} = \sigma' \tan \phi_{ult} \tag{9.26}$$

(i.e., the *c* component is 0). For triaxial tests

$$\phi_{ult} = \sin^{-1}\left(\frac{\sigma_1' - \sigma_3'}{\sigma' + \sigma_3'}\right)_{residual} \tag{9.27}$$

where $\sigma_1' = \sigma_3' + \Delta\sigma_{ult}$.

The residual friction angle in clays is of importance in subjects such as the long-term stability of slopes.

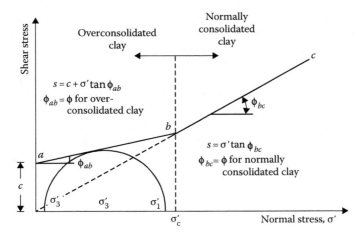

Figure 9.25 Failure envelope of a clay with preconsolidation pressure of σ_c'.

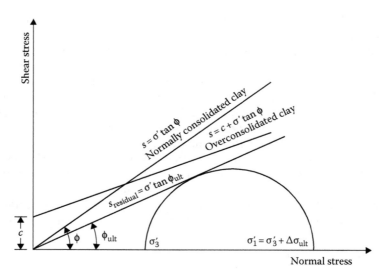

Figure 9.26 Residual shear strength of clay.

The consolidated drained triaxial test procedure described earlier is of the conventional type. However, failure in the soil specimens can be produced by any one of the methods of axial compression or axial extension as described in Section 9.3 (with reference to Figure 9.11), allowing full drainage condition.

9.9.2 Consolidated undrained test

In the consolidated undrained test, the soil specimen is first consolidated by a chamber-confining pressure σ_3; full drainage from the specimen is allowed. After complete dissipation of excess pore water pressure, u_c, generated by the confining pressure, the deviator stress $\Delta\sigma$ is increased to cause failure of the specimen. During this phase of loading, the drainage line from the specimen is closed. Since drainage is not permitted, the pore water pressure (pore water pressure due to deviator stress u_d) in the specimen increases. Simultaneous measurements of $\Delta\sigma$ and u_d are made during the test. Figure 9.27 shows the nature of the variation of $\Delta\sigma$ and u_d with axial strain; also shown is the nature of the variation of the pore water pressure parameter A ($A = u_d/\Delta\sigma$; see Equation 5.11) with axial strain. The value of A at failure, A_f, is positive for normally consolidated clays and becomes negative for overconsolidated clays (also see Table 5.2). Thus, A_f is dependent on the overconsolidation ratio (OCR). The OCR for triaxial test conditions may be defined as

$$\text{OCR} = \frac{\sigma_c'}{\sigma_3} \tag{9.28}$$

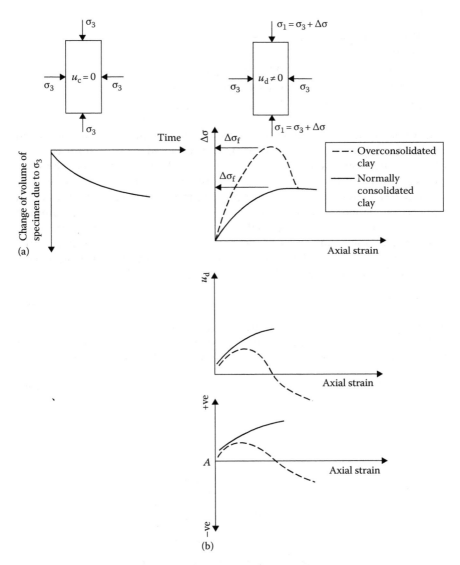

Figure 9.27 Consolidated undrained triaxial test: (a) application of confining pressure and (b) application of deviator stress.

where $\sigma'_c = \sigma_c$ is the maximum chamber pressure at which the specimen is consolidated and then allowed to rebound under a chamber pressure of σ_3.

The typical nature of the variation of A_f with the OCR for Weald clay is shown in Figure 5.11.

At failure

Total major principal stress $= \sigma_1 = \sigma_3 + \Delta\sigma_f$
Total minor principal stress $= \sigma_3$
Pore water pressure at failure $= u_{d(failure)} = A_f \Delta\sigma_f$
Effective major principal stress $= \sigma_1 - A_f \Delta\sigma_f = \sigma_1'$
Effective minor principal stress $= \sigma_3 - A_f \Delta\sigma_f = \sigma_3'$

Consolidated undrained tests on a number of specimens can be conducted to determine the shear strength parameters of a soil, as shown for the case of a normally consolidated clay in Figure 9.28. The total-stress Mohr's circles (circles A and B) for two tests are shown by dashed lines. The effective-stress Mohr's circles C and D correspond to the total-stress circles A and B, respectively. Since C and D are effective-stress circles at failure, a common tangent drawn to these circles will give the Mohr–Coulomb failure envelope given by the equation $s = \sigma' \tan \phi$. If we draw a common tangent to the total-stress circles, it will be a straight line passing through the origin. This is the total-stress failure envelope, and it may be given by

$$s = \sigma \tan \phi_{cu} \tag{9.29}$$

where ϕ_{cu} is the consolidated undrained angle of friction.

The total-stress failure envelope for an overconsolidated clay will be of the nature shown in Figure 9.29 and can be given by the relation

$$s = c_{cu} + \sigma \tan \phi_{cu} \tag{9.30}$$

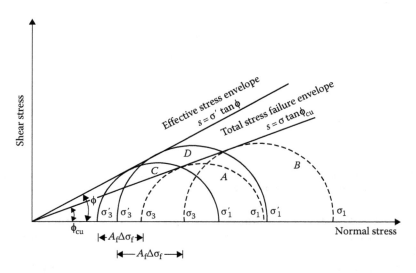

Figure 9.28 Consolidated undrained test results—normally consolidated clay.

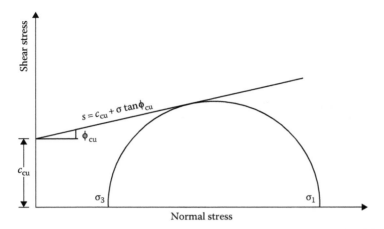

Figure 9.29 Consolidated undrained test—total stress envelope for overconsolidated clay.

where c_{cu} is the intercept of the total-stress failure envelope along the shear stress axis.

The shear strength parameters for overconsolidated clay based on effective stress, that is, c and ϕ, can be obtained by plotting the effective-stress Mohr's circle and then drawing a common tangent.

As in consolidated drained tests, shear failure in the specimen can be produced by axial compression or extension by changing the loading conditions.

9.9.3 Unconsolidated undrained test

In unconsolidated undrained triaxial tests, drainage from the specimen is not allowed at any stage. First, the chamber-confining pressure σ_3 is applied, after which the deviator stress $\Delta\sigma$ is increased until failure occurs. For these tests,

Total major principal stress $= \sigma_3 + \Delta\sigma_f = \sigma_1$
Total minor principal stress $= \sigma_3$

Tests of this type can be performed quickly, since drainage is not allowed. For a saturated soil, the deviator stress failure, $\Delta\sigma_f$, is practically the same, irrespective of the confining pressure σ_3 (Figure 9.30). So the total-stress failure envelope can be assumed to be a horizontal line, and $\phi = 0$. The undrained shear strength can be expressed as

$$s = S_u = \frac{\Delta\sigma_f}{2} \tag{9.31}$$

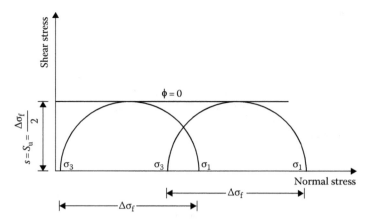

Figure 9.30 Unconsolidated undrained triaxial test.

This is generally referred to as the shear strength based on the $\phi = 0$ concept.

The fact that the strength of saturated clays in unconsolidated und-rained loading conditions is the same, irrespective of the confining pressure σ_3 can be explained with the help of Figure 9.31. If a saturated clay specimen A is consolidated under a chamber-confining pressure of σ_3 and then sheared to failure under undrained conditions, Mohr's circle at failure will be represented by circle no. 1. The effective-stress Mohr's circle

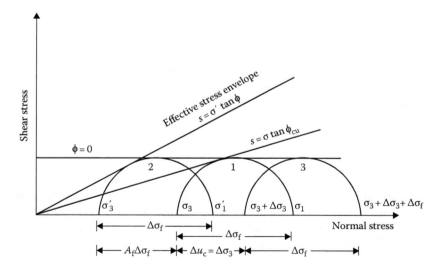

Figure 9.31 Effective and total stress Mohr's circles for unconsolidated undrained tests.

corresponding to circle no. 1 is circle no. 2, which touches the effective-stress failure envelope. If a similar soil specimen B, consolidated under a chamber-confining pressure of σ_3, is subjected to an additional confining pressure of $\Delta\sigma_3$ without allowing drainage, the pore water pressure will increase by Δu_c. We saw in Chapter 5 that $\Delta u_c = B\Delta\sigma_3$ and, for saturated soils, $B = 1$. So, $\Delta u_c = \Delta\sigma_3$.

Since the effective confining pressure of specimen B is the same as specimen A, it will fail with the same deviator stress, $\Delta\sigma_f$. The total-stress Mohr's circle for this specimen (i.e., B) at failure can be given by circle no. 3. So, at failure, for specimen B,

Total minor principal stress $= \sigma_3 + \Delta\sigma_3$
Total minor principal stress $= \sigma_3 + \Delta\sigma_3 + \Delta\sigma_f$

The effective stresses for the specimen are as follows:

$$\text{Effective major principal stress} = (\sigma_3 + \Delta\sigma_3 + \Delta\sigma_f) - (\Delta u_c + A_f\Delta\sigma_f)$$

$$= (\sigma_3 + \Delta\sigma_f) - A_f\Delta\sigma_f$$

$$= \sigma_1 - A_f\Delta\sigma_f = \sigma_1'$$

$$\text{Effective minor principal stress} = (\sigma_3 + \Delta\sigma_3) - (\Delta u_c + A_f\Delta\sigma_f)$$

$$= \sigma_3 - A_f\Delta\sigma_f = \sigma_3'$$

The aforementioned principal stresses are the same as those we had for specimen A. Thus, the effective-stress Mohr's circle at failure for specimen B will be the same as that for specimen A, that is, circle no. 1.

The value of $\Delta\sigma_3$ could be of any magnitude in specimen B; in all cases, $\Delta\sigma_f$ would be the same.

Example 9.1

Consolidated drained triaxial tests on two specimens of a soil gave the following results:

Test no.	Confining pressure σ_3 (kN/m²)	Deviator stress at failure $\Delta\sigma_f$ (kN/m²)
1	70	440.4
2	92	474.7

Determine the values of c and ϕ for the soil.

Solution

From Equation 9.25, $\sigma_1 = \sigma_3 \tan^2 (45° + \phi/2) + 2c \tan (45° + \phi/2)$.
For test 1, $\sigma_3 = 70 \text{ kN/m}^2$; $\sigma_1 = \sigma_3 + \Delta\sigma_f = 70 + 440.4 = 510.4 \text{ kN/m}^2$. So,

$$510.4 = 70 \tan^2 \left(45° + \frac{\phi}{2} \right) + 2c \tan \left(45° + \frac{\phi}{2} \right) \tag{E9.1}$$

Similarly, for test 2, $\sigma_3 = 92 \text{ kN/m}^2$; $\sigma_1 = 92 + 474.7 = 566.7 \text{ kN/m}^2$.
Thus

$$566.7 = 92 \tan^2 \left(45° + \frac{\phi}{2} \right) + 2c \tan \left(45° + \frac{\phi}{2} \right) \tag{E9.2}$$

Subtracting Equation E9.1 from Equation E9.2 gives

$$56.3 = 22 \tan^2 \left(45° + \frac{\phi}{2} \right)$$

$$\phi = 2 \left[\tan^{-1} \left(\frac{56.3}{22} \right)^{1/2} - 45° \right] = 26°$$

Substituting $\phi = 26°$ in Equation E9.1 gives

$$c = \frac{510.4 - 70 \tan^2 (45° + 26/2)}{2 \tan (45° + 25/2)} = \frac{510.4 - 70(2.56)}{2(1.6)} = 103.5 \text{ kN/m}^2$$

Example 9.2

A normally consolidated clay specimen was subjected to a consolidated undrained test. At failure, $\sigma_3 = 100 \text{ kN/m}^2$, $\sigma_1 = 204 \text{ kN/m}^2$, and $u_d = 50 \text{ kN/m}^2$. Determine ϕ_{cu} and ϕ.

Solution

Referring to Figure 9.32

$$\sin\phi_{cu} = \frac{\overline{ab}}{Oa} = \frac{(\sigma_1 - \sigma_3)/2}{(\sigma_1 + \sigma_3)/2} = \frac{\sigma_1 - \sigma_3}{\sigma_1 + \sigma_3} = \frac{204 - 100}{204 + 100} = \frac{104}{304}$$

Hence

$$\phi_{cu} = 20°$$

Again

$$\sin\phi = \frac{\overline{cd}}{Oc} = \frac{\sigma'_1 - \sigma'_3}{\sigma'_1 + \sigma'_3}$$

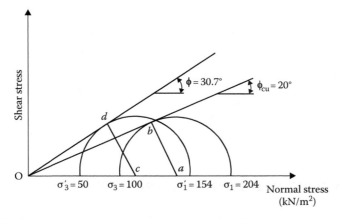

Figure 9.32 Total- and effective-stress Mohr's circles.

$$\sigma_3' = 100 - 50 = 50 \text{ kN/m}^2$$

$$\sigma_1' = 204 - 50 = 154 \text{ kN/m}^2$$

So

$$\sin \phi = \frac{154 - 50}{154 + 54} = \frac{104}{204}$$

Hence

$$\phi = 30.7°$$

Example 9.3

Consider the normally consolidated clay described in Example 9.2 (i.e., $\phi_{cu} = 20°$ and $\phi = 30.7°$). If a consolidated undrained triaxial test is conducted on such a clay specimen with $\sigma_3 = 80 \text{ kN/m}^2$, what will be σ_1 and u_d at failure?

Solution

$$\sin \phi_{cu} = \frac{\sigma_1 - \sigma_3}{\sigma_1 + \sigma_3}$$

or

$$\sin 20 = \frac{\sigma_1 - 80}{\sigma_1 + 80}; \ \sigma_1 = 163.16 \text{ kN/m}^2$$

Again,

$$\sin\phi = \frac{\sigma_1' - \sigma_3'}{\sigma_1' + \sigma_3'} = \frac{(\sigma_1 - u_d) - (\sigma_3 - u_d)}{(\sigma_1 - u_d) + (\sigma_3 - u_d)} = \frac{\sigma_1 - \sigma_3}{\sigma_1 + \sigma_3 - 2u_d}$$

Hence,

$$\sin 30.7° = \frac{163.16 - 80}{(163.16 + 80) - 2u_d}$$

$$u_d = 41.71\,\text{kN/m}^2$$

9.10 UNCONFINED COMPRESSION TEST

The unconfined compression test is a special case of the unconsolidated undrained triaxial test. In this case, no confining pressure to the specimen is applied (i.e., $\sigma_3 = 0$). For such conditions, for saturated clays, the pore water pressure in the specimen at the beginning of the test is negative (capillary pressure). Axial stress on the specimen is gradually increased until the specimen fails (Figure 9.33). At failure, $\sigma_3 = 0$ and so

$$\sigma_1 = \sigma_3 + \Delta\sigma_f = \Delta\sigma_f = q_u \tag{9.32}$$

where q_u is the unconfined compression strength.

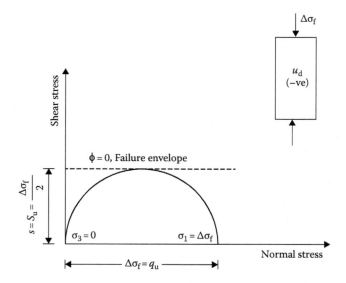

Figure 9.33 Unconfined compression strength.

Table 9.2 Consistency and unconfined
compression strength of clays

Consistency	q_u (kN/m²)
Very soft	0–24
Soft	24–48
Medium	48–96
Stiff	96–192
Very stiff	192–383
Hard	>383

Theoretically, the value of $\Delta\sigma_f$ of a saturated clay should be the same as that obtained from unconsolidated undrained tests using similar specimens. Thus, $s = S_u = q_u/2$. However, this seldom provides high-quality results.

The general relation between consistency and unconfined compression strength of clays is given in Table 9.2.

9.11 MODULUS OF ELASTICITY AND POISSON'S RATIO FROM TRIAXIAL TESTS

For calculation of soil settlement and distribution of stress in a soil mass, it may be required to know the magnitudes of the modulus of elasticity and Poisson's ratio of soil. These values can be determined from a triaxial test. Figure 9.34 shows a plot of $\sigma_1' - \sigma_3'$ versus axial strain $\in$ for a triaxial test, where σ_3 is kept constant. The definitions of the initial tangent modulus E_i and

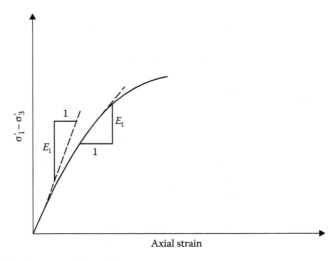

Figure 9.34 Definition of E_i and E_t.

the tangent modulus E_t at a certain stress level are also shown in the figure. Janbu (1963) showed that the initial tangent modulus can be estimated as

$$E_i = Kp_a \left(\frac{\sigma_3'}{p_a} \right)^n \tag{9.33}$$

where
σ_3' is the minor effective principal stress
p_a is the atmospheric pressure (same pressure units E_i and σ_3')
K is the modulus number
n is the exponent determining the rate of variation of E_i with σ_3'

For a given soil, the magnitudes of K and n can be determined from the results of a number of triaxial tests and then plotting E_i versus σ_3' on log–log scales. The magnitude of K for various soils usually falls in the range of 300–2000. Similarly, the range of n is between 0.3 and 0.6.

The tangent modulus E_t can be determined as

$$E_t = \frac{\partial(\sigma_1' - \sigma_3')}{\partial \in} \tag{9.34}$$

Duncan and Chang (1970) showed that

$$E_t = \left[1 - \frac{R_f(1 - \sin\phi)(\sigma_1' - \sigma_3')}{2c\cos\phi + 2\sigma_3'\sin\phi} \right]^2 Kp_a \left(\frac{\sigma_3'}{p_a} \right)^n \tag{9.35}$$

where R_f is the failure ratio. For most soils, the magnitude of R_f falls between 0.75 and 1.

For *drained conditions*, Trautman and Kulhawy (1987) suggested that

$$K \approx 300 + 900 \left(\frac{\phi° - 25°}{20°} \right) \tag{9.36}$$

The approximate values of n and R_f are as follows (Kulhawy et al., 1983):

Soil type	n	R_f
GW	1/3	0.7
SP	1/3	0.8
SW	1/2	0.7
SP	1/2	0.8
ML	2/3	0.8

The value of Poisson's ratio (ν) can be determined by the same type of triaxial test (i.e., σ_3 constant) as

$$\nu = \frac{\Delta\epsilon_a - \Delta\epsilon_v}{2\Delta\epsilon_a} \tag{9.37}$$

where
$\Delta\epsilon_a$ is the increase in axial strain
$\Delta\epsilon_v$ is the volumetric strain $= \Delta\epsilon_a + 2\Delta\epsilon_r$
$\Delta\epsilon_r$ is the lateral strain

So

$$\nu = \frac{\Delta\epsilon_a - (\Delta\epsilon_a + 2\Delta\epsilon_r)}{2\Delta\epsilon_a} = -\frac{\Delta\epsilon_r}{\Delta\epsilon_a} \tag{9.38}$$

For undrained loading of saturated cohesive soil

$\nu = 0.5$

For drained conditions, Poisson's ratio may be approximated as (Trautman and Kulhawy, 1987)

$$\nu = 0.1 + 0.3\left(\frac{\phi^\circ - 25^\circ}{20^\circ}\right) \tag{9.39}$$

9.12 FRICTION ANGLES ϕ AND ϕ_{ult}

The drained angle of friction ϕ of normally consolidated clays generally increases with the plasticity index (PI) of soil. This fact is illustrated in Figure 9.35 for a number of clays from data compiled by Sorensen and Okkels (2013). From this plot,

$$\phi = 43 - 10\log PI \quad \text{(mean)} \tag{9.40}$$

and

$$\phi = 39 - 11\log PI \quad \text{(lower bound)} \tag{9.41}$$

Based on tests conducted over 30 years on clays obtained in Denmark, Sorensen and Okkels (2013) gave the following correlations for overconsolidated clays.

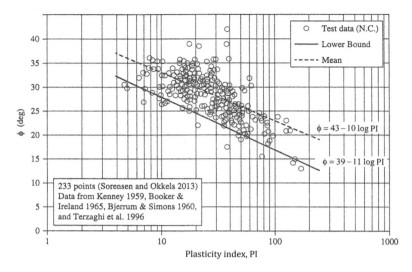

Figure 9.35 Variation of φ with plasticity index (PI) for several normally consolidated clays. (Based on Sorensen, K. K. and N. Okkels, Correlation between drained shear strength and plasticity index of undisturbed overconsolidated clays, Proc., *18th Int. Conf. Soil Mech. Geotech. Eng.*, Paris, Presses des Ponts, 1, 423–428, 2013.)

Mean value of φ:

$$\phi \text{ (deg)} = 45 - 14 \log PI \text{ (for } 4 < PI < 50) \tag{9.42}$$

$$\phi \text{ (deg)} = 26 - 3 \log PI \text{ (for } 50 \leq PI < 150) \tag{9.43}$$

Lower bound value of φ:

$$\phi \text{ (deg)} = 44 - 14 \log PI \text{ (for } 4 < PI < 50) \tag{9.44}$$

$$\phi \text{ (deg)} = 30 - 6 \log PI \text{ (for } 50 \leq PI < 150) \tag{9.45}$$

Lower bound value of *c*:

$$c \text{ (kN/m}^2) = 30 \text{ (for } 7 < PI < 30) \tag{9.46}$$

$$c \text{ (kN/m}^2) = 48 - 0.6 \, PI \text{ (for } 30 \leq PI < 80) \tag{9.47}$$

$$c \text{ (kN/m}^2) = 0 \text{ (for } PI > 80) \tag{9.48}$$

Castellanos and Brandon (2013) showed that φ for *undisturbed riverine and lacustrine alluvial deposits* determined from a consolidated undrained

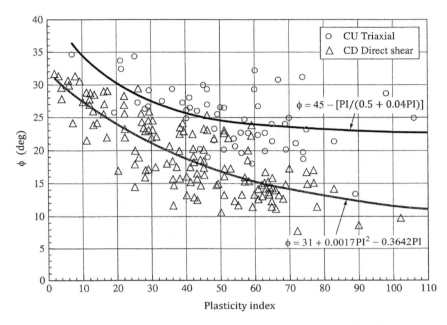

Figure 9.36 Variation of ϕ for undisturbed riverine and lacustrine alluvial soils determined from triaxial and direct shear tests against plasticity index. (Based
on Castellanos, B. A. and T. L. Brandon, A comparison between the shear
strength measured with direct shear and triaxial devices on undisturbed and
remoulded soils, *Proc. 18th Int. Conf. Soil Mech. Geotech. Eng.*, Paris, Presses
des Ponts, 1, 317–320, 2013.)

triaxial test is significantly greater than ϕ determined from a consolidated
drained direct shear test (Figure 9.36).

Figure 9.37 shows the variation of the magnitude of ϕ_{ult} for several clays
with the percentage of clay-size fraction present. ϕ_{ult} gradually decreases
with the increase of clay-size fraction. At very high clay content, ϕ_{ult}
approached the value of ϕ_μ (angle of sliding friction) for sheet minerals.
For highly plastic sodium montmorillonites, the value of ϕ_{ult} can be as low
as 3°–4°.

Stark and Eid (1994) evaluated the residual friction angle of 32 clays and
clay shales using the torsional ring shear tests. Based on those tests, the
effect of clay mineralogy on ϕ_{ult} is shown in Figure 9.38a. It can be seen that
ϕ_{ult} decreases with increasing liquid limit; also ϕ_{ult} decreases with increasing activity. Figure 9.38a also shows that the drained residual failure envelope can be nonlinear. The relation between ϕ_{ult} (secant residual friction
angle) and liquid limit of clays with varying clay-size fractions is shown
in Figure 9.38b. From this figure, it appears that there is a definite relation
between ϕ_{ult}, liquid limit, and the clay-size fraction.

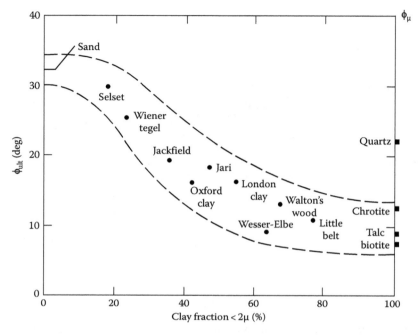

Figure 9.37 Variation of ϕ_{ult} with percentage of clay content. (After Skempton, A. W., *Geotechnique*, 14, 77, 1964.)

9.13 EFFECT OF RATE OF STRAIN ON THE UNDRAINED SHEAR STRENGTH

Casagrande and Wilson (1949, 1951) studied the problem of the effect of rate of strain on the undrained shear strength of saturated clays and clay shales. The time of loading ranged from 1 to 10^4 min. Using a time of loading of 1 min as the reference, the undrained strength of some clays decreased by as much as 20%. The nature of the variation of the undrained shear strength and time to cause failure, t, can be approximated by a straight line in a plot of S_u versus $\log t$, as shown in Figure 9.39. Based on this, Hvorslev (1960) gave the following relation:

$$S_{u(t)} = S_{u(a)}\left[1 - \rho_a \log\left(\frac{t}{t_a}\right)\right] \tag{9.49}$$

where

$S_{u(t)}$ is the undrained shear strength with time, t, to cause failure
$S_{u(a)}$ is the undrained shear strength with time, t_a, to cause failure
ρ_a is the coefficient for decrease of strength with time

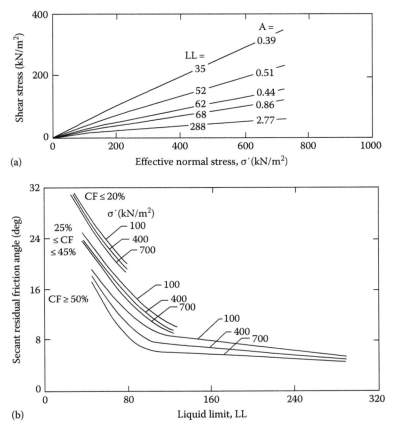

Figure 9.38 (a) Drained failure envelopes; (b) plot of secant residual friction angle versus liquid limit. *Note:* LL, liquid limit; A, activity; CF, clay size fraction. [Redrawn after Stark, T. D. and Eid, H. T., *J. Geotech. Eng. Div.*, ASCE, 120(5), 856, 1994.]

In view of the time duration, Hvorslev suggested that the reference time be taken as 1000 min. In that case

$$S_{u(t)} = S_{u(m)} \left[1 - \rho_m \log\left(\frac{t \text{ min}}{1000 \text{ min}} \right) \right] \tag{9.50}$$

where
$S_{u(m)}$ is the undrained shear strength at time 1000 min
ρ_m is the coefficient for decrease of strength with reference time of 1000 min

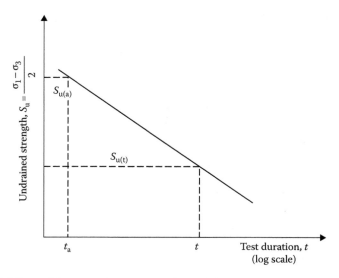

Figure 9.39 Effect of the rate of strain on undrained shear strength.

The relation between ρ_a in Equation 9.49 and ρ_m in Equation 9.50 can be given by

$$\rho_m = \frac{\rho_a}{1 - \rho_a \log[(1000 \text{ min})/(t_a \text{ min})]} \tag{9.51}$$

For $t_a = 1$ min, Equation 9.38 gives

$$\rho_m = \frac{\rho_1}{1 - 3\rho_1} \tag{9.52}$$

Hvorslev's analysis of the results of Casagrande and Wilson (1951) yielded the following results: general range $\rho_1 = 0.04-0.09$ and $\rho_m = 0.05-0.13$; Cucaracha clay-shale $\rho_1 = 0.07-0.19$ and $\rho_m = 0.09-0.46$. The study of the strength–time relation of Bjerrum et al. (1958) for a normally consolidated marine clay (consolidated undrained test) yielded a value of ρ_m in the range 0.06–0.07.

9.14 EFFECT OF TEMPERATURE ON THE UNDRAINED SHEAR STRENGTH

A number of investigations have been conducted to determine the effect of temperature on the shear strength of saturated clay. Most studies indicate

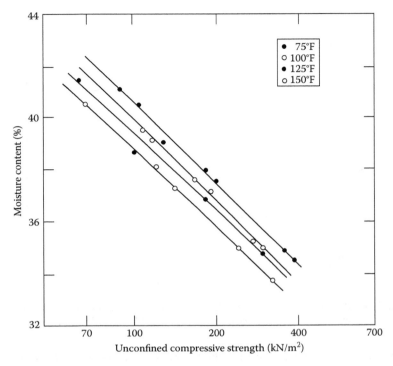

Figure 9.40 Unconfined compression strength of kaolinite—effect of temperature. (After Sherif, M. A. and Burrous, C. M. Temperature effect on the unconfined shear strength of saturated cohesive soils, Special Report 103, Highway Research Board, 267–272, 1969.)

that a decrease in temperature will cause an increase in shear strength. Figure 9.40 shows the variation of the unconfined compression strength ($q_u = 2S_u$) of kaolinite with temperature. Note that for a given moisture content, the value of q_u decreases with increase of temperature. A similar trend has been observed for San Francisco Bay mud (Mitchell, 1964), as shown in Figure 9.41. The undrained shear strength ($S_u = (\sigma_1 - \sigma_3)/2$) varies linearly with temperature. The results are for specimens with equal mean effective stress and similar structure. From these tests

$$\frac{dS_u}{dT} \approx 0.59 \text{ kN/(m}^2 \text{ °C)} \tag{9.53}$$

Kelly (1978) also studied the effect of temperature on the undrained shear strength of some undisturbed marine clay samples and commercial illite and montmorillonite. Undrained shear strengths at 4°C and 20°C

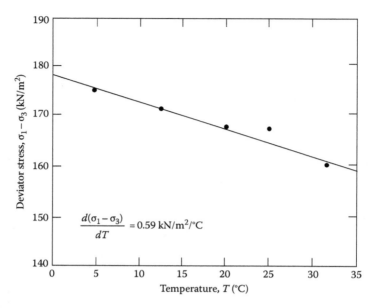

Figure 9.41 Effect of temperature on shear strength of San Francisco Bay mud. (After Mitchell, J. K., *J. Soil Mech. Found. Eng. Div.*, ASCE, 90(SM1), 29, 1964.)

were determined. Based on the laboratory test results, Kelly proposed the following correlation:

$$\frac{\Delta S_u}{\Delta T} = 0.0102 + 0.00747 S_{u(average)} \tag{9.54}$$

where

$$S_{u(average)} = (S_{u(4°C)} + S_{u(20°C)})/2 \text{ in } kN/m^2 \tag{9.55}$$

T is the temperature in °C

Example 9.4

The following are the results of an unconsolidated undrained test: $\sigma_3 = 70$ kN/m², $\sigma_1 = 210$ kN/m². The temperature of the test was 12°C. Estimate the undrained shear strength of the soil at a temperature of 20°C.

Solution

$$S_{u(12°C)} = \frac{\sigma_1 - \sigma_3}{2} = \frac{210 - 70}{2} = 70 \text{ kN/m}^2$$

From Equation 9.54

$$\Delta S_u = \Delta T[0.0102 + 0.00747 S_{u(average)}]$$

Now

$$\Delta T = 20 - 12 = 8°C$$

and

$$\Delta S_u = 8[0.0102 + 0.00747(70)] = 4.26 \text{ kN/m}^2$$

Hence

$$S_{u(20°C)} = 70 - 4.26 = 65.74 \text{ kN/m}^2$$

9.15 CORRELATION FOR UNDRAINED SHEAR STRENGTH OF REMOLDED CLAY

Several correlations have been suggested in the past for the *undrained shear strength of remolded clay* $(S_u = S_{ur})$, and some are given in Table 9.3. It is important to point out that these relationships should be used as an approximation only. O'Kelly (2013) has also shown that S_{ur} at a moisture content w can be estimated as

$$\log S_{ur} = (1 - W_{LN})\left[\log\left(\frac{S_{ur(A)}}{S_{ur(B)}}\right)\right] + \log S_{ur(B)} \tag{9.56}$$

where

$S_{ur(A)}$ = undrained shear strength at moisture content w_A
$S_{ur(B)}$ = undrained shear strength at moisture content w_B

$$W_{LN} = \frac{\log w - \log w_A}{\log w_B - \log w_A} \tag{9.57}$$

Table 9.3 Correlations for S_{ur} (kN/m²)

Investigator	Relationship
Leroueil et al. (1983)	$S_{ur} = \dfrac{1}{[(LI) - 0.21]^2}$
Hirata et al. (1990)	$S_{ur} = \exp[-3.36(LI) + 0.376]$
Terzaghi et al. (1996)	$S_{ur} = 2(LI)^{-2.8}$
Yang et al. (2006)	$S_{ur} = 159.6\exp[-3.97(LI)]$

Note: LI = liquidity index

Example 9.5

Consider a saturated remolded clay soil. Given:

Liquid limit = 48
Plastic limit = 23
Moisture content of soil = 43%

Estimate the undrained shear strength S_{ur} using the equations of Leroueil et al. (1983) and Terzaghi et al. (1996) given in Table 9.3.

Solution

The liquidity index,

$$LI = \frac{w - PL}{LL - PL} = \frac{43 - 23}{48 - 23} = 0.8$$

• Leroueil et al. (1983):

$$S_{ur} = \frac{1}{[(LI) - 0.21]^2} = \frac{1}{(0.8 - 0.21)^2} = 2.87 \text{ kN/m}^2$$

• Terzaghi et al. (1996):

$$S_{ur} = 2(LI)^{-2.8} = (1)(0.8)^{-2.8} = 3.74 \text{ kN/m}^2$$

Example 9.6

Consider a remolded saturated clay. Given:

Moisture content, w (%)	Undrained shear strength, S_{ur} (kN/m²)
68	4.68
54	10.68

Estimate the undrained shear strength S_{ur} when the moisture content is 40%. Use Equation 9.56.

Solution

Given:

• @ $w_A = 68\%$, the value of $S_{ur(A)} = 4.86$ kN/m²
• @ $w_B = 54\%$, the value of $S_{ur(B)} = 10.68$ kN/m²
• $w = 40\%$

From Equation 9.57

$$W_{LN} = \frac{\log w - \log w_A}{\log w_B - \log w_A} = \frac{\log(40) - \log(68)}{\log(54) - \log(68)} = \frac{-0.231}{-0.101} = 2.287$$

From Equation 9.56

$$\log S_{ur} = (1 - 2.287)\left[\log\left(\frac{4.86}{10.68}\right)\right] + \log(10.68) = \textbf{29.44 kN/m}^2$$

9.16 STRESS PATH

Results of triaxial tests can be represented by diagrams called *stress paths*. A stress path is a line connecting a series of points, each point representing a successive stress state experienced by a soil specimen during the progress of a test. There are several ways in which the stress path can be drawn, two of which are discussed later.

9.16.1 Rendulic plot

A Rendulic plot is a plot representing the stress path for triaxial tests originally suggested by Rendulic (1937) and later developed by Henkel (1960). It is a plot of the state of stress during triaxial tests on a plane $Oabc$, as shown in Figure 9.42.

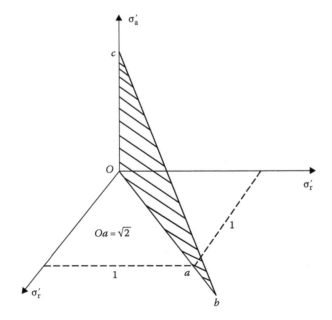

Figure 9.42 Rendulic plot.

Along Oa, we plot $\sqrt{2}\sigma_r'$, and along Oc, we plot σ_a' (σ_r' is the effective radial stress and σ_a' the effective axial stress). Line Od in Figure 9.43 represents the isotropic stress line. The direction cosines of this line are $1/\sqrt{3}$, $1/\sqrt{3}$, $1/\sqrt{3}$. Line Od in Figure 9.43 will have slope of 1 vertical to $\sqrt{2}$ horizontal. Note that the trace of the octahedral plane ($\sigma_1' + \sigma_2' + \sigma_3' = $ const) will be at right angles to the line Od.

In triaxial equipment, if a soil specimen is hydrostatically consolidated (i.e., $\sigma_a' = \sigma_r'$), it may be represented by point 1 on the line Od. If this specimen is subjected to a drained axial compression test by increasing σ_a' and keeping σ_r' constant, the stress path can be represented by the line 1–2. Point 2 represents the state of stress at failure. Similarly

Line 1–3 will represent a drained axial compression test conducted by keeping σ_a' constant and reducing σ_r'.

Line 1–4 will represent a drained axial compression test where the mean principal stress (or $J = \sigma_1' + \sigma_2' + \sigma_3'$) is kept constant.

Line 1–5 will represent a drained axial extension test conducted by keeping σ_r' constant and reducing σ_a'.

Line 1–6 will represent a drained axial extension test conducted by keeping σ_a' constant and increasing σ_r'.

Line 1–7 will represent a drained axial extension test with $J = \sigma_1' + \sigma_2' + \sigma_3'$ constant (i.e., $J = \sigma_a' + 2\sigma_r'$ constant).

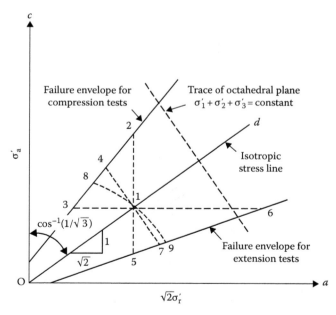

Figure 9.43 Rendulic diagram.

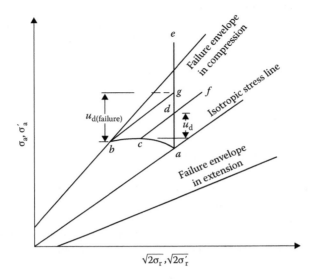

Figure 9.44 Determination of pore water pressure in a Rendulic plot.

Curve 1–8 will represent an undrained compression test.
Curve 1–9 will represent an undrained extension test.

Curves 1–8 and 1–9 are independent of the total stress combination, since the pore water pressure is adjusted to follow the stress path shown.

If we are given the effective stress path from a triaxial test in which failure of the specimen was caused by loading in an undrained condition, the pore water pressure at a given state during the loading can be easily determined. This can be explained with the aid of Figure 9.44. Consider a soil specimen consolidated with an encompassing pressure σ_r and with failure caused in the undrained condition by increasing the axial stress σ_a. Let *acb* be the effective stress path for this test. We are required to find the excess pore water pressures that were generated at points *c* and *b* (i.e., at failure). For this type of triaxial test, we know that the *total stress path* will follow a vertical line such as *ae*. To find the excess pore water pressure at *c*, we draw a line *cf* parallel to the isotropic stress line. Line *cf* intersects line *ae* at *d*. The pore water pressure u_d at *c* is the *vertical distance* between points *c* and *d*. The pore water pressure $u_{d(failure)}$ at *b* can similarly be found by drawing *bg* parallel to the isotropic stress line and measuring the vertical distance between points *b* and *g*.

9.16.2 Lambe's stress path

Lambe (1964) suggested another type of stress path in which are plotted the successive effective normal and shear stresses on a plane making an angle of 45° to the major principal plane. To understand what a stress path is, consider a normally consolidated clay specimen subjected to a consolidated drained triaxial test (Figure 9.45a). At any time during the test, the stress condition in the specimen can be represented by Mohr's circle (Figure 9.45b). Note here that, in a drained test, total stress is equal to effective stress. So

$$\sigma_3 = \sigma_3' \text{ (minor principal stress)}$$

$$\sigma_1 = \sigma_3 + \Delta\sigma = \sigma_1' \text{ (major principal stress)}$$

At failure, Mohr's circle will touch a line that is the Mohr–Coulomb failure envelope; this makes an angle ϕ with the normal stress axis (ϕ is the soil friction angle).

We now consider the effective normal and shear stresses on a plane making an angle of 45° with the major principal plane. Thus

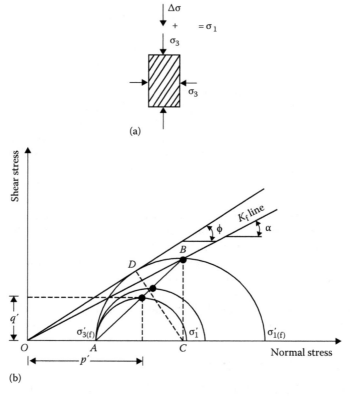

Figure 9.45 Definition of stress path: (a) sample under loading; (b) definition of K_f line.

Effective normal stress, $p' = \dfrac{\sigma_1' + \sigma_3'}{2}$ $\qquad$ (9.58)

Shear stress, $q' = \dfrac{\sigma_1' - \sigma_3'}{2}$ $\qquad$ (9.59)

The point on Mohr's circle having coordinates p' and q' is shown in Figure 9.45b. If the points with p' and q' coordinates of all Mohr's circles are joined, this will result in the line AB. This line is called a stress path. The straight line joining the origin and point B will be defined here as the K_f line. The K_f line makes an angle α with the normal stress axis. Now

$$\tan \alpha = \frac{BC}{OC} = \frac{(\sigma_{1(f)}' - \sigma_{3(f)}')/2}{(\sigma_{1(f)}' + \sigma_{3(f)}')/2} \qquad (9.60)$$

where $\sigma_{1(f)}'$ and $\sigma_{3(f)}'$ are the effective major and minor principal stresses at failure.
Similarly

$$\sin \phi = \frac{DC}{OC} = \frac{(\sigma_{1(f)}' - \sigma_{3(f)}')/2}{(\sigma_{1(f)}' + \sigma_{3(f)}')/2} \qquad (9.61)$$

From Equations 9.60 and 9.61, we obtain

$$\tan \alpha = \sin \phi \qquad (9.62)$$

For a consolidated undrained test, consider a clay specimen consolidated under an isotropic stress $\sigma_3 = \sigma_3'$ in a triaxial test. When a deviator stress $\Delta \sigma$ is applied on the specimen and drainage is not permitted, there will be an increase in the pore water pressure, Δu (Figure 9.46a):

$$\Delta u = A \Delta \sigma \qquad (9.63)$$

where A is the pore water pressure parameter.
At this time, the effective major and minor principal stresses can be given by

Minor effective principal stress $= \sigma_3' = \sigma_3 - \Delta u$

Major effective principal stress $= \sigma_1' = \sigma_1 - \Delta u = (\sigma_3 + \Delta \sigma) - \Delta u$

Mohr's circles for the total and effective stress at any time of deviator stress application are shown in Figure 9.46b. (Mohr's circle no. 1 is for total stress

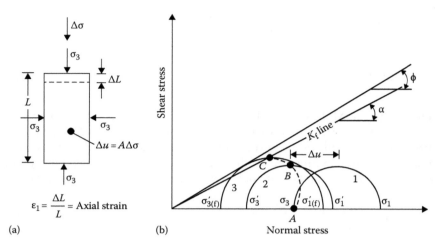

Figure 9.46 Stress path for consolidated undrained triaxial test: (a) sample under loading; (b) effective-stress path.

and no. 2 for effective stress.) Point B on the effective-stress Mohr's circle has the coordinates p' and q'. If the deviator stress is increased until failure occurs, the effective-stress Mohr's circle at failure will be represented by circle no. 3, as shown in Figure 9.46b, and the effective-stress path will be represented by the curve ABC.

The general nature of the effective-stress path will depend on the value of $A = \Delta u / \Delta \sigma$. Figure 9.47 shows the stress path in a p' versus q' plot for Lagunilla clay (Lambe, 1964). In any particular problem, if a stress path

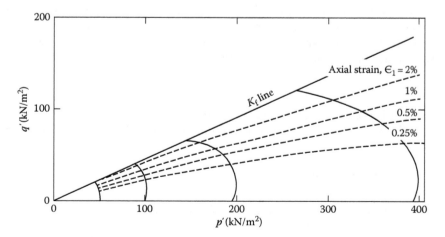

Figure 9.47 Stress path for Lagunilla clay. [After Lambe, T. W., *Soil Mech. Found. Div.*, ASCE, 90(5), 43, 1964.]

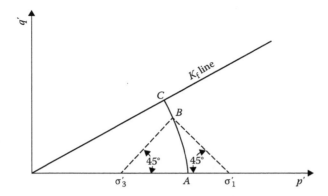

Figure 9.48 Determination of major and minor principal stresses for a point on a stress path.

is given in a p' versus q' plot, we should be able to determine the values of the major and minor effective principal stresses for any given point on the stress path. This is demonstrated in Figure 9.48, in which ABC is an effective stress path.

From Figure 9.47, two important aspects of effective stress path can be summarized as follows:

1. The stress paths for a given normally consolidated soil are geometrically similar.
2. The axial strain in a CU test may be defined as $\epsilon_1 = \Delta L/L$, as shown in Figure 9.46a. For a given soil, if the points representing equal strain in a number of stress paths are joined, they will be approximately straight lines passing through the origin. This is also shown in Figure 9.47.

Example 9.7

Given here are the loading conditions of a number of consolidated *drained* triaxial tests on a remolded clay ($\phi = 25°$, $c = 0$).

Test no.	Consolidation pressure (kN/m²)	Type of loading applied to cause failure
1	400	σ_a increased; σ_r constant
2	400	σ_a constant; σ_r increased
3	400	σ_a decreased; σ_r constant
4	400	σ_a constant; σ_r decreased
5	400	$\sigma_a + 2\sigma_r$ constant; increased σ_d and decreased σ_r
6	400	$\sigma_a + 2\sigma_r$ constant; decreased σ_d and increased σ_r

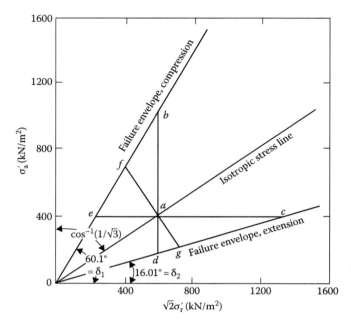

Figure 9.49 Stress paths for tests 1–6 in Example 9.7.

  a. Draw the isotropic stress line.
  b. Draw the failure envelopes for compression and extension tests.
  c. Draw the stress paths for tests 1–6.

Solution

Part a: The isotropic stress line will make an angle $\theta = \cos^{-1} 1/\sqrt{3}$ with the σ_a' axis, so $\theta = 54.8°$. This is shown in Figure 9.49 as line $0a$.
Part b:

$$\sin\phi = \left(\frac{\sigma_1' - \sigma_3'}{\sigma_1' + \sigma_3'}\right)_{\text{failure}} \quad \text{or} \quad \left(\frac{\sigma_1'}{\sigma_3'}\right)_{\text{failure}} = \frac{1 + \sin\phi}{1 - \sin\phi}$$

where σ_1' and σ_3' are the major and minor principal stresses. For *compression tests*, $\sigma_1' = \sigma_a'$ and $\sigma_3' = \sigma_r'$. Thus

$$\left(\frac{\sigma_a'}{\sigma_r'}\right)_{\text{failure}} = \frac{1 + \sin 25°}{1 - \sin 25°} = 2.46$$

  or $\quad \left(\sigma_a'\right)_{\text{failure}} = 2.46 \left(\sigma_r'\right)_{\text{failure}}$

The slope of the failure envelope is

$$\tan\delta_1 = \frac{\sigma'_a}{\sqrt{2}\sigma'_r} = \frac{2.46\sigma'_r}{\sqrt{2}\sigma'_r} = 1.74$$

Hence, $\delta_1 = 60.1°$. The failure envelope for the compression tests is shown in Figure 9.49.

For *extension tests*, $\sigma'_1 = \sigma'_r$ and $\sigma'_3 = \sigma'_a$. So

$$\left(\frac{\sigma'_a}{\sigma'_r}\right)_{failure} = \frac{1-\sin 25}{1+\sin 25} = 0.406 \quad\text{or}\quad \sigma'_a = 0.406\sigma'_r$$

The slope of the failure envelope for extension tests is

$$\tan\delta_2 = \frac{\sigma'_a}{\sqrt{2}\sigma'_r} = \frac{0.406\sigma'_r}{\sqrt{2}\sigma'_r} = 0.287$$

Hence, $\delta_2 = 16.01°$. The failure envelope is shown in Figure 9.49.

Part c: Point a on the isotropic stress line represents the point where $\sigma'_a = \sigma'_r$(or $\sigma'_1 = \sigma'_2 = \sigma'_3$). The stress paths of the test are plotted in Figure 9.49.

Test no.	Stress path in Figure 9.49
1	ab
2	ac
3	ad
4	ae
5	af
6	ag

Example 9.8

The stress path for a normally consolidated clay is shown in Figure 9.50 (Rendulic plot). The stress path is for a consolidated undrained triaxial test where failure was caused by increasing the axial stress while keeping the radial stress constant. Determine

   a. ϕ for the soil,
   b. The pore water pressure at A,
   c. The pore water pressure at failure, and
   d. The value of A_f.

Solution

Refer to Figure 9.51.

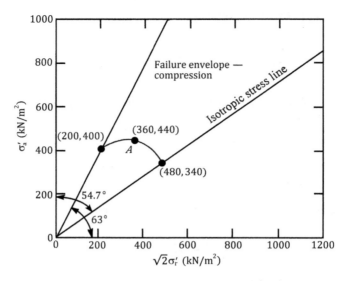

Figure 9.50 Stress path for a normally consolidated clay (consolidated undrained test).

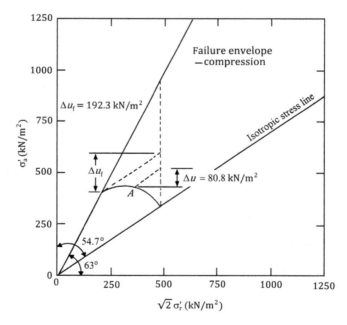

Figure 9.51 Determination of pore water pressures from stress path.

a. For compression, $\delta_1 = 63°$

$$\frac{\sigma_a'}{\sigma_r'} = \sqrt{2}\tan\delta_1 = \sqrt{2}\tan 63 = 2.776$$

$$\frac{\sigma_a'}{\sigma_r'} = 2.776 = \frac{1+\sin\phi}{1-\sin\phi}; \quad \phi = 28°$$

b. From the graph, at A, $\Delta u = 80.8$ kN/m²
c. From the graph, at failure, $\Delta u = 192.3$ kN/m²
d. At failure, $\sigma_a' = 400$ kN/m²; $\sigma_r' = 200/\sqrt{2} = 141.4$ kN/m²

$$A_f = \frac{192.3}{400-141.4} = 0.744$$

Example 9.9

The results of a consolidated undrained test, in which $\sigma_3 = 392$ kN/m², on a normally consolidated clay are given below.

Axial strain (%)	$\Delta\sigma$ (kN/m²)	u_d (kN/m²)
0	0	0
0.5	156	99
0.75	196	120
1	226	132
1.3	235	147
2	250	161
3	245	170
4	240	173
4.5	235	175

Draw the K_f line in a p' versus q' diagram. Also draw the stress path for this test in the diagram and determine α.

Solution

$\sigma_3 = 392$ kN/m²

Axial strain (%)	$\Delta\sigma$ (kN/m²)	u_d (kN/m²)	p' (kN/m²)	q' (kN/m²)
0	0	0	392	0
0.5	156	99	371	78
0.75	196	120	370	98
1	226	132	373	113
1.3	235	147	362.5	117.5

(Continued)

Axial strain (%)	$\Delta\sigma$ (kN/m²)	u_d (kN/m²)	p' (kN/m²)	q' (kN/m²)
2	250	161	356	125
3	245	170	344.5	122.5
4	240	173	339	120
4.5	235	175	334.5	117.5

The p' versus q' plot is shown (Figure 9.52). From the K_f line, $\alpha \approx 20°$.

Example 9.10

For a saturated clay soil, the following are the results of some consolidated drained triaxial tests at failure:

Test no.	$p' = \dfrac{\sigma_1' + \sigma_3'}{2}$ (kN/m²)	$q' = \dfrac{\sigma_1' - \sigma_3'}{2}$ (kN/m²)
1	420	179.2
2	630	255.5
3	770	308.0
4	1260	467.0

Draw a p' versus q' diagram, and from that, determine c and ϕ for the soil.

Solution

The diagram of q' versus p' is shown in Figure 9.53; this is a straight line, and its equation may be written in the form

$$q' = m + p' \tan \alpha \qquad (E9.3)$$

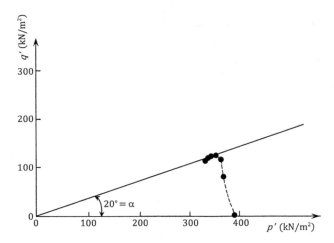

Figure 9.52 Plot of q' versus p' for the saturated clay (undrained triaxial test).

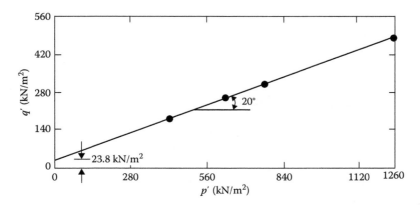

Figure 9.53 Plot of q' versus p' diagram.

Also

$$\frac{\sigma_1' - \sigma_3'}{2} = c\cos\phi + \frac{\sigma_1' + \sigma_3'}{2}\sin\phi \qquad\qquad \text{(E9.4)}$$

Comparing Equations E9.3 and E9.4, we find $m = c\cos\phi$ or $c = m/\cos\phi$ and $\tan\alpha = \sin\phi$. From Figure 9.53, $m = 23.8$ kN/m² and $\alpha = 20°$. So

$$\phi = \sin^{-1}(\tan 20°) = 21.34°$$

and

$$c = \frac{m}{\cos\alpha} = \frac{23.8}{\cos 21.34°} = 25.55 \text{ kN/m}^2$$

9.17 HVORSLEV'S PARAMETERS

Considering cohesion to be the result of physicochemical bond forces (thus the interparticle spacing and hence void ratio), Hvorslev (1937) expressed the shear strength of a soil in the form

$$s = c_e + \sigma'\tan\phi_e \qquad\qquad (9.64)$$

where c_e and ϕ_e are "true cohesion" and "true angle of friction," respectively, which are dependent on the void ratio.

The procedure for determination of the aforementioned parameters can be explained with the aid of Figure 9.54, which shows the relation of the moisture content (i.e., void ratio) with effective consolidation

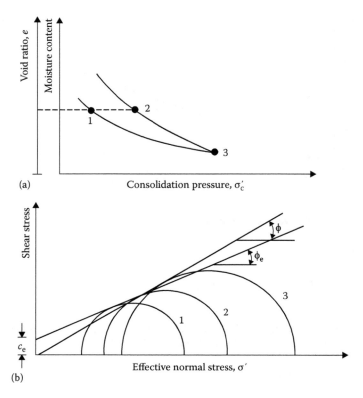

Figure 9.54 Determination of c_e and ϕ_e: (a) plot of e, and moisture content vs. σ_c'; (b) effective stress Mohr's circles.

pressure. Points 2 and 3 represent normally consolidated stages of a soil, and point 1 represents the overconsolidation stage. We now test the soil specimens represented by points 1, 2, and 3 in an undrained condition. The effective-stress Mohr's circles at failure are given in Figure 9.54b.

The soil specimens at points 1 and 2 in Figure 9.54a have the same moisture content and hence the same void ratio. If we draw a common tangent to Mohr's circles 1 and 2, the slope of the tangent will give ϕ_e, and the intercept on the shear stress axis will give c_e.

Gibson (1953) found that ϕ_e varies slightly with void ratio. The *true angle of internal friction* decreases with the plasticity index of soil, as shown in Figure 9.55. The variation of the effective cohesion c_e with void ratio may be given by the relation (Hvorslev, 1960)

$$c_e = c_0 \exp(-Be) \tag{9.65}$$

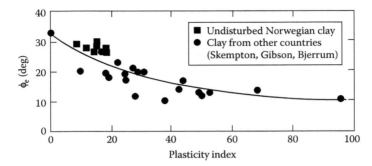

Figure 9.55 Variation of true angle of friction with plasticity index. (After Bjerrum, L. and Simons, N. E., Comparison of shear strength characteristics of normally consolidated clay, in *Proc. Res. Conf. Shear Strength Cohesive Soils*, ASCE, 711–726, 1960.)

where

c_0 is the true cohesion at zero void ratio

e is the void ratio at failure

B is the slope of plot of ln c_e versus void ratio at failure

Example 9.11

A clay soil specimen was subjected to confining pressures $\sigma_3 = \sigma'_3$ in a triaxial chamber. The moisture content versus σ'_3 relation is shown in Figure 9.56a.

A *normally consolidated* specimen of the same soil was subjected to a consolidated undrained triaxial test. The results are as follows: $\sigma_3 = 440$ kN/m²; $\sigma_1 = 840$ kN/m²; moisture content at failure, 27%; $u_d = 240$ kN/m².

An overconsolidated specimen of the same soil was subjected to a consolidated undrained test. The results are as follows: overconsolidation pressure, $\sigma'_c = 550$ kN/m²; $\sigma_3 = 100$ kN/m²; $\sigma_1 = 434$ kN/m²; $u_d = -18$ kN/m²; initial and final moisture content, 27%.

Determine ϕ_e, c_e for a moisture content of 27%; also determine ϕ.

Solution

For the normally consolidated specimen,

$$\sigma'_3 = 440 - 240 = 200 \text{ kN/m}^2$$

$$\sigma'_1 = 840 - 240 = 600 \text{ kN/m}^2$$

$$\phi = \sin^{-1}\left(\frac{\sigma'_1 - \sigma'_3}{\sigma'_1 + \sigma'_3}\right) = \sin^{-1}\left(\frac{600 - 200}{600 + 200}\right) = 30°$$

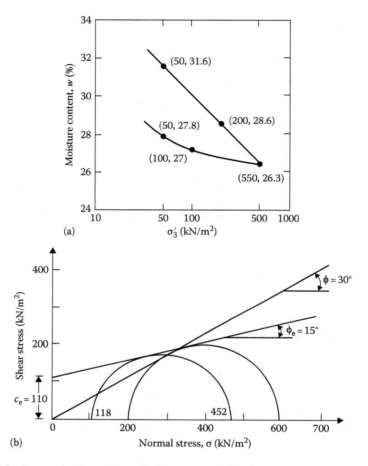

Figure 9.56 Determination of Hvorslev's parameters: (a) plot of moisture content vs. σ_3'; (b) plot of Mohr's circles (w = 27%).

The failure envelope is shown in Figure 9.56b.
For the overconsolidated specimen

$$\sigma_3' = 100 - (-18) = 118 \text{ kN/m}^2$$

$$\sigma_1' = 434 - (-18) = 452 \text{ kN/m}^2$$

Mohr's circle at failure is shown in Figure 9.56b; from this

$$c_e = 110 \text{ kN/m}^2 \quad \phi_e = 15°$$

9.18 RELATIONS BETWEEN MOISTURE CONTENT, EFFECTIVE STRESS, AND STRENGTH FOR CLAY SOILS

9.18.1 Relations between water content and strength

The strength of a soil at failure (i.e., $(\sigma_1 - \sigma_3)_{failure}$ or $(\sigma_1' - \sigma_3')_{failure}$) is dependent on the moisture content at failure. Henkel (1960) pointed out that there is a unique relation between the moisture content w at failure and the strength of a clayey soil. This is shown in Figures 9.57 and 9.58 for Weald clay.

For normally consolidated clays, the variation of w versus $\log(\sigma_1 - \sigma_3)_{failure}$ is approximately linear. For overconsolidated clays, this relation is not linear but lies slightly below the relation of normally consolidated specimens. The curves merge when the strength approaches the overconsolidation pressure. Also note that slightly different relations for w versus $\log(\sigma_1 - \sigma_3)_{failure}$ are obtained for axial compression and axial extension tests.

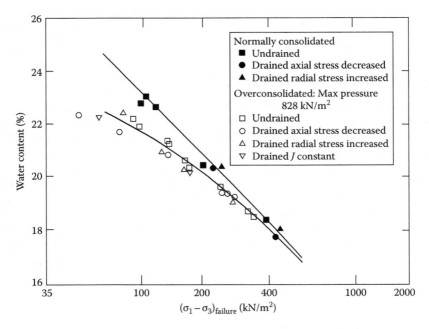

Figure 9.57 Water content versus $(\sigma_1 - \sigma_3)_{failure}$ for Weald clay—extension tests. (After Henkel, D. J., The shearing strength of saturated remolded clays, in Proc. Res. Conf. Shear Strength of Cohesive Soils, ASCE, 533–554, 1960.)

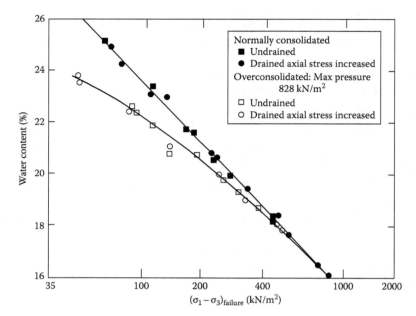

Figure 9.58 Water content versus $(\sigma_1 - \sigma_3)_{\text{failure}}$ for Weald clay—compression tests. (After Henkel, D. J., The shearing strength of saturated remolded clays, in *Proc. Res. Conf. Shear Strength of Cohesive Soils*, ASCE, 533–554, 1960.)

9.18.2 Unique effective stress failure envelope

When Mohr's envelope is used to obtain the relation for normal and shear stress at failure, from triaxial test results, separate envelopes need to be drawn for separate preconsolidation pressures, σ'_c, as shown in Figure 9.59. For a soil with a preconsolidation pressure of σ'_{c_1}, $s = c_1 + \sigma' \tan \phi_{c(1)}$; similarly, for a preconsolidation pressure of σ'_{c_2}, $s = c_2 + \sigma' \tan \phi_{c(2)}$.

Henkel (1960) showed that a single, general failure envelope for normally consolidated and preconsolidated (irrespective of preconsolidation pressure) soils can be obtained by plotting the ratio of the major to minor effective stress at failure against the ratio of the maximum consolidation pressure to the average effective stress at failure. This fact is demonstrated in Figure 9.60, which gives results of triaxial compression tests for Weald clay. In Figure 9.60

$$J_m = \text{maximum consolidation pressure} = \sigma'_c$$

$$J_f = \text{average effective stress at failure}$$

$$= \frac{\sigma'_{1(\text{failure})} + \sigma'_{2(\text{failure})} + \sigma'_{3(\text{failure})}}{3}$$

$$= \frac{\sigma'_a + 2\sigma'_r}{3}$$

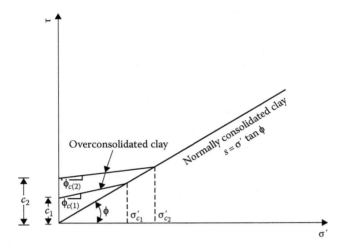

Figure 9.59 Mohr's envelope for overconsolidated clay.

The results shown in Figure 9.60 are obtained from normally consolidated specimens and overconsolidated specimens having a maximum preconsolidation pressure of 828 kN/m². Similarly, a unique failure envelope can be obtained from extension tests. Note, however, that the failure envelopes for compression tests and extension tests are slightly different.

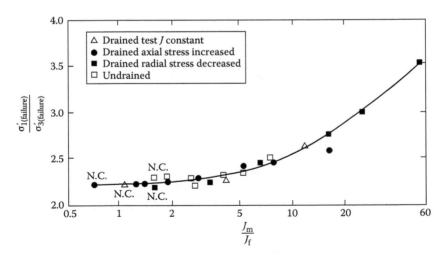

Figure 9.60 Plot of $\sigma'_{1(failure)}/\sigma'_{3(failure)}$ against J_m/J_f for Weald clay—compression tests. (After Henkel, D. J., The shearing strength of saturated remolded clays, in *Proc. Res. Conf. Shear Strength of Cohesive Soils*, ASCE, 533–554, 1960.)

9.18.3 Unique relation between water content and effective stress

There is a unique relation between the water content of a soil and the effective stresses to which it is being subjected, provided that normally consolidated specimens and specimens with common maximum consolidation pressures are considered separately. This can be explained with the aid of Figure 9.61, in which a Rendulic plot for a normally consolidated clay is shown. Consider several specimens consolidated at various confining pressures in a triaxial chamber; the states of stress of these specimens are represented by the points a, c, e, g, etc., located on the isotropic stress lines. When these specimens are sheared to failure by drained compressions, the corresponding stress paths will be represented by lines such as ab, cd, ef, and gh. During drained tests, the moisture contents of the specimens change. We can determine the moisture contents of the specimens during the tests, such as w_1, w_2, ..., as shown in Figure 9.61. If these points of equal moisture contents on the drained stress paths are joined, we obtain

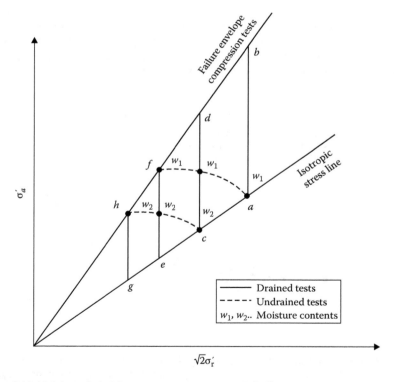

Figure 9.61 Unique relation between water content and effective stress.

contours of stress paths of equal moisture contents (for moisture contents $w_1, w_2, \ldots$).

Now, if we take a soil specimen and consolidate it in a triaxial chamber under a state of stress as defined by point a and shear it to failure in an undrained condition, it will follow the effective stress path af, since the moisture content of the specimen during shearing is w_1. Similarly, a specimen consolidated in a triaxial chamber under a state of stress represented by point c (moisture content w_2) will follow a stress path ch (which is the stress contour of moisture content w_2) when sheared to failure in an undrained state. This means that a unique relation exists between water content and effective stress.

Figures 9.62 and 9.63 show the stress paths for equal water contents for normally consolidated and overconsolidated Weald clay. Note the similarity of shape of the stress paths for normally consolidated clay in Figure 9.63. For overconsolidated clay, the shape of the stress path gradually changes, depending on the OCR.

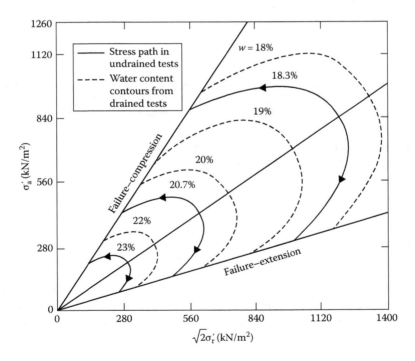

Figure 9.62 Weald clay—normally consolidated. (After Henkel, D. J., The shearing strength of saturated remolded clays, in *Proc. Res. Conf. Shear Strength of Cohesive Soils*, ASCE, 533–554, 1960.)

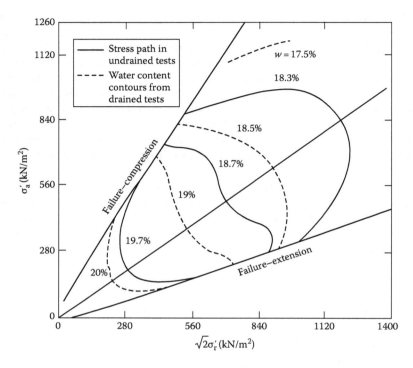

Figure 9.63 Weald clay—overconsolidated; maximum consolidation pressure = 828 kN/m². (After Henkel, D. J., The shearing strength of saturated remolded clays, in *Proc. Res. Conf. Shear Strength of Cohesive Soils*, ASCE, 533–554, 1960.)

9.19 CORRELATIONS FOR EFFECTIVE STRESS FRICTION ANGLE

It is difficult in practice to obtain undisturbed samples of sand and gravelly soils to determine the shear strength parameters. For that reason, several approximate correlations were developed over the years to determine the soil friction angle based on field test results, such as standard penetration number (N) and cone penetration resistance (q_c). In granular soils, N and q_c are dependent on the effective-stress level. Schmertmann (1975) provided a correlation between the standard penetration resistance, drained triaxial friction angle obtained from axial compression tests ($\phi = \phi_{tc}$), and the vertical effective stress (σ_0'). This correlation can be approximated as (Kulhawy and Mayne, 1990)

$$\phi_{tc} = \tan^{-1}\left[\frac{N}{12.2 + 20.3(\sigma_0'/p_a)}\right]^{0.34} \quad \text{(for granular soils)} \qquad (9.66)$$

where p_a is atmospheric pressure (in the same units as σ'_0). In a similar manner, the correlation between ϕ_{tc}, σ'_0, and q_c was provided by Robertson and Campanella (1983), which can be approximated as (Kulhawy and Mayne, 1990)

$$\phi_{tc} = \tan^{-1}\left[0.9 + 0.38\log\left(\frac{q_c}{\sigma'_0}\right)\right] \quad \text{(for granular soils)} \tag{9.67}$$

Kulhawy and Mayne (1990) also provided the approximate relations between the triaxial drained friction angle $\langle\phi_{tc}\rangle$ obtained from triaxial compression tests with the drained friction angle obtained from other types of tests for cohesionless and cohesive soils. Their findings are summarized in Table 9.4.

Following are some other correlations generally found in the recent literature.

- Wolff (1989)

$$\phi_{tc} = 27.1 + 0.3N_1 - 0.00054(N_1)^2 \quad \text{(for granular soil)} \tag{9.68}$$

- Hatanaka and Uchida (1996)

$$\phi_{tc} = \sqrt{15.4N_1} + 20 \quad \text{(for granular soil)} \tag{9.69}$$

$$\text{where } N_1 = \sqrt{\frac{98}{\sigma'_o}}N \tag{9.70}$$

  = standard penetration number corrected to a standard value of σ'_o equal to one atmospheric pressure

(*Note*: σ'_o is vertical stress in kN/m².)

Table 9.4 Relative values of drained friction angle

Test type	Drained friction angle	
	Cohesionless soil	*Cohesive soil*
Triaxial compression	$1.0\phi_{tc}$	$1.0\phi_{tc}$
Triaxial extension	$1.12\phi_{tc}$	$1.22\phi_{tc}$
Plane strain compression	$1.12\phi_{tc}$	$1.10\phi_{tc}$
Plane strain extension	$1.25\phi_{tc}$	$1.34\phi_{tc}$
Direct shear	$\tan^{-1}[\tan(1.12\phi_{tc})\cos\phi_{cv}]$	$\tan^{-1}[\tan(1.1\phi_{tc})\cos\phi_{ult}]$

Source: Compiled from Kulhawy, F. H. and Mayne, P. W., *Manual on Estimating Soil Properties in Foundation Design*, Electric Power Research Institute, Palo Alto, CA, 1990.

- Ricceri et al. (2002)

$$\phi_{tc} = \tan^{-1}\left[0.38 + 0.27\log\left(\frac{q_c}{\sigma_o'}\right)\right]\left(\begin{array}{c}\text{for silt with low plasticity,} \\ \text{poorly graded sand, and silty} \\ \text{sand}\end{array}\right) \quad (9.71)$$

- Ricceri et al. (2002)

$$\phi_{tc} = 31 + \frac{K_D}{0.236 + 0.066K_D}\left(\begin{array}{c}\text{for silt with low plasticity, poorly} \\ \text{graded sand, and silty sand}\end{array}\right) \quad (9.72)$$

where K_D is the horizontal stress index in the dilatometer test.

9.20 ANISOTROPY IN UNDRAINED SHEAR STRENGTH

Owing to the nature of the deposition of cohesive soils and subsequent consolidation, clay particles tend to become oriented perpendicular to the direction of the major principal stress. Parallel orientation of clay particles could cause the strength of the clay to vary with direction, or in other words, the clay could be anisotropic with respect to strength. This fact can be demonstrated with the aid of Figure 9.64, in which V and H are vertical and horizontal directions that coincide with lines perpendicular and parallel to the bedding planes of a soil deposit. If a soil specimen with its axis inclined at an angle i with the horizontal is collected

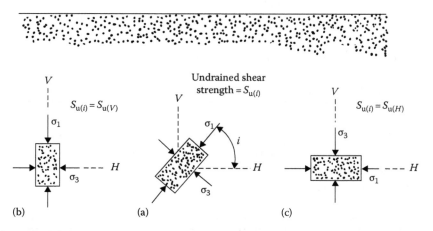

Figure 9.64 Strength anisotropy in clay with direction of major principal stress: (a) $i = 90°$; (b) $i = i$; (c) $i = 0°$.

and subjected to an undrained test, the undrained shear strength can be given by

$$S_{u(i)} = \frac{\sigma_1 - \sigma_3}{2} \tag{9.73}$$

where $S_{u(i)}$ is the undrained shear strength when the major principal stress makes an angle i with the horizontal.

Let the undrained shear strength of a soil specimen with its axis vertical (i.e., $S_{u(i = 90°)}$] be referred to as $S_{u(V)}$ (Figure 9.64a); similarly, let the undrained shear strength with its axis horizontal (i.e., $S_{u(i = 0°)}$] be referred to as $S_{u(H)}$ (Figure 9.64c). If $S_{u(V)} = S_{u(i)} = S_{u(H)}$, the soil is isotropic with respect to strength, and the variation of undrained shear strength can be represented by a circle in a polar diagram, as shown by curve a in Figure 9.65. However, if the soil is anisotropic, $S_{u(i)}$ will change with direction. Casagrande and Carrillo (1944) proposed the following equation for the directional variation of the undrained shear strength:

$$S_{u(i)} = S_{u(H)} + [S_{u(V)} - S_{u(H)}]\sin^2 i \tag{9.74}$$

When $S_{u(V)} > S_{u(H)}$, the nature of variation of $S_{u(i)}$ can be represented by curve b in Figure 9.65. Again, if $S_{u(V)} < S_{u(H)}$, the variation of $S_{u(i)}$ is given by curve c. The coefficient of anisotropy can be defined as

$$K = \frac{S_{u(V)}}{S_{u(H)}} \tag{9.75}$$

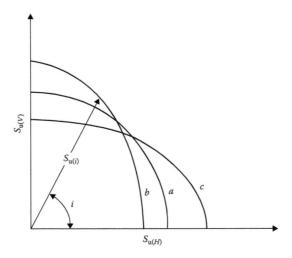

Figure 9.65 Directional variation of undrained strength of clay.

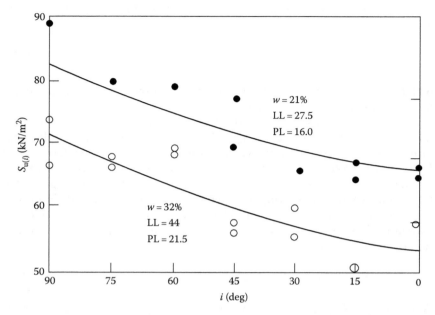

Figure 9.66 Directional variation of undrained shear strength of Welland Clay, Ontario, Canada. [After Lo, K. Y., Stability of slopes in anisotropic soils, *J. Soil Mech. Found. Eng. Div.*, ASCE, 91(SM4), 85, 1965.]

In the case of natural soil deposits, the value of K can vary from 0.75 to 2.0. K is generally less than 1 in overconsolidated clays. An example of the directional variation of the undrained shear strength $S_{u(i)}$ for a clay is shown in Figure 9.66.

Richardson et al. (1975) made a study regarding the anisotropic strength of a soft deposit of marine clay in Thailand. The undrained strength was determined by field vane shear tests. Both rectangular and triangular vanes were used for this investigation. Based on the experimental results (Figure 9.67), Richardson et al. concluded that $S_{u(i)}$ can be given by the following relation:

$$S_{u(i)} = \frac{S_{u(H)}S_{u(V)}}{\sqrt{S_{u(H)}^2 \sin^2 i + S_{u(V)}^2 \cos^2 i}} \tag{9.76}$$

Example 9.12

For an anisotropic clay deposit, the results from unconfined compression tests were $S_{u(i=30°)} = 101.5$ kN/m^2 and $S_{u(i=60°)} = 123$ kN/m^2. Find the anisotropy coefficient K of the soil based on the Casagrande-Carillo equation (Equation 9.74).

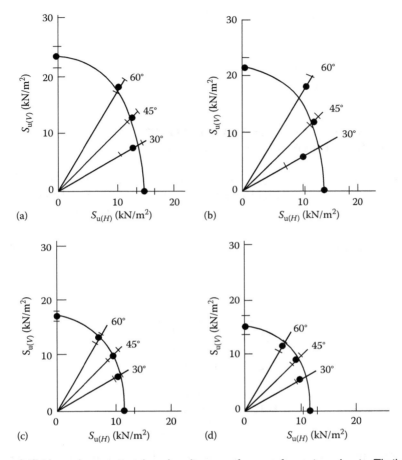

Figure 9.67 Vane shear strength polar diagrams for a soft marine clay in Thailand. (a) Depth = 1 m; (b) depth = 2 m; (c) depth = 3 m; (d) depth = 4 m. (After Richardson, A. M. et al., In situ determination of anisotropy of a soft clay, in *Proc. Conf. In Situ Meas. Soil Prop.*, vol. I, ASCE, 336–349, 1975.)

Solution

Equation 9.74:

$$S_{u(i)} = S_{u(H)} + [S_{u(V)} - S_{u(H)}]\sin^2 i$$

$$101.5 = S_{u(H)} + [S_{u(V)} - S_{u(H)}]\sin^2 30° \tag{a}$$

$$123 = S_{u(H)} + [S_{u(V)} - S_{u(H)}]\sin^2 60° \tag{b}$$

$$\frac{101.5/S_{u(H)}}{123/S_{u(H)}} = \frac{1+(K-1)\sin^2 30°}{1+(K-1)\sin^2 60°}$$

$$0.825 = \frac{1+(K-1)(0.25)}{1+(K-1)(0.75)}$$

$$K = 1.47$$

9.21 SENSITIVITY AND THIXOTROPIC CHARACTERISTICS OF CLAYS

Most undisturbed natural clayey soil deposits show a pronounced reduction of strength when they are remolded. This characteristic of saturated cohesive soils is generally expressed quantitatively by a term referred to as *sensitivity*. Thus

$$\text{Sensitivity} = \frac{S_{u(\text{undisturbed})}}{S_{u(\text{remolded})}} \tag{9.77}$$

The classification of clays based on sensitivity is as follows:

Sensitivity	Clay
≈1	Insensitive
1–2	Low sensitivity
2–4	Medium sensitivity
4–8	Sensitive
8–16	Extra sensitive
>16	Quick

The sensitivity of most clays generally falls in a range 1–8. However, sensitivity as high as 150 for a clay deposit at St Thurible, Canada, was reported by Peck et al. (1951).

The loss of strength of saturated clays may be due to the breakdown of the original structure of natural deposits and thixotropy. *Thixotropy* is defined as an isothermal, reversible, time-dependent process that occurs under constant composition and volume, whereby a material softens as a result of remolding and then gradually returns to its original strength when allowed to rest. This is shown in Figure 9.68. A general review of the thixotropic nature of soils is given by Seed and Chan (1959).

Figure 9.69, which is based on the work of Moretto (1948), shows the thixotropic strength regain of a Laurentian clay with a liquidity index of 0.99 (i.e., the natural water content was approximately equal to the liquid limit). In Figure 9.70, the acquired sensitivity is defined as

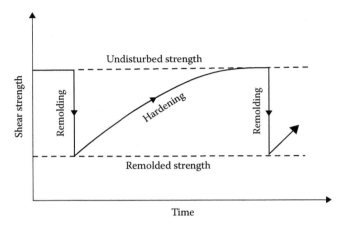

Figure 9.68 Thixotropy of a material.

$$\text{Acquired sensitivity} = \frac{S_{u(t)}}{S_{u(\text{remolded})}} \tag{9.78}$$

where $S_{u(t)}$ is the undrained shear strength after a time t from remolding.

Acquired sensitivity generally decreases with the liquidity index (i.e., the natural water content of soil), and this is demonstrated in Figure 9.70. It can also be seen from this figure that the acquired sensitivity of clays with

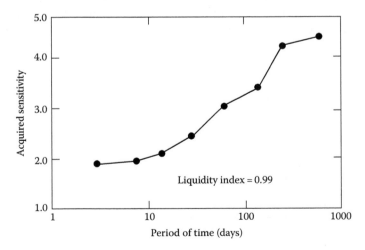

Figure 9.69 Acquired sensitivity for Laurentian clay. (After Seed, H. B. and Chan, C. K., Trans., ASCE, 24, 894, 1959.)

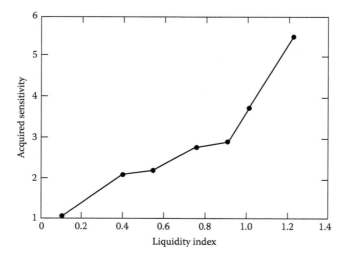

Figure 9.70 Variation of sensitivity with liquidity index for Laurentian clay. (After Seed, H. B. and Chan, C. K., *Trans.*, ASCE, 24, 894, 1959.)

a liquidity index approaching zero (i.e., natural water content equal to the plastic limit) is approximately one. Thus, thixotropy in the case of overconsolidated clay is very small.

There are some clays that show that sensitivity cannot be entirely accounted for by thixotropy (Berger and Gnaedinger, 1949). This means that only a part of the strength loss due to remolding can be recovered by hardening with time. The other part of the strength loss is due to the breakdown of the original structure of the clay. The general nature of the strength regain of a partially thixotropic material is shown in Figure 9.71.

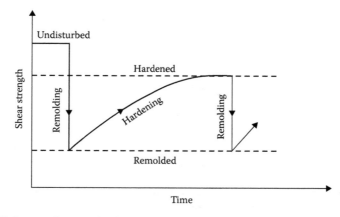

Figure 9.71 Regained strength of a partially thixotropic material.

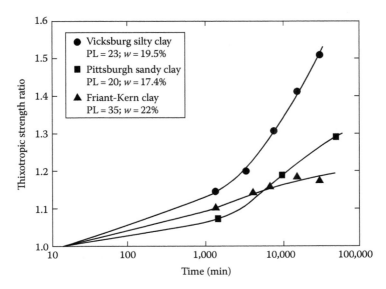

Figure 9.72 Increase of thixotropic strength with time for three compacted clays. (After Seed, H. B. and Chan, C. K., *Trans.*, ASCE, 24, 894, 1959.)

Seed and Chan (1959) conducted several tests on three compacted clays with a water content near or below the plastic limit to study their thixotropic strength-regain characteristics. Figure 9.72 shows their thixotropic strength ratio with time. The thixotropic strength ratio is defined as follows:

$$\text{Thixotropic strength ratio} = \frac{S_{u(t)}}{S_{u(\text{compacted at } t=0)}} \tag{9.79}$$

where $S_{u(t)}$ is the undrained strength at time t after compaction.

These test results demonstrate that thixotropic strength regain is also possible for soils with a water content at or near the plastic limit.

Figure 9.73 shows a general relation between sensitivity, liquidity index, and effective vertical pressure for natural soil deposits.

9.22 VANE SHEAR TEST

The field vane shear test is another method of obtaining the undrained shear strength of cohesive soils. The common shear vane usually consists of four thin steel plates of equal size welded to a steel torque rod (Figure 9.74a). To perform the test, the vane is pushed into the soil and torque is applied at the top of the torque rod. The torque is gradually increased until the cylindrical

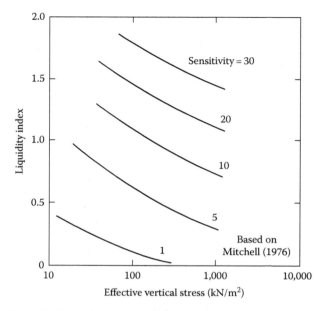

Figure 9.73 General relation between sensitivity, liquidity index, and effective vertical stress.

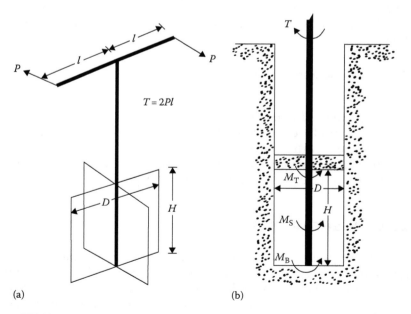

Figure 9.74 Vane shear test: (a) vane shear apparatus; (b) test in soil.

soil of height H and diameter D fails (Figure 9.74b). The maximum torque T applied to cause failure is the sum of the resisting moment at the top, M_T, and bottom, M_B, of the soil cylinder, plus the resisting moment at the sides of the cylinder, M_S. Thus

$$T = M_S + M_T + M_B \tag{9.80}$$

However

$$M_S = \pi D H \frac{D}{2} S_u \quad \text{and} \quad M_T = M_B = \frac{\pi D^2}{4} \frac{2}{3} \frac{D}{2} S_u$$

(assuming uniform undrained shear strength distribution at the ends; see Carlson [1948]). So

$$T = \pi S_u \left[\left(\pi D H \frac{D}{2} \right) + 2 \left(\frac{\pi D^2}{4} \frac{2}{3} \frac{D}{2} \right) \right]$$

or

$$S_u = \frac{T}{\pi (D^2 H/2 + D^3/6)} \tag{9.81}$$

If only one end of the vane (i.e., the bottom) is engaged in shearing the clay, $T = M_S + M_B$. So

$$S_u = \frac{T}{\pi (D^2 H/2 + D^3/12)} \tag{9.82}$$

Standard vanes used in field investigations have $H/D = 2$. In such cases, Equation 9.81 simplifies to the form

$$S_u = 0.273 \frac{T}{D^3} \tag{9.83}$$

The American Society for Testing and Materials (2002) recommends the following dimensions for field vanes:

D (mm)	H (mm)	Thickness of blades (mm)
38.1	76.2	1.6
50.8	101.6	1.6
63.5	127.0	3.2
92.1	184.2	3.2

In some cases, tapered vanes (Figure 9.75) are also used in the field. If both ends of the vane are engaged and $H/D = 2$, Equation 9.81 can be modified as

$$S_u = \frac{T}{K} \tag{9.84}$$

where

$$K = \frac{\pi D^2}{12}\left(\frac{D}{\cos i_T} + \frac{D}{\cos i_B} + 6H\right) \tag{9.85}$$

If the undrained shear strength is different in the vertical $[S_{u(V)}]$ and horizontal $[S_{u(H)}]$ directions, then Equation 9.81 translates to

$$T = \pi D^2 \left[\frac{H}{2}S_{u(V)} + \frac{D}{6}S_{u(H)}\right] \tag{9.86}$$

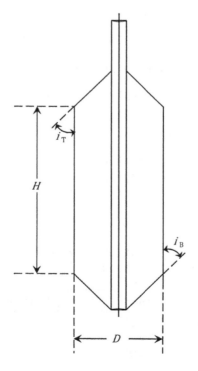

Figure 9.75 Tapered vanes.

In addition to rectangular vanes, triangular vanes can be used in the field (Richardson et al., 1975) to determine the directional variation of the undrained shear strength. Figure 9.76a shows a triangular vane. For this vane

$$S_{u(i)} = \frac{T}{\frac{4}{3}\pi L^3 \cos^2 i}$$

(9.87)

The term $S_{u(i)}$ was defined in Equation 9.73.

Silvestri and Tabib (1992) analyzed elliptical vanes (Figure 9.76b). For uniform shear stress distribution,

$$S_u = C\frac{T}{8a^3}$$

(9.88)

where $C = f(a/b)$. The variation of C with a/b is shown in Figure 9.77.

Bjerrum (1972) studied a number of slope failures and concluded that the undrained shear strength obtained by vane shear is too high. He proposed that the vane shear test results obtained from the field should be corrected for the actual design. Thus

$$S_{u(design)} = \lambda S_{u(field\ vane)}$$

(9.89)

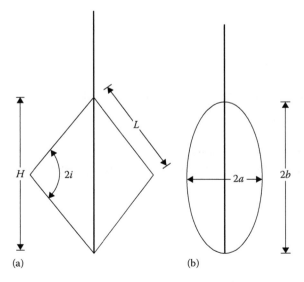

Figure 9.76 (a) Triangular vane and (b) elliptical vane.

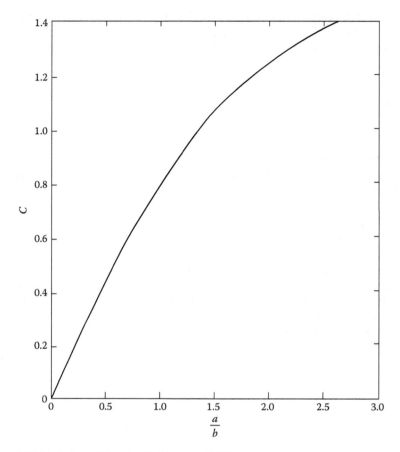

Figure 9.77 Variation of *C* with *a/b* (Equation 9.88).

where λ is a correction factor, which may be expressed as

$$\lambda = 1.7 - 0.54 \log(\text{PI}) \tag{9.90}$$

where PI is the plasticity index (%).

Morris and Williams (1994) gave the following correlations of λ:

$$\lambda = 1.18e^{-0.08(\text{PI})} + 0.57 \qquad \text{PI} > 5 \tag{9.91}$$

and

$$\lambda = 7.01e^{-0.08(\text{LL})} + 0.57 \qquad \text{LL} > 20 \tag{9.92}$$

where LL is the liquid limit (%).

9.22.1 Correlations with field vane shear strength

The field vane shear strength has been correlated with the preconsolidation pressure and the OCR of the clay. Using 343 data points, Mayne and Mitchell (1988) derived the following empirical relationship for estimating the preconsolidation pressure of a natural clay deposit:

$$\sigma'_c = 7.04[S_{u(\text{field vane})}]^{0.83} \tag{9.93}$$

where

σ'_c is the preconsolidation pressure (kN/m^2)
$S_{u(\text{field vane})}$ is the field vane shear strength (kN/m^2)

The OCR can also be correlated to $S_{u(\text{field vane})}$ according to the equation

$$OCR = \beta \frac{S_{u(\text{field vane})}}{\sigma'_o} \tag{9.94}$$

where σ'_o is the effective overburden pressure.

The magnitudes of β developed by various investigators are given later (also see Chapter 8)

- Mayne and Mitchell (1988)

$$\beta = 22[PI(\%)]^{-0.48} \tag{9.95}$$

  where PI is the plasticity index.
- Hansbo (1957):

$$\beta = \frac{222}{w(\%)} \tag{9.96}$$

  where w is the natural moisture content.
- Larsson (1980):

$$\beta = \frac{1}{0.08 + 0.0055[(PI)\%]} \tag{9.97}$$

9.23 RELATION OF UNDRAINED SHEAR STRENGTH (S_u) AND EFFECTIVE OVERBURDEN PRESSURE (p')

A relation between S_u, p', and the drained friction angle can also be derived as follows. Referring to Figure 9.78a, consider a soil specimen at A. The major and minor effective principal stresses at A can be given by p' and $K_o p'$, respectively (where K_o is the coefficient of at-rest earth pressure). Let this soil specimen be subjected to a UU triaxial test. As shown in Figure 9.78b, at failure the *total major* principal stress is $\sigma_1 = p' + \Delta\sigma_1$; the *total minor* principal stress is $\sigma_3 = K_o p' + \Delta\sigma_3$; and the *excess* pore water pressure is Δu. So, the *effective* major and minor principal stresses can be given by $\sigma_1' = \sigma_1 - \Delta u$ and $\sigma_3' = \sigma_3 - \Delta u$, respectively. The total- and effective-stress Mohr's circles for this test, at failure, are shown in Figure 9.78c. From this, we can write

$$\frac{S_u}{c\cot\phi + (\sigma_1' + \sigma_3')/2} = \sin\phi$$

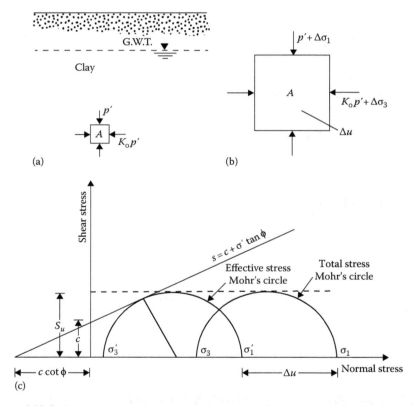

Figure 9.78 Relation between the undrained strength of clay and the effective overburden pressure: (a) soil specimen at A with major and minor principal stresses; (b) specimen at A subjected to a UU triaxial test; (c) total and effective stress Mohr's circles.

where ϕ is the drained friction angle, or

$$S_u = c\cos\phi + \frac{\sigma_1' + \sigma_3'}{2}\sin\phi$$

$$= c\cos\phi + \left(\frac{\sigma_1' + \sigma_3'}{2} - \sigma_3'\right)\sin\phi + \sigma_3'\sin\phi$$

However

$$\frac{\sigma_1' + \sigma_3'}{2} - \sigma_3' = \frac{\sigma_1' - \sigma_3'}{2} = S_u$$

So,

$$S_u = c\cos\phi + S_u\sin\phi + \sigma_3'\sin\phi$$

$$S_u(1 - \sin\phi) = c\cos\phi + \sigma_3'\sin\phi \tag{9.98}$$

$$\sigma_3' = \sigma_3 - \Delta u = K_o p' + \Delta\sigma_3 - \Delta u \tag{9.99}$$

However (Chapter 5)

$$\Delta u = B\Delta\sigma_3 + A_f(\Delta\sigma_1 - \Delta\sigma_3)$$

For saturated clays, $B = 1$. Substituting the preceding equation into Equation 9.99

$$\sigma_3' = K_o p' + \Delta\sigma_3 - [\Delta\sigma_3 + A_f(\Delta\sigma_1 - \Delta\sigma_3)]$$

$$= K_o p' - A_f(\Delta\sigma_1 - \Delta\sigma_3) \tag{9.100}$$

Again,

$$S_u = \frac{\sigma_1 - \sigma_3}{2} = \frac{(p' + \Delta\sigma_1) - (K_o p' + \Delta\sigma_3)}{2}$$

$$\text{or} \quad 2S_u = (\Delta\sigma_1 - \Delta\sigma_3) + (p' - K_o p')$$

$$\text{or} \quad (\Delta\sigma_1 - \Delta\sigma_3) = 2S_u - (p' - K_o p') \tag{9.101}$$

Table 9.5 Empirical equations related to S_u and p'

Reference	Relation	Remarks
Skempton (1957)	$S_{u(VST)}/p' = 0.11 + 0.0037\ PI$	For normally consolidated clay
Chandler (1988)	$S_{u(VST)}/p'_c = 0.11 + 0.0037\ PI$	Can be used in overconsoildated soil; accuracy ±25%; not valid for sensitive and fissured clays
Jamiolkowski et al. (1985)	$S_u/p'_c = 0.23 \pm 0.04$	For low overconsolidated clays
Mesri (1989)	$S_u/p'_c = 0.22$	
Bjerrum and Simons (1960)	$S_u/p' = f(LI)$	See Figure 9.79; for normally consolidated clays
Ladd et al. (1977)	$\dfrac{(S_u/p')_{\text{overconsolidated}}}{(S_u/p')_{\text{normally consolidated}}} = (OCR)^{0.8}$	

Notes: PI, plasticity index (%); $S_{u(VST)}$, undrained shear strength from vane shear test; p'_c, preconsolidation pressure; LI, liquidity index; and OCR, overconsolidation ratio.

Substituting Equation 9.101 into Equation 9.100, we obtain

$$\sigma'_3 = K_o p' - 2 S_u A_f + A_f p'(1 - K_o) \tag{9.102}$$

Substituting of Equation 9.102 into the right-hand side of Equation 9.98 and simplification yields

$$S_u = \frac{c \cos\phi + p' \sin\phi \, [K_o + A_f(1 - K_o)]}{1 + (2A_f - 1)\sin\phi} \tag{9.103}$$

For normally consolidated clays, $c = 0$; hence, Equation 9.103 becomes

$$\frac{S_u}{p'} = \frac{\sin\phi \, [K_o + A_f(1 - K_o)]}{1 + (2A_f - 1)\sin\phi} \tag{9.104}$$

There are also several empirical relations between S_u and p' suggested by various investigators. These are given in Table 9.5 and Figure 9.79.

Example 9.13

A specimen of clay was collected from the field from a depth of 15 m. Given, $p' = 183.43$ kN/m². A consolidated undrained triaxial test yielded the following results: $\phi = 32°$ and $A_f = 0.8$. Estimate the undrained shear strength S_u of the clay.

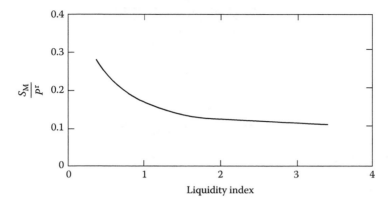

Figure 9.79 Variation of S_u/p' with liquidity index (see Table 9.5 for Bjerrum and Simon' relation).

Solution

From Equation 9.104,

$$\frac{S_u}{p'} = \frac{\sin\phi[K_o + A_f(1-K_o)]}{1+(2A_f-1)\sin\phi}$$

$$\sin\phi = \sin 32° = 0.53$$
$$K_o = 1-\sin\phi = 1-0.53 = 0.47$$

$$\frac{S_u}{183.43} = \frac{(0.53)[0.47+0.8(1-0.47)]}{1+[(2)(0.8)-1]0.53}$$

$$S_u = 65.9 \text{ kN/m}^2$$

Example 9.14

A soil profile is shown in Figure 9.80. From a laboratory consolidation test, the preconsolidation pressure of a soil specimen obtained from a depth of 8 m below the ground surface was found to be 140 kN/m². Estimate the undrained shear strength of the clay at that depth. Use Skempton's and Ladd et al.'s relations from Table 9.5 and Equation 9.90.

Solution

$$\gamma_{\text{sat(clay)}} = \frac{G_s\gamma_w + wG_s\gamma_w}{1+wG_s} = \frac{(2.7)(9.81)(1+0.3)}{1+0.3(2.7)}$$

$$= 19.02 \text{ kN/m}^3$$

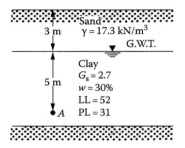

Figure 9.80 Undrained shear strength of a clay deposit.

The effective overburden pressure at A is

$$p' = 3\,(17.3) + 5\,(19.02 - 9.81) = 51.9 + 46.05 = 97.95 \text{ kN/m}^2$$

Thus, the OCR is

$$\text{OCR} = \frac{140}{97.95} = 1.43$$

From Table 9.5 (Ladd et al.'s relationship)

$$\left(\frac{S_u}{p'}\right)_{OC} = \left(\frac{S_u}{p'}\right)_{NC} (\text{OCR})^{0.8} \qquad (E9.5)$$

However, from Table 9.5 (Skempton's relationship)

$$\left(\frac{S_{u(VST)}}{p'}\right)_{NC} = 0.11 + 0.037 \text{ PI} \qquad (E9.6)$$

From Equation 9.90

$$S_u = \lambda S_{u(VST)} = [1.7 - 0.54 \log(\text{PI})]S_{u(VST)}$$
$$= [1.7 - (0.54)\log(52 - 31)]S_{u(VST)} = 0.986 S_{u(VST)}$$

$$S_{u(VST)} = \frac{S_u}{0.986} \qquad (E9.7)$$

Combining Equations E9.6 and E9.7

$$\left(\frac{S_u}{0.986 p'}\right)_{NC} = 0.11 + 0.0037 \text{ PI}$$

$$\left(\frac{S_u}{p'}\right)_{NC} = (0.986)[0.11 + 0.0037(52 - 31)] = 0.185 \qquad (E9.8)$$

From Equations E9.5 and E9.8

$$S_{u(OC)} = (0.185)(1.43)^{0.8}(97.95) = 24.12 \text{ kN/m}^2$$

9.24 CREEP IN SOILS

Like metals and concrete, most soils exhibit creep, that is, continued deformation under a sustained loading (Figure 9.81). In order to understand Figure 9.81, consider several similar clay specimens subjected to standard undrained loading. For specimen no. 1, if a deviator stress $(\sigma_1 - \sigma_3)_1 <$ $(\sigma_1 - \sigma_3)_{failure}$ is applied, the strain versus time ($\in$ versus t) relation will be similar to that shown by curve 1. If specimen no. 2 is subjected to a deviator stress $(\sigma_1 - \sigma_3)_2$ such that $(\sigma_1 - \sigma_3)_{failure} > (\sigma_1 - \sigma_3)_2 > (\sigma_1 - \sigma_3)_1$, the strain versus time relation may be similar to that shown curve 2. After the occurrence of a large strain, creep failure will take place in the specimen.

In general, the strain versus time plot for a given soil can be divided into three parts: primary, secondary, and tertiary. The primary part is the transient stage; this is followed by a steady state, which is secondary creep. The tertiary part is the stage during which there is a rapid strain, which results in failure. These three steps are shown in Figure 9.81. Although the

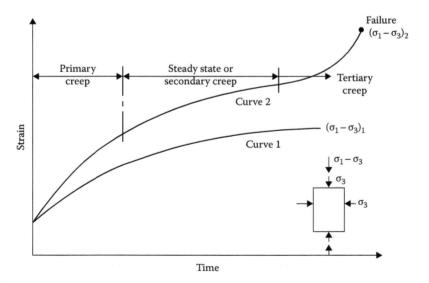

Figure 9.81 Creep in soils.

secondary stage is referred to as steady-state creep, in reality a true steady-state creep may not really exist (Singh and Mitchell, 1968).

It was observed by Singh and Mitchell (1968) that for most soils (i.e., sand, clay—dry, wet, normally consolidated, and overconsolidated) the logarithm of strain rate has an approximately linear relation with the logarithm of time. This fact is illustrated in Figure 9.82 for remolded San Francisco Bay mud. The strain rate is defined as

$$\dot{\epsilon} = \frac{\Delta \varepsilon}{\Delta t} \tag{9.105}$$

where
$\dot{\epsilon}$ is the strain rate
ε is the strain
t is the time

From Figure 9.82, it is apparent that the slope of the log $\dot{\epsilon}$ versus log t plot for a given soil is constant irrespective of the level of the deviator stress. When the failure stage due to creep at a given deviator stress level is reached, the log $\dot{\epsilon}$ versus log t plot will show a reversal of slope as shown in Figure 9.83.

Figure 9.84 shows the nature of the variation of the creep strain rate with deviator stress $D = \sigma_1 - \sigma_3$ at a given time t after the start of the creep.

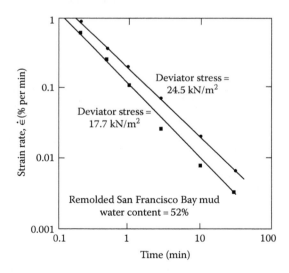

Figure 9.82 Plot of log $\dot{\epsilon}$ versus log t during undrained creep of remolded San Francisco Bay mud. [After Singh, A. and Mitchell, J. K., *J. Soil Mech. Found. Eng. Div.*, ASCE, 94(SMI), 21, 1968.]

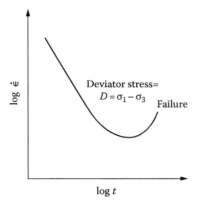

Figure 9.83 Nature of variation of log $\dot{\varepsilon}$ versus log t for a given deviator stress showing the failure stage at large strains.

For small values of the deviator stress, the curve of log $\dot{\varepsilon}$ versus D is convex upward. Beyond this portion, log $\dot{\varepsilon}$ versus D is approximately a straight line. When the value of D approximately reaches the strength of the soil, the curve takes an upward turn, signaling impending failure.

For a mathematical interpretation of the variation of strain rate with the deviator stress, several investigators (e.g., Christensen and Wu, 1964; Mitchell et al., 1968) have used the *rate-process theory*. Christensen and Das (1973) also used the rate-process theory to predict the rate of erosion of cohesive soils.

The fundamentals of the rate-process theory can be explained as follows. Consider the soil specimen shown in Figure 9.85. The deviator stress on the specimen is $D = \sigma_1 - \sigma_3$. Let the shear stress along a plane AA in the specimen be equal to τ. The shear stress is resisted by the bonds at the points of contact of the particles along AA. Due to the shear stress τ the weaker

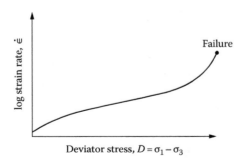

Figure 9.84 Variation of the strain rate $\dot{\varepsilon}$ with deviator stress at a given time t after the start of the test.

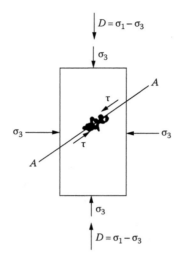

Figure 9.85 Fundamentals of rate-process theory.

bonds will be overcome, with the result that shear displacement occurs at these localities. As this displacement proceeds, the force carried by the weaker bonds is transmitted partly or fully to stronger bonds. The effect of applied shear stress can thus be considered as making some flow units cross the energy barriers as shown in Figure 9.86, in which ΔF is equal to the activation energy (in cal/mol of flow unit). The frequency of activation of the flow units to overcome the energy barriers can be given by

$$k' = \frac{kT}{h} \exp\left(-\frac{\Delta F}{RT}\right) = \frac{kT}{h} \exp\left(-\frac{\Delta F}{NkT}\right) \tag{9.106}$$

where
  k' is the frequency of activation
  k is Boltzmann's constant = 1.38×10^{-16} erg/K = 3.29×10^{-24} cal/K

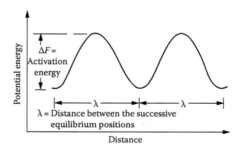

Figure 9.86 Definition of activation energy.

T is the absolute temperature

h is Plank's constant = 6.624×10^{-27} erg/s

ΔF is the free energy of activation, cal/mol

R is the universal gas constant

N is Avogadro's number = 6.02×10^{23}

Now, referring to Figure 9.87 when a force f is applied across a flow unit, the energy-barrier height is reduced by $f\lambda/2$ in the direction of the force and increased by $f\lambda/2$ in the opposite direction. By this, the frequency of activation in the direction of the force is

$$\underset{\rightarrow}{k'} = \frac{kT}{h}\exp\left(-\frac{\Delta F/N - f\lambda/2}{kT}\right) \qquad (9.107)$$

and, similarly, the frequency of activation in the opposite direction becomes

$$\underset{\leftarrow}{k'} = \frac{kT}{h}\exp\left(-\frac{\Delta F/N + \lambda f/2}{kT}\right) \qquad (9.108)$$

where λ is the distance between successive equilibrium positions.

So, the net frequency of activation in the direction of the force is equal to

$$\underset{\rightarrow}{k'} - \underset{\leftarrow}{k'} = \frac{kT}{h}\left[\exp\left(-\frac{\Delta F/N - f\lambda/2}{kT}\right) - \exp\left(-\frac{\Delta F/N + f\lambda/2}{kT}\right)\right]$$

$$= \frac{2kT}{h}\exp\left(-\frac{\Delta F}{RT}\right)\sinh\left(\frac{\lambda f}{2kT}\right) \qquad (9.109)$$

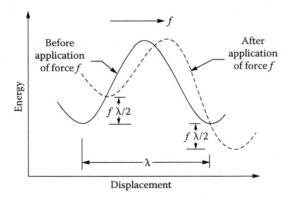

Figure 9.87 Derivation of Equation 9.111.

The rate of strain in the direction of the applied force can be given by

$$\dot{\epsilon} = x\left(\underset{\rightarrow}{k'} - \underset{\leftarrow}{k'}\right) \tag{9.110}$$

where x is a constant depending on the successful barrier crossings. So

$$\dot{\epsilon} = 2x\frac{kT}{h}\exp\left(-\frac{\Delta F}{RT}\right)\sinh\left(\frac{f\lambda}{2kT}\right) \tag{9.111}$$

In the previous equation

$$f = \frac{\tau}{S} \tag{9.112}$$

where
  τ is the shear stress
  S is the number of flow units per unit area

For triaxial shear test conditions as shown in Figure 9.85

$$\tau_{\max} = \frac{D}{2} = \frac{\sigma_1 - \sigma_3}{2} \tag{9.113}$$

Combining Equations 9.112 and 9.113

$$f = \frac{D}{2S} \tag{9.114}$$

Substituting Equation 9.114 into Equation 9.111, we get

$$\dot{\epsilon} = 2x\frac{kT}{h}\exp\left(-\frac{\Delta F}{RT}\right)\sinh\left(\frac{D\lambda}{4kST}\right) \tag{9.115}$$

For large stresses to cause significant creep—that is, $D > 0.25$ [$D_{\max} = 0.25$ (Mitchell et al., 1968)] the magnitude of $D\lambda/4kST$ is greater than 1. So, in that case

$$\sinh\frac{D\lambda}{4kST} \approx \frac{1}{2}\exp\left(\frac{D\lambda}{4kST}\right) \tag{9.116}$$

Hence, from Equations 9.115 and 9.116

$$\dot{\epsilon} = x\frac{kT}{h}\exp\left(-\frac{\Delta F}{RT}\right)\exp\left(\frac{D\lambda}{4kST}\right) \tag{9.117}$$

$$\dot{\epsilon} = A \exp(BD) \tag{9.118}$$

where

$$A = x \frac{kT}{h} \exp\left(-\frac{\Delta F}{RT}\right) \tag{9.119}$$

and

$$B = \frac{\lambda}{4kST} \tag{9.120}$$

The quantity A is likely to vary with time because of the variation of x and ΔF with time. B is a constant for a given value of the effective consolidation pressure.

Figure 9.88 shows the variation of the undrained creep rate $\dot{\epsilon}$ with the deviator stress D for remolded illite at elapsed times t equal to 1, 10, 100, and 1000 min. From this, note that at any given time the following apply:

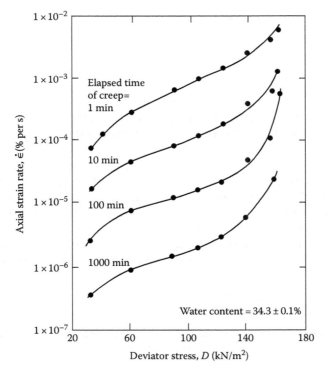

Figure 9.88 Variation of strain rate with deviator stress for undrained creep of remolded illite. [After Mitchell, J. K. et al., J. Soil Mech. Found. Eng. Div., ASCE, 95(SM5), 1219, 1969.]

Table 9.6 Values of ΔF for some soils

Soil	ΔF (kcal/mol)
Saturated, remolded illite; water content 30%–43%	25–40
Dried illite, samples air-dried from saturation, then evacuated over desiccant	37
Undisturbed San Francisco Bay mud	25–32
Dry Sacramento River sand	~25

Source: After Mitchell, J. K. et al., *J. Soil Mech. Found. Eng. Div.,* ASCE, 95(SM5), 1219, 1969.

1. For $D < 49$ kN/m², the log $\dot{\epsilon}$ versus D plot is convex upward following the relation given by Equation 9.115, $\dot{\epsilon} = 2A \sinh(BD)$. For this case, $D\lambda/4SkT < 1$.
2. For 128 kN/m² $> D > 49$ kN/m², the log $\dot{\epsilon}$ versus D plot is approximately a straight line following the relation given by Equation 9.118, $\dot{\epsilon} = Ae^{BD}$. For this case, $D\lambda/4SkT > 1$.
3. For $D > 128$ kN/m², the failure stage is reached when the strain rate rapidly increases; this stage cannot be predicted by Equations 9.115 or 9.118.

Table 9.6 gives the values of the experimental activation energy ΔF for four different soils.

9.25 OTHER THEORETICAL CONSIDERATIONS: YIELD SURFACES IN THREE DIMENSIONS

Comprehensive failure conditions or yield criteria were first developed for metals, rocks, and concrete. In this section, we will examine the application of these theories to soil and determine the yield surfaces in the principal stress space. The notations σ_1', σ_2', and σ_3' will be used for effective principal stresses without attaching an order of magnitude—that is, σ_1', σ_2', and σ_3' are not necessarily major, intermediate, and minor principal stresses, respectively.

Von Mises (1913) proposed a simple yield function, which may be stated as

$$F = \left(\sigma_1' - \sigma_2'\right)^2 + \left(\sigma_2' - \sigma_3'\right)^2 + \left(\sigma_3' - \sigma_1'\right)^2 - 2Y^2 = 0 \tag{9.121}$$

where Y is the yield stress obtained in axial tension. However, the octahedral shear stress can be given by the relation

$$\tau_{\text{oct}} = \frac{1}{3}\sqrt{\left(\sigma_1' - \sigma_2'\right)^2 + \left(\sigma_2' - \sigma_3'\right)^2 + \left(\sigma_3' - \sigma_1'\right)^2}$$

Thus, Equation 9.121 may be written as

$$3\tau_{oct}^2 = 2Y^2$$

$$\text{or} \quad \tau_{oct} = \sqrt{\frac{2}{3}}\, Y \tag{9.122}$$

Equation 9.122 means that failure will take place when the octahedral shear stress reaches a constant value equal to $\sqrt{2/3}Y$. Let us plot this on the octahedral plane $(\sigma_1' + \sigma_2' + \sigma_3' = \text{const})$, as shown in Figure 9.89. The locus will be a circle with a radius equal to $\tau_{oct} = \sqrt{2/3}Y$ and with its center at point a. In Figure 9.89a, Oa is the octahedral normal stress $(\sigma_1' + \sigma_2' + \sigma_3')/3 = \sigma_{oct}'$; also, $ab = \tau_{oct}$ and $Ob = \sqrt{\sigma_{oct}'^2 + \tau_{oct}^2}$. Note that the locus is unaffected by the value of σ_{oct}'. Thus, various values of σ_{oct}' will generate a circular cylinder coaxial with the hydrostatic axis, which is a yield surface (Figure 9.89b).

Another yield function suggested by Tresca (1868) can be expressed in the form

$$\sigma_{max} - \sigma_{min} = 2k \tag{9.123}$$

Equation 9.123 assumes that failure takes place when the maximum shear stress reaches a constant critical value. The factor k of Equation 9.123 is defined for the case of simple tension by Mohr's circle shown in Figure 9.90. Note that for soils this is actually the $\phi = 0$ condition. In Figure 9.90, the yield function is plotted on the octahedral plane

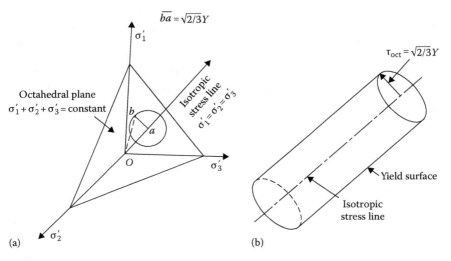

Figure 9.89 Yield surface—Von Mises criteria: (a) plot of Equation 9.122 on the octahedral plane; (b) yield surface.

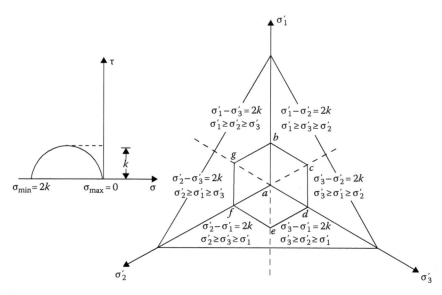

Figure 9.90 Yield surface—Tresca criterion.

$(\sigma_1' + \sigma_2' + \sigma_3' = \text{const})$. The locus is a regular hexagon. Point a is the point of intersection of the *hydrostatic axis* or *isotropic stress line* with octahedral plane, and so it represents the octahedral normal stress. Point b represents the failure condition in compression for $\sigma_1' > \sigma_2' = \sigma_3'$, and point e represents the failure condition in extension with $\sigma_2' = \sigma_3' > \sigma_1'$. Similarly, point d represents the failure condition for $\sigma_3' > \sigma_1' = \sigma_2'$, point g for $\sigma_1' = \sigma_2' > \sigma_3'$, point f for $\sigma_2' > \sigma_3' = \sigma_1'$, and point c for $\sigma_3' = \sigma_1' > \sigma_2'$. Since the locus is unaffected by the value of σ_{oct}', the yield surface will be a hexagonal cylinder.

We have seen from Equation 9.24 that, for the Mohr–Coulomb condition of failure, $(\sigma_1' - \sigma_3') = 2c\cos\phi + (\sigma_1' + \sigma_3')\sin\phi$, or $(\sigma_1' - \sigma_3')^2 = [2c\cos\phi + (\sigma_1' + \sigma_3')\sin\phi]^2$. In its most general form, this can be expressed as

$$\left\{(\sigma_1' - \sigma_2')^2 - \left[2c\cos\phi + (\sigma_1' + \sigma_2')\sin\phi\right]\right\}^2$$
$$\times \left\{(\sigma_2' - \sigma_3')^2 - \left[2c\cos\phi + (\sigma_2' + \sigma_3')\sin\phi\right]\right\}^2$$
$$\times \left\{(\sigma_3' - \sigma_1') - \left[2c\cos\phi + (\sigma_3' + \sigma_1')\sin\phi\right]\right\}^2 = 0 \tag{9.124}$$

When the yield surface defined by Equation 9.124 is plotted on the octahedral plane, it will appear as shown in Figure 9.91. This is an irregular hexagon

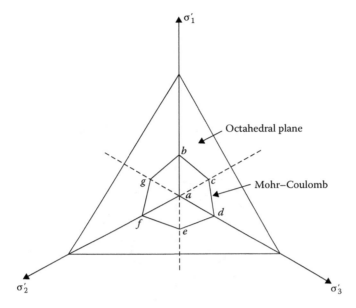

Figure 9.91 Mohr–Coulomb failure criterion.

in section with nonparallel sides of equal length. Point a in Figure 9.91 is the point of intersection of the hydrostatic axis with the octahedral plane. Thus, the yield surface will be a hexagonal cylinder coaxial with the isotropic stress line.

Figure 9.92 shows a comparison of the three yield functions described previously. In a Rendulic-type plot, the failure envelopes will appear in a manner shown in Figure 9.92b. At point a, $\sigma_1' = \sigma_2' = \sigma_3' = \sigma'$ (say). At point b, $\sigma_1' = \sigma' + \overline{ba'} = \sigma' + ab\sin\theta$, where $\theta = \cos^{-1}(1/\sqrt{3})$. Thus

$$\sigma_1' = \sigma' + \sqrt{\frac{2}{3}}\,\overline{ab} \tag{9.125}$$

$$\sigma_2' = \sigma_3' = \sigma' - \frac{aa'}{\sqrt{2}} = \sigma' - \frac{\overline{ab}\cos\theta}{\sqrt{2}} = \sigma' - \frac{1}{\sqrt{6}}\,\overline{ab} \tag{9.126}$$

For the Mohr–Coulomb failure criterion, $\sigma_1' - \sigma_3' = 2c\cos\phi + (\sigma_1' + \sigma_3')\sin\phi$. Substituting Equations 9.125 and 9.126 in the preceding equation, we obtain

$$\left(\sigma' + \sqrt{\frac{2}{3}}\,\overline{ab} - \sigma' + \frac{1}{\sqrt{6}}\,\overline{ab}\right) = 2c\cos\phi + \left(\sigma' + \sqrt{\frac{2}{3}}\,\overline{ab} + \sigma' - \frac{1}{\sqrt{6}}\,\overline{ab}\right)\sin\phi$$

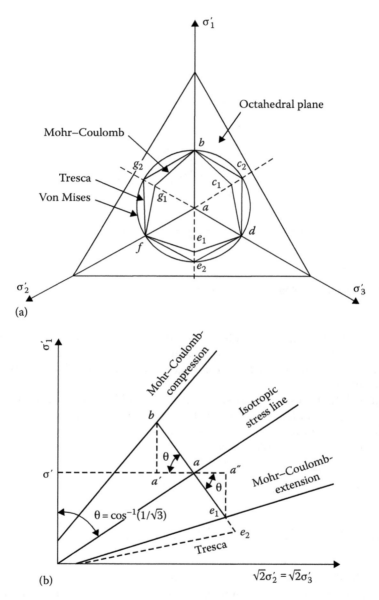

Figure 9.92 (a) Comparison of Von Mises, Tresca, and Mohr–Coulomb yield functions; (b) failure envelopes in a Rendulic type plot.

or

$$\overline{ab}\left[\left(\sqrt{\frac{2}{3}}+\frac{1}{\sqrt{6}}\right)-\left(\sqrt{\frac{2}{3}}-\frac{1}{\sqrt{6}}\right)\sin\phi\right]=2(c\cos\phi+\sigma'\sin\phi)$$

or $$\overline{ab}\frac{3}{\sqrt{6}}\left(1-\frac{1}{3}\sin\phi\right)=2(c\cos\phi+\sigma'\sin\phi) \tag{9.127}$$

Similarly, for *extension* (i.e., at point e_1)

$$\sigma'_1=\sigma'-\overline{e_1a''}=\sigma'-\overline{ae_1}\sin\theta=\sigma'-\sqrt{\frac{2}{3}}\overline{ae_1} \tag{9.128}$$

$$\sigma'_2=\sigma'_3=\sigma'+\frac{\overline{aa''}}{\sqrt{2}}=\sigma'+\frac{\overline{ae_1}\cos\theta}{\sqrt{2}}=\sigma'\frac{1}{\sqrt{6}}\overline{ae_1} \tag{9.129}$$

Now $\sigma'_3-\sigma'_1=2c\cos\phi+(\sigma'_3+\sigma'_1)\sin\phi$. Substituting Equations 9.128 and 9.129 into the preceding equation, we get

$$\overline{ae_1}\left[\left(\sqrt{\frac{2}{3}}+\frac{1}{\sqrt{6}}\right)+\left(\sqrt{\frac{2}{3}}-\frac{1}{\sqrt{6}}\right)\sin\phi\right]=2(c\cos\phi+\sigma'\sin\phi) \tag{9.130}$$

or

$$\overline{ae_1}\frac{3}{\sqrt{6}}\left(1+\frac{1}{3}\sin\phi\right)=2(c\cos\phi+\sigma'\sin\phi) \tag{9.131}$$

Equating Equations 9.127 and 9.131

$$\frac{\overline{ab}}{\overline{ae_1}}=\frac{1+\frac{1}{3}\sin\phi}{1-\frac{1}{3}\sin\phi} \tag{9.132}$$

Table 9.7 gives the ratios of $\overline{ab}$ to $\overline{ae_1}$ for various values of ϕ. Note that this ratio is not dependent on the value of cohesion, c.

It can be seen from Figure 9.92a that the Mohr–Coulomb and the Tresca yield functions coincide for the case $\phi=0$.

Table 9.7 Ratio of $\overline{ab}$ to $\overline{ae_1}$
(Equation 9.132)

ϕ	$\overline{ab}/\overline{ae_1}$
40	0.647
30	0.715
20	0.796
10	0.889
0	1.0

Von Mises' yield function (Equation 9.121) can be modified to the form

$$\left(\sigma_1' - \sigma_2'\right)^2 + \left(\sigma_2' - \sigma_1'\right)^2 + \left(\sigma_3' - \sigma_1'\right)^2 = \left[c + \frac{k_2}{3}\left(\sigma_1' + \sigma_2' + \sigma_3'\right)\right]^2$$

$$\text{or} \quad \left(\sigma_1' - \sigma_2'\right)^2 + \left(\sigma_2' - \sigma_3'\right)^2 + \left(\sigma_3' - \sigma_1'\right)^2 = (c + k_2\sigma_{oct}')^2 \qquad (9.133)$$

where
k_2 is a function of sin ϕ
c is cohesion

Equation 9.133 is called the extended Von Mises' yield criterion.
Similarly, Tresca's yield function (Equation 9.123) can be modified to the form

$$\left[\left(\sigma_1' - \sigma_2'\right)^2 - (c + k_3\sigma_{oct}')^2\right]$$
$$\times\left[\left(\sigma_2' - \sigma_3'\right) - (c + k_3\sigma_{oct}')^2\right]$$
$$\times\left[\left(\sigma_3' - \sigma_1'\right)^2 - (c + k_3\sigma_{oct}')^2\right] \qquad (9.134)$$

where
k_3 is a function of sin ϕ
c is cohesion

Equation 9.134 is generally referred to as the extended Tresca criterion.

9.26 EXPERIMENTAL RESULTS TO COMPARE THE YIELD FUNCTIONS

Kirkpatrick (1957) devised a special shear test procedure for soils, called the *hollow cylinder test*, which provides the means for obtaining the variation in the three principal stresses. The results from this test can be used to compare the validity of the various yield criteria suggested in the preceding section.

A schematic diagram of the laboratory arrangement for the hollow cylinder test is shown in Figure 9.93a. A soil specimen in the shape of a hollow cylinder is placed inside a test chamber. The specimen is encased by both an inside and an outside membrane. As in the case of a triaxial test, radial pressure on the soil specimen can be applied through water. However, in this

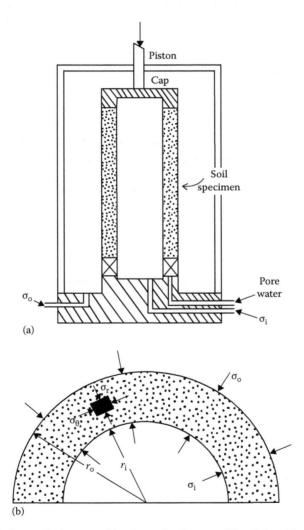

Figure 9.93 Hollow cylinder test: (a) schematic diagram; (b) relationship for principal stresses in the soil specimen.

type of test, the pressures applied to the inside and outside of the specimen can be controlled separately. Axial pressure on the specimen is applied by a piston. In the original work of Kirkpatrick, the axial pressure was obtained from load differences applied to the cap by the fluid on top of the specimen (i.e., piston pressure was not used; see Equation 9.141).

The relations for the principal stresses in the soil specimen can be obtained as follows (see Figure 9.93b). Let σ_o and σ_i be the outside and inside fluid pressures, respectively. For *drained tests*, the total stresses σ_o and σ_i are equal to the effective stresses, σ'_o and σ'_i. For an axially symmetrical case, the equation of continuity for a given point in the soil specimen can be given by

$$\frac{d\sigma'_r}{dr} + \frac{\sigma'_r - \sigma'_\theta}{r} = 0 \tag{9.135}$$

where
σ'_r and σ'_θ are the radial and tangential stresses; respectively
r is the radial distance from the center of the specimen to the point

We will consider a case where the failure in the specimen is caused by increasing σ'_i and keeping σ'_o constant. Let

$$\sigma'_\theta = \lambda \sigma'_r \tag{9.136}$$

Substituting Equation 9.136 in Equation 9.135, we get

$$\frac{d\sigma'_r}{dr} + \frac{\sigma'_r(1-\lambda)}{r} = 0$$

$$\text{or} \quad \frac{1}{\lambda-1}\int \frac{d\sigma'_r}{\sigma'_r} = \int \frac{dr}{r}$$

$$\sigma'_r = Ar^{\lambda-1} \tag{9.137}$$

where A is a constant.

However, $\sigma'_r = \sigma'_o$ at $r = r_o$, which is the outside radius of the specimen. So

$$A = \frac{\sigma'_o}{r_o^{\lambda-1}} \tag{9.138}$$

Combining Equations 9.137 and 9.138

$$\sigma'_r = \sigma'_o \left(\frac{r}{r_o} \right)^{\lambda-1} \tag{9.139}$$

Again, from Equations 9.136 and 9.139

$$\sigma'_\theta = \lambda \sigma'_o \left(\frac{r}{r_o} \right)^{\lambda-1} \tag{9.140}$$

The effective axial stress σ'_a can be given by the equation

$$\sigma'_a = \frac{\sigma'_o \left(\pi r_o^2 \right) - \sigma'_i \left(\pi r_i^2 \right)}{\pi r_o^2 - \pi r_i^2} = \frac{\sigma'_o r_o^2 - \sigma'_i r_i^2}{r_o^2 - r_i^2} \tag{9.141}$$

where r_i is the inside radius of the specimen.

At failure, the radial and tangential stresses at the inside face of the specimen can be obtained from Equations 9.139 and 9.140:

$$\sigma'_{r(inside)} = \left(\sigma'_i \right)_{failure} = \sigma'_o \left(\frac{r_i}{r_o} \right)^{\lambda-1} \tag{9.142}$$

or

$$\left(\frac{\sigma'_i}{\sigma'_o} \right)_{failure} = \left(\frac{r_i}{r_o} \right)^{\lambda-1} \tag{9.143}$$

$$\sigma'_{\theta(inside)} = \left(\sigma'_\theta \right)_{failure} = \lambda \sigma'_o \left(\frac{r_i}{r_o} \right)^{\lambda-1} \tag{9.144}$$

To obtain σ'_a at failure, we can substitute Equation 9.142 into Equation 9.141

$$\begin{aligned}
\left(\sigma'_a \right)_{failure} &= \frac{\sigma'_o \left[(r_o/r_i)^2 - (\sigma'_i/\sigma'_o) \right]}{(r_o/r_i)^2 - 1} \\
&= \frac{\sigma'_o [(r_o/r_i)^2 - (r_o/r_i)^{1-\lambda}]}{(r_o/r_i)^2 - 1}
\end{aligned} \tag{9.145}$$

From the earlier derivations, it is obvious that for this type of test (i.e., increasing σ'_i to cause failure and keeping σ'_o constant) the major and minor principal stresses are σ_r' and σ_θ'. The intermediate principal stress is σ_a'. For granular soils the value of the cohesion c is 0, and from the Mohr–Coulomb failure criterion

$$\left(\frac{\text{Minor principal stress}}{\text{Major principal stress}}\right)_{\text{failure}} = \frac{1-\sin\phi}{1+\sin\phi}$$

$$\text{or} \quad \left(\frac{\sigma_\theta'}{\sigma_r'}\right)_{\text{failure}} = \frac{1-\sin\phi}{1+\sin\phi} \tag{9.146}$$

Comparing Equations 9.136 and 9.146

$$\frac{1-\sin\phi}{1+\sin\phi} = \tan^2\left(45° - \frac{\phi}{2}\right) = \lambda \tag{9.147}$$

The results of some hollow cylinder tests conducted by Kirkpatrick (1957) on a sand are given in Table 9.8, together with the calculated values of λ, $(\sigma_a')_{\text{failure}}$, $(\sigma_r')_{\text{failure}}$, and $(\sigma_\theta')_{\text{failure}}$.

A comparison of the yield functions on the octahedral plane and the results of Kirkpatrick is given in Figure 9.94. The results of triaxial compression and extension tests conducted on the same sand by Kirkpatrick are also shown in Figure 9.94. The experimental results indicate that the

Table 9.8 Results of Kirkpatrick's hollow cylinder test on a sand

Test no.	$(\sigma_i')_{\text{failure}}$[a] (kN/m²)	σ_o'[b] (kN/m²)	λ (from Equation 9.143)[c]	$\sigma_{\theta(\text{inside})}'$ at failure[d] (kN/m²)	$\sigma_{\theta(\text{outside})}'$ at failure[e] (kN/m²)	σ_a' (from Equation 9.141) (kN/m²)
1	146.3	99.4	0.196	28.7	19.5	72.5
2	187.5	129.0	0.208	39.0	26.8	91.8
3	304.2	211.1	0.216	65.7	45.6	153.9
4	384.2	265.7	0.215	82.5	57.1	192.9
5	453.7	316.0	0.192	87.0	60.7	222.9
6	473.5	330.6	0.198	93.8	65.4	234.9
7	502.9	347.1	0.215	107.8	74.6	247.7
8	532.4	372.7	0.219	116.6	81.6	268.4
9	541.2	378.1	0.197	106.3	74.5	263.6

[a] $(\sigma_i')_{\text{failure}} = \sigma_{r(\text{inside})}'$ at failure.
[b] $(\sigma_o') = \sigma_{r(\text{outside})}'$ at failure.
[c] For these tests, $r_o = 50.8$ mm and $r_i = 31.75$ mm.
[d] $\sigma_{\theta(\text{inside})}' = \lambda(\sigma_i')_{\text{failure}}$.
[e] $\sigma_{o(\text{outside})}' = \lambda(\sigma_o')_{\text{failure}}$.

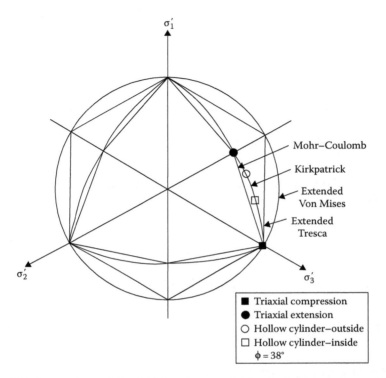

Figure 9.94 Comparison of the yield functions on the octahedral plane along with the results of Kirkpatrick.

Mohr–Coulomb criterion gives a better representation for soils than the extended Tresca and Von Mises criteria. However, the hollow cylinder tests produced slightly higher values of ϕ than those from the triaxial tests.

Wu et al. (1963) also conducted a type of hollow cylinder shear test with sand and clay specimens. In these tests, failure was produced by increasing the inside, outside, and axial stresses on the specimens in various combinations. The axial stress increase was accomplished by the application of a force P on the cap through the piston as shown in Figure 9.93. Triaxial compression and extension tests were also conducted. Out of a total of six series of tests, there were two in which failure was caused by increasing the outside pressure. For those two series of tests, $\sigma'_\theta > \sigma'_a > \sigma'_r$. Note that this is opposite to Kirkpatrick's tests, in which $\sigma'_r > \sigma'_a > \sigma'_\theta$. Based on the Mohr–Coulomb criterion, we can write (see Equation 9.25) $\sigma'_{max} = \sigma'_{min} N + 2cN^{1/2}$. So, for the case where $\sigma'_\theta > \sigma'_a > \sigma'_r$,

$$\sigma'_\theta = \sigma'_r N + 2cN^{1/2} \tag{9.148}$$

The value of N in the previous equation is $\tan^2(45° + \phi/2)$, and so the λ in Equation 9.136 is equal to $1/N$. From Equation 9.135

$$\frac{d\sigma'_r}{dr} = \frac{\sigma'_\theta - \sigma'_r}{r}$$

Combining the preceding equation and Equation 9.148, we get

$$\frac{d\sigma'_r}{dr} = \frac{1}{r}\left[\sigma'_r(N-1) + 2cN^{1/2}\right] \tag{9.149}$$

Using the boundary condition that at $r = r_i$, $\sigma'_r = \sigma'_i$, Equation 9.149 gives the following relation:

$$\sigma'_r = \left(\sigma'_i + \frac{2cN^{1/2}}{N-1}\right)\left(\frac{r}{r_i}\right)^{N-1} - \frac{2cN^{1/2}}{N-1} \tag{9.150}$$

Also, combining Equations 9.148 and 9.150

$$\sigma'_\theta = \left(\sigma'_i + \frac{2cN^{3/2}}{N-1}\right)\left(\frac{r}{r_i}\right)^{N-1} - \frac{2cN^{1/2}}{N-1} \tag{9.151}$$

At failure, $\sigma'_{r(outside)} = \left(\sigma'_o\right)_{failure}$. So

$$\left(\sigma'_o\right)_{failure} = \left(\sigma'_i + \frac{2cN^{1/2}}{N-1}\right)\left(\frac{r_o}{r_i}\right)^{N-1} - \frac{2cN^{1/2}}{N-1} \tag{9.152}$$

For granular soils and normally consolidated clays, $c = 0$. So, at failure, Equations 9.150 and 9.151 simplify to the form

$$\left(\sigma'_r\right)_{outside\,at\,failure} = \left(\sigma'_o\right)_{failure} = \sigma'_i\left(\frac{r_o}{r_i}\right)^{N-1} \tag{9.153}$$

and $$\left(\sigma'_\theta\right)_{outside\,at\,failure} = \sigma'_i N\left(\frac{r_o}{r_i}\right)^{N-1} \tag{9.154}$$

Hence $$\left(\frac{\sigma'_r}{\sigma'_\theta}\right)_{failure} = \frac{\text{Minor principal effective stress}}{\text{Major principal effective stress}} = \frac{1}{N} = \lambda \tag{9.155}$$

Compare Equations 9.136 and 9.155.

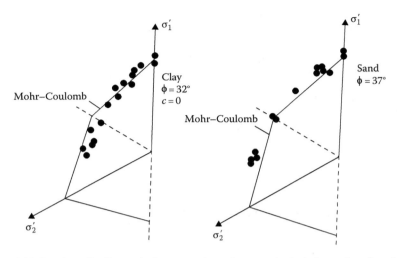

Figure 9.95 Results of hollow cylinder tests plotted on octahedral plane $\sigma_1' + \sigma_2' + \sigma_3' = 1$. [After Wu, T. H. et al., *J. Soil Mech. Found. Eng. Div.*, ASCE, 89(SMI), 145, 1963.]

Wu et al. also derived equations for σ_r' and σ_θ' for the case $\sigma_a' > \sigma_\theta' > \sigma_r'$. Figure 9.95 shows the results of Wu et al. plotted on the octahedral plane $\sigma_1' + \sigma_2' + \sigma_3' = 1$. The Mohr–Coulomb yield criterion has been plotted by using the triaxial compression and extension test results. The results of other hollow cylinder tests are plotted as points. In general, there is good agreement between the experimental results and the yield surface predicted by the Mohr–Coulomb theory. However, as in Kirkpatrick's test, hollow cylinder tests indicated somewhat higher values of ϕ than triaxial tests in the case of sand. In the case of clay, the opposite trend is generally observed.

Example 9.15

A sand specimen was subjected to a drained shear test using hollow cylinder test equipment. Failure was caused by increasing the inside pressure while keeping the outside pressure constant. At failure, $\sigma_o = 28$ kN/m² and $\sigma_i = 38.3$ kN/m². The inside and outside radii of the specimen were 40 and 60 mm, respectively.

    a. Calculate the soil friction angle.
    b. Calculate the axial stress on the specimen at failure.

Solution

    a. From Equation 9.143:

$$\left(\frac{\sigma_i'}{\sigma_o'}\right)_f = \left(\frac{r_i}{r_o}\right)^{\lambda-1}$$

$$\log\left(\frac{38.3}{28}\right) = (\lambda - 1)\log\left(\frac{40}{60}\right)$$

$$\lambda - 1 = \frac{0.136}{-0.176} = -0.773$$

Equation 9.147:

$$\lambda = \tan^2(45 - \phi/2) = 0.227$$

$$\phi = 39°$$

b. Equation 9.141:

$$\sigma_a' = \frac{\sigma_o' r_o^2 - \sigma_i' r_i^2}{r_o^2 - r_i^2} = \frac{(28)(60)^2 - (38.3)(40)^2}{60^2 - 40^2} = 19.76\,\text{kN/m}^2$$

REFERENCES

American Society for Testing and Materials, *Annual Book of ASTM Standards*, vol. 04.08, ASTM, Philadelphia, PA, 2002.

Berger, L. and J. Gnaedinger, Strength regain of clays, ASTM *Bull.*, 160, 72–78, Sept. 1949.

Bishop, A. W. and L. Bjerrum, The relevance of the triaxial test to the solution of stability problems, *Proc. Res. Conf. Shear Strength Cohesive Soils*, Boulder, Colarado, ASCE, 437–501, 1960.

Bishop, A. W. and D. J. Henkel, *The Measurement of Soil Properties in the Triaxial Test*, 2nd edn., Edward Arnold, London, U.K., 1969.

Bjerrum, L., Embankments on soft ground, *Proc. Spec. Conf. Earth and Earth-Supported Struct.*, Purdue University, Indiana, 2, ASCE, 1–54, 1972.

Bjerrum, L. and N. E. Simons, Comparison of shear strength characteristics of normally consolidated clay, *Proc. Res. Conf. Shear Strength Cohesive Soils*, Boulder, Colorado. ASCE, 711–726, 1960.

Bjerrum, L., Simons, and I. Torblaa, The Effect of Time on Shear Strength of a Soft Marine Clay, *Proc. Brussels Conf. Earth Press. Prob.*, 1, 148–158, 1958.

Bolten, M. D., The strength and dilatancy of sands, *Geotechnique*, 36(1), 65–75, 1986.

Booker, W. E. and H. O. Ireland, Earth pressures at rest related to stress history, *Can. Geotech. J.*, 2(1), 1–15, 1965.

Carlson, L., Determination of the in situ shear strength of undisturbed clay by means of a rotating auger, *Proc. Second Int. Conf. Soil Mech. Found. Eng.*, Rotterdam, Netherlands, 1, 265–270, 1948.

Casagrande, A. and N. Carrillo, Shear failure of anisotropic materials, in *Contribution to Soil Mechanics 1941–1953*, Boston Society of Civil Engineers, Boston, MA, 1944.

Casagrande, A. and S. D. Wilson, Investigation of the effects of the long-time loading on the shear strength of clays and shales at constant water content, Report to the U.S. Waterways Experiment Station, Harvard University, 1949.

Casagrande, A. and S. D. Wilson, Effect of the rate of loading on the strength of clays and shales at constant water content, *Geotechnique*, 1, 251–263, 1951.

Castellanos, B. A. and T. L. Brandon, A comparison between the shear strength measured with direct shear and triaxial devices on undisturbed and remoulded soils, *Proc. 18th Int. Conf. Soil Mech. Geotech. Eng.*, Paris, Presses des Ponts, 1, 317–320, 2013.

Chandler, R. J., The in situ measurement of the undrained shear strength of clays using the field vane, in A. F. Richards (Ed.), *STP1014, Vane Shear Strength Testing in Soils: Field and Laboratory Studies*, ASTM, Philadelphia, PA, 13–44, 1988.

Christensen, R. W. and B. M. Das, Hydraulic erosion of remolded cohesive soils, Highway Research Board, Special Report 135, 9–19, 1973.

Christensen, R. W. and T. H. Wu, Analysis of clay deformation as a rate process, *J. Soil Mech. Found. Eng. Div.*, ASCE, 90(SM6), 125–157, 1964.

Coulomb, C. A., Essai Sur une Application des regles des Maximis et Minimis a Quelques Problemes des Statique Relatifs a L'Architecture, *Mem. Acad. R. Pres. Divers Savants*, Paris, France, 3, 38, 1776.

Duncan, J. M. and C. Y. Chang, Nonlinear analysis of stress and strain in soils, *J. Soil Mech. Pound. Eng. Div.*, ASCE, 96(5), 1629–1653, 1970.

Gibson, R. E., Experimental determination of true cohesion and true angle of internal friction in clay, *Proc. Third Int. Conf. Soil Mech. Found. Eng.*, Zurich, Switzerland, 1, 126, 1953.

Hansbo, S., A new approach to the determination of shear strength of clay by the fall cone method, Swedish Geotechnical Institute Report No. 114, 1957.

Hatanaka, M. and A. Uchida, Empirical correlation between penetration resistance and internal friction angle of sandy soils, *Soils Found.*, 36(4), 1–10, 1996.

Henkel, D. J., The shearing strength of saturated remolded clays, *Proc. Res. Conf. Shear Strength Cohesive Soils*, ASCE, Boulder, Colorado, 533–554, 1960.

Horne, H. M. and D. U. Deere, Frictional characteristics of minerals, *Geotechnique*, 12, 319–335, 1962.

Hvorslev, J., Physical component of the shear strength of saturated clays, in *Proc. Res. Conf. Shear Strength Cohesive Soils*, Boulder, Colorado, ASCE, 169–273, 1960.

Hvorslev, M. J., Uber Die Festigheitseigen-schaften Gestorter Bindinger Boden, in *Ingeniorvidenskabelige Skrifter*, no. 45, Danmarks Naturvidenskabelige Samfund, Kovenhavn, 1937.

Jamiolkowski, M., C. C. Ladd, J. T. Germaine, and R. Lancellotta, New developments in field and laboratory testing of soils, in *Proc. 11th Int. Conf. Soil Mech. Found. Eng.*, San Francisco, 1, 57–153, 1985.

Janbu, N., Soil compressibility as determined by oedometer and triaxial tests, in *Proc. Eur. Conf. Soil Mech. Found. Eng.*, Wiesbaden, Germany, 1, 259–263, 1963.

Kelly, W. E., Correcting shear strength for temperature, *J. Geotech. Eng. Div.*, ASCE, 104(GT5), 664–667, 1978.

Kenney, T. C., Discussion, *Proc.*, ASCE, 85(SM3), 67–79, 1959.

Kirkpatrick, W. M., The condition of failure in sands, *Proc. Fourth Int. Conf. Soil Mech. Found. Eng.*, London, 1, 172–178, 1957.

Koerner, R. M., Effect of particle characteristics on soil strength, *J. Soil Mech. Found Div.*, ASCE, 96(4), 1221–1234, 1970.

Kulhawy, F. H. and P. W. Mayne, *Manual on Estimating Soil Properties in Foundation Design*, Electric Power Research Institute, Palo Alto, CA, 1990.

Kulhawy, F. H., C. H. Traumann, J. F. Beech, T. D. O'Rourke, W. McGuire, W. A. Wood, and C. Capano, Transmission line structure foundations for uplift-compression loading, Report EL-2870, Electric Power Research Institute, Palo Alto, CA, 1983.

Ladd, C. C., R. Foote, K. Ishihara, F. Schlosser, and H. G. Poulos, Stress-deformation and strength characteristics, *Proc. Ninth Int. Conf. Soil Mech. Found. Eng.*, Tokyo, Japan, 2, 421–494, 1977.

Lade, P. V. and K. L. Lee, Engineering properties of soils, Report UCLA-ENG-7652, 1976.

Lambe, T. W., Methods of estimating settlement, *Soil Mech. Found. Div.*, ASCE, 90(5), 43, 1964.

Larsson, R., Undrained shear strength in stability calculation of embankments and foundations on clay, *Can. Geotech. J.*, 41(5), 591–602, 1980.

Lee, I. K., Stress-dilatency performance of feldspar, *J. Soil Mech. Found. Div.*, ASCE, 92(SM2), 79–103, 1966.

Lee, K. L., Comparison of plane strain and triaxial tests on sand, *J. Soil Mech. Found. Div.*, ASCE, 96(SM3), 901–923, 1970.

Lee, K. L. and H. B. Seed, Drained strength characteristics of sands, *J. Soil Mech. Found. Div.*, ASCE, 83(6), 117–141, 1968.

Lo, K. Y., Stability of slopes in anisotropic soils, *J. Soil Mech. Found. Eng. Div.*, ASCE, 91(SM4), 85–106, 1965.

Mayne, P. W. and J. K. Mitchell, Profiling of overconsolidation ratio in clays by field vane, *Can. Geotech. J.*, 25(1), 150–158, 1988.

Mesri, G., A re-evaluation of $s_{u(mob)} \approx 0.22\sigma_p$ using laboratory shear tests, *Can. Geotech. J.*, 26(1), 162–164, 1989.

Meyerhof, G. G., Some recent research on the bearing capacity of foundations, *Can. Geotech. J.*, l(l), 16–26, 1963.

Mitchell, J. K., Shearing resistance of soils as a rate process, *J. Soil Mech. Found. Eng. Div.*, ASCE, 90(SM1), 29–61, 1964.

Mitchell, J. K., R. G. Campanella, and A. Singh, Soil creep as a rate process, *J. Soil Mech. Found. Eng. Div.*, ASCE, 94(SM1), 231–253, 1968.

Mitchell, J. K., A. Singh, and R. G. Campanella, Bonding, effective stresses, and strength of soils, *J. Soil Mech. Found. Eng. Div.*, ASCE, 95(SM5), 1219–1246, 1969.

Mohr, O., Welche Umstände Bedingen die Elastizitätsgrenze und den Bruch eines Materiales? *Z. Ver. Deut. Ing.*, 44, 1524–1530, 1572–1577, 1900.

Moretto, O., Effect of natural hardening on the unconfined strength of remolded clays, *Proc. Second Int. Conf. Soil Mech. Found. Eng.*, Harvard University, Boston, 1, 218–222, 1948.

Morris, P. M. and D. J. Williams, Effective stress vane shear strength correction factor correlations, *Can. Geotech. J.*, 31(3), 335–342, 1994.

O'Kelly, B. C., Atterberg limit and remolded shear strength—water content relationship, *Geotech. Testing J.*, ASTM, 36(6), 939–947, 2013.

Olson, R. E., Shearing strength of kaolinite, illite, and montmorillonite, *J. Geotech. Eng. Div.*, ASCE, 100(GT11), 1215–1230, 1974.

Peck, R. B., H. O. Ireland, and T. S. Fry, Studies of soil characteristics: The earth flows of St. Thuribe, Quebec, Soil Mechanics Series No. 1, University of Illinois, Urbana, IL, 1951.

Rendulic, L., Ein Grundgesetzder tonmechanik und sein experimentaller beweis, *Bauingenieur*, 18, 459–467, 1937.

Ricceri, G., P. Simonini, and S. Cola, Applicability of piezocone and dilatometer to characterize the soils of the Venice lagoon, *Geotech. Geol. Eng.*, 20(2), 89–121, 2002.

Richardson, A. M., E. W. Brand, and A. Menon, In situ determination of anisotropy of a soft clay, *Proc. Conf. in situ Meas. Soil Prop.*, Raleigh, North Calorina, 1, ASCE, 336–349, 1975.

Robertson, P. K. and R. G. Campanella, Interpretation of cone penetration tests. Part I: sand, *Can. Geotech. J.*, 20(4), 718–733, 1983.

Roscoe, K. H., A. N. Schofield, and A. Thurairajah, An evaluation of test data for selecting a yield criterion for soils, *STP 361, Proc. Symp. Lab. Shear Test. Soils*, Am. Soc. Testing and Materials (ASTM) Philadelphia, 111–129, 1963.

Rowe, P. W., The stress-dilatency relation for static equilibrium of an assembly of particles in contact, *Proc. R. Soc. Lond. Ser. A*, 269, 500–527, 1962.

Schmertmann, J. H., Measurement of in situ shear strength, *Proc. Spec. Conf. In Situ Meas. Soil Prop.*, Raleigh, North Calorina, 2, ASCE, 57–138, 1975.

Seed, H. B. and C. K. Chan, Thixotropic characteristics of compacted clays, *Trans. ASCE*, 24, 894–916, 1959.

Seed, H. B. and K. L. Lee, Liquefaction of saturated sands during cyclic loading, *J. Soil Mech. Found. Eng. Div.*, ASCE, 92(SM6), 105–134, 1966.

Sherif, M. A. and C. M. Burrous, Temperature effect on the unconfined shear strength of saturated cohesive soils, Special Report 103, Highway Research Board, 267–272, 1969.

Silvestri, V. and C. Tabib, Analysis of elliptic vanes, *Can. Geotech. J.*, 29(6), 993–997, 1992.

Simons, N. E., The effect of overconsolidation on the shear strength characteristics of an undisturbed Oslo clay, *Proc. Res. Conf. Shear Strength Cohesive Soils*, Boulder, Colorado, ASCE, 747–763, 1960.

Singh, A. and J. K. Mitchell, General stress-strain-time functions for soils, *J. Soil Mech. Found. Eng. Div.*, ASCE, 94(SM1), 21–46, 1968.

Skempton, A. W., Correspondence, *Geotechnique*, 10(4), 186, 1960.

Skempton, A. W., Discussion: The planning and design of New Hong Kong Airport, *Proc. Inst. Civil Eng.*, 7, 305–307, 1957.

Skempton, A. W., Effective stress in soils, concrete and rock, in *Pore Pressure and Suction in Soils*, Butterworths, London, U.K., 4–16, 1961.

Skempton, A. W., Long-term stability of clay slopes, *Geotechnique*, 14, 77, 1964.

Sorensen, K. K. and N. Okkels, Correlation between drained shear strength and plasticity index of undisturbed overconsolidated clays, *Proc., 18th Int. Conf. Soil Mech. Geotech. Eng.*, Paris, Presses des Ponts, 1, 423–428, 2013.

Stark, T. D. and H. T. Eid, Drained residual strength of cohesive soils, *J. Geotech. Eng. Div.*, ASCE, 120(5), 856–871, 1994.

Terzaghi, K., R. B. Peck, and G. Mesri, *Soil Mechanics in Engineering Practice*, 3rd Edn., Wiley, New York, 1996.

Trautmann, C. H. and F. H. Kulhawy, CUFAD—A computer program for compression and uplift foundation analysis and design, Report EL-4540-CCM, Electric Power Research Institute, Palo Alto, CA, 16, 1987.

Tresca, H., Memoire sur L'Ecoulement des corps solids, *Mem. Pres. Par Div. Savants*, 18, 733–799, 1868.

U.S. Navy, *Soil Mechanic-Design Manual 7.1*, Department of the Navy, Naval facilities Engineering Command, U.S. Government Printing Press, Washington, DC, 1986.

Von Mises, R., Mechanik der festen Korper in Plastichdeformablen Zustand, *Goettinger-Nachr. Math-Phys. Kl.*, 1, 582–592, 1913.

Whitman, R. V. and K. A. Healey, Shear strength of sand during rapid loadings, *Trans.* ASCE, 128, 1553–1594, 1963.

Wolff, T. F., Pile capacity predicting using parameter functions, in *Predicted and Observed Axial Behavior of Piles, Results of a Pile Prediction Symposium*, sponsored by the Geotechnical Engineering Division, ASCE, Evanston, IL, June 1989, Geotech. Spec. Pub. 23, ASCE, 96–106, 1989.

Wu, T. H., A. K. Loh, and L. E. Malvern, Study of failure envelope of soils, *J. Soil Mech. Found. Eng. Div.*, ASCE, 89(SM1), 145–181, 1963.

Chapter 10

Elastic settlement
of shallow foundations

10.1 INTRODUCTION

The increase of stress in soil layers due to the load imposed by various structures at the foundation level will always be accompanied by some strain, which will result in the settlement of the structures.

In general, the total settlement S of a foundation can be given as

$$S = S_e + S_p + S_s$$

where

S_e is the elastic settlement
S_p is the primary consolidation settlement
S_s is the secondary consolidation settlement

In granular soils elastic settlement is the predominant part of the settlement, whereas in saturated inorganic silts and clays the primary consolidation settlement probably predominates. The secondary consolidation settlement forms the major part of the total settlement in highly organic soils and peats. In this chapter, the procedures for estimating elastic settlement will be discussed in detail. Consolidation settlement calculation procedures will be discussed in Chapter 11.

10.2 ELASTIC SETTLEMENT OF FOUNDATIONS ON SATURATED CLAY (POISSON'S RATIO ν = 0.5)

Janbu et al. (1956) proposed a generalized equation for average elastic settlement for uniformly loaded flexible foundation supported by a saturated clay soil in the form

$$S_e(\text{average}) = \mu_1\mu_0 \frac{qB}{E} \quad (\nu = 0.5) \tag{10.1}$$

585

where

μ_1 is the correction factor for *finite thickness of elastic soil layer H*, as shown in Figure 10.1

μ_0 is the correction factor for *depth of embedment of foundation D_f*, as shown in Figure 10.1

B is the width of rectangular loaded foundation or diameter of circular loaded foundation

E is the modulus of elasticity of the clay soil

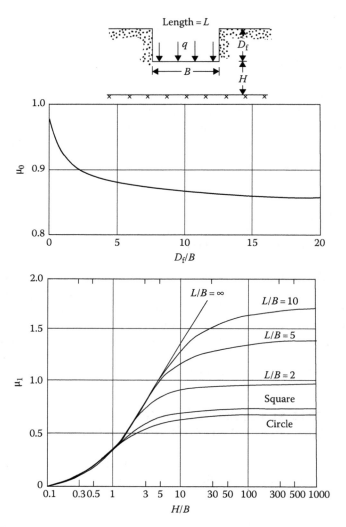

Figure 10.1 Variation of μ_0 and μ_1 for use in Equation 10.1. [Based on Christian, J. T. and Carrier, III, W. D., *Can. Geotech. J.*, 15(1), 124, 1978.]

Table 10.1 Variation of β with plasticity index and overconsolidation ratio

OCR	PI range	Range of β
1	PI < 30	1500–600
	30 < PI < 50	600–300
	PI > 50	300–125
2	PI < 30	1450–575
	30 < PI < 50	575–275
	PI > 50	275–115
4	PI < 30	975–400
	30 < PI < 50	400–185
	PI > 50	185–70
6	PI < 30	600–250
	30 < PI < 50	250–115
	PI > 50	115–60

Source: Compiled from Duncan, J. M., and Buchignani, A. N., Department of Civil Engineering University of California, Berkley, 1976.

Christian and Carrier (1978) made a critical evaluation of Equation 10.1, the details of which will not be presented here. However, they suggested that for Poisson's ratio $\nu = 0.5$, Equation 10.1 could be retained for elastic settlement calculations with a modification of the values of μ_1 and μ_0. The modified values of μ_1 are based on the work of Giroud (1972), and those for μ_0 are based on the work of Burland (1970). These are shown in Figure 10.1.

The *undrained secant modulus E* of clay soils can generally be expressed as

$$E = \beta S_u \tag{10.2}$$

where S_u is undrained shear strength. Duncan and Buchignani (1976) compiled the results of the variation of β with plasticity index PI and overconsolidation ratio OCR for a number of soils. Table 10.1 gives a summary of these results.

Example 10.1

A flexible shallow foundation measures $0.67 \, \text{m} \times 1.34 \, \text{m}$ $(B \times L)$ in plan and is subjected to a uniform load of $200 \, \text{kN/m}^2$. It is located at a depth of $0.67 \, \text{m}$ below the ground surface in a clay layer. The undrained shear strength of the clay is $150 \, \text{kN/m}^2$. A rock layer is located at a depth of $5 \, \text{m}$ from the bottom of the foundation. Estimate the average elastic settlement of the foundation. Use $\beta = 600$ in Equation 10.2.

Solution

Referring to Figure 10.1,

$$\frac{L}{B} = \frac{1.34}{0.67} = 2$$

$$\frac{D_f}{B} = \frac{0.67}{0.67} = 1$$

$$\frac{H}{B} = \frac{5}{0.67} = 7.46$$

$$E = \beta S_u = (600)(150) = 90,000 \text{ kN/m}^2$$

$$\mu_0 \approx 0.92$$

$$\mu_1 \approx 0.9$$

From Equation 10.1,

$$S_{e(average)} = \mu_1 \mu_0 \frac{qB}{E} = (0.9)(0.92)\frac{(200)(0.67)}{90,000}$$
$$= 0.00123 \text{ m} = 1.23 \text{ mm}$$

10.3 ELASTIC SETTLEMENT OF FOUNDATIONS ON GRANULAR SOIL

Various methods available at the present time to calculate the elastic settlement of foundations on granular soil can be divided into three general categories. They are as follows:

1. Methods based on observed settlement of structures and full-scale prototypes: These methods are empirical in nature and are correlated with the results of the standard in situ tests such as the standard penetration test (SPT) and the cone penetration test (CPT). They include, for example, procedures developed by Terzaghi and Peck (1948, 1967), Meyerhof (1956, 1965), Peck and Bazaraa (1969), and Burland and Burbidge (1985).
2. Semi-empirical methods: These methods are based on a combination of field observations and some theoretical studies. They include, for example, the procedures outlined by Schmertmann (1970), Schmertmann et al. (1978), and Akbas and Kulhawy (2009).
3. Methods based on theoretical relationships derived from the theory of elasticity: The relationships for settlement calculation available in this category contain the term modulus of elasticity E and Poisson's ratio v.

The general outlines for some of these methods are given in the following sections.

10.4 SETTLEMENT CALCULATION OF FOUNDATIONS ON GRANULAR SOIL USING METHODS BASED ON OBSERVED SETTLEMENT OF STRUCTURES AND FULL-SCALE PROTOTYPES

The methods suggested by Terzaghi and Peck (1948, 1967), Meyerhof (1965), and Burland and Burbidge (1985) are elaborated upon in the following sections.

10.4.1 Terzaghi and Peck's method

Terzaghi and Peck (1948) proposed the following empirical relationship between the settlement (S_e) of a prototype foundation measuring $B \times B$ in plan and the settlement of a test plate $[S_{e(1)}]$ measuring $B_1 \times B_1$ loaded to the same intensity:

$$\frac{S_e}{S_{e(1)}} = \frac{4}{[1 + (B_1/B)]^2} \tag{10.3}$$

Although a full-sized footing can be used for a load test, the normal practice is to employ a plate with B_1 in the order of 0.3–1 m.

Terzaghi and Peck (1948, 1967) proposed a correlation for the allowable bearing capacity, field standard penetration number N_{60}, and the width of the foundation B corresponding to a 25 mm settlement based on the observation given by Equation 10.3. This correlation is shown in Figure 10.2 (for depth of foundation equal to zero). The curves shown in Figure 10.2 can be approximated by the relation

$$S_e \, (\text{mm}) = \frac{3q}{N_{60}} \left(\frac{B}{B + 0.3} \right)^2 \tag{10.4}$$

where
   N_{60} is the field standard penetration number for an energy ratio of 60%
   q is the bearing pressure (kN/m²)
   B is the width of foundation (m)

If corrections for ground water table location and depth of embedment are included, then Equation 10.4 takes the form

$$S_e(\text{mm}) = C_W \, C_D \frac{3q}{N_{60}} \left(\frac{B}{B + 0.3} \right)^2 \tag{10.5}$$

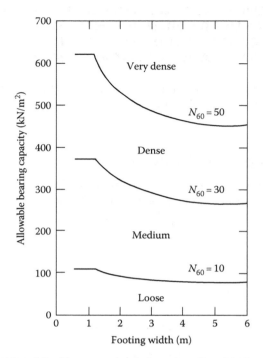

Figure 10.2 Terzaghi and Peck's recommendations for allowable bearing capacity for 25 mm settlement variation with B and N_{60}.

where

C_W is the ground water table correction
C_D is the correction for depth of embedment = $1 - (D_f/4B)$
D_f is the depth of embedment

The magnitude of C_W is equal to 1.0 if the depth of water table is greater than or equal to $2B$ below the foundation, and it is equal to 2.0 if the depth of water table is less than or equal to B below the foundation. The N_{60} value that is to be used in Equations 10.4 and 10.5 should be the average value of N_{60} up to a depth of about $3B$ to $4B$ measured from the bottom of the foundation.

10.4.2 Meyerhof's method

In 1956, Meyerhof proposed relationships for the elastic settlement of foundations on granular soil similar to Equation 10.4. In 1965, he compared the predicted (by the relationships proposed in 1956) and observed settlements of eight structures and suggested that the allowable pressure q for a desired magnitude of S_e can be increased by 50% compared to what he recommended in 1956. The revised relationships including the correction factors for water table location C_W and depth of embedment C_D can be expressed as

$$S_e(\text{mm}) = C_W \, C_D \, \frac{1.25q}{N_{60}} \quad (\text{for } B \leq 1.22 \, \text{m}) \tag{10.6}$$

and

$$S_e(\text{mm}) = C_W \, C_D \, \frac{2q}{N_{60}} \left(\frac{B}{B+0.3} \right)^2 \quad (\text{for } B > 1.22 \, \text{m}) \tag{10.7}$$

$$C_W = 1.0 \tag{10.8}$$

and

$$C_D = 1.0 - \frac{D_f}{4B} \tag{10.9}$$

10.4.3 Method of Peck and Bazaraa

Peck and Bazaraa (1969) recognized that the original Terzaghi and Peck method (see Section 10.4.1) was overly conservative and revised Equation 10.5 to the following form:

$$S_e(\text{mm}) = C_W \, C_D \, \frac{2q}{N'_{60}} \left(\frac{B}{B+0.3} \right)^2 \tag{10.10}$$

where
  S_e is in mm
  q is in kN/m^2
  B is in m
  N'_{60} is the standard penetration number, N_{60}, corrected to a standard
    effective overburden pressure of 75 kN/m^2

$$C_W = \frac{\sigma_0 \text{ at } 0.5 \, B \text{ below the bottom of the foundation}}{\sigma'_0 \text{ at } 0.5 \, B \text{ below the bottom of the foundation}} \tag{10.11}$$

where

σ_0 is the total overburden pressure
σ_0' is the effective overburden pressure

$$C_D = 1.0 - 0.4 \left(\frac{\gamma D_f}{q} \right)^{0.5} \tag{10.12}$$

where γ is the unit weight of soil.
The relationships for N_{60}' are as follows:

$$N_{60}' = \frac{4 N_{60}}{1 + 0.04 \sigma_0'} \quad \text{(for } \sigma_0' \leq 75 \text{ kN/m}^2\text{)} \tag{10.13}$$

and

$$N_{60}' = \frac{4 N_{60}}{3.25 + 0.01 \sigma_0'} \quad \text{(for } \sigma_0' > 75 \text{ kN/m}^2\text{)} \tag{10.14}$$

where σ_0' is the effective overburden pressure (kN/m^2).

10.4.4 Method of Burland and Burbidge

Burland and Burbidge (1985) proposed a method for calculating the elastic settlement of sandy soil using the field standard penetration number N. The method can be summarized as follows:

Step 1: Determination of variation of standard penetration number with depth

Obtain the field penetration numbers N_{60} with depth at the location of the foundation. The following adjustments of N_{60} may be necessary, depending on the field conditions:

For gravel or sandy gravel

$$N_{60(a)} \approx 1.25 N_{60} \tag{10.15}$$

For fine sand or silty sand below the ground water table and $N_{60} > 15$

$$N_{60(a)} \approx 15 + 0.5(N_{60} - 15) \tag{10.16}$$

where $N_{60(a)}$ is the adjusted N value.

Step 2: Determination of depth of stress influence (z')

In determining the depth of stress influence, the following three cases may arise:

Case I: If N_{60} [or $N_{60(a)}$] is approximately constant with depth, calculate z' from

$$\frac{z'}{B_R} = 1.4\left(\frac{B}{B_R}\right)^{0.75} \tag{10.17}$$

where

B_R is the reference width = 0.3 m
B is the width of the actual foundation (m)

Case II: If N_{60} [or $N_{60(a)}$] is increasing with depth, use Equation 10.17 to calculate z'.

Case III: If N_{60} [or $N_{60(a)}$] is decreasing with depth, calculate $z' = 2B$ and z' = distance from the bottom of the foundation to the bottom of the soft soil layer (=z''). Use $z' = 2B$ or $z' = z''$ (whichever is smaller).

Step 3: Determination of depth of stress influence correction factor α

The correction factor α is given as

$$\alpha = \frac{H}{z'}\left(2 - \frac{H}{z'}\right) \le 1 \tag{10.18}$$

where H is the thickness of the compressible layer.

Step 4: Calculation of elastic settlement

The elastic settlement of the foundation S_e can be calculated as

a. For normally consolidated soil

$$\frac{S_e}{B_R} = 0.14\alpha\left\{\frac{1.71}{\left[\bar{N}_{60} \text{ or } \bar{N}_{60(a)}\right]^{1.4}}\right\}\left[\frac{1.25(L/B)}{0.25 + (L/B)}\right]^2\left(\frac{B}{B_R}\right)^{0.7}\left(\frac{q}{p_a}\right) \tag{10.19}$$

where

L is the length of the foundation
p_a is the atmospheric pressure (≈ 100 kN/m^2)
$\bar{N}_{60}$ or $\bar{N}_{60(a)}$ is the average value of N_{60} or $N_{60(a)}$ in the depth of stress increase

b. For overconsolidated soil ($q \le \sigma'_c$; where σ'_c is the overconsolidation pressure)

$$\frac{S_e}{B_R} = 0.047\alpha \left\{ \frac{0.57}{\left[\bar{N}_{60} \text{ or } \bar{N}_{60(a)} \right]^{1.4}} \right\} \left[\frac{1.25(L/B)}{0.25 + (L/B)} \right]^2 \left(\frac{B}{B_R} \right)^{0.7} \left(\frac{q}{p_a} \right) \quad (10.20)$$

c. For overconsolidated soil ($q > \sigma'_c$)

$$\frac{S_e}{B_R} = 0.14\alpha \left\{ \frac{0.57}{[\bar{N}_{60} \text{ or } \bar{N}_{60(a)}]^{1.4}} \right\} \left[\frac{1.25(L/B)}{0.25 + (L/B)} \right]^2 \left(\frac{B}{B_R} \right)^{0.7} \left(\frac{q - 0.67\sigma'_c}{p_a} \right)$$
$$(10.21)$$

Example 10.2

A shallow foundation measuring 1.75 m × 1.75 m is to be constructed over a layer of sand. Given $D_f = 1$ m; N_{60} is generally increasing with depth, $\bar{N}_{60}$ in the depth of stress influence = 10; $q = 120$ kN/m². The sand is normally consolidated. Estimate the elastic settlement of the foundation. Use the Burland and Burbidge method.

Solution

From Equation 10.17

$$\frac{z'}{B_R} = 1.4 \left(\frac{B}{B_R} \right)^{0.75}$$

Depth of stress influence

$$z' = 1.4 \left(\frac{B}{B_R} \right)^{0.75} \quad z' = (1.4)(0.3) \left(\frac{1.75}{0.3} \right)^{0.75} \approx 1.58 \, \text{m}$$

From Equation 10.19

$$\frac{S_e}{B_R} = 0.14\alpha \left\{ \frac{1.71}{[\bar{N}_{60} \text{ or } \bar{N}_{60(a)}]^{1.4}} \right\} \left[\frac{1.25(L/B)}{0.25 + (L/B)} \right]^2 \left(\frac{B}{B_R} \right)^{0.7} \left(\frac{q}{p_a} \right)$$

For this case, $\alpha = 1$

$$\frac{1.71}{\bar{N}_{60}} = \frac{1.71}{(10)^{1.4}} = 0.068$$

Hence

$$\frac{S_e}{0.3} = (0.14)(1)(0.068)\left[\frac{(1.25)(1.75/1.75)}{0.25+(1.75/1.75)}\right]^2\left(\frac{1.75}{0.3}\right)^{0.7}\left(\frac{120}{100}\right)$$

$S_e \approx 0.0118 \text{ m} = 11.8 \text{ mm}$

Example 10.3

Solve Example 10.2 using Meyerhof's method.

Solution

From Equation 10.7

$$S_e = C_W C_D \frac{2q}{N_{60}}\left(\frac{B}{B+0.3}\right)^2$$

$$C_W = 1$$

$$C_D = 1.0 - \frac{D_f}{4B} = 1 - \frac{1}{(4)(1.75)} = 0.857$$

$$S_e = (1)(0.857)\left[\frac{(2)(120)}{10}\right]\left(\frac{1.75}{1.75+0.3}\right)^2 = 14.99 \text{ mm} \approx 15 \text{ mm}$$

Example 10.4

Consider a shallow foundation 1 m × 1 m in plan in a granular soil. Given: $D_f = 1$ m; unit weight of granular soil, $\gamma = 17$ kN/m³; uniform load on the foundation, $q = 200$ kN/m². Following are the results of standard penetration tests taken at the site.

Depth from the ground surface (m)	N_{60}
1.5	8
3.0	10
4.5	12
6.0	14

Estimate the elastic settlement based on the Peck and Bazaara method. *Note*: Ground water table is at a depth of 9 m from the ground surface.

Solution

Determination of N'_{60}. The following table can be prepared.

Depth (m)	N_{60}	σ'_0 (kN/m²)	N'_{60}
1.5	8	25.5	15.8[a]
3.0	10	51.0	13.16[a]
4.5	12	76.5	11.96[b]
6.0	14	102.0	13.1[b]
			Av. ≈ 13.51 ≈ 14

[a] Equation 10.13
[b] Equation 10.14

From Equation 10.11, $C_w = 1$.
From Equation 10.12,

$$C_D = 2 - 0.4\left(\frac{\gamma D_f}{q}\right)^{0.5} = 1 - 0.4\left[\frac{(17)(1)}{200}\right]^{0.5} = 0.88$$

Equation 10.10:

$$S_c = C_W C_D \frac{2q}{N'_{60}}\left(\frac{B}{B+0.3}\right)^2 = (1)(0.88)\frac{2 \times 200}{14}\left(\frac{1}{1+0.3}\right)^2 = 14.88 \text{ mm}$$

10.5 SEMI-EMPIRICAL METHODS FOR SETTLEMENT CALCULATION OF FOUNDATIONS ON GRANULAR SOIL

In the following sections, we will discuss the strain influence factor method suggested by Schmertmann et al. (1978); Terzaghi et al. (1996); and the L_1–L_2 method developed by Akbas and Kulhawy (2009) for calculation of settlement of foundations on granular soil.

10.5.1 Strain influence factor method (Schmertmann et al. 1978)

Based on the theory of elasticity, the equation for vertical strain ε_z at a depth below the center of a flexible circular load of diameter B has been expressed as (Equation 4.40)

$$\varepsilon_z = \frac{q(1+\nu)}{E}[(1-2\nu)A' + B']$$

(4.40)

or

$$I_z = \frac{\varepsilon_z E}{q} = (1+\nu)[(1-2\nu)A' + B']$$

(10.22)

where
A' and B' = f(z/B)
q is the load per unit area
E is the modulus of elasticity
ν is the Poisson's ratio
I_z is the strain influence factor

Schmertmann et al. (1978) proposed a simple variation of I_z with depth below a shallow foundation that is supported by a granular soil. This variation of I_z is shown in Figure 10.3. Referring to this figure.

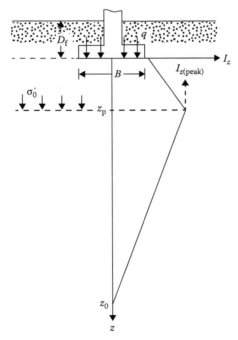

Figure 10.3 Nature of strain influence diagram suggested by Schmertmann et al. (1978).

- For square or circular foundation:

$$I_z = 0.1 \quad \text{at } z = 0$$

$$I_{z(\text{peak})} \quad \text{at } z = z_p = 0.5B$$

$$I_z = 0 \quad \text{at } z = z_0 = 2B$$

- For foundation with $L/B \geq 10$:

$$I_z = 0.2 \quad \text{at } z = 0$$

$$I_{z(\text{peak})} \quad \text{at } z = z_p = B$$

$$I_z = 0 \quad \text{at } z = z_0 = 4B$$

where L is the length of foundation. For L/B between 1 and 10, interpolation can be done. Also

$$I_{z(\text{peak})} = 0.5 + 0.1 \left(\frac{q}{\sigma_0'} \right)^{0.5} \tag{10.23}$$

The value of σ_0' in Equation 10.23 is the effective overburden pressure at a depth where $I_{z(\text{peak})}$ occurs. Salgado (2008) gave the following interpolation for I_z at $z = 0$, z_p, and z_0 (for $L/B = 1$ to $L/B \geq 10$):

$$I_{z(\text{at } z=0)} = 0.1 + 0.0111 \left(\frac{L}{B} - 1 \right) \leq 0.2 \tag{10.24}$$

$$\frac{z_p}{B} = 0.5 + 0.0555 \left(\frac{L}{B} - 1 \right) \leq 1 \tag{10.25}$$

$$\frac{z_0}{B} = 2 + 0.222 \left(\frac{L}{B} - 1 \right) \leq 4 \tag{10.26}$$

The total elastic settlement of the foundation can now be calculated as

$$S_e = C_1 C_2 q \sum_0^{2B} \frac{I_z}{E} \Delta z \tag{10.27}$$

where

q is the *net* effective pressure applied at the level of the foundation

q_0 is the effective overburden pressure at the level of the foundation

C_1 is the correction factor for embedment of foundation $= 1 - 0.5(q_0/q)$

$$(10.28)$$

C_2 is the correction factor to account for creep in soil $= 1 + 0.2\log(t/0.1)$

$$(10.29)$$

t is the time, in years

Noting that stiffness is about 40% larger for plane strain compared to axisymmetric loading, Schmertmann et al. (1978) recommended that

$$E = 2.5q_c \quad \text{(for square and circular foundations)} \qquad (10.30)$$

and

$$E = 3.5q_c \quad \text{(for strip foundation)} \qquad (10.31)$$

where q_c is the cone penetration resistance.

For rectangular foundation with $L \times B$ plan, Terzaghi et al. (1996) suggested that

$$\frac{E_{(L/B)}}{E_{(L/B=1)}} = 1 + 0.4\log\left(\frac{L}{B}\right) \le 1.4 \qquad (10.32)$$

Example 10.5

Consider a rectangular foundation 2 m × 4 m in plan at a depth of 1.2 m in a sand deposit as shown in Figure 10.4a. Given $\gamma = 17.5$ kN/m³; $q = 124$ k/m²; and the following approximated variation of q_c with z:

z (m)	q_c (kN/m²)
0–0.5	2250
0.5–2.5	3430
2.5–5.0	2950

Estimate the elastic settlement of the foundation using the strain influence factor method.

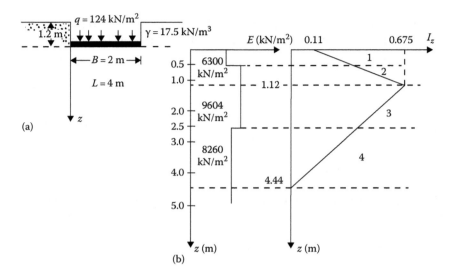

Figure 10.4 (a) Rectangular foundation in a sand deposit; (b) variation of E and I_z with depth.

Solution

From Equation 10.25

$$\frac{z_p}{B} = 0.5 + 0.0555\left(\frac{L}{B} - 1\right) = 0.5 + 0.0555\left(\frac{4}{2} - 1\right) \approx 0.56$$

$$z_p = (0.56)(2) = 1.12 \text{ m}$$

From Equation 10.26

$$\frac{z_0}{B} = 2 + 0.222\left(\frac{L}{B} - 1\right) = 2 + 0.222(2 - 1) = 2.22$$

$$z_0 = (2.22)(2) = 4.44 \text{ m}$$

From Equation 10.24, at $z = 0$

$$I_z = 0.1 + 0.0111\left(\frac{L}{B} - 1\right) = 0.1 + 0.0111\left(\frac{4}{2} - 1\right) \approx 0.11$$

From Equation 10.23

$$I_{z(peak)} = 0.5 + 0.1\left(\frac{q}{\sigma_0'}\right)^{0.5} = 0.5 + 0.1\left[\frac{124}{(1.2 + 1.12)(17.5)}\right]^{0.5} = 0.675$$

The plot of I_z versus z is shown in Figure 10.4b. Again, from Equation 10.32

$$E_{(L/B)} = \left[1 + 0.4\log\frac{L}{B}\right](E_{(L/B=1)}) = \left[1 + 0.4\log\left(\frac{4}{2}\right)\right](2.5 \times q_c) = 2.8q_c$$

Thus, the approximate variation of E with z is as follows:

z (m)	q_c (kN/m²)	E (kN/m²)
0–0.5	2250	6300
0.5–2.5	3430	9604
2.5–5.0	2950	8260

The plot of E versus z is shown in Figure 10.4b.
The soil layer is divided into four layers as shown in Figure 10.4b. Now the following table can be prepared.

Layer no.	Δz (m)	E (kN/m²)	I_z at middle of layer	$\dfrac{I_z}{E}\Delta z$ (m³/kN)
1	0.50	6300	0.236	1.87×10^{-5}
2	0.62	9604	0.519	3.35×10^{-5}
3	1.38	9604	0.535	7.68×10^{-5}
4	1.94	8260	0.197	4.62×10^{-5}
			Total	17.52×10^{-5}

$$S_e = C_1 C_2(q) \sum \frac{I_z}{E}\Delta z$$

$$C_1 = 1 - 0.5\left(\frac{q_0}{q}\right) = 1 - 0.5\left(\frac{17.5 \times 1.2}{124}\right) = 0.915$$

Assume the time for creep is 10 years. So

$$C_2 = 1 + 0.2\log\left(\frac{10}{0.1}\right) = 1.4$$

Hence

$$S_e = (0.915)(1.4)(124)(17.52 \times 10^{-5}) = 2783 \times 10^{-5}\,\text{m} = 27.83\,\text{mm}$$

10.5.2 Strain influence factor method (Terzaghi et al. 1996)

Terzaghi et al. (1966) proposed a slightly different form of the strain influence factor diagram, as shown in Figure 10.3. According to Terzaghi et al. (1996) and referring to Figure 10.3,

At $z = 0$, $I_z = 0.2$ (for all L/B values)

At $z = z_p = 0.5B$, $I_z = 0.6$ (for all L/B values)

At $z = z_0 = 2B$, $I_z = 0$ (for $L/B = 1$)

At $z = z_0 = 4B$, $I_z = 0$ (for $L/B \geq 10$)

For L/B between 1 and 10 (or >10),

$$\frac{z_0}{B} = 2\left[1 + \log\left(\frac{L}{B}\right)\right] \tag{10.33}$$

The elastic settlement can be given as

$$S_e = C_d(q)\sum_0^{z_0}\frac{I_z}{E}\Delta z + \underbrace{0.02\left[\frac{0.1}{\dfrac{\Sigma(q_c\Delta z)}{z_0}}\right]z_0\log\left(\frac{t\text{ days}}{1\text{ day}}\right)}_{\text{Post-construction settlement}} \tag{10.34}$$

In Equation 10.34, q_c is in MN/m². The relationships for E are

$$E = 3.5q_c \text{ (for square and circular foundations)} \tag{10.35}$$

and

$$E_{\text{rectangular}} = \left[1 + 0.4\left(\frac{L}{B}\right)\right]E_{\text{square}} \quad \text{(for } L/B \geq 10\text{)} \tag{10.36}$$

In Equation 10.34, C_d is the depth factor. Table 10.2 gives the interpolated values of C_d for values of D_f/B.

Table 10.2 Variation of C_d with D_f/B^a

D_f/B	C_d
0.1	1
0.2	0.96
0.3	0.92
0.5	0.86
0.7	0.82
1.0	0.77
2.0	0.68
2.0	0.65

[a] Based on data from Terzaghi et al. (1996)

Example 10.6

Solve Example 10.5 using the method of Terzaghi et al. (1996).

Solution

Given: $L/B = 4/2 = 2$

Figure 10.5 shows the plot of I_z with depth below the foundation. Note that

$$\frac{z_0}{B} = 2\left[1 + \log\left(\frac{L}{B}\right)\right] = 2[1 + \log(2)] = 2.6$$

or

$$z_0 = (2.6)(B) = (2.6)(2) = 5.2 \text{ m}$$

Also, from Equations 10.35 and 10.36,

$$E = \left[1 + 0.4\left(\frac{L}{B}\right)\right](3.5q_c) = \left[1 + 0.4\left(\frac{4}{2}\right)\right](3.5q_c)] = 6.3q_c$$

Also, $D_f/B = 1.2/2 = 0.6$. From Table 10.2, $C_d \approx 0.85$.

The following is the table to calculate $\sum_0^{z_0} \frac{I_z}{E}\Delta z$.

Layer No.	Δz (m)	E (kN/m²)	I_z at the middle of the layer	$\frac{I_z}{E}\Delta z$ (m²/kN)
1	0.5	14,175	0.3	1.058×10^{-5}
2	0.5	21,609	0.3	1.157×10^{-5}
3	1.5	21,609	0.493	3.422×10^{-5}
4	2.7	18,585	0.193	2.804×10^{-5}
				$\Sigma 8.441 \times 10^{-5}$

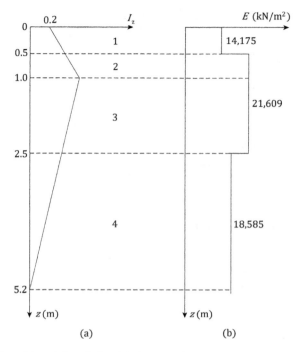

Figure 10.5 Plot of I_z and E with depth below the foundation.

Thus,

$$C_d(q) \sum_0^{z_0} \frac{I_z}{E} \Delta z = (0.85)(124)(8.441 \times 10^{-5}) = 889.68 \times 10^{-5}\,\text{m}$$

Post-construction creep is

$$0.02 \left[\frac{0.1}{\dfrac{\sum(q_c \Delta z)}{z_0}} \right] z_0 \log\left(\frac{t \text{ days}}{1 \text{ day}} \right)$$

$$\frac{\sum(q_c \Delta z)}{z_0} = \frac{(2250 \times 0.5) + (3430 \times 2) + (2950 \times 2.7)}{3.2}$$

$$= 3067.3 \text{ kN/m}^2 \approx 3.07 \text{ MN/m}^2$$

Hence, the elastic settlement is

$$S_e = 889.6 \times 10^{-5} + 0.02 \left[\frac{0.1}{3.07} \right] (5.2) \log\left(\frac{10 \times 365 \text{ days}}{1 \text{ day}} \right)$$

$$= 2096.68 \times 10^{-5} \text{ m}$$

$$\approx 20.97 \text{ mm}$$

10.5.3 Field tests on load–settlement behavior: L_1–L_2 method

Akbas and Kulhawy (2009) evaluated 167 load–settlement relationships obtained from field tests. Figure 10.6 shows a generalized relationship of load Q versus settlement S_e from these field tests, which they referred to as the L_1–L_2 method. From this figure, note that (a) Q_{L_1} is the load at settlement level $S_{e(L_1)}$; (b) Q_T is the load at settlement level $S_{e(T)}$; and (c) Q_{L_2} is the load at settlement level $S_{e(L_2)}$, which is the ultimate load ($\approx Q_u$).

The field test results yielded the mean value of $S_{e(L_1)}$ to be 0.23% of the width of the foundation, B. Similarly, the mean value of $S_{e(L_2)}$s was 5.39% of B. The final analysis showed a nondimensional load–settlement relationship as given in Figure 10.7. The mean plot can be expressed as

$$\frac{Q}{Q_{L_2}} = \frac{S_e/B}{0.69(S_e/B)+1.68} \tag{10.37}$$

In order to find Q for a given settlement level, one needs to know that $Q_{L_2} = Q_u$ = ultimate bearing capacity for which the following is recommended.

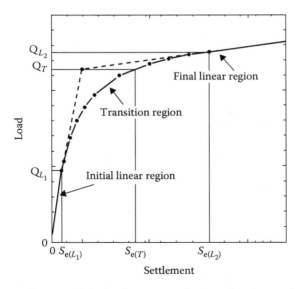

Figure 10.6 General nature of the load versus settlement plot observed from the field (L_1–L_2 method).

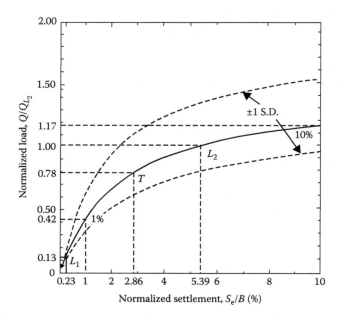

Figure 10.7 Nondimensional plot of Q/Q_{L_2} versus S_e/B (Equation 10.37).

1. For $B > 1$ m:

$$Q_{L_2} = Q_u^\gamma + Q_u^q = \underbrace{\frac{1}{2}\gamma B\, N_\gamma\, F_{\gamma s}\, F_{\gamma d}\, F_{\gamma c}}_{Q_u^\gamma} + \underbrace{q\, N_q\, F_{qs}\, F_{qd}\, F_{qc}}_{Q_u^q} \qquad (10.38)$$

where
  N_γ, N_q are the bearing capacity factors
  $F_{\gamma s}$, F_{qs} are the shape factors
  $F_{\gamma d}$, F_{qd} are the depth factors
  $F_{\gamma c}$, F_{qc} are the compressibility factors

2. For $B \leq 1$ m:

$$Q_{L_2} = \frac{Q_u^\gamma}{B} + Q_u^q \qquad (10.39)$$

In order to determine Q_u^γ and Q_u^q, see Vesic (1973) or a foundation engineering book (e.g., Das, 2011).

Equation 10.37 was further refined by Akbas and Kulhawy (2009) based on the modulus of elasticity E_{L_1} of soil. E_{L_1} is based on Q_{L_1} and $S_{e(L_1)}$, which is the initial linear region shown in Figure 10.6.

For $\dfrac{E_{L_1}}{p_a} > 500$,

$$\frac{Q}{Q_{L_2}} = \frac{S_e/B}{0.68(S_e/B)+1.18} \tag{10.40}$$

For $500 > \dfrac{E_{L_1}}{p_a} > 250$,

$$\frac{Q}{Q_{L_2}} = \frac{S_e/B}{0.72(S_e/B)+1.59} \tag{10.41}$$

For $\dfrac{E_{L_1}}{p_a} < 250$,

$$\frac{Q}{Q_{L_2}} = \frac{S_e/B}{0.75(S_e/B)+1.95} \tag{10.42}$$

In Equations 10.40 through 10.42, p_a is atmospheric pressure ($\approx 100 \ kN/m^2$).

Example 10.7

Consider a shallow foundation of granular soil. Given:

Foundation: $B = 1.5$ m; $D_f = 1$ m

Unit weight of soil: $\gamma = 16.5 \ kN/m^3$

Equation 10.38 was solved to determine Q_{L_2} as

$$Q_{L_2} = Q_u^\gamma + Q_u^q$$

with $Q_u^\gamma = 333.8$ kN and $Q_u^q = 1136.6$ kN.
Estimate the following:

a. $S_{e(L_1)}$
b. $S_{e(L_2)}$
c. Settlement S_e with application of load $Q = 800$ kN

Solution

a. $S_{e(L_1)} = 0.23B \ (\%) = \dfrac{(0.23)(1.5 \times 1000)}{100} = 3.45 \ mm$

b. $S_{e(L_2)} = 5.39B \ (\%) = \dfrac{(5.39)(1.5 \times 1000)}{100} = 80.85 \ \text{mm}$

c. $B > 1$ m. So $Q_{L_2} = 333.8 + 1136.6 = 1467.4$ kN. From Equation 10.37,

$$1467.4 = \frac{S_e/B \ (\%)}{0.69[S_e/B \ (\%)]+1.68}; \quad \frac{S_e}{B} = 1.467\%$$

$$S_e = (1.457)\left(\frac{1.5 \times 1000}{100}\right) \approx 22 \ \text{m}$$

10.6 SETTLEMENT DERIVED FROM THEORY OF ELASTICITY

The following sections describe two methods of elastic settlement calculation derived from the theory of elasticity. They are

- Schleicher's (1926), Steinbrenner's (1934) and Fox's (1948) theories
- Theory of Mayne and Poulos (1999)

10.6.1 Settlement based on theories of Schleicher (1926), Steinbrenner (1934) and Fox (1948)

Consider a foundation measuring $L \times B$ (L = length; B = width) located at a depth D_f below the ground surface (Figure 10.8). A rigid layer is located at a depth H below the ground surface.

Now we will consider three cases as follows:

- Case I: Flexible surface foundation on elastic half-space (i.e., $D_f = 0$ and $H = \infty$)

The elastic settlement under a point on the foundation can be given by

$$S_{e(\text{flexible})} = \frac{qB}{E}(1 - v^2)I \tag{10.43}$$

where I = influence factor that depends on the location of the point of interest on the foundation; q = net pressure applied by the foundation to the underlying soil; E = modulus of elasticity of the soil, and v = Poisson's ratio of the soil.

Schleicher (1926) expressed the influence factor for the *corner of a flexible foundation* as

$$I_{\text{corner}} = \frac{1}{\pi}\left[m' \ln\left(\frac{1 + \sqrt{m'^2 + 1}}{m'}\right) + \ln\left(m' + \sqrt{m'^2 + 1}\right) \right] \tag{10.44}$$

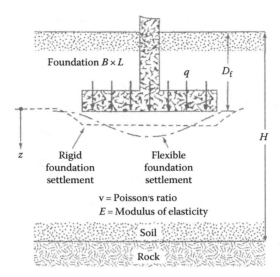

Figure 10.8 Elastic settlement of flexible and rigid foundations.

where $m' = L/B$. The influence factors for the other locations on the foundation can be determined by dividing the foundation into four rectangles and using the principle of superposition. It can be deduced that the influence factor for the center is twice that of the corner. Das and Sivakugan (2019) have noted that the variation of the influence factor with m' can be approximated as,

Corner: $I = 0.7283 \log m' + 0.5469$ (10.45)

Center: $I = 1.4566 \log m' + 1.0939$ (10.46)

Flexible foundations apply uniform pressure and settle nonuniformly. Rigid foundations apply nonuniform pressure and settle uniformly. Some influence factors for estimating the average values of the settlements of flexible and rigid foundations, as reported in the literature, are given in Table 10.3.

Bowles (1987) suggested that the settlement of the rigid foundation can be *estimated as 93% of the settlement computed for a flexible foundation under the center.*

- Case II: Effect of a rigid layer on the settlement of surface foundation (i.e., $D_f = 0$, $H \neq \infty$)

Table 10.3 Influence factors to compute average settlement of flexible and rigid foundation

$m' = L/B$	Flexible	Rigid
Circle	0.85	0.79
1	0.95	0.82
1.5	1.20	1.07
2	1.30	1.21
3	1.52	1.42
5	1.82	1.60
10	2.24	2.00
100	2.96	3.40

Theoretically, if the foundation is perfectly flexible (Bowles, 1987), the settlement may be expressed as

$$S_{e(flexible)} = q(\alpha B')\frac{1 - v^2}{E} I_s \tag{10.47}$$

where

q is the net applied pressure on the foundation

v is the Poisson's ratio of soil

E is the average modulus of elasticity of the soil under the foundation, measured from $z = 0$ to about $z = 4B$

$B' = B/2$ for center of foundation

$\quad = B$ for corner of foundation

$$I_s = \text{Shape factor (Steinbrenner, 1934)} = F_1 + \frac{1 - 2v}{1 - v}F_2 \tag{10.48}$$

$$F_1 = \frac{1}{\pi}(A_0 + A_1) \tag{10.49}$$

$$F_2 = \frac{n'}{2\pi}\tan^{-1} A_2 \tag{10.50}$$

$$A_0 = m'\ln\frac{\left(1 + \sqrt{m'^2 + 1}\right)\sqrt{m'^2 + n'^2}}{m'\left(1 + \sqrt{m'^2 + n'^2 + 1}\right)} \tag{10.51}$$

$$A_1 = \ln\frac{\left(m' + \sqrt{m'^2 + 1}\right)\sqrt{1 + n'^2}}{m' + \sqrt{m'^2 + n'^2 + 1}} \tag{10.52}$$

$$A_2 = \frac{m'}{n' + \sqrt{m'^2 + n'^2 + 1}} \tag{10.53}$$

α is a factor that depends on the location on the foundation where settlement is being calculated.

To calculate settlement at the *center* of the foundation, we use

$$\alpha = 4$$

$$m' = \frac{L}{B}$$

and

$$n' = \frac{H}{(B/2)}$$

To calculate settlement at a *corner* of the foundation, use

$$\alpha = 1$$

$$m' = \frac{L}{B}$$

and

$$n' = \frac{H}{B}$$

The variations of F_1 and F_2 with m' and n' are given in Tables 10.4 through 10.7, respectively. Furthermore, to rigid foundations,

$$S_{e(\text{rigid})} = 0.93\, S_{e(\text{flexible – center})} \tag{10.54}$$

- Case III: Effect of foundation embedment (i.e., $D_f \neq 0$, $H \neq \infty$)

When the foundation base is located at some depth beneath the ground level, the embedment reduces the settlement further. Fox (1948) proposed reduction factor I_f to account for this reduction in settlement. The settlement computed using Equation 10.43 or Equation 10.47 would have to be modified as

$$S_{e(\text{flexible})} = \frac{qB}{E}(1 - v^2)II_f \quad \text{(modified Equation 10.43)} \tag{10.55}$$

and

$$S_{e(\text{flexible})} = q(\alpha B')\frac{1 - v^2}{E}I_s I_f \quad \text{(modified Equation 10.47)} \tag{10.56}$$

The variation of I_f with D_f/B, L/B, and v is shown in Figure 10.9.

Table 10.4 Variation of F_1 with m' and n'

	m'									
n'	1.0	1.2	1.4	1.6	1.8	2.0	2.5	3.0	3.5	4.0
0.25	0.014	0.013	0.012	0.011	0.011	0.011	0.010	0.010	0.010	0.010
0.50	0.049	0.046	0.044	0.042	0.041	0.040	0.038	0.038	0.037	0.037
0.75	0.095	0.090	0.087	0.084	0.082	0.080	0.077	0.076	0.074	0.074
1.00	0.142	0.138	0.134	0.130	0.127	0.125	0.121	0.118	0.116	0.115
1.25	0.186	0.183	0.179	0.176	0.173	0.170	0.165	0.161	0.158	0.157
1.50	0.224	0.224	0.222	0.219	0.216	0.213	0.207	0.203	0.199	0.197
1.75	0.257	0.259	0.259	0.258	0.255	0.253	0.247	0.242	0.238	0.235
2.00	0.285	0.290	0.292	0.292	0.291	0.289	0.284	0.279	0.275	0.271
2.25	0.309	0.317	0.321	0.323	0.323	0.322	0.317	0.313	0.308	0.305
2.50	0.330	0.341	0.347	0.350	0.351	0.351	0.348	0.344	0.340	0.336
2.75	0.348	0.361	0.369	0.374	0.377	0.378	0.377	0.373	0.369	0.365
3.00	0.363	0.379	0.389	0.396	0.400	0.402	0.402	0.400	0.396	0.392
3.25	0.376	0.394	0.406	0.415	0.420	0.423	0.426	0.424	0.421	0.418
3.50	0.388	0.408	0.422	0.431	0.438	0.442	0.447	0.447	0.444	0.441
3.75	0.399	0.420	0.436	0.447	0.454	0.460	0.467	0.458	0.466	0.464
4.00	0.408	0.431	0.448	0.460	0.469	0.476	0.484	0.487	0.486	0.484
4.25	0.417	0.440	0.458	0.472	0.481	0.484	0.495	0.514	0.515	0.515
4.50	0.424	0.450	0.469	0.484	0.495	0.503	0.516	0.521	0.522	0.522
4.75	0.431	0.458	0.478	0.494	0.506	0.515	0.530	0.536	0.539	0.539
5.00	0.437	0.465	0.487	0.503	0.516	0.526	0.543	0.551	0.554	0.554
5.25	0.443	0.472	0.494	0.512	0.526	0.537	0.555	0.564	0.568	0.569
5.50	0.448	0.478	0.501	0.520	0.534	0.546	0.566	0.576	0.581	0.584
5.75	0.453	0.483	0.508	0.527	0.542	0.555	0.576	0.588	0.594	0.597
6.00	0.457	0.489	0.514	0.534	0.550	0.563	0.585	0.598	0.606	0.609
6.25	0.461	0.493	0.519	0.540	0.557	0.570	0.594	0.609	0.617	0.621
6.50	0.465	0.498	0.524	0.546	0.563	0.577	0.603	0.618	0.627	0.632
6.75	0.468	0.502	0.529	0.551	0.569	0.584	0.610	0.627	0.637	0.643
7.00	0.471	0.506	0.533	0.556	0.575	0.590	0.618	0.635	0.646	0.653
7.25	0.474	0.509	0.538	0.561	0.580	0.596	0.625	0.643	0.655	0.662
7.50	0.477	0.513	0.541	0.565	0.585	0.601	0.631	0.650	0.663	0.671
7.75	0.480	0.516	0.545	0.569	0.589	0.606	0.637	0.658	0.671	0.680
8.00	0.482	0.519	0.549	0.573	0.594	0.611	0.643	0.664	0.678	0.688
8.25	0.485	0.522	0.552	0.577	0.598	0.615	0.648	0.670	0.685	0.695
8.50	0.487	0.524	0.555	0.580	0.601	0.619	0.653	0.676	0.692	0.703
8.75	0.489	0.527	0.558	0.583	0.605	0.623	0.658	0.682	0.698	0.710
9.00	0.491	0.529	0.560	0.587	0.609	0.627	0.663	0.687	0.705	0.716
9.25	0.493	0.531	0.563	0.589	0.612	0.631	0.667	0.693	0.710	0.723
9.50	0.495	0.533	0.565	0.592	0.615	0.634	0.671	0.697	0.716	0.719
9.75	0.496	0.536	0.568	0.595	0.618	0.638	0.675	0.702	0.721	0.735
10.00	0.498	0.537	0.570	0.597	0.621	0.641	0.679	0.707	0.726	0.740
20.00	0.529	0.575	0.614	0.647	0.677	0.702	0.756	0.797	0.830	0.858
50.00	0.548	0.598	0.640	0.678	0.711	0.740	0.803	0.853	0.895	0.931
100.00	0.555	0.605	0.649	0.688	0.722	0.753	0.819	0.872	0.918	0.956

Table 10.5 Variation of F_1 with m' and n'

n'	4.5	5.0	6.0	7.0	8.0	9.0	10.0	25.0	50.0	100.0
0.25	0.010	0.010	0.010	0.010	0.010	0.010	0.010	0.010	0.010	0.010
0.50	0.036	0.036	0.036	0.036	0.036	0.036	0.036	0.036	0.036	0.036
0.75	0.073	0.073	0.072	0.072	0.072	0.072	0.071	0.071	0.071	0.071
1.00	0.114	0.113	0.112	0.112	0.112	0.111	0.111	0.110	0.110	0.110
1.25	0.155	0.154	0.153	0.152	0.152	0.151	0.151	0.150	0.150	0.150
1.50	0.195	0.194	0.192	0.191	0.190	0.190	0.189	0.188	0.188	0.188
1.75	0.233	0.232	0.229	0.228	0.227	0.226	0.225	0.223	0.223	0.223
2.00	0.269	0.267	0.264	0.262	0.261	0.260	0.259	0.257	0.256	0.256
2.25	0.302	0.300	0.296	0.294	0.293	0.291	0.291	0.287	0.287	0.287
2.50	0.333	0.331	0.327	0.324	0.322	0.321	0.320	0.316	0.315	0.315
2.75	0.362	0.359	0.355	0.352	0.350	0.348	0.347	0.343	0.342	0.342
3.00	0.389	0.386	0.382	0.378	0.376	0.374	0.373	0.368	0.367	0.367
3.25	0.415	0.412	0.407	0.403	0.401	0.399	0.397	0.391	0.390	0.390
3.50	0.438	0.435	0.430	0.427	0.424	0.421	0.420	0.413	0.412	0.411
3.75	0.461	0.458	0.453	0.449	0.446	0.443	0.441	0.433	0.432	0.432
4.00	0.482	0.479	0.474	0.470	0.466	0.464	0.462	0.453	0.451	0.451
4.25	0.516	0.496	0.484	0.473	0.471	0.471	0.470	0.468	0.462	0.460
4.50	0.520	0.517	0.513	0.508	0.505	0.502	0.499	0.489	0.487	0.487
4.75	0.537	0.535	0.530	0.526	0.523	0.519	0.517	0.506	0.504	0.503
5.00	0.554	0.552	0.548	0.543	0.540	0.536	0.534	0.522	0.519	0.519
5.25	0.569	0.568	0.564	0.560	0.556	0.553	0.550	0.537	0.534	0.534
5.50	0.584	0.583	0.579	0.575	0.571	0.568	0.585	0.551	0.549	0.548
5.75	0.597	0.597	0.594	0.590	0.586	0.583	0.580	0.565	0.583	0.562
6.00	0.611	0.610	0.608	0.604	0.601	0.598	0.595	0.579	0.576	0.575
6.25	0.623	0.623	0.621	0.618	0.615	0.611	0.608	0.592	0.589	0.588
6.50	0.635	0.635	0.634	0.631	0.628	0.625	0.622	0.605	0.601	0.600
6.75	0.646	0.647	0.646	0.644	0.641	0.637	0.634	0.617	0.613	0.612
7.00	0.656	0.658	0.658	0.656	0.653	0.650	0.647	0.628	0.624	0.623
7.25	0.666	0.669	0.669	0.668	0.665	0.662	0.659	0.640	0.635	0.634
7.50	0.676	0.679	0.680	0.679	0.676	0.673	0.670	0.651	0.646	0.645
7.75	0.685	0.688	0.690	0.689	0.687	0.684	0.681	0.661	0.656	0.655
8.00	0.694	0.697	0.700	0.700	0.698	0.695	0.692	0.672	0.666	0.665
8.25	0.702	0.706	0.710	0.710	0.708	0.705	0.703	0.682	0.676	0.675
8.50	0.710	0.714	0.719	0.719	0.718	0.715	0.713	0.692	0.686	0.684
8.75	0.717	0.722	0.727	0.728	0.727	0.725	0.723	0.701	0.695	0.693
9.00	0.725	0.730	0.736	0.737	0.736	0.735	0.732	0.710	0.704	0.702
9.25	0.731	0.737	0.744	0.746	0.745	0.744	0.742	0.719	0.713	0.711
9.50	0.738	0.744	0.752	0.754	0.754	0.753	0.751	0.728	0.721	0.719
9.75	0.744	0.751	0.759	0.762	0.762	0.761	0.759	0.737	0.729	0.727
10.00	0.750	0.758	0.766	0.770	0.770	0.770	0.768	0.745	0.738	0.735
20.00	0.878	0.896	0.925	0.945	0.959	0.969	0.977	0.982	0.965	0.957
50.00	0.962	0.989	1.034	1.070	1.100	1.125	1.146	1.265	1.279	1.261
100.00	0.990	1.020	1.072	1.114	1.150	1.182	1.209	1.408	1.489	1.499

Table 10.6 Variation of F_2 with m' and n'

					m'					
n'	1.0	1.2	1.4	1.6	1.8	2.0	2.5	3.0	3.5	4.0
0.25	0.049	0.050	0.051	0.051	0.051	0.052	0.052	0.052	0.052	0.052
0.50	0.074	0.077	0.080	0.081	0.083	0.084	0.086	0.086	0.0878	0.087
0.75	0.083	0.089	0.093	0.097	0.099	0.101	0.104	0.106	0.107	0.108
1.00	0.083	0.091	0.098	0.102	0.106	0.109	0.114	0.117	0.119	0.120
1.25	0.080	0.089	0.096	0.102	0.107	0.111	0.118	0.122	0.125	0.127
1.50	0.075	0.084	0.093	0.099	0.105	0.110	0.118	0.124	0.128	0.130
1.75	0.069	0.079	0.088	0.095	0.101	0.107	0.117	0.123	0.128	0.131
2.00	0.064	0.074	0.083	0.090	0.097	0.102	0.114	0.121	0.127	0.131
2.25	0.059	0.069	0.077	0.085	0.092	0.098	0.110	0.119	0.125	0.130
2.50	0.055	0.064	0.073	0.080	0.087	0.093	0.106	0.115	0.122	0.127
2.75	0.051	0.060	0.068	0.076	0.082	0.089	0.102	0.111	0.119	0.125
3.00	0.048	0.056	0.064	0.071	0.078	0.084	0.097	0.108	0.116	0.122
3.25	0.045	0.053	0.060	0.067	0.074	0.080	0.093	0.104	0.112	0.119
3.50	0.042	0.050	0.057	0.064	0.070	0.076	0.089	0.100	0.109	0.116
3.75	0.040	0.047	0.054	0.060	0.067	0.073	0.086	0.096	0.105	0.113
4.00	0.037	0.044	0.051	0.057	0.063	0.069	0.082	0.093	0.102	0.110
4.25	0.036	0.042	0.049	0.055	0.061	0.066	0.079	0.090	0.099	0.107
4.50	0.034	0.040	0.046	0.052	0.058	0.063	0.076	0.086	0.096	0.104
4.75	0.032	0.038	0.044	0.050	0.055	0.061	0.073	0.083	0.093	0.101
5.00	0.031	0.036	0.042	0.048	0.053	0.058	0.070	0.080	0.090	0.098
5.25	0.029	0.035	0.040	0.046	0.051	0.056	0.067	0.078	0.087	0.095
5.50	0.028	0.033	0.039	0.044	0.049	0.054	0.065	0.075	0.084	0.092
5.75	0.027	0.032	0.037	0.042	0.047	0.052	0.063	0.073	0.082	0.090
6.00	0.026	0.031	0.036	0.040	0.045	0.050	0.060	0.070	0.079	0.087
6.25	0.025	0.030	0.034	0.039	0.044	0.048	0.058	0.068	0.077	0.085
6.50	0.024	0.029	0.033	0.038	0.042	0.046	0.056	0.066	0.075	0.083
6.75	0.023	0.028	0.032	0.036	0.041	0.045	0.055	0.064	0.073	0.080
7.00	0.022	0.027	0.031	0.035	0.039	0.043	0.053	0.062	0.071	0.078
7.25	0.022	0.026	0.030	0.034	0.038	0.042	0.051	0.060	0.069	0.076
7.50	0.021	0.025	0.029	0.033	0.037	0.041	0.050	0.059	0.067	0.074
7.75	0.020	0.024	0.028	0.032	0.036	0.039	0.048	0.057	0.065	0.072
8.00	0.020	0.023	0.027	0.031	0.035	0.038	0.047	0.055	0.063	0.071
8.25	0.019	0.023	0.026	0.030	0.034	0.037	0.046	0.054	0.062	0.069
8.50	0.018	0.022	0.026	0.029	0.033	0.036	0.045	0.053	0.060	0.067
8.75	0.018	0.021	0.025	0.028	0.032	0.035	0.043	0.051	0.059	0.066
9.00	0.017	0.021	0.024	0.028	0.031	0.034	0.042	0.050	0.057	0.064
9.25	0.017	0.020	0.024	0.027	0.030	0.033	0.041	0.049	0.056	0.063
9.50	0.017	0.020	0.023	0.026	0.029	0.033	0.040	0.048	0.055	0.061
9.75	0.016	0.019	0.023	0.026	0.029	0.032	0.039	0.047	0.054	0.060
10.00	0.016	0.019	0.022	0.025	0.028	0.031	0.038	0.046	0.052	0.059
20.00	0.008	0.010	0.011	0.013	0.014	0.016	0.020	0.024	0.027	0.031
50.00	0.003	0.004	0.004	0.005	0.006	0.006	0.008	0.010	0.011	0.013
100.00	0.002	0.002	0.002	0.003	0.003	0.003	0.004	0.005	0.006	0.006

Table 10.7 Variation of F_2 with m' and n'

n'	m'									
	4.5	5.0	6.0	7.0	8.0	9.0	10.0	25.0	50.0	100.0
0.25	0.053	0.053	0.053	0.053	0.053	0.053	0.053	0.053	0.053	0.053
0.50	0.087	0.087	0.088	0.088	0.088	0.088	0.088	0.088	0.088	0.088
0.75	0.109	0.109	0.109	0.110	0.110	0.110	0.110	0.111	0.111	0.111
1.00	0.121	0.122	0.123	0.123	0.124	0.124	0.124	0.125	0.125	0.125
1.25	0.128	0.130	0.131	0.132	0.132	0.133	0.133	0.134	0.134	0.134
1.50	0.132	0.134	0.136	0.137	0.138	0.138	0.139	0.140	0.140	0.140
1.75	0.134	0.136	0.138	0.140	0.141	0.142	0.142	0.144	0.144	0.145
2.00	0.134	0.136	0.139	0.141	0.143	0.144	0.145	0.147	0.147	0.148
2.25	0.133	0.136	0.140	0.142	0.144	0.145	0.146	0.149	0.150	0.150
2.50	0.132	0.135	0.139	0.142	0.144	0.146	0.147	0.151	0.151	0.151
2.75	0.130	0.133	0.138	0.142	0.144	0.146	0.147	0.152	0.152	0.153
3.00	0.127	0.131	0.137	0.141	0.144	0.145	0.147	0.152	0.153	0.154
3.25	0.125	0.129	0.135	0.140	0.143	0.145	0.147	0.153	0.154	0.154
3.50	0.122	0.126	0.133	0.138	0.142	0.144	0.146	0.153	0.155	0.155
3.75	0.119	0.124	0.131	0.137	0.141	0.143	0.145	0.154	0.155	0.155
4.00	0.116	0.121	0.129	0.135	0.139	0.142	0.145	0.154	0.155	0.156
4.25	0.113	0.119	0.127	0.133	0.138	0.141	0.144	0.154	0.156	0.156
4.50	0.110	0.116	0.125	0.131	0.136	0.140	0.143	0.154	0.156	0.156
4.75	0.107	0.113	0.123	0.130	0.135	0.139	0.142	0.154	0.156	0.157
5.00	0.105	0.111	0.120	0.128	0.133	0.137	0.140	0.154	0.156	0.157
5.25	0.102	0.108	0.118	0.126	0.131	0.136	0.139	0.154	0.156	0.157
5.50	0.099	0.106	0.116	0.124	0.130	0.134	0.138	0.154	0.156	0.157
5.75	0.097	0.103	0.113	0.122	0.128	0.133	0.136	0.154	0.157	0.157
6.00	0.094	0.101	0.111	0.120	0.126	0.131	0.135	0.153	0.157	0.157
6.25	0.092	0.098	0.109	0.118	0.124	0.129	0.134	0.153	0.157	0.158
6.50	0.090	0.096	0.107	0.116	0.122	0.128	0.132	0.153	0.157	0.158
6.75	0.087	0.094	0.105	0.114	0.121	0.126	0.131	0.153	0.157	0.158
7.00	0.085	0.092	0.103	0.112	0.119	0.125	0.129	0.152	0.157	0.158
7.25	0.083	0.090	0.101	0.110	0.117	0.123	0.128	0.152	0.157	0.158
7.50	0.081	0.088	0.099	0.108	0.115	0.121	0.126	0.152	0.156	0.158
7.75	0.079	0.086	0.097	0.106	0.114	0.120	0.125	0.151	0.156	0.158
8.00	0.077	0.084	0.095	0.104	0.112	0.118	0.124	0.151	0.156	0.158
8.25	0.076	0.082	0.093	0.102	0.110	0.117	0.122	0.150	0.156	0.158
8.50	0.074	0.080	0.091	0.101	0.108	0.115	0.121	0.150	0.156	0.158
8.75	0.072	0.078	0.089	0.099	0.107	0.114	0.119	0.150	0.156	0.158
9.00	0.071	0.077	0.088	0.097	0.105	0.112	0.118	0.149	0.156	0.158
9.25	0.069	0.075	0.086	0.096	0.104	0.110	0.116	0.149	0.156	0.158
9.50	0.068	0.074	0.085	0.094	0.102	0.109	0.115	0.148	0.156	0.158
9.75	0.066	0.072	0.083	0.092	0.100	0.107	0.113	0.148	0.156	0.158
10.00	0.065	0.071	0.082	0.091	0.099	0.106	0.112	0.147	0.156	0.158
20.00	0.035	0.039	0.046	0.053	0.059	0.065	0.071	0.124	0.148	0.156
50.00	0.014	0.016	0.019	0.022	0.025	0.028	0.031	0.071	0.113	0.142
100.00	0.007	0.008	0.010	0.011	0.013	0.014	0.016	0.039	0.071	0.113

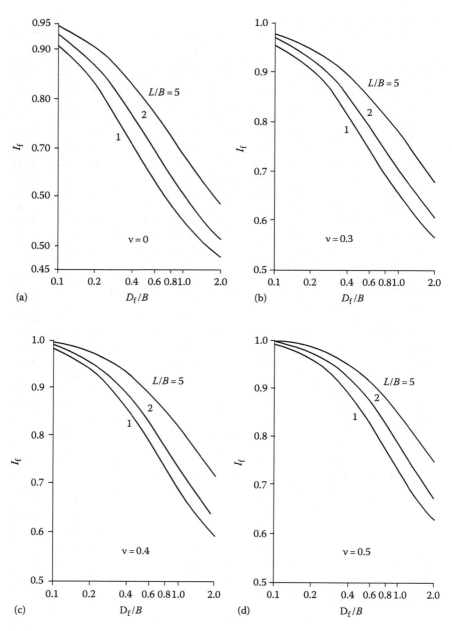

Figure 10.9 Variation of I_f with D_f/B, L/B, and ν: (a) $\nu = 0$; (b) $\nu = 0.3$; (c) $\nu = 0.4$; (d) $\nu = 0.5$.

Due to the nonhomogeneous nature of soil deposits, the magnitude of E may vary with depth. For that reason, Bowles (1987) recommended using a weighted average of E in Equation 10.55, or

$$E = \frac{\sum E_{(i)} \Delta z}{\bar{z}} \qquad (10.57)$$

where

$E_{(i)}$ is the soil modulus of elasticity within a depth Δz
$\bar{z}$ is the depth $(H - D_f)$ or $5B$, whichever is smaller

For a rigid foundation

$$S_{e(rigid)} \approx 0.93 S_{e(flexible, center)}$$

Example 10.8

Consider a rigid surface foundation $1\,m \times 2\,m$ in plan on a granular soil deposit ($D_f = 0$ and $H = \infty$ in Figure 10.8). Given: $v = 0.3$; $q = 150\,kN/m^2$; and $E = 10,400\,kN/m^2$. Determine the elastic settlement.

Solution

From Equation 10.43,

$$S_{e(flexible)} = \frac{qB}{E}(1 - v^2)I$$

From Equation 10.46,

$$I = 1.4566 \log m' + 1.0939 = (1.4566)\left[\log\left(\frac{2}{1}\right)\right] + 1.0939 = 1.532$$

$$S_{e(flexible)} = \frac{(150)(1)}{10,400}(1 - 0.3^2)(1.532) = 0.02\,m = 20\,mm$$

So,

$$S_{e(rigid)} = (0.93)(20) = 18.6\,mm$$

Example 10.9

A rigid shallow foundation $1\,m \times 2\,m$ is shown in Figure 10.10. Calculate the elastic settlement at the center of the foundation.

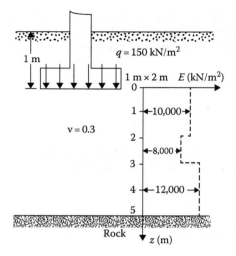

Figure 10.10 Elastic settlement for a rigid shallow foundation.

Solution

Given $B = 1$ m and $L = 2$ m. Note that $\bar{z} = 5$ m $= 5B$. From Equation 10.57

$$E = \frac{\sum E_{(i)} \Delta z}{\bar{z}} = \frac{(10,000)(2) + (8,000)(1) + (12,000)(2)}{5} = 10,400 \text{ kN/m}^2$$

For the center of the foundation

$\alpha = 4$

$$m' = \frac{L}{B} = \frac{2}{1} = 1$$

and

$$n' = \frac{H}{(B/2)} = \frac{5}{(1/2)} = 10$$

From Tables 10.4 and 10.6, $F_1 = 0.641$ and $F_2 = 0.031$. From Equation 10.48

$$I_s = F_1 + \frac{1 - 2v}{1 - v} F_2 = 0.641 + \frac{2 - 0.3}{1 - 0.3}(0.031) = 0.716$$

Again, $D_f/B = 1/1 = 1$, $L/B = 2$, and $v = 0.3$. From Figure 10.9b, $I_f = 0.709$. Hence

$$S_{e(\text{flexible})} = q(\alpha B') \frac{1-v^2}{E} I_s I_f$$

$$= (150)\left(4 \times \frac{1}{2}\right)\left(\frac{1-0.3^2}{10,400}\right)(0.716)(0.709)$$

$$= 0.0133\,\text{m} = 13.3\,\text{m}$$

Since the foundation is rigid, we obtain

$$S_{e(\text{rigid})} = (0.93)(13.3) = 12.4 \text{ mm}$$

10.6.2 Improved equation for elastic settlement

Mayne and Poulos (1999) presented an improved formula for calculating the elastic settlement of foundations. The formula takes into account the rigidity of the foundation, the depth of embedment of the foundation, the increase in the modulus of elasticity of the soil with depth, and the location of rigid layers at a limited depth. To use Mayne and Poulos's equation, one needs to determine the equivalent diameter B_e of a rectangular foundation, or

$$B_e = \sqrt{\frac{4BL}{\pi}} \tag{10.58a}$$

where
   B is the width of foundation
   L is the length of foundation

For circular foundations

$$B_e = B \tag{10.58b}$$

where B is the diameter of foundation.

Figure 10.11 shows a foundation with an equivalent diameter B_e located at a depth D_f below the ground surface. Let the thickness of the foundation be t and the modulus of elasticity of the foundation material be E_F. A rigid layer is located at a depth H below the bottom of the foundation. The modulus of elasticity of the compressible soil layer can be given as

$$E = E_o + kz \tag{10.59}$$

With the preceding parameters defined, the elastic settlement below the center of the foundation is

$$S_e = \frac{q B_e I_G I_F I_E}{E_o}(1 - v^2) \tag{10.60}$$

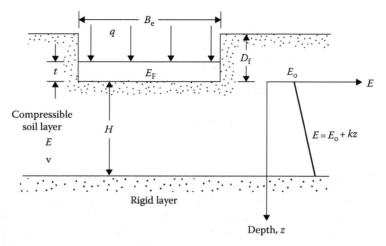

Figure 10.11 Improved equation for calculating elastic settlement—general parameters.

where I_G is the influence factor for the variation of E with depth

$$= f\left(\beta' = \frac{E_o}{kB_e}, \frac{H}{B_e}\right)$$

I_F is the foundation rigidity correction factor
I_E is the foundation embedment correction factor

Figure 10.12 shows the variation of I_G with $\beta' = E_o/kB_e$ and H/B_e. The foundation rigidity correction factor can be expressed as

$$I_F = \frac{\pi}{4} + \frac{1}{4.6 + 10(E_f/[E_o + (B_e/2)k])(2t/B_e)^3} \tag{10.61}$$

Similarly, the embedment correction factor is

$$I_E = 1 - \frac{1}{3.5\exp(1.22v - 0.4)[(B_e/D_f) + 1.6]} \tag{10.62}$$

Figures 10.13 and 10.14 show the variation of I_E and I_F with terms expressed in Equations 10.61 and 10.62.

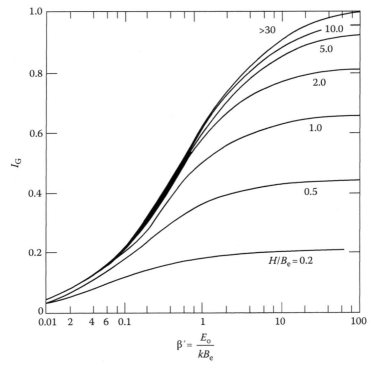

Figure 10.12 Variation of I_G with β'.

Example 10.10

For a shallow foundation supported by a silty clay as shown in Figure 10.11,

Length = L = 1.5 m
Width = B = 1 m
Depth of foundation = D_f = 1 m
Thickness of foundation = t = 0.23 m
Load per unit area = q = 190 kN/m²
$E_F = 15 \times 10^6$ kN/m²

The silty clay soil has the following properties:

H = 2 m
v = 0.3
E_o = 9000 kN/m²
k = 500 kN/m²/m

Estimate the elastic settlement of the foundation.

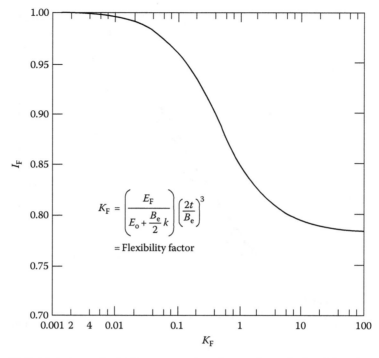

Figure 10.13 Variation of rigidity correction factor I_F with flexibility factor K_F (Equation 10.61).

Solution

From Equation 10.58, the equivalent diameter is

$$B_e = \sqrt{\frac{4BL}{\pi}} = \sqrt{\frac{(4)(1.5)(1)}{\pi}} = 1.38\,\text{m}$$

so

$$\beta' = \frac{E_o}{kB_e} = \frac{9000}{(500)(1.38)} = 13.04$$

and

$$\frac{H}{B_e} = \frac{2}{1.38} = 1.45$$

From Figure 10.12, for $\beta' = 13.04$ and $H/B_e = 1.45$, the value of $I_G \approx 0.74$.

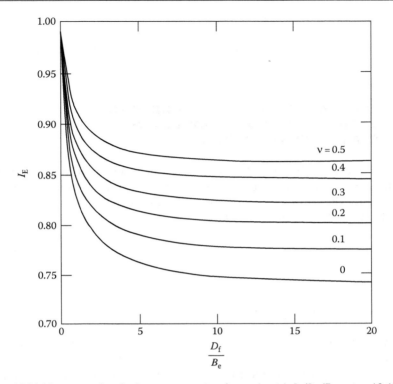

Figure 10.14 Variation of embedment correction factor I_E with D_f/B_e (Equation 10.62).

From Equation 10.61

$$I_F = \frac{\pi}{4} + \frac{1}{4.6 + 10(E_f/[E_o + (B_e/2)k])(2t/B_e)^3}$$

$$= \frac{\pi}{4} + \frac{1}{4.6 + 10[15 \times 10^6/9000 + (1.38/2)(500)][(2)(0.23)/1.38]^3} = 0.787$$

From Equation 10.50

$$I_E = 1 - \frac{1}{3.5 \exp(1.22v - 0.4)[(B_e/D_f) + 1.6]}$$

$$= 1 - \frac{1}{3.5 \exp[(1.22)(0.3) - 0.4][(1.38/1) + 1.6]} = 0.907$$

From Equation 10.60

$$S_e = \frac{qB_e I_G I_F I_E}{E_o}(1 - v^2)$$

So, with $q = 190 \text{ kN/m}^2$, it follows that

$$S_c = \frac{(190)(1.38)(0.74)(0.787)(0.907)}{9000}(1 - 0.3^2) = 0.014 \text{ m} \approx 14 \text{ mm}$$

10.7 ELASTIC SETTLEMENT CONSIDERING VARIATION OF SOIL MODULUS OF ELASTICITY WITH STRAIN

Berardi and Lancellotta (1991) proposed a method to estimate the elastic settlement that takes into account the variation of the modulus of elasticity of soil with the strain level. This method is also described by Berardi et al. (1991). According to this procedure,

$$S_e = I_F \frac{qB}{E_s} \tag{10.63}$$

where
I_F = influence factor for a rigid foundation

This is based on the work of Tsytovich (1951). The variation of I_F for $v = 0.15$ is given in Table 10.8. In this table, H_i is the depth of influence.

Berardi et al. (1991) noted that, for square foundations, $H_i \approx B$; and, for strip foundations, $H_i \approx 2B$. For rectangular foundations,

$$H_i \approx \left[1 + \log\left(\frac{L}{B}\right)\right]B \tag{10.64}$$

where
L = length of the foundation

(Note: Equation 10.64 is for $L/B \le 10$).

Table 10.8. Variation of I_F

L/B	Depth of influence H_i/B			
	0.5	1.0	1.5	2.0
1	0.35	0.56	0.63	0.69
2	0.39	0.65	0.76	0.88
3	0.40	0.67	0.81	0.96
5	0.41	0.68	0.84	0.89
10	0.42	0.71	0.89	1.06

The modulus of elasticity E in Equation 10.64 can be expressed as,

$$E = K_E p_a \left(\frac{\sigma_0' + 0.5\Delta\sigma'}{p_a} \right)^{0.5} \tag{10.65}$$

where

p_a = atmospheric pressure (≈ 100 kN/m^2)

σ_0' and $\Delta\sigma'$ = effective overburden stress and net effective stress increase due to the foundation loading, respectively, at the center of the influence zone below the foundation

K_E = nondimensional modulus number

Based on Lancellotta (2009) (also see Das and Sivakugan, 2017), at 0.1% strain level (i.e., S_e/B)

$$K_{E,0.1\%} = 9.1D_r + 92.5 \quad \text{(for } H_i = B) \tag{10.66}$$

and

$$K_{E,0.1\%} = 11.44D_r - 76.5 \quad \text{(for } H_i = 2B) \tag{10.67}$$

where D_r = relative density of sand (%).

It is important to note that, at $D_r = 60\%$, Equations 10.66 and 10.67 will give values for $K_{E,0.1\%}$ as 638.5 and 609.9, respectively. Similarly, at $D_r = 80\%$, Equations 10.66 and 10.67 will give values for $K_{E,0.1\%}$ as 820.5 and 838.7, respectively. These values of $K_{E,0.1\%}$ are relatively close. Most of the foundations are analyzed within a range of $D_r = 60\%$ to 80%. So, $K_{E,0.1\%}$ values for the range of $H_i = B$ and $2B$ can be reasonably interpolated.

The magnitude of D_r can be estimated as (Skempton, 1986),

$$D_r = \left[\frac{(N_1)_{60}}{60} \right]^{0.5} \tag{10.68}$$

where

$(N_1)_{60}$ = average corrected (to an effective overburden pressure of $p_a = 100$ kN/m^2) standard penetration resistance in the zone of influence

The modulus number K_E, as any other strain level, can be estimated as (Berardi et al. 1991; Das and Sivakugan, 2017)

$$\frac{K_{E(S_e/B\%)}}{K_{E,0.1\%}} = 0.008 \left(\frac{S_e}{B} \right)^{-0.7} \tag{10.69}$$

It has been suggested by Lancellotta (2009) that

$$\frac{E_{s(S_e/B\%)}}{E_{s(0.1\%)}} = 0.008\left(\frac{S_e}{B}\right)^{-0.7}$$

(10.70)

where $E_{s(0.1\%)}$ is the modulus of elasticity of sand when the vertical strain level $\varepsilon_v = S_e/B = 0.1\%$. The value of $K_{E,0.1\%}$ determined from Equations 10.66 and 10.67 can be substituted into Equation 10.65 for estimation of $E_{s(0.1\%)}$.

Again, from Equations 10.63 and 10.70,

$$\left(\frac{S_e}{B}\right)^{0.3} = \frac{125q(I_F)}{E_{s(0.1\%)}}$$

(10.71)

Example 10.11

Consider a square foundation 2 m × 2 m in plan. Given:

- $D_f = 0.5$ m
- Load on the foundation = 150 kN/m²
- Unit weight of sand = 19 kN/m³
- $(N_1)_{60} = 28$
- Poisson's ratio, $v = 0.15$

Estimate S_e.

Solution

From Equation 10.68,

$$D_r = \left[\frac{(N_1)_{60}}{60}\right]^{0.5} = \left[\frac{28}{60}\right]^{0.5} = 0.683 = 68.3\%$$

For a square foundation $H_i = B$. So the center of the influence zone will be 1.5 m from the ground surface. Hence, from Equation 10.65

$$\sigma'_o = (1.5)(19) = 28.5 \text{ kN/m}^2$$

For estimating $\Delta\sigma'$, we use Table 4.13. For this case,

$$\frac{z}{B/2} = \frac{(1.5/2)}{(1.5/2)} = 1; \quad \frac{L}{B} = \frac{1.5}{1.5} = 1$$

Hence,

$$\frac{\Delta\sigma}{q} = 0.701; \quad \Delta\sigma = (0.701)(150) = 105.15 \text{ kN/m}^2$$

From Table 10.8 for $L/B = 1$ and $H_i/B = 1$, the value of $I_F = 0.56$.

$$\left(\frac{\sigma_0' + 0.5\Delta\sigma'}{p_a}\right)^{0.5} = \left(\frac{28.5 + 0.5 \times 105.15}{100}\right)^{0.5} = 0.9$$

From Equation 10.66,

$$K_{E,0.1\%} = 9.1D_r + 92.5 = (9.1)(68.3) + 92.5 = 714.03$$

$$E_{s(0.1\%)} = K_{E,0.1\%}p_a\left(\frac{\sigma_0' + 0.5\Delta\sigma'}{p_a}\right)^{0.5} = (714.03)(100)(0.9) \approx 62,263 \text{ kN/m}^2$$

From Equation 10.71,

$$\left(\frac{S_e}{B}\right)^{0.3} = \frac{(125)(150)(0.56)}{64,263} = 0.163$$
$$S_e = 0.00476 \text{ m} = 4.76 \text{ mm}$$

10.8 EFFECT OF GROUND WATER TABLE RISE ON ELASTIC SETTLEMENT OF GRANULAR SOIL

Any future rise in the water table can reduce the ultimate bearing capacity. Similarly, future water table rise in the vicinity of the foundations in granular soil can reduce the soil stiffness and produce additional settlement. Terzaghi (1943) concluded that, when the water table rises from very deep to the foundation level, the settlement will be doubled in granular soil. Shahriar et al. (2015) recently conducted several laboratory model tests and related numerical modeling based on the concept of strain influence factor as shown in Figure 10.15 to show the settlement correction factor due to rise in water table can be expressed as

$$C_w = \frac{S_{e(\text{water table})}}{S_e} \tag{10.72}$$

where

C_w = settlement correction factor
$S_{e(\text{water table})}$ = total settlement with the water table rising
S_e = elastic settlement in dry sand

The magnitude of C_w when the water table is located at a depth z below the foundation may be given as,

$$C_w = 1 + [C_{w(\text{max})} - 1]\left(\frac{A_w}{A_t}\right)^n \tag{10.73}$$

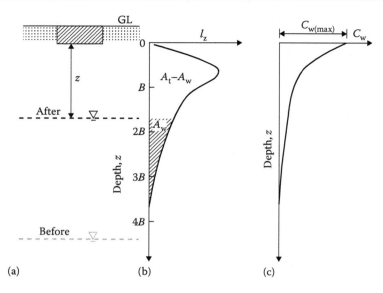

Figure 10.15 Schematic diagram of (a) shallow foundation with rising groundwater table; (b)straininfluencefactorunderthefoundation;(c)groundwatertablecorrectionfactor, C_w.

where

A_t = total area of the strain influence factor

A_w = area of the strain influence factor covered by water

$C_{w(max)}$ = maximum value of C_w at the base of the foundation

n = a variable depending on the denseness of sand

For loose sand:

$C_{w(max)} = 6.3$ and $n = 0.85$

and for dense sand:

$C_{w(max)} = 3.4$ and $n = 1.1$

The strain influence factor used in this study was developed using FLAC and FLAC^{CD} with $E = 30$ MN/m²; $\nu = 0.2$; and q (load per unit area of the foundation) = 100 kN/m². The influence factor diagram is shown in Figure 10.16 (Shahriar et al. 2012). Based on Figure 10.16, the variation of A_w/A_t with z/B and B/L can be calculated. These values are shown in Table 10.9.

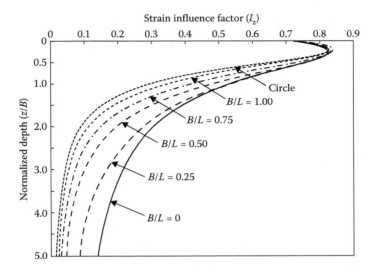

Strain influence factor (I_z)

Figure 10.16 Shahriar et al.'s (2012) strain influence factor diagram.

Table 10.9. A_w/A_t values at different depths for various foundation shapes

Water table depth (z/B)	Foundation shape					
	Circular	Square (B/L = 1)	Rectangular (B/L = 0.75)	Rectangular (B/L = 0.50)	Rectangular (B/L = 0.25)	Strip (B/L = 0)
0	I	I	I	I	I	I
0.5	0.573	0.612	0.658	0.703	0.757	0.785
I	0.33	0.368	0.416	0.475	0.562	0.314
2	0.149	0.171	0.2	0.241	0.327	0.399
3	0.08	0.094	0.11	0.135	0.196	0.264
4	0.044	0.051	0.06	0.075	0.113	0.163
5	0.019	0.023	0.027	0.034	0.051	0.078
6	0	0	0	0	0	0

Example 10.12

Consider the shallow foundation described in Example 10.5. If the ground water table rises to a level of $z = 6$ m, estimate the settlement of the foundation. Consider the sand to be dense.

Solution

$$C_w = 1 + [C_{w(max)} - 1]\left(\frac{A_w}{A_t}\right)^n$$

For dense sand, $C_{w(max)} = 3.4$ and $n = 1.1$. $B/L = 2/4 = 0.5$; $z/B = 6/2 = 3$. From Table 10.9, $A_w/A_t = 0.135$. So

$$C_w = 1 + (3.4 - 1)(0.135)^{1.1} = 1.265$$

$$\frac{S_{e(water\ table)}}{S_e} = 1.265$$

$$S_{e(water\ table)} = (1.265)(27.83) = 35.2\ \mathbf{mm}$$

REFERENCES

Akbas, S. O. and F. H. Kulhawy, Axial compression of footings in cohesionless soils. 1: Load settlement behavior, *J. Geotech. Geoenviron. Eng.*, ASCE, 123(11), 1562–1574, 2009.

Berardi, R., M. Jamiolkowski, and R. Lancellotta. Settlement of shallow foundations in sands: Selection of stiffness on the basis of penetration resistance, in *Geotechnical Engineering Congress 1991*, Geotech. Special Pub. 27, ASCE, 185–200, 1991.

Berardi, R. and R. Lancellotta. Stiffness of granular soil from field performance, *Geotechnique*, 41(1), 149–157, 1991.

Bowles, J. E., Elastic foundation settlement on sand deposits, *J. Geotech. Eng.*, ASCE, 113(8), 846–860, 1987.

Burland, J. B., Discussion, Session A, *Proc. Conf. in situ Invest. Soil Rocks*, British Geotechnical Society, London, U.K., 61–62, 1970.

Burland, J. B. and M. C. Burbidge, Settlement of foundations on sand and gravel, *Proc. Inst. Civ. Eng.*, 78(1), 1325–1381, 1985.

Christian, J. T. and W. D. Carrier III, Janbu, Bjerrum and Kjaernsli's chart reinterpreted, *Can. Geotech. J.*, 15(1), 124–128, 1978.

Das, B. M., *Principles of Foundation Engineering*, Cengage, Stamford, CT, 2011.

Das, B. and N. Sivakugan. *Fundamentals of Geotechnial Engineering*, 5th edn., Cengage, Stamford, CT, 2017.

Das, B. and N. Sivakugan. *Principles of Foundation Engineering*, 9th edn., Cengage, Stamford, CT, 2019.

Duncan, J. M. and A. N. Buchignani, *An Engineering Manual for Settlement Studies*, Department of Civil Engineering, Univ. of California, Berkeley, 1976.

Fox, E. N., The mean elastic settlement of a uniformly loaded area at a depth below the ground surface, *Proc. 2nd Int. Conf. Soil Mech. Found. Eng.*, Rotterdam, the Netherlands, vol. 1, 129–132, 1948.

Giroud, J. L., Settlement of rectangular foundations on soil layer, *J. Soil Mech. Found. Div.*, ASCE, 98(1), 149–154, 1972.

Janbu, N., L. Bjerrum, and B. Kjaernsli, Veiledning ved Losning av Fundamentering soppgaver, Publ. 16, Norwegian Geotechnical Institute, Oslo, Norway, 30–32, 1956.

Lancellotta, R., *Geotechnical Engineering*, 2nd edn. London: Taylor & Francis, 2009.

Mayne, P. W. and H. G. Poulos, Approximate displacement influence factors for elastic shallow foundations, *J. Geotech. Geoenviron. Eng.*, ASCE, 125(6), 453–460, 1999.

Meyerhof, G. G., Penetration tests and bearing capacity of cohesionless soils, *J. Soil Mech. Found. Div.*, ASCE, 82(1), 1–19, 1956.

Meyerhof, G. G., Shallow foundations, *J. Soil Mech. Found. Div.*, ASCE, 91(2), 21–31, 1965.

Peck, R. B. and A. R. S. S. Bazaraa, Discussion of paper by D'Appolonia et al., *J. Soil Mech. Found. Div.*, ASCE, 95(3), 305–309, 1969.

Salgado, R., *The Engineering of Foundations*, Mcgraw-Hill, New York, 2008.

Schleicher, F., Zur Theorie des Baugrundes, *Bauingenieur*, 7, 931–935, 949–954, 1926.

Schmertmann, J. H., Static cone to compute static settlement over sand, *J. Soil Mech. Found. Div.*, ASCE, 96(3), 1011–1043, 1970.

Schmertmann, J. H., J. P. Hartman, and P. R. Brown, Improved strain influence factor diagrams, *J. Geotech. Eng. Div.*, ASCE, 104(8), 1131–1135, 1978.

Shahriar, M. A., N. Sivakugan, and B. M. Das, Strain influence factors for footings on an elastic medium, *Proc. XI Australia-New Zealand Conf. on Geomechanics*, Melbourne, Australian Geomechanics Society and New Zealand Geotechnical Society, 131–136, 2012.

Shahriar, M. A., N. Sivakugan, B. M. Das, A. Urquhart, and M. Tapiolas, Water table correction factors for settlement of shallow foundations in granular soil, *Int. J. Geomechanics*, ASCE, 15(1), 76, 2015.

Skempton, A. W., Standard penetration test procedures and the effects in sand of overburden pressure, relative density, particle size, ageing and overconsolidation, *Geotechnique*, 36(3), 425–447, 1986.

Steinbrenner, W., Tafeln zur Setzungsberechnung, *Die Strasse*, 1, 121–124, 1934.

Terzaghi, K., *Theoretical Solil Mechanics*, Wiley, New York, 1943.

Terzaghi, K. and R. B. Peck, *Soil Mechanics in Engineering Practice*, 1st edn., Wiley, New York, 1948.

Terzaghi, K. and R. B. Peck, *Soil Mechanics in Engineering Practice*, 2nd edn., Wiley, New York, 1967.

Terzaghi, K., R. B. Peck, and G. Mesri, *Soil Mechanics in Engineering Practice*, 3rd edn., Wiley, New York, 1996.

Tsytovich, N. A., *Soil Mechanics*, Moscow: Stroitielstvo I. Archiketura (in Russian), 1951.

Vesic, A. S. Analysis of ultimate loads of shallow foundations, *J. Soil Mech. Found. Div.*, ASCE, 99(1), 45–73, 1973.

Chapter 11

Consolidation settlement of shallow foundations

11.1 INTRODUCTION

As mentioned in Chapter 10, the total settlement of a shallow foundation, in general, is the sum of the elastic settlement (S_e) and the consolidation settlement (S_c) of the soil supporting the foundation. The procedures for estimating the elastic settlement were treated in Chapter 10. It was also pointed out that the total consolidation settlement is the sum of the settlements occurring from primary consolidation (S_p) and secondary consolidation (S_s). Or

$$S_c = S_p + S_s \tag{11.1}$$

In this chapter, we will consider the procedures for estimation of S_p and S_s.

11.2 ONE-DIMENSIONAL PRIMARY CONSOLIDATION SETTLEMENT CALCULATION

Based on Equation 8.98 in Section 8.9, the settlement for one-dimensional consolidation can be given by

$$S_p = \Delta H_t = \frac{\Delta e}{1 + e_0} H_t \tag{8.98}$$

where

$$\Delta e = C_c \log \frac{\sigma_0' + \Delta\sigma}{\sigma_0'} \quad \text{(for normally consolidated clays)} \tag{8.99}$$

$$\Delta e = C_r \log \frac{\sigma_0' + \Delta\sigma}{\sigma_0'} \quad \text{(for overconsolidated clays, } \sigma_0' + \Delta\sigma \le \sigma_c') \tag{8.100}$$

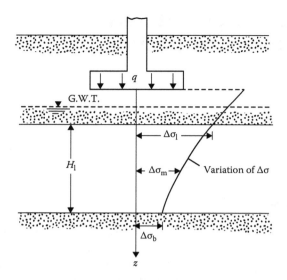

Figure 11.1 Calculation of consolidation settlement—method A.

$$\Delta e = C_r \log \frac{\sigma'_c}{\sigma'_0} + C_c \log \frac{\sigma'_0 + \Delta\sigma}{\sigma'_c} \quad \text{(for } \sigma'_0 < \sigma'_c < \sigma'_0 + \Delta\sigma) \tag{8.101}$$

where σ'_c is the preconsolidation pressure.

When a load is applied over a limited area, the increase of pressure due to the applied load will decrease with depth, as shown in Figure 11.1. So, for a more realistic settlement prediction, we can use the following methods.

Method A

1. Calculate the average effective pressure σ'_0 on the clay layer before application of the load under consideration.
2. Calculate the increase of stress due to the applied load at the top, middle, and bottom of the clay layer below the center of the foundation. This can be done by using theories developed in Chapters 3 and 4. The average increase of stress below the center of the foundation in the clay layer can be estimated by Simpson's rule

$$\Delta\sigma_{av} = \frac{1}{6}(\Delta\sigma_t + 4\Delta\sigma_m + \Delta\sigma_b) \tag{11.2}$$

where $\Delta\sigma_t$, $\Delta\sigma_m$, and $\Delta\sigma_b$ are stress increases at the top, middle, and bottom of the clay layer, respectively.

The magnitude of $\Delta\sigma_{av}$ below a rectangular foundation can also be determined by using Griffiths's influence factor I_a (Griffiths, 1984). The procedure is described below.

From Equation 4.47,

$$\Delta\sigma_z = qI_9 \tag{11.3}$$

where $\Delta\sigma_z$ = stress increase at a depth z below the corner of a rectangular foundation; q = load per unit area on the foundation; I_9 = influence factor which is a function of

$$m = \frac{B}{z} \tag{11.4}$$

and

$$n = \frac{L}{z} \tag{11.5}$$

If it is required to find the average stress increase, $\Delta\sigma_z$, below the corner of a uniformly loaded rectangular area with limits of $z = 0$ to $z = H$ as shown in Figure 11.2 (point A), then

$$\Delta\sigma_{av} = \frac{1}{H}\int_0^H (qI_9)dz = qI_a$$

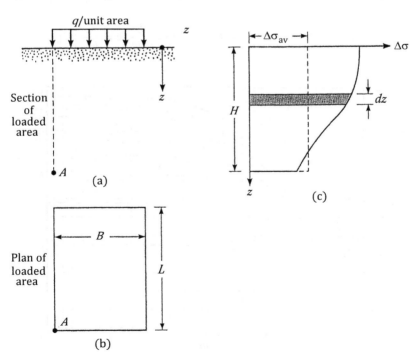

Figure 11.2 Average vertical stress increase below the corner of a rectangular flexible.

where

$$I_a = f(m, n) \tag{11.6}$$

$$m = \frac{B}{H} \tag{11.7}$$

$$n = \frac{L}{H} \tag{11.8}$$

The variation of I_a with m and n as proposed by Griffiths (1984) is shown in Figure 11.3.

For estimating the consolidation settlement, the average vertical stress increase in a clay layer; that is, between $z = H_1$ and $z = H_2$ as shown in Figure 11.4, can be done as (Griffiths, 1984)

$$\Delta\sigma_{av(H_2/H_1)} = q\left[\frac{H_2 I_{a(H_2)} - H_1 I_{a(H_1)}}{H_2 - H_1}\right] \tag{11.9}$$

where $\Delta\sigma_{av(H_2/H_1)}$ = average stress increase below the corner of a uniformly loaded rectangular area between depths $z = H_1$ and $z = H_2$

$$I_{a(H_2)} = I_a \text{ for } z = 0 \text{ to } z = H_2 = f\left(m = \frac{B}{H_2}, n = \frac{L}{H_2}\right)$$

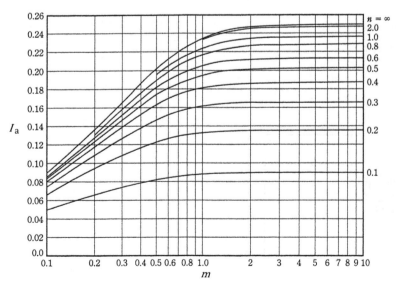

Figure 11.3 Griffiths' influence factor I_a.

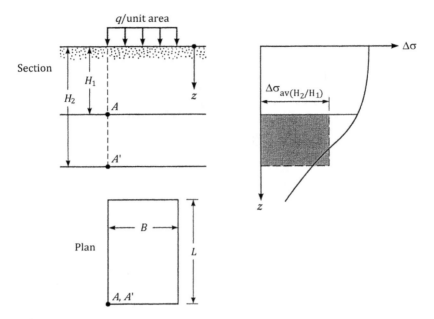

Figure 11.4 Average stress increase between $z = H_1$ and $z = H_2$ below the corner of a uniformly loaded rectangular area.

$$I_{a(H_1)} = I_a \text{ for } z = 0 \text{ to } z = H_1 = f\left(m = \frac{B}{H_1}, n = \frac{L}{H_1}\right)$$

The procedure for using Figure 11.3 is shown in Example 11.1.

For circular foundations, the magnitude of the average stress increase $\Delta\sigma_{av}$ can also be obtained by a method developed by Saikia (2012) using this procedure. Figure 11.5 gives the variation of $\Delta\sigma_{av}/q$ with z_1/R and z_2/R, where z_1 and z_2 are the vertical distances between the bottom of the foundation and the top and bottom of the clay layer, respectively.

3. Using σ'_0 and σ_{av} calculated earlier, obtain Δe from Equations 8.99, 8.100, or 8.101, whichever is applicable.
4. Calculate the settlement by using Equation 8.98.

Method B

1. Better results in settlement calculation may be obtained by dividing a given clay layer into n layers as shown in Figure 11.6.
2. Calculate the effective stress $\sigma'_{0(i)}$ at the middle of each layer.
3. Calculate the increase of stress at the middle of each layer $\Delta\sigma_i$ due to the applied load.

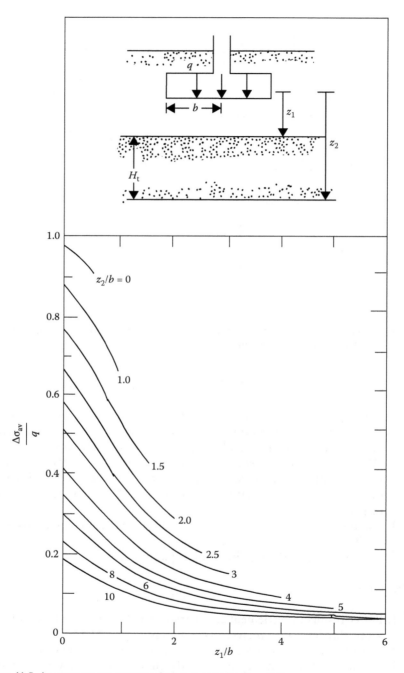

Figure 11.5 Average stress increase below a circular foundation.

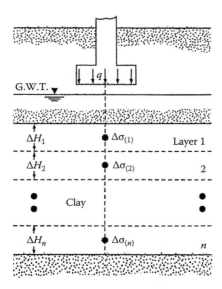

Figure 11.6 Calculation of consolidation settlement—method B.

4. Calculate Δe_i for each layer from Equations 8.99, 8.100, or 8.101, whichever is applicable.
5. Total settlement for the entire clay layer can be given by

$$S_{\mathrm{p}} = \sum_{i=1}^{i=n} \Delta S_{\mathrm{p}} = \sum_{i=1}^{n} \frac{\Delta e_i}{1 + e_0} \Delta H_i \qquad (11.10)$$

Example 11.1

Refer to Figure 11.7. Determine the *average* stress increase below the center of the loaded area between $z = 3$ m to $z = 5$ m (that is, between points A and A').

Solution

Refer to Figure 11.7. The loaded area can be divided into four rectangular areas, each measuring 1.5 m × 1.5 m ($L \times B$). Using Equation 11.9, the average stress increase (between the required depths) below the corner of each rectangular area can be given as

$$\Delta\sigma_{\mathrm{av}(H_2/H_1)} = q\left[\frac{H_2 I_{\mathrm{a}(H_2)} - H_1 I_{\mathrm{a}(H_1)}}{H_2 - H_1}\right] = 100\left[\frac{(5)I_{\mathrm{a}(H_2)} - (3)I_{\mathrm{a}(H_1)}}{5 - 3}\right]$$

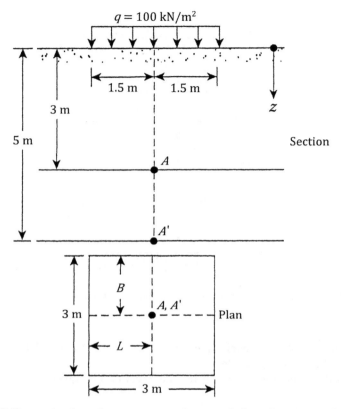

Figure 11.7 Determination of average stress increase below the center of a rectangular area.

For $I_{a(H_2)}$,

$$m_2 = \frac{B}{H_2} = \frac{1.5}{5} = 0.3$$

$$n_2 = \frac{L}{H_2} = \frac{1.5}{5} = 0.3$$

Referring to Figure 11.3 for $m_2 = 0.3$ and $n_2 = 0.3$, $I_{a(H_2)} = 0.126$.
For $I_{a(H_1)}$,

$$m_2 = \frac{B}{H_1} = \frac{1.5}{3} = 0.5$$

$$n_2 = \frac{L}{H_1} = \frac{1.5}{3} = 0.5$$

Referring to Figure 11.3, $I_{a(H_1)} = 0.175$. So

$$\Delta\sigma_{av(H_2/H_1)} = 100\left[\frac{(5)(0.126)-(3)(0.175)}{5-3}\right] = 5.25 \text{ kN/m}^2$$

The stress increase between $z = 3$ m to $z = 5$ m below the center of the loaded area is equal to

$$4\Delta\sigma_{av(H_2/H_1)} = (4)(5.25) = 21 \text{ kN/m}^2$$

Example 11.2

Solve Example 1.1 using Equation 4.47 and Equation 11.2.

Solution

The loaded area has been divided into four rectangular areas in Figure 11.7. Determination of $\Delta\sigma_t$, $\Delta\sigma_m$, and $\Delta\sigma_b$ below the corner of each rectangular area can be done as follows:

- $\Delta\sigma_t$ determination

$$m_t = \frac{B}{z_t} = \frac{B}{H_1} = \frac{1.5}{3} = 0.5$$

$$n_t = \frac{L}{z_t} = \frac{L}{H_1} = \frac{1.5}{3} = 0.5$$

For $m_t = n_t = 0.5$, from Table 4.12, $I_{9(1)} = 0.0840$. So
$\Delta\sigma_t = qI_{9(1)} = (100)(0.084) = 8.4 \text{ kN/m}^2$
- $\Delta\sigma_m$ determination

$$m_m = \frac{B}{z_m} = \frac{1.5}{4} = 0.375$$

$$n_m = \frac{L}{z_m} = \frac{1.5}{4} = 0.375$$

From Table 4.12, for $m_m = n_m = 0.375$, the value of $I_{9(m)}$ is about 0.05. Hence,

$$\Delta\sigma_m = qI_{9(m)} = (100)(0.05) = 5 \text{ kN/m}^2$$

- $\Delta\sigma_b$ determination

$$m_b = \frac{B}{z_b} = \frac{B}{H_2} = \frac{1.5}{5} = 0.3$$

$$n_b = \frac{L}{z_b} = \frac{L}{H_2} = \frac{1.5}{5} = 0.3$$

From Table 4.12, $I_{9(b)} = 0.0374$. So

$$\Delta\sigma_b = (100)(0.0374) = 3.74 \text{ kN/m}^2$$

The stress increase, $\Delta\sigma_{av}$, below the *corner* of each rectangular area is

$$\Delta\sigma_{av} = \frac{1}{6}[8.4 + (4 \times 5) + 3.74] = 5.36 \text{ kN/m}^2$$

Hence, $\Delta\sigma_{av}$ below the center of the entire rectangular area = $(5.36)(4) = 21.44 \text{ kN/m}^2$

Example 11.3

A circular foundation 2 m in diameter is shown in Figure 11.8a. A normally consolidated clay layer 5 m thick is located below the foundation. Determine the primary consolidation settlement of the clay. Use method B (Section 11.2).

Solution

We divide the clay layer into five layers, each 1 m thick. *Calculation of* $\sigma'_{0(i)}$: The effective stress at the middle of layer 1 is

$$\sigma'_{0(1)} = 17(1.5) + (19 - 9.81)(0.5) + (18.5 - 9.81)(0.5) = 34.44 \text{ kN/m}^2$$

The effective stress at the middle of the second layer is

$$\sigma'_{0(2)} = 34.44 + (18.5 - 9.81)(1) = 34.44 + 8.69 = 43.13 \text{ kN/m}^2$$

Similarly

$$\sigma'_{0(3)} = 43.13 + 8.69 = 51.81 \text{ kN/m}^2$$

$$\sigma'_{0(4)} = 51.82 + 8.69 = 60.51 \text{ kN/m}^2$$

$$\sigma'_{0(5)} = 60.51 + 8.69 = 69.2 \text{ kN/m}^2$$

Calculation of $\Delta\sigma_i$: For a circular loaded foundation, the increase of stress below the center is given by Equation 4.35, and so

$$\Delta\sigma_i = q\left\{1 - \frac{1}{[(b/z)^2 + 1]^{3/2}}\right\}$$

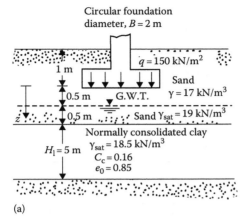

(a)

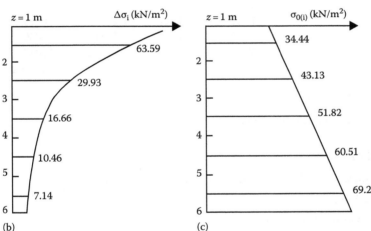

(b)   (c)

Figure 11.8 Consolidation settlement calculation from layers of finite thickness: (a) soil profile; (b) variation of $\Delta\sigma_i$ with depth; (c) variation of $\sigma_{0(i)}$ with depth.

where b is the radius of the circular foundation, 1 m. Hence

$$\Delta\sigma_1 = 150\left\{1 - \frac{1}{[(1/1.5)^2 + 1]^{3/2}}\right\} = 63.59 \text{ kN/m}^2$$

$$\Delta\sigma_2 = 150\left\{1 - \frac{1}{[(1/2.5)^2 + 1]^{3/2}}\right\} = 29.93 \text{ kN/m}^2$$

$$\Delta\sigma_3 = 150\left\{1 - \frac{1}{[(1/3.5)^2 + 1]^{3/2}}\right\} = 16.66 \text{ kN/m}^2$$

$$\Delta\sigma_4 = 150\left\{1 - \frac{1}{[(1/4.5)^2 + 1]^{3/2}}\right\} = 10.46 \text{ kN/m}^2$$

$$\Delta\sigma_5 = 150\left\{1 - \frac{1}{[(1/5.5)^2 + 1]^{3/2}}\right\} = 7.14 \text{ kN/m}^2$$

Calculation of primary consolidation settlement: The steps in the calculation are given in the following table (see also Figure 11.8b and c):

Layer	ΔH_i (m)	$\sigma'_{0(i)}$ (kN/m²)	$\Delta\sigma_i$ (kN/m²)	Δe^a	$\dfrac{\Delta e}{1 + e_0}\Delta H_i$ (m)
1	1	34.44	63.59	0.0727	0.0393
2	1	43.13	29.93	0.0366	0.0198
3	1	51.82	16.66	0.0194	0.0105
4	1	60.51	10.46	0.0111	0.0060
5	1	69.2	7.14	0.00682	0.0037
					$\Sigma = 0.0793$

^{a} $\Delta e = C_c \log \dfrac{\sigma'_{0(i)} + \Delta\sigma_i}{\sigma'_{0(i)}}$; $C_c = 0.16$

So, $S_p = 0.0793$ m $= 79.3$ mm.

Example 11.4

Solve Example 11.3 using Method A and Equation 11.2.

Solution

From Equation 4.35

$$\Delta\sigma = q\left\{1 - \frac{1}{[(b/z)^2 + 1]^{3/2}}\right\}$$

Hence

$$\Delta\sigma_t = 150\left\{1 - \frac{1}{[(1/1)^2 + 1]^{3/2}}\right\} = 96.97 \text{ kN/m}^2$$

$$\Delta\sigma_m = 150\left\{1 - \frac{1}{[(1/3.5)^2 + 1]^{3/2}}\right\} = 16.66 \text{ kN/m}^2$$

$$\Delta\sigma_b = 150\left\{1 - \frac{1}{[(1/6)^2 + 1]^{3/2}}\right\} = 6.04 \text{ kN/m}^2$$

$$\Delta\sigma_{av} = \frac{1}{6}(\Delta\sigma_t + 4\Delta\sigma_m + \Delta\sigma_b)$$

$$= \frac{1}{6}[96.97 + (4)(16.66) + 6.04] = 28.28 \text{ kN/m}^2$$

Also

$$\sigma_0' = (1.5)(17) + (0.5)(19 - 9.81) + \left(\frac{5}{2}\right)(18.5 - 9.81) = 51.82 \text{ kN/m}^2$$

Combining Equations 8.98 and 8.99

$$S_p = \frac{C_c H}{1 + e_0} \log\left(\frac{\sigma_0' + \Delta\sigma}{\sigma_0'}\right)$$

$$= \frac{(0.16)(5000 \text{ mm})}{1 + 0.85} \log\left(\frac{51.82 + 28.28}{51.82}\right) = 81.79 \text{ mm}$$

Example 11.5

Refer to Example 11.4. Calculate $\Delta\sigma_{av}$ using Figure 11.5.

Solution

For this case

$$\frac{z_1}{b} = \frac{1}{1} = 1$$

$$\frac{z_2}{b} = \frac{6}{1} = 6$$

From Figure 11.5 for $z_1/b = 1$ and $z_2/b = 6$, the value of $\Delta\sigma_{av}/q \approx 0.175$. So

$$\Delta\sigma_{av} = (0.175)(150) = 26.25 \text{ kN/m}^2 \text{ (close to that in Example 11.4)}$$

Example 11.6

Calculate the primary consolidation settlement of the 3 m thick clay layer (Figure 11.9) that will result from the load carried by a 1.5 m square footing. The clay is normally consolidated. Use Equation 11.2.

Solution

For normally consolidated clay, combining Equations 8.98 and 8.99, we have

$$S_p = \frac{C_c H}{1 + e_0} \log\left(\frac{\sigma_0' + \Delta\sigma}{\sigma_0'}\right)$$

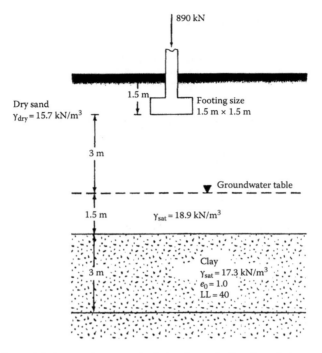

Figure 11.9 Consolidation settlement calculation for a shallow foundation.

where

$$C_c = 0.009(LL - 10) = 0.009(40 - 10) = 0.27$$
$$H = 3000 \text{ mm}$$
$$e_0 = 1.0$$

$$\sigma_0' = 4.5 \times \gamma_{dry(sand)} + 1.5[\gamma_{sat(sand)} - 9.81] + \frac{3}{2}[\gamma_{sat(clay)} - 9.81]$$

$$= 4.5 \times 15.7 + 1.5(18.9 - 9.81) + 1.5(17.3 - 9.81) = 95.52 \text{ kN/m}^2$$

In order to calculate $\Delta\sigma$, we can prepare the following table:

z (m)	$m_i' = \dfrac{L}{B}$ [a]	B (m)	$n_i' = \dfrac{z}{B/2}$ [b]	I_{I0} (Table 4.13)	$\Delta\sigma = qI_{I0}$ [c] (kN/m^2)
4.5	1	1.5	6	0.051	20.17
6.0	1	1.5	8	0.029	11.47
7.5	1	1.5	10	0.019	7.52

[a] Equation 4.50.
[b] Equation 4.51.
[c] $q = \dfrac{890}{2.25} = 395.6 \text{ kN/m}^2$.

We calculate (Equation 11.2)

$$\Delta\sigma_{av} = \frac{20.17 + (4)(11.47) + 7.52}{6} = 12.26 \text{ kN/m}^2$$

Substituting these values into the settlement equation gives

$$S_p = \frac{(0.27)(3000)}{1+1} \log\left(\frac{95.52 + 12.26}{95.52}\right) = 21.3 \text{ mm}$$

11.3 SKEMPTON–BJERRUM MODIFICATION FOR CALCULATION OF CONSOLIDATION SETTLEMENT

In one-dimensional consolidation tests, there is no lateral yield of the soil specimen and the ratio of the minor to major principal effective stresses, K_o, remains constant. In that case, the increase of pore water pressure due to an increase of vertical stress is equal in magnitude to the latter; or

$$\Delta u = \Delta\sigma \tag{11.11}$$

where

Δu is the increase of pore water pressure
$\Delta\sigma$ is the increase of vertical stress

However, in reality, the final increase of major and minor principal stresses due to a given loading condition at a given point in a clay layer does not maintain a ratio equal to K_o. This causes a lateral yield of soil. The increase of pore water pressure at a point due to a given load is (Figure 11.10) (see Chapter 5)

$$\Delta u = \Delta\sigma_3 + A(\Delta\sigma_1 - \Delta\sigma_3)$$

Skempton and Bjerrum (1957) proposed that the vertical compression of a soil element of thickness dz due to an increase of pore water pressure Δu may be given by

$$dS_p = m_\upsilon \Delta u dz \tag{11.12}$$

where m_υ is coefficient of volume compressibility (Section 8.2), or

$$dS_p = m_\upsilon[\Delta\sigma_3 + A(\Delta\sigma_1 - \Delta\sigma_3)]dz = m_\upsilon\Delta\sigma_1\left[A + \frac{\Delta\sigma_3}{\Delta\sigma_1}(1 - A)\right]dz$$

The preceding equation can be integrated to obtain the total primary consolidation settlement:

$$S_p = \int_0^{H_t} m_\upsilon\Delta\sigma_1\left[A + \frac{\Delta\sigma_3}{\Delta\sigma_1}(1 - A)\right]dz \tag{11.13}$$

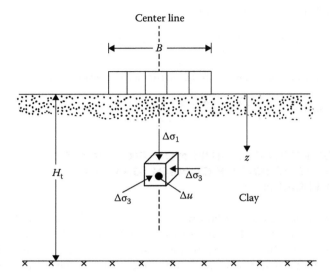

Figure 11.10 Development of excess pore water pressure below the centerline of a circular loaded foundation.

For conventional one-dimensional consolidation (K_o condition),

$$S_{p(oed)} = \int_0^{H_t} \frac{\Delta e}{1 + e_0} dz = \int_0^{H_t} \frac{\Delta e}{\Delta \sigma_1} \frac{1}{1 + e_0} \Delta \sigma_1 dz = \int_0^{H_t} m_v \Delta \sigma_1 dz \qquad (11.14)$$

(Note that Equation 11.14 is the same as that used for settlement calculation in Section 11.2.) Thus

$$\text{Settlement ratio, } \rho_{circle} = \frac{S_p}{S_{p(oed)}}$$

$$= \frac{\int_0^{H_t} m_v \Delta \sigma_1 [A + (\Delta \sigma_3 / \Delta \sigma_1)(1 - A)]dz}{\int_0^{H_t} m_v \Delta \sigma_1 dz}$$

$$= A + (1 - A)\frac{\int_0^{H_t} \Delta \sigma_3 dz}{\int_0^{H_t} \Delta \sigma_1 dz}$$

$$= A + (1 - A)M_1 \qquad (11.15)$$

where

$$M_1 = \frac{\int_0^{H_t} \Delta\sigma_3 dz}{\int_0^{H_t} \Delta\sigma_1 dz} \qquad (11.16)$$

We can also develop an expression similar to Equation 11.15 for consolidation under the center of a strip load (Scott, 1963) of width B. From Chapter 5,

$$\Delta u = \Delta\sigma_3 + \left[\frac{\sqrt{3}}{2}\left(A - \frac{1}{3}\right) + \frac{1}{2}\right](\Delta\sigma_1 - \Delta\sigma_3) \quad \upsilon = 0.5$$

$$\text{So,} \quad S_c = \int_0^{H_t} m_\upsilon \Delta u \, dz = \int_0^{H_t} m_\upsilon \Delta\sigma_1 \left[N + (1-N)\frac{\Delta\sigma_3}{\Delta\sigma_1}\right] dz \qquad (11.17)$$

where

$$N = \frac{\sqrt{3}}{2}\left(A - \frac{1}{3}\right) + \frac{1}{2}$$

Hence

$$\text{Settlement ratio, } \rho_{\text{strip}} = \frac{S_p}{S_{p(\text{oed})}}$$

$$= \frac{\int_0^{H_t} m_\upsilon \Delta\sigma_1 [N + (1-N)(\Delta\sigma_3/\Delta\sigma_1)] dz}{\int_0^{H_t} m_\upsilon \Delta\sigma_1 dz}$$

$$= N + (1-N)M_2 \qquad (11.18)$$

where

$$M_2 = \frac{\int_0^{H_t} \Delta\sigma_3 dz}{\int_0^{H_t} \Delta\sigma_1 dz} \qquad (11.19)$$

The values of ρ_{circle} and ρ_{strip} for different values of the pore pressure parameter A are given in Figure 11.11.

It must be pointed out that the settlement ratio obtained in Equations 11.15 and 11.18 can only be used for settlement calculation along the axes of symmetry. Away from the axes of symmetry, the principal stresses are no longer in vertical and horizontal directions.

It is also important to know that the settlement ratios given in Figure 11.11 are for normally consolidated clays. Leonards (1976) considered the settlement ratio $\rho_{cir(OC)}$ for three-dimensional consolidation effect in the field for a circular foundation located over *overconsolidated clay*. Referring to Figure 11.10

$$S_p = \rho_{circle(OC)}S_{p(oed)} \tag{11.20}$$

where

$$\rho_{circle(OC)} = f\left(OCR, \frac{B}{H_t}\right) \tag{11.21}$$

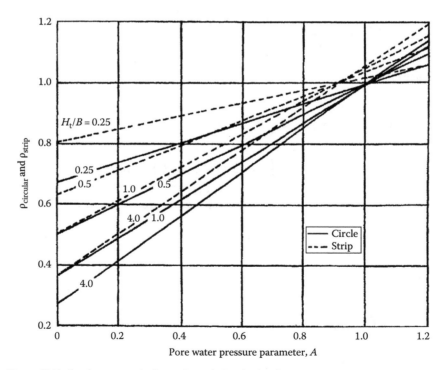

Figure 11.11 Settlement ratio for strip and circular loading.

$$OCR = \text{overconsolidation ratio} = \frac{\sigma'_c}{\sigma'_0} \qquad (11.22)$$

σ'_c is the preconsolidation pressure
σ'_0 is the present effective overburden pressure

The interpolated values of $\rho_{circle(OC)}$ from the work of Leonards (1976) are given in Table 11.1.

The settlement ratio equations (Equations 11.15, 11.18, and 11.20) have been developed assuming that the foundation is located on the top of the clay layer. In most practical situations, this may not be true. So, an approximate procedure needs to be adopted to obtain an equivalent foundation on the clay layers so that the settlement ratio relationships can be used. This approximate procedure of load distribution is usually referred to as the 2:1 stress distribution procedure. The 2:1 stress distribution procedure can be explained using Figure 11.12 as follows. When a foundation measuring $B \times L$ is subjected to a stress increase of q, the load spreads out along planes 2V:1H. Thus, the load at a depth z on the top of a clay layer will be distributed over an area $B' \times L'$.

$$B' = B + z \qquad (11.23)$$

Table 11.1 Variation of $\rho_{circle(OC)}$ with OCR and B/H_t

OCR	$\rho_{circle(OC)}$		
	$B/H_t = 4.0$	$B/H_t = 1.0$	$B/H_t = 0.2$
1	1	1	1
2	0.986	0.957	0.929
3	0.972	0.914	0.842
4	0.964	0.871	0.771
5	0.950	0.829	0.707
6	0.943	0.800	0.643
7	0.929	0.757	0.586
8	0.914	0.729	0.529
9	0.900	0.700	0.493
10	0.886	0.671	0.457
11	0.871	0.643	0.429
12	0.864	0.629	0.414
13	0.857	0.614	0.400
14	0.850	0.607	0.386
15	0.843	0.600	0.371
16	0.843	0.600	0.357

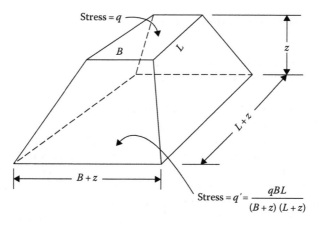

Figure 11.12 2:1 Stress distribution.

and

$$L' = L + z \tag{11.24}$$

So, the load per unit area q' at a depth z will be

$$q' = \frac{qBL}{(B+z)(L+z)} \tag{11.25}$$

The application of this 2V:1H load distribution concept is shown in Examples 11.7 and 11.8.

Example 11.7

The average representative value of the pore water pressure parameter A (as determined from triaxial tests on undisturbed samples) for the clay layer shown in Figure 11.13 is about 0.6. Estimate the consolidation settlement of the circular tank.

Solution

The average effective overburden pressure for the 6 m thick clay layer is $\sigma_0' = (6/2)(19.24 - 9.81) = 28.29 \, \text{kN/m}^2$. We will use Equation 11.2 to obtain the average pressure increase:

$$\Delta\sigma_{av} = \frac{1}{6}(\Delta\sigma_t + 4\Delta\sigma_m + \Delta\sigma_b)$$

$$\Delta\sigma_t = 100 \, \text{kN/m}^2$$

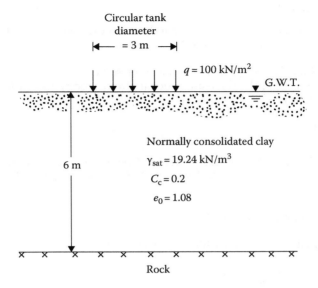

Figure 11.13 Consolidation settlement under a circular tank.

From Equation 4.35

$$\Delta\sigma_m = 100\left\{1 - \frac{1}{[(1.5/3)^2 + 1]^{3/2}}\right\} = 28.45 \text{ kN/m}^2$$

$$\Delta\sigma_b = 100\left\{1 - \frac{1}{[(1.5/6)^2 + 1]^{3/2}}\right\} = 8.69 \text{ kN/m}^2$$

$$\Delta\sigma_{av} = \frac{1}{6}[100 + 4(28.45) + 8.69] = 37.1 \text{ kN/m}^2$$

$$\Delta e = C_c \log \frac{\sigma'_0 + \Delta\sigma_{av}}{\sigma'_0} = 0.2 \log\left(\frac{28.29 + 37.1}{28.29}\right) = 0.073$$

$$e_0 = 1.08$$

$$S_{p(oed)} = \frac{\Delta e H_t}{1 + e_0} = \frac{0.073 \times 6}{1 + 1.08} = 0.21 \text{ m} = 210 \text{ mm}$$

From Figure 11.11, the settlement ratio $\rho_{circular}$ is approximately 0.73 (note that $H_t/B = 2$), so

$$S_p = \rho_{circular} S_{p(oed)} = 0.73 (210) = 153.3 \text{ mm}$$

Example 11.8

Refer to Example 11.6. Assume that the clay is overconsolidated. Given OCR = 3, swell index $(C_r) \approx 1/4 C_c$.

a. Calculate the primary consolidation settlement S_p.
b. Assuming the three-dimensional effect, modify the settlement calculated in Part a.

Solution

Part a:

From Example 11.6, $\sigma_0' = 95.52$ kN/m². Since OCR = 3, the preconsolidation pressure $\sigma_c' = (\text{OCR})\,(\sigma_0') = (3)(95.52) = 286.56$ kN/m². For this case

$$\sigma_0' + \Delta\sigma_{av} = 95.52 + 12.26 < \sigma_c'$$

So, Equations 8.98 and 8.100 may be used. Or

$$S_p = \frac{C_r H}{1 + e_0} \log\left(\frac{\sigma_0' + \Delta\sigma_{av}'}{\sigma_0'}\right)$$

$$= \frac{(0.27/4)(3000)}{1+1} \log\left(\frac{95.52 + 12.25}{95.52}\right) = 5.3 \text{ mm}$$

Part b:

Assuming that the 2:1 method of stress increase holds good, the area of distribution of stress at the top of the clay layer will have dimensions of

$$B' = \text{width} = B + z = 1.5 + 4.5 = 6 \text{ m}$$

$$L' = \text{width} = L + z = 1.5 + 4.5 = 6 \text{ m}$$

The diameter of an equivalent circular area B_{eq} can be given as

$$\frac{\pi}{4} B_{eq}^2 = B'L'$$

$$B_{eq} = \sqrt{\frac{4B'L'}{\pi}} = \sqrt{\frac{(4)(6)(6)}{\pi}} = 6.77 \text{ m}$$

$$\frac{B_{eq}}{H_t} = \frac{6.77}{3} = 2.26$$

From Table 11.1, for OCR = 3 and $B_{eq}/H_t = 2.26$, $\rho_{circle(OC)} \approx 0.95$. Hence

$$S_p = \rho_{circle(OC)}\, S_{p(oed)} = (0.95)(5.3) = 5.04 \text{ mm}$$

11.4 SETTLEMENT CALCULATION USING STRESS PATH

Lambe's (1964) stress path was explained in Section 9.16. Based on Figure 9.47, it was also concluded that (1) the stress paths for a given *normally consolidated* clay are geometrically similar, and (2) when the points representing equal axial strain (ϵ_1) are joined, they will be approximate straight lines passing through the origin.

Let us consider a case where a soil specimen is subjected to an oedometer (one-dimensional consolidation) type of loading (Figure 11.14). For this case, we can write

$$\sigma'_3 = K_o \sigma'_1 \tag{11.26}$$

where K_o is the at-rest earth pressure coefficient and can be given by the expression (Jaky, 1944)

$$K_o = 1 - \sin\phi \tag{11.27}$$

For Mohr's circle shown in Figure 11.14, the coordinates of point E can be given by

$$q' = \frac{\sigma'_1 - \sigma'_3}{2} = \frac{\sigma'_1(1 - K_o)}{2}$$

$$p' = \frac{\sigma'_1 + \sigma'_3}{2} = \frac{\sigma'_1(1 + K_o)}{2}$$

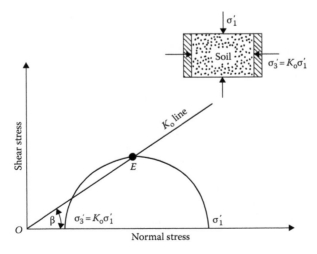

Figure 11.14 Determination of the slope of the K_o line.

Thus

$$\beta = \tan^{-1}\left(\frac{q'}{p'}\right) = \tan^{-1}\left(\frac{1-K_o}{1+K_o}\right) \tag{11.28}$$

where β is the angle that the line OE (K_o line) makes with the normal stress axis.

Figure 11.15 shows a q' versus p' plot for a soil specimen in which the K_o line has also been incorporated. Note that the K_o line also corresponds to a certain value of ϵ_1.

To obtain a general idea of the nature of distortion in soil specimens derived from the application of an axial stress, we consider a soil specimen. If $\sigma'_1 = \sigma'_3$ (i.e., hydrostatic compression) and the specimen is subjected to a hydrostatic stress increase of $\Delta\sigma$ under drained conditions (i.e., $\Delta\sigma = \Delta\sigma'$), then the drained stress path would be EF, as shown in Figure 11.16. There would be uniform strain in all directions. If $\sigma'_3 = K_o\sigma'_1$ (at-rest pressure) and the specimen is subjected to an axial stress increase of $\Delta\sigma$ under drained conditions (i.e., $\Delta\sigma = \Delta\sigma'$), the specimen deformation would depend on the stress path it follows. For stress path AC, which is along the K_o line, there will be axial deformation only and no lateral deformation. For stress path AB, there will be lateral expansion, and so the axial strain at B will be greater than that at C. For stress path AD, there will be some lateral compression, and the axial strain at D will be more than at F but less than that at C. Note that the axial strain is gradually increasing as we go from F to B.

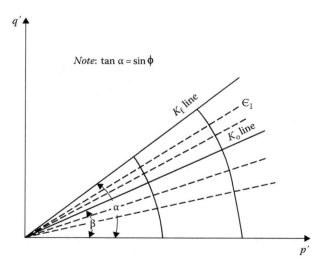

Figure 11.15 Plot of p' versus q' with K_o and K_f lines.

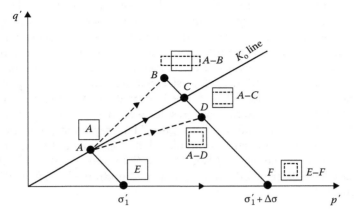

Figure 11.16 Stress path and specimen distortion.

In all cases, the effective major principal stress is $\sigma'_1 + \Delta\sigma'$. However, the lateral strain is compressive at F and zero at C, and we get lateral expansion at B. This is due to the nature of the lateral effective stress to which the specimen is subjected during the loading.

In the calculation of settlement from stress paths, it is assumed that, for normally consolidated clays, the volume change between any two points on a p' versus q' plot is independent of the path followed. This is explained in Figure 11.17. For a soil specimen, the volume changes between stress paths AB, GH, CD, and CI, for example, are all the same. However, the axial strains will be different. With this basic assumption, we can now proceed to determine the settlement.

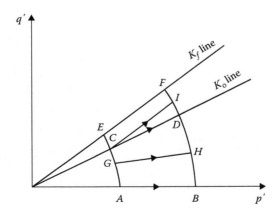

Figure 11.17 Volume change between two points of a p' versus q' plot.

For ease in understanding, the procedure for settlement calculation will be explained with the aid of an example. For settlement calculation in a normally consolidated clay, undisturbed specimens from representative depths are obtained. Consolidated undrained triaxial tests on these specimens at several confining pressures, σ_3, are conducted, along with a standard one-dimensional consolidated test. The stress–strain contours are plotted on the basis of the consolidated undrained triaxial test results. The standard one-dimensional consolidation test results will give us the values of compression index C_c. For example, let Figure 11.18 represent the stress–strain contours for a given normally consolidated clay specimen obtained from an *average depth* of a clay layer. Also, let $C_c = 0.25$ and $e_0 = 0.9$. The drained friction angle ϕ (determined from consolidated undrained tests) is 30°. From Equation 11.28

$$\beta = \tan^{-1}\left(\frac{1 - K_o}{1 + K_o}\right)$$

and $K_o = 1 - \sin\phi = 1 - \sin 30° = 0.5$. So

$$\beta = \tan^{-1}\left(\frac{1 - 0.5}{1 + 0.5}\right) = 18.43°$$

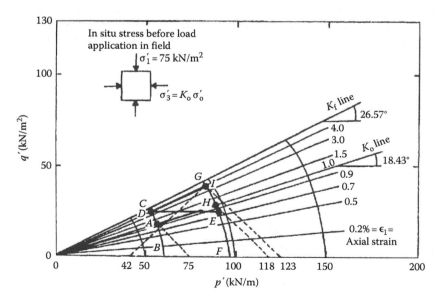

Figure 11.18 Use of stress path to calculate settlement.

Knowing the value of β, we can now plot the K_o line in Figure 11.18. Also note that $\tan \alpha = \sin\phi$. Since $\phi = 30°$, $\tan \alpha = 0.5$. So $\alpha = 26.57°$. Let us calculate the settlement in the clay layer for the following conditions (Figure 11.18):

1. In situ average effective overburden pressure $= \sigma_1' = 75 \text{ kN/m}^2$.
2. Total thickness of clay layer $= H_t = 3$ m.

Owing to the construction of a structure, the increase of the total major and minor principal stresses at an average depth are

$$\Delta\sigma_1 = 40 \text{ kN/m}^2$$

$$\Delta\sigma_3 = 25 \text{ kN/m}^2$$

(assuming that the load is applied instantaneously). The in situ minor principal stress (at-rest pressure) is $\sigma_3 = \sigma_3' = K_o\sigma_1' = 0.5(75) = 37.5 \text{ kN/m}^2$.
   So, before loading

$$p' = \frac{\sigma_1' + \sigma_3'}{2} = \frac{75 + 37.5}{2} = 56.25 \text{ kN/m}^2$$

$$q' = \frac{\sigma_1' - \sigma_3'}{2} = \frac{75 - 37.5}{2} = 18.75 \text{ kN/m}^2$$

The stress conditions before loading can now be plotted in Figure 11.18 from the previously mentioned values of p' and q'. This is point A.
   Since the stress paths are geometrically similar, we can plot BAC, which is the stress path through A. Also, since the loading is instantaneous (i.e., undrained), the stress conditions in clay, represented by the p' versus q' plot immediately after loading, will fall on the stress path BAC. Immediately after loading

$$\sigma_1 = 75 + 40 = 115 \text{ kN/m}^2 \text{ and } \sigma_3 = 37.5 + 25 = 62.5 \text{ kN/m}^2$$

$$\text{So, } q' = \frac{\sigma_1' - \sigma_3'}{2} = \frac{\sigma_1 - \sigma_3}{2} = \frac{115 - 62.5}{2} = 26.25 \text{ kN/m}^2$$

With this value of q', we locate point D. At the end of consolidation

$$\sigma_1' = \sigma_1 = 115 \text{ kN/m}^2 \quad \sigma_3' = \sigma_3 = 62.5 \text{ kN/m}^2$$

$$\text{So, } p' = \frac{\sigma_1' + \sigma_3'}{2} = \frac{115 + 62.5}{2} = 88.75 \text{ kN/m}^2$$

and $q' = 26.25 \text{ kN/m}^2$

The preceding values of p' and q' are plotted as point E. FEG is a geometrically similar stress path drawn through E. ADE is the effective stress path that a soil element, at average depth of the clay layer, will follow. AD represents the elastic settlement, and DE represents the consolidation settlement.

For elastic settlement (stress path A to D)

$$S_e = [(\epsilon_1 \text{ at } D) - (\epsilon_1 \text{ at } A)]H_t = (0.04 - 0.01)3 = 0.09 \text{ m}$$

For consolidation settlement (stress path D to E), based on our previous assumption, the volumetric strain between D and E is the same as the volumetric strain between A and H. Note that H is on the K_o line. For point A, $\sigma_1' = 75 \text{ kN/m}^2$, and for point H, $\sigma_1' = 118 \text{ kN/m}^2$. So, the volumetric strain, ϵ_v, is

$$\epsilon_v = \frac{\Delta e}{1 + e_0} = \frac{C_c \log(118/75)}{1 + 0.9} = \frac{0.25 \log(118/75)}{1.9} = 0.026$$

The axial strain ϵ_1 along a horizontal stress path is about one-third the volumetric strain along the K_o line, or

$$\epsilon_1 = \frac{1}{3}\epsilon_v = \frac{1}{3}(0.026) = 0.0087$$

So, the primary consolidation settlement is

$$S_p = 0.0087H_t = 0.0087(3) = 0.0261 \text{ m}$$

and hence the total settlement is

$$S_e + S_p = 0.09 + 0.0261 = 0.116 \text{ m}$$

Another type of loading condition is also of some interest. Suppose that the stress increase at the average depth of the clay layer was carried out in two steps: (1) instantaneous load application, resulting in stress increases of $\Delta\sigma_1 = 40 \text{ kN/m}^2$ and $\Delta\sigma_3 = 25 \text{ kN/m}^2$ (stress path AD), followed by (2) a gradual load increase, which results in a stress path DI (Figure 11.18). As mentioned earlier, the undrained shear along stress path AD will produce an axial strain of 0.03. The volumetric strains for stress paths DI and AH will be the same; so $\epsilon_v = 0.026$. The axial strain ϵ_1 for the stress path DI can be given by the relation (based on the theory of elasticity)

$$\frac{\epsilon_1}{\epsilon_0} = \frac{1 + K_o - 2KK_o}{(1 - K_o)(1 + 2K)} \tag{11.29}$$

where $K = \sigma_3'/\sigma_1'$ for the point I. In this case, $\sigma_3' = 42 \text{ kN/m}^2$ and $\sigma_1' = 123 \text{ kN/m}^2$. So

$$K = \frac{42}{123} = 0.341$$

$$\frac{\epsilon_1}{\epsilon_v} = \frac{\epsilon_1}{0.026} = \frac{1+0.5-2(0.341)(0.5)}{(1-0.5)[1+2(0.341)]} = 1.38$$

or

$$\epsilon_1 = (0.026)(1.38) = 0.036$$

Hence, the total settlement due to the loading is equal to

$$S = [(\epsilon_1 \text{ along } AD) + (\epsilon_1 \text{ along } DI)]H_t$$

$$= (0.03+0.036)H_t = 0.066H_t$$

11.5 COMPARISON OF PRIMARY CONSOLIDATION SETTLEMENT CALCULATION PROCEDURES

It is of interest at this point to compare the primary settlement calculation procedures outlined in Sections 11.2 and 11.3 with the stress path technique described in Section 11.4 (Figure 11.19).

Based on the one-dimensional consolidation procedure outlined in Sections 11.2, essentially we calculate the settlement along the stress path AE, that is, along the K_o line. A is the initial at-rest condition of the soil, and E is the final stress condition (at rest) of soil at the end of consolidation. According to the Skempton–Bjerrum modification, the consolidation settlement is calculated for stress path DE. AB is the elastic settlement.

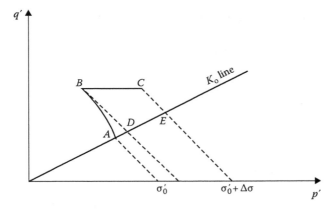

Figure 11.19 Comparison of consolidation settlement calculation procedures.

However, Lambe's stress path method gives the consolidation settlement for stress path BC. AB is the elastic settlement. Although the stress path technique provides us with a better insight into the fundamentals of settlement calculation, it is more time-consuming because of the elaborate laboratory tests involved.

A number of works have been published that compare the observed and the predicted settlements of various structures. Terzaghi and Peck (1967) pointed out that the field consolidation settlement is approximately one-dimensional when a comparatively thin layer of clay is located between two stiff layers of soil. Peck and Uyanik (1955) analyzed the settlement of eight structures in Chicago located over thick deposits of soft clay. The settlements of these structures were predicted by the method outlined in Section 11.2. Elastic settlements were not calculated. For this investigation, the ratio of the settlements observed to that calculated had an average value of 0.85. Skempton and Bjerrum (1957) also analyzed the settlements of four structures in the Chicago area (auditorium, Masonic temple, Monadnock block, Isle of Grain oil tank) located on overconsolidated clays. The predicted settlements included the elastic settlements and the consolidation settlements (by the method given in Section 11.3). The ratio of the observed to the predicted settlements varied from 0.92 to 1.17. Settlement analysis of Amuya Dam, Venezuela (Lambe, 1963), by the stress path method, showed very good agreement with the observed settlement.

However, there are several instances where the predicted settlements vary widely from the observed settlements. The discrepancies can be attributed to deviation of the actual field conditions from those assumed in the theory, difficulty in obtaining undisturbed samples for laboratory tests, and so forth.

11.6 SECONDARY CONSOLIDATION SETTLEMENT

The coefficient of secondary consolidation C_α was defined in Section 8.7 as

$$C_\alpha = \frac{\Delta H_t / H_t}{\Delta \log t}$$

where
  t is the time
  H_t is the thickness of the clay layer

It has been reasonably established that C_α decreases with time in a logarithmic manner and is directly proportional to the total thickness of the clay layer at the beginning of secondary consolidation. Thus, secondary consolidation settlement can be given by

$$S_s = C_\alpha H_{ts} \log \frac{t}{t_p}$$ (11.30)

where

H_{ts} is the thickness of the clay layer at the beginning of secondary consolidation = $H_t - S_c$

t is the time at which secondary compression is required

t_p is the time at the end of primary consolidation

Actual field measurements of secondary settlements are relatively scarce. However, good agreement of measured and estimated settlements has been reported by some observers, for example, Horn and Lambe (1964), Crawford and Sutherland (1971), and Su and Prysock (1972).

11.7 PRECOMPRESSION FOR IMPROVING FOUNDATION SOILS

In instances when it appears that too much consolidation settlement is likely to occur due to the construction of foundations, it may be desirable to apply some surcharge loading before foundation construction in order to eliminate or reduce the postconstruction settlement. This technique has been used with success in many large construction projects (Johnson, 1970). In this section, the fundamental concept of surcharge application for elimination of primary consolidation of compressible clay layers is presented.

Let us consider the case where a given construction will require a permanent uniform loading of intensity σ_f, as shown in Figure 11.20. The total primary consolidation settlement due to loading is estimated to be equal to $S_{p(f)}$. If we want to eliminate the expected settlement due to primary consolidation, we will have to apply a total uniform load of intensity $\sigma = \sigma_f + \sigma_s$. This load will cause a faster rate of settlement of the underlying compressible layer; when a total settlement of $S_{p(f)}$ has been reached, the surcharge can be removed for actual construction.

For a quantitative evaluation of the magnitude of σ_s and the time it should be kept on, we need to recognize the nature of the variation of the degree of consolidation at any time after loading for the underlying clay layer, as shown in Figure 11.21. The degree of consolidation U_z will vary with depth and will be minimum at midplane, that is, at $z = H$. If the average degree of consolidation U_{av} is used as the criterion for surcharge load removal, then after removal of the surcharge, the clay close to the midplane will continue

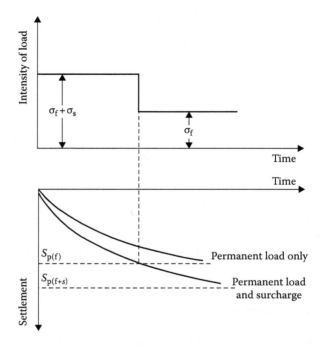

Figure 11.20 Concept of precompression technique.

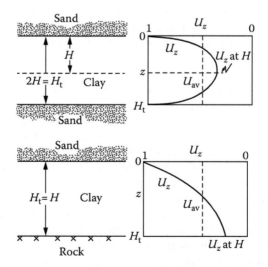

Figure 11.21 Choice of degree of consolidation for calculation of precompression.

to settle, and the clay close to the previous layer(s) will tend to swell. This will probably result in a net consolidation settlement. To avoid this problem, we need to take a more conservative approach and use the midplane degree of consolidation $U_{z\,=\,H}$ as the criterion for our calculation. Using the procedure outlined by Johnson (1970)

$$S_{p(f)} = \left(\frac{H_t}{1+e_0}\right) C_c \log\left(\frac{\sigma_0' + \sigma_f}{\sigma_0'}\right) \tag{11.31}$$

and

$$S_{p(f+s)} = \left(\frac{H_t}{1+e_0}\right) C_c \log\left(\frac{\sigma_0' + \sigma_f + \sigma_s}{\sigma_0'}\right) \tag{11.32}$$

where
   σ_0' is the initial average in situ effective overburden pressure
   $S_{p(f)}$ and $S_{p(f+s)}$ are the primary consolidation settlements due to load intensities of σ_f and $\sigma_f + \sigma_s$, respectively

However

$$S_{p(f)} = U_{(f+s)} S_{p(f+s)} \tag{11.33}$$

where $U_{(f+s)}$ is the degree of consolidation due to the loading of $\sigma_f + \sigma_s$. As explained earlier, this is conservatively taken as the midplane ($z = H$) degree of consolidation. Thus

$$U_{(f+s)} = \frac{S_{p(f)}}{S_{p(f+s)}} \tag{11.34}$$

Combining Equations 11.31, 11.32, and 11.34

$$U_{(f+s)} = \frac{\log[1 + (\sigma_f/\sigma_0')]}{\log\{1 + (\sigma_f/\sigma_0')[1 + (\sigma_s/\sigma_f)]\}} \tag{11.35}$$

The values of $U_{(f+s)}$ for several combinations of σ_f/σ_0' and σ_s/σ_f are given in Figure 11.22. Once $U_{(f+s)}$ is known, we can evaluate the nondimensional time factor T_v from Figure 8.7. (Note that $U_{(f+s)} = U_z$ at $z = H$ of Figure 8.7,

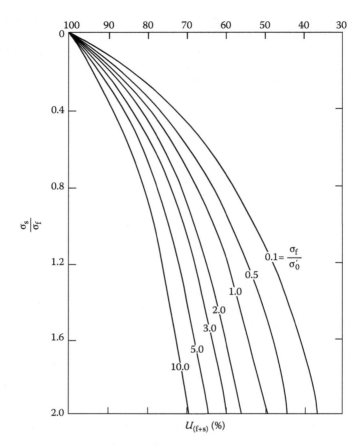

Figure 11.22 Variation of $U_{(f+s)}$ with σ_s/σ_f and σ_f/σ_0'.

based on our assumption.) For convenience, a plot of $U_{(f+s)}$ versus T_v is given in Figure 11.23. So, the time for surcharge load removal $t = t'$ is

$$t' = \frac{T_v H^2}{C_v} \tag{11.36}$$

where
   C_v is the coefficient of consolidation
   H is the length of the maximum drainage path

A similar approach may be adopted to estimate the intensity of the surcharge fill and the time for its removal to eliminate or reduce postconstruction settlement due to secondary consolidation.

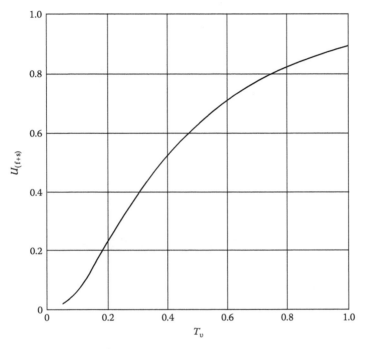

Figure 11.23 Plot of $U_{(f+s)}$ versus T_v.

Example 11.9

The soil profile shown in Figure 11.24 is in an area where an airfield is to be constructed. The entire area has to support a permanent surcharge of 58 kN/m² due to the fills that will be placed. It is desired to eliminate all the primary consolidation in 6 months by precompression before the start of construction. Estimate the total surcharge $(q = q_s + q_f)$ that will be required for achieving the desired goal.

Solution

$$t = t' = \frac{T_v H^2}{C_v} \quad \text{or} \quad T_v = \frac{t' C_v}{H^2}$$

For two-way drainage

$$H = \frac{H_t}{2} = 2.25 \text{ m} = 225 \text{ cm}$$

We are given that

$$t' = 6 \times 30 \times 24 \times 60 \text{ min}$$

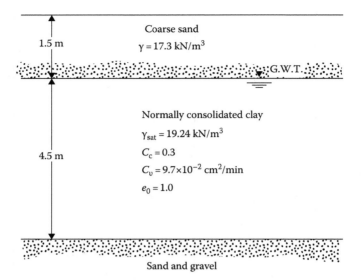

Figure 11.24 Soil profile for precompression.

So

$$T_\upsilon = \frac{(6\times30\times24\times60)(9.7\times10^{-2})}{(225)^2} = 0.497$$

From Figure 11.23, for $T_\upsilon = 0.497$ and $U_{(f+s)} \approx 0.62$

$$\sigma_0' = 17.3(1.5) + 2.25(19.24 - 9.81) = 47.17 \text{ kN/m}^2$$

$$\sigma_f = 58 \text{ kN/m}^2 \text{ (given)}$$

So

$$\frac{\sigma_f}{\sigma_0'} = \frac{58}{47.17} = 1.23$$

From Figure 11.22, for $U_{(f+s)} = 0.62$ and $\sigma_f/\sigma_0' = 1.23$,

$$\frac{\sigma_s}{\sigma_f} = 1.17$$

So

$$\sigma_s = 1.17\sigma_f = 1.17(58) = 67.86 \text{ kN/m}^2$$

Thus

$$\sigma = \sigma_f + \sigma_s = 58 + 67.86 = 125.86 \text{ kN/m}^2$$

Example 11.10

Refer to Example 11.9. If $\sigma_s = 35$ kN/m², determine the time needed to eliminate all of the primary consolidation by precompression before the start of construction.

Solution

Given: $\sigma_f = 58$ kN/m²; $\sigma_s = 35$ kN/m²
From Equation 11.35,

$$U_{(f+s)} = \frac{\log\left[1 + \dfrac{\sigma_f}{\sigma_0'}\right]}{\log\left\{1 + \left(\dfrac{\sigma_f}{\sigma_0'}\right)\left[1 + \left(\dfrac{\sigma_s}{\sigma_f}\right)\right]\right\}}$$

From Example 11.9, $\sigma_0' = 47.17$ kN/m².

$$U_{(f+s)} = \frac{\log\left[1 + \dfrac{58}{47.17}\right]}{\log\left\{1 + \left(\dfrac{58}{47.17}\right)\left[1 + \left(\dfrac{35}{58}\right)\right]\right\}} = 0.736$$

From Figure 11.23, for $U_{(f+s)} = 0.736$, the magnitude of T_v is about 0.66. Hence, from Equation 11.36,

$$t' = \frac{T_v H^2}{C_v} = \frac{(0.736)(225 \text{ cm})^2}{9.7 \times 10^{-2}} = 384,124 \text{ min} \approx 8.89 \text{ months}$$

11.8 PRECOMPRESSION WITH SAND DRAINS

The concept of accelerating consolidation settlement by including sand drains was presented in Section 8.14. In the presence of sand drains, to determine the surcharge that needs to be applied at the ground surface, we refer to Figure 11.20 and Equation 11.35. In a modified form, for a combination of vertical and radial drainage, Equation 11.35 can be rewritten as

$$U_{v,r} = \frac{\log[1 + (\sigma_f/\sigma_0')]}{\log\{1 + (\sigma_f/\sigma_0')[1 + (\sigma_s/\sigma_f)]\}} \tag{11.37}$$

The notations σ_f, σ_0', and σ_s are the same as those in Equation 11.35; however, the left-hand side of Equation 11.37 is the average degree of consolidation instead of the degree of consolidation at midplane. Both *radial* and *vertical* drainage contribute to the average degree of consolidation. If $U_{v,r}$ can be determined at any time $t = t'$ (see Figure 11.20), the total surcharge $\sigma_f + \sigma_s$ may be easily obtained from Figure 11.22.

For a given surcharge and duration $t = t'$, the average degree of consolidation can be obtained from Equation 8.200. Or

$$U_{v,r} = 1 - (1 - U_v)(1 - U_r)$$

The magnitude of U_v can be obtained from Equations 8.42 and 8.43, and the magnitude of U_r can be obtained from Equation 8.178.

Example 11.11

Redo Example 11.9 with the addition of sand drains. Assume $r_w = 0.1$ m, $d_e = 3$ m (see Figure 8.51), $C_v = C_{v,r}$, and the surcharge is applied instantaneously. Also assume that this is a no-smear case.

Solution

From Example 11.9, $T_v = 0.497$. Using Equation 8.35, we obtain

$$U_v = \sqrt{\frac{4T_v}{\pi}} \times 100 = \sqrt{\frac{(4)(0.497)}{\pi}} \times 100 = 79.5\%$$

Also

$$n = \frac{d_e}{2r_w} = \frac{3}{(2)(0.1)} = 15$$

Again (Equation 8.170)

$$T_r = \frac{C_{vr}t'}{d_e^2} = \frac{(9.7 \times 10^{-2} \text{ cm}^2/\text{min})(6 \times 30 \times 24 \times 60 \text{ min})}{(300 \text{ cm})^2} \approx 0.28$$

From Table 8.9, for $n = 15$ and $T_r = 0.28$, the value of U_r is about 68%. Hence

$$U_{v,r} = 1 - (1 - U_v)(1 - U_r) = 1 - (1 - 0.795)(1 - 0.68) = 0.93 = 93\%$$

From Example 11.9

$$\frac{\sigma_f}{\sigma'_0} = 1.23$$

For $\sigma_f/\sigma'_0 = 1.23$ and $U_{v,r} = 93\%$, the value of $\sigma_s/\sigma_f \approx 0.124$ (Figure 11.22). Hence

$$\sigma_s = 0.124\sigma_f = 0.124(58) = 7.19 \text{ kN/m}^2$$

So

$$\sigma = \sigma_f + \sigma_s = 58 + 7.19 = 65.19 \text{ kN/m}^2$$

REFERENCES

Crawford, C. B. and J. G. Sutherland, The Empress Hotel, Victoria, British Columbia, Sixty-five years of foundation settlements, *Can. Geotech. J.*, 8(1), 77–93, 1971.

Griffiths, D. V., A chart to estimate the average vertical stress increase in an elastic foundation below a uniformly loaded vertical area, *Can. Geotech. J.*, 21(4), 710–713, 1984.

Horn, H. M. and T. W. Lambe, Settlement of buildings on the MIT campus, *J. Soil Mech. Found. Div.*, ASCE, 90(SM5), 181–195, 1964.

Jaky, J., The coefficient of earth pressure at rest, *J. Soc. Hung. Arch. Eng.*, 355–358, 1944.

Johnson, S. J., Precompression for improving foundation soils, *J. Soil Mech. Found. Div.*, ASCE, 96(SM1), 111–144, 1970.

Lambe, T. W., An earth dam for storage of fuel oil, *Proc. II Pan Am. Conf. Soil Mech. Found. Eng.*, Sao Paulo, Brazil, 1, 257, 1963.

Lambe, T. W., Methods of estimating settlement, *J. Soil Mech. Found. Div.*, ASCE, 90(SM5), 43, 1964.

Leonards, G. A., Estimating consolidation settlement of shallow foundations on overconsolidated clay, Special Report 163, Transportation Research Board, Washington, DC, 13–16, 1976.

Peck, R. B. and M. E. Uyanik, Observed and computed settlements of structures in chicago, Bulletin 429, University of Illinois Engineering Experiment Station, 1955.

Saikia, A., Average vertical stress under uniformly loaded spread footing: An analytical cum graphical solution, *Int. J. Geotech. Eng.*, 6(3), 359–363, 2012.

Scott, R. F., *Principles of Soil Mechanics*, Addison-Wesley, Reading, MA, 1963.

Skempton, A. W. and L. Bjerrum, A contribution to settlement analysis of foundations in clay, *Geotechnique*, 7, 168, 1957.

Su, H. H. and R. H. Prysock, Settlement analysis of two highway embankments, *Proc. Spec. Conf. Perform. Earth Supported Struct.*, American Society of Civil Engineers, Reston, VA, vol. 1, 465–488, 1972.

Terzaghi, K. and R. B. Peck, *Soil Mechanics in Engineering Practice*, 2nd edn., Wiley, New York, 1967.

Appendix: Calculation of stress at the interface of a three-layered flexible system

The procedure for calculating vertical stress at the interface of a three-layer flexible system was discussed in Section 4.14. The related variables K_1, K_2, A, H, ZZ_1 and ZZ_2 were also defined there. Figures A.1 through A.32 give the plots of the variation of ZZ_1 and ZZ_2 with A, H, K_1 and K_2 (Peattie, 1962).

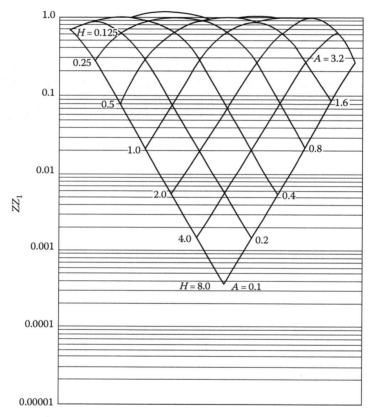

Figure A.1 Values of ZZ_1 for $K_1 = 0.2$ and $K_2 = 0.2$. (After Peattie, K.R., Stress and strain factors for three-layer systems, *Bulletin 342*, Highway Research Board, 1962.)

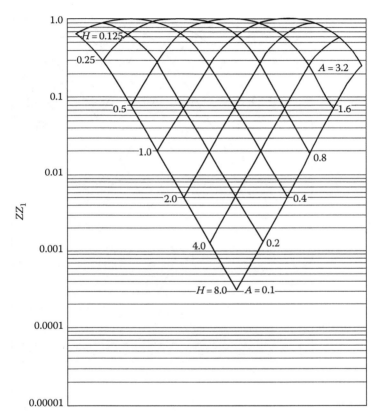

Figure A.2 Values of ZZ_1 for $K_1 = 0.2$ and $K_2 = 2.0$. (After Peattie, K.R., Stress and strain factors for three-layer systems, *Bulletin 342*, Highway Research Board, 1962.)

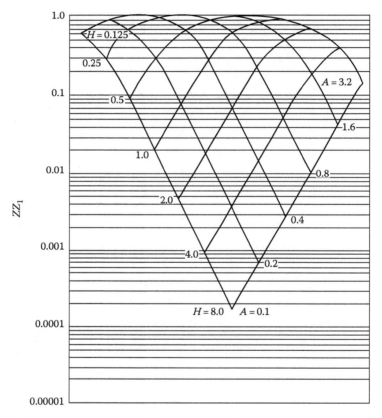

Figure A.3 Values of ZZ_1 for K_1 = 0.2 and K_2 = 20.0. (After Peattie, K.R., Stress and strain factors for three-layer systems, *Bulletin 342*, Highway Research Board, 1962.)

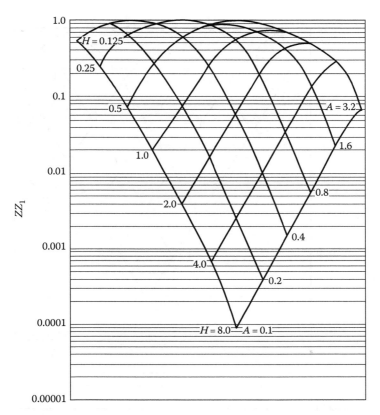

Figure A.4 Values of ZZ_1 for $K_1 = 0.2$ and $K_2 = 200.0$. (After Peattie, K.R., Stress and strain factors for three-layer systems, *Bulletin 342*, Highway Research Board, 1962.)

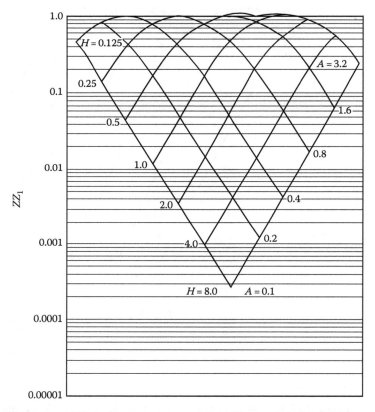

Figure A.5 Values of ZZ_1 for $K_1 = 2.0$ and $K_2 = 0.2$. (After Peattie, K.R., Stress and strain factors for three-layer systems, *Bulletin 342*, Highway Research Board, 1962.)

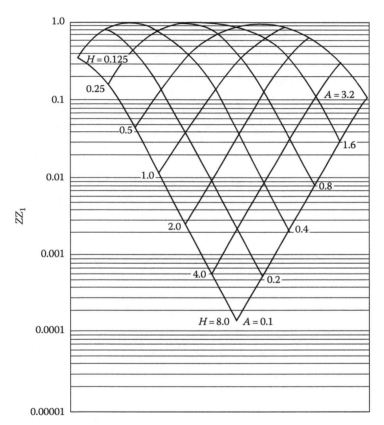

Figure A.6 Values of ZZ_1 for $K_1 = 2.0$ and $K_2 = 2.0$. (After Peattie, K.R., Stress and strain factors for three-layer systems, *Bulletin 342*, Highway Research Board, 1962.)

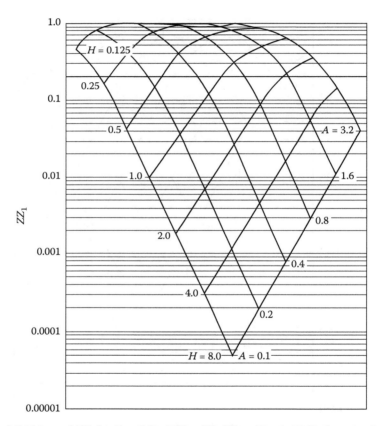

Figure A.7 Values of ZZ_1 for $K_1 = 2.0$ and $K_2 = 20$. (After Peattie, K.R., Stress and strain factors for three-layer systems, *Bulletin 342*, Highway Research Board, 1962.)

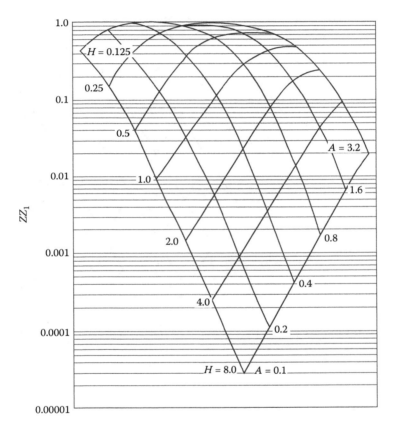

Figure A.8 Values of ZZ_1 for $K_1 = 2.0$ and $K_2 = 200.0$. (After Peattie, K.R., Stress and strain factors for three-layer systems, *Bulletin 342*, Highway Research Board, 1962.)

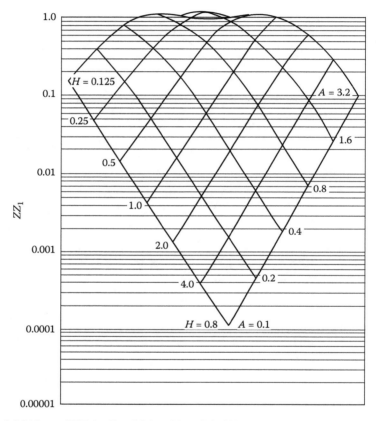

Figure A.9 Values of ZZ_1 for $K_1 = 20.0$ and $K_2 = 0.2$. (After Peattie, K.R., Stress and strain factors for three-layer systems, *Bulletin 342*, Highway Research Board, 1962.)

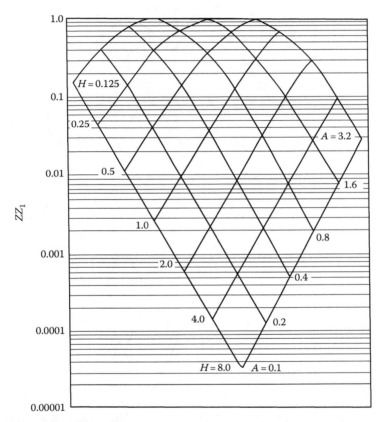

Figure A.10 Values of ZZ_1 for $K_1 = 20$ and $K_2 = 2.0$. (After Peattie, K.R., Stress and strain factors for three-layer systems, *Bulletin 342*, Highway Research Board, 1962.)

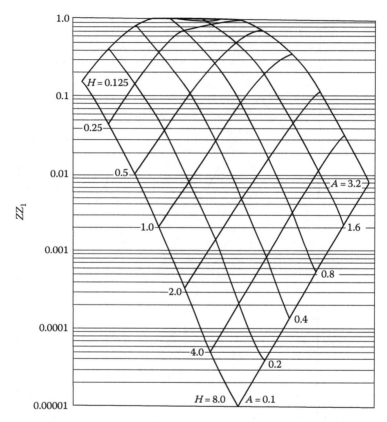

Figure A.11 Values of ZZ_1 for $K_1 = 20.0$ and $K_2 = 20.0$. (After Peattie, K.R., Stress and strain factors for three-layer systems, *Bulletin 342*, Highway Research Board, 1962.)

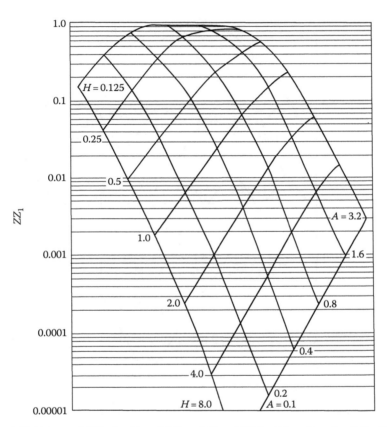

Figure A.12 Values of ZZ_1 for $K_1 = 20.0$ and $K_2 = 200.0$. (After Peattie, K.R., Stress and strain factors for three-layer systems, *Bulletin 342*, Highway Research Board, 1962.)

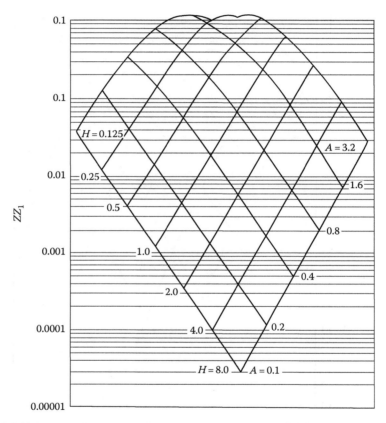

Figure A.13 Values of ZZ_1 for K_1 = 200.0 and K_2 = 0.2. (After Peattie, K.R., Stress and strain factors for three-layer systems, *Bulletin 342*, Highway Research Board, 1962.)

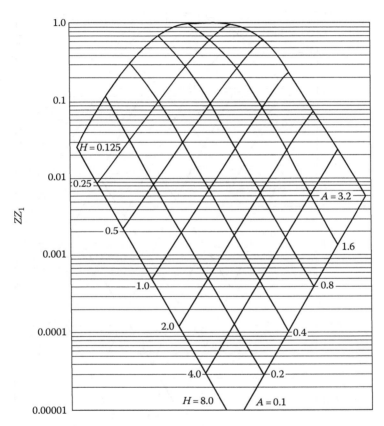

Figure A.14 Values of ZZ_1 for $K_1 = 200.0$ and $K_2 = 2.0$. (After Peattie, K.R., Stress and strain factors for three-layer systems, *Bulletin 342*, Highway Research Board, 1962.)

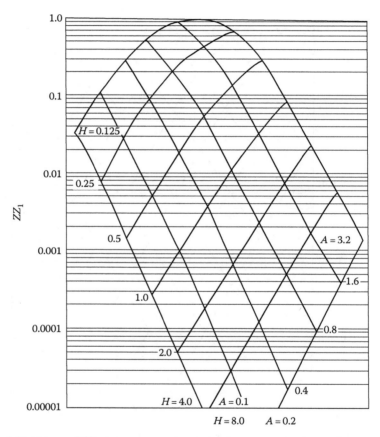

Figure A.15 Values of ZZ_1 for K_1 = 200.0 and K_2 = 20. (After Peattie, K.R., Stress and strain factors for three-layer systems, *Bulletin 342*, Highway Research Board, 1962.)

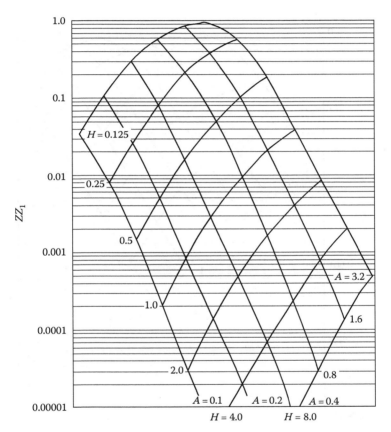

Figure A.16 Values of ZZ_1 for K_1 = 200.0 and K_2 = 200.0. (After Peattie, K.R., Stress and strain factors for three-layer systems, *Bulletin 342*, Highway Research Board, 1962.)

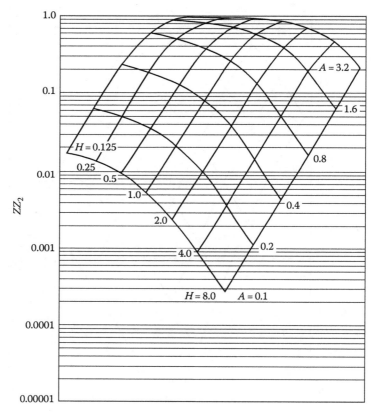

Figure A.17 Values of ZZ_2 for $K_1 = 0.2$ and $K_2 = 0.2$. (After Peattie, K.R., Stress and strain factors for three-layer systems, *Bulletin 342*, Highway Research Board, 1962.)

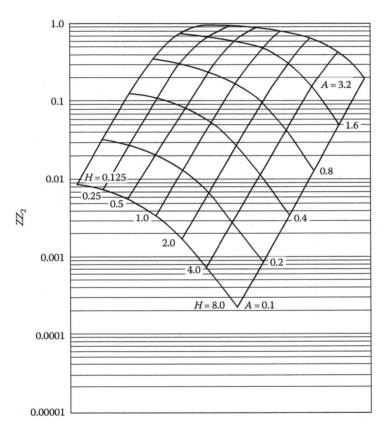

Figure A.18 Values of ZZ_2 for $K_1 = 0.2$ and $K_2 = 2.0$. (After Peattie, K.R., Stress and strain factors for three-layer systems, *Bulletin 342*, Highway Research Board, 1962.)

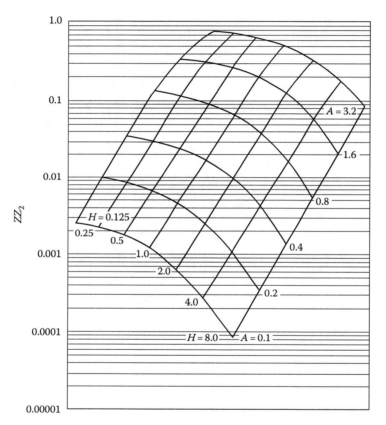

Figure A.19 Values of ZZ_2 for $K_1 = 0.2$ and $K_2 = 20.0$. (After Peattie, K.R., Stress and strain factors for three-layer systems, *Bulletin 342*, Highway Research Board, 1962.)

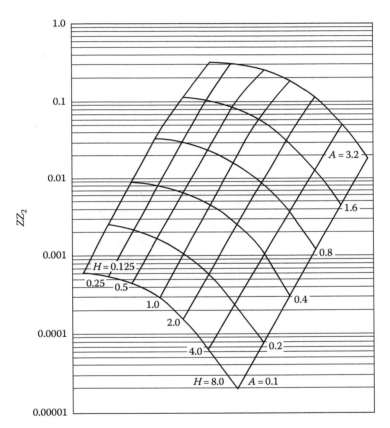

Figure A.20 Values of ZZ_2 for $K_1 = 0.2$ and $K_2 = 200.0$. (After Peattie, K.R., Stress and strain factors for three-layer systems, *Bulletin 342*, Highway Research Board, 1962.)

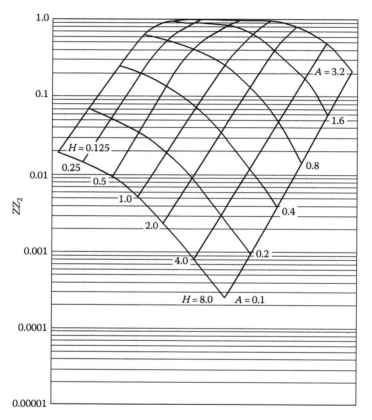

Figure A.21 Values of ZZ_2 for $K_1 = 2.0$ and $K_2 = 0.2$. (After Peattie, K.R., Stress and strain factors for three-layer systems, *Bulletin 342*, Highway Research Board, 1962.)

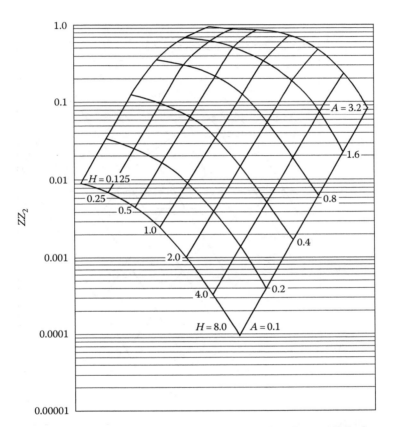

Figure A.22 Values of ZZ_2 for $K_1 = 2.0$ and $K_2 = 2.0$. (After Peattie, K.R., Stress and strain factors for three-layer systems, *Bulletin 342*, Highway Research Board, 1962.)

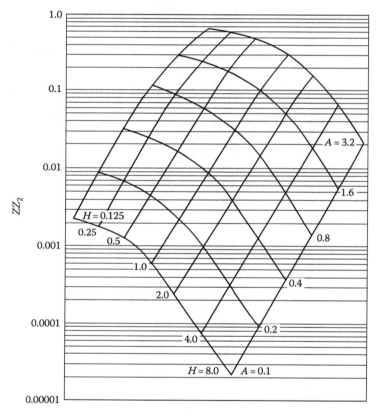

Figure A.23 Values of ZZ_2 for $K_1 = 2.0$ and $K_2 = 20.0$. (After Peattie, K.R., Stress and strain factors for three-layer systems, *Bulletin 342*, Highway Research Board, 1962.)

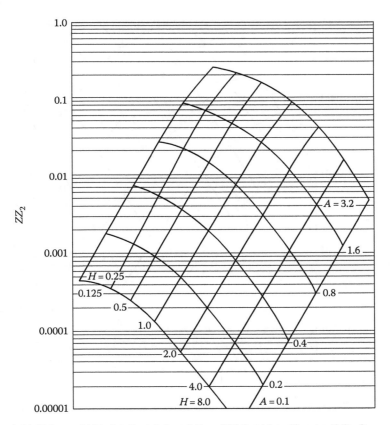

Figure A.24 Values of ZZ_2 for $K_1 = 2.0$ and $K_2 = 200.0$. (After Peattie, K.R., Stress and strain factors for three-layer systems, *Bulletin 342*, Highway Research Board, 1962.)

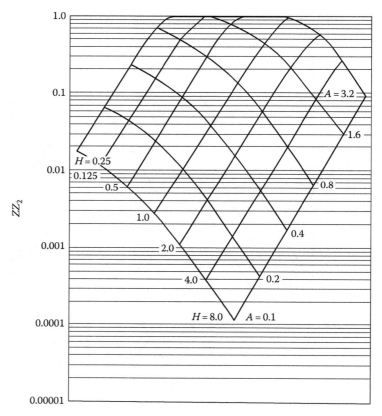

Figure A.25 Values of ZZ_2 for $K_1 = 20.0$ and $K_2 = 0.2$. (After Peattie, K.R., Stress and strain factors for three-layer systems, *Bulletin 342*, Highway Research Board, 1962.)

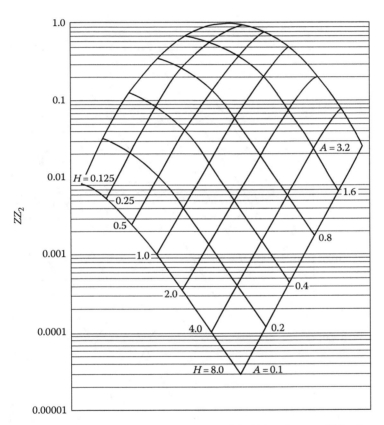

Figure A.26 Values of ZZ_2 for $K_1 = 20.0$ and $K_2 = 2.0$. (After Peattie, K.R., Stress and strain factors for three-layer systems, *Bulletin 342*, Highway Research Board, 1962.)

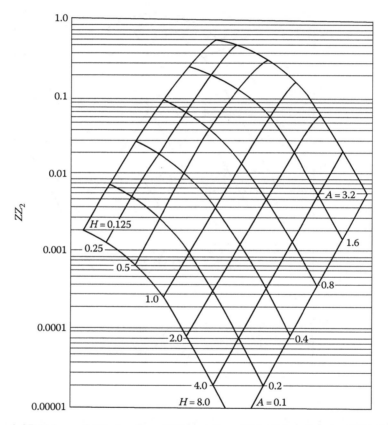

Figure A.27 Values of ZZ$_2$ for K_1 = 20.0 and K_2 = 20.0. (After Peattie, K.R., Stress and strain factors for three-layer systems, *Bulletin 342*, Highway Research Board, 1962.)

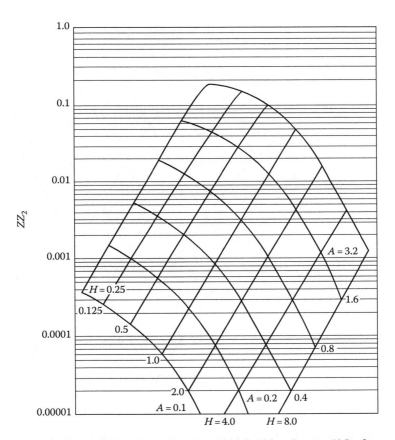

Figure A.28 Values of ZZ_2 for $K_1 = 20$ and $K_2 = 200.0$. (After Peattie, K.R., Stress and strain factors for three-layer systems, *Bulletin 342*, Highway Research Board, 1962.)

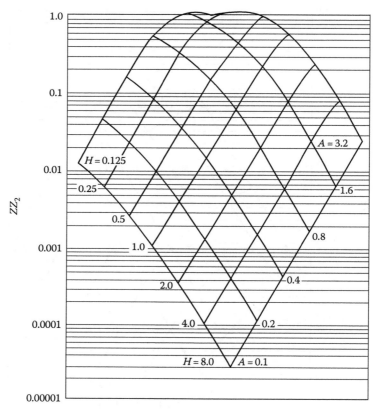

Figure A.29 Values of ZZ_2 for K_1 = 200.0 and K_2 = 0.2. (After Peattie, K.R., Stress and strain factors for three-layer systems, *Bulletin 342*, Highway Research Board, 1962.)

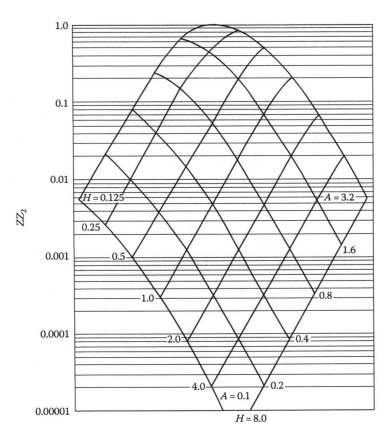

Figure A.30 Values of ZZ_2 for $K_1 = 200.0$ and $K_2 = 2.0$. (After Peattie, K.R., Stress and strain factors for three-layer systems, *Bulletin 342*, Highway Research Board, 1962.)

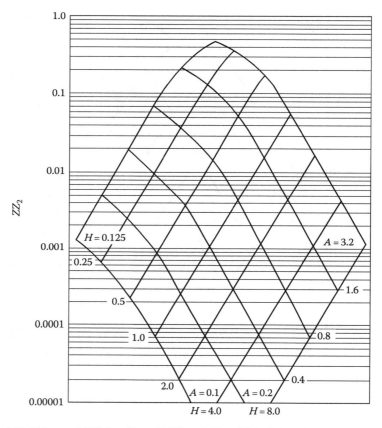

Figure A.31 Values of ZZ_2 for $K_1 = 200.0$ and $K_2 = 20.0$. (After Peattie, K.R., Stress and strain factors for three-layer systems, *Bulletin 342*, Highway Research Board, 1962.)

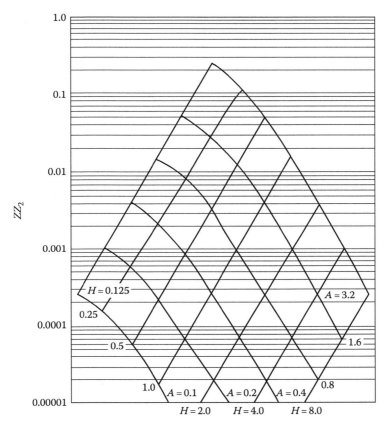

Figure A.32 Values of ZZ_2 for K_1 = 200.0 and K_2 = 200.0. (After Peattie, K.R., Stress and strain factors for three-layer systems, *Bulletin 342*, Highway Research Board, 1962.)

REFERENCE

Peattie, K. R., Stress and strain factors for three-layer systems, *Bulletin 342*, Highway Research Board, pp. 215–253, 1962.

Index